Springer Polar Sciences

Springer Polar Sciences

Springer Polar Sciences is an interdisciplinary book series that is dedicated to research on the Arctic and sub-Arctic regions and Antarctic. The series aims to present a broad platform that will include both the sciences and humanities and to facilitate exchange of knowledge between the various polar science communities.

Topics and perspectives will be broad and will include but not be limited to climate change impacts, environmental change, polar ecology, governance, health, economics, indigenous populations, tourism and resource extraction activities.

Books published in the series will have ready appeal to scientists, students and policy makers.

More information about this series at http://www.springer.com/series/15180

Luke Copland • Derek Mueller
Editors

Arctic Ice Shelves and Ice Islands

Springer

Editors
Luke Copland
Geography, Environment and Geomatics
University of Ottawa
Ottawa, ON, Canada

Derek Mueller
Geography and Environmental Studies
Carleton University
Ottawa, ON, Canada

ISSN 2510-0475 ISSN 2510-0483 (electronic)
Springer Polar Sciences
ISBN 978-94-024-1099-0 ISBN 978-94-024-1101-0 (eBook)
DOI 10.1007/978-94-024-1101-0

Library of Congress Control Number: 2017940395

Printed on acid-free paper

This Springer imprint is published by Springer Nature
The registered company is Springer Science+Business Media B.V.
The registered company address is: Van Godewijckstraat 30, 3311 GX Dordrecht, The Netherlands

Preface

This book provides an overview of our current state of knowledge of Arctic ice shelves and related features. Ice shelves are permanent areas of ice which float on the ocean surface while attached to the coast, and typically occur in very cold environments where perennial sea ice builds up to great thickness, and/or where glaciers flow off the land and are preserved on the ocean surface. These landscape features are relatively poorly studied in the Arctic, yet they are potentially highly sensitive indicators of climate change because they respond to changes in atmospheric, oceanic and glaciological conditions. Recent fracturing and breakup events of ice shelves in the Canadian High Arctic (e.g., Ward Hunt Ice Shelf in 2002; Ayles Ice Shelf in 2005; Serson Ice Shelf in 2008) have attracted large scientific and public attention, particularly in light of the rapid recent reductions in sea ice that have occurred across the Arctic. Much has been published about Antarctic ice shelves, which fringe ~55% of the Antarctic coastline and are far more common than Arctic ice shelves. However, to date there hasn't been a dedicated book about Arctic ice shelves or ice islands. This volume fills that gap.

Part I provides an overview of the general characteristics and distribution of Arctic ice shelves. Chapter 1 defines what ice shelves are, how and where they form, and the differences between Arctic and Antarctic ice shelves. Features that are closely related to Arctic ice shelves, such as ice islands, epishelf lakes, ice rises and multiyear landfast sea ice are also defined and their significance explained. This is followed by three chapters describing the distribution of all known Northern Hemisphere ice shelves in the Canadian Arctic (Chap. 2), Eurasian Arctic (Chap. 3) and Greenland (Chap. 4).

Part II begins with reviews of the physical processes that are important for ice shelf growth and retreat. Chapter 5 examines the changes that have occurred in Canadian Arctic ice shelves over the past century, while Chap. 6 concentrates on the surface mass balance of the Ward Hunt Ice Shelf, the largest remaining in the Northern Hemisphere. Following this, Chaps. 7 and 8 explore two approaches to determining the Holocene history of ice shelves, along with evidence for Arctic ice shelf presence and absence over this period. Chapter 9 focuses on biological aspects of Arctic ice shelves, and the unique communities of zooplankton found in epishelf lakes. Ice shelves were once considered as strictly abiotic environments, but are now known to be a habitat for diverse microbial communities. These organisms are tolerant of their extreme environment and serve as an analogue for how life survived and evolved during widespread glaciations on Earth.

Part III focuses on the causes and patterns of Arctic ice shelf calving events, and the primary features that are produced from these calving events, ice islands. A discussion of the main mechanisms driving recent Arctic ice shelf losses is the focus of Chap. 10. This is an area which is currently poorly documented, yet is important to understand in light of the rapid climate changes occurring in the Arctic. Chapter 11 provides a review of the tracks that previous ice islands have taken across the Arctic over the last 60 years, and the mechanisms that have driven these drift patterns. Chapter 12 provides detailed insights into recent changes in sea ice conditions along northern Ellesmere Island, with a focus on semi-permanent ice plugs that once filled the inter-island channels of the northern Queen Elizabeth Islands and formed incipient ice shelves. Chapter 13 addresses the strategic military importance that ice shelves and ice islands had during the Cold War due to their ability to be used as forward operating bases across the Arctic Ocean. Chapter 14 explores the Russian historical perspective on ice islands and includes a review of the long-term use of ice islands as platforms for scientific research stations. Finally, in Chap. 15 a review of the engineering and risk concerns for the installation and design of oil rigs and other infrastructure in iceberg-infested waters is presented.

In summary, the book provides the first comprehensive overview of Arctic ice shelves. The few previous publications in this area have only considered ice shelves from a single perspective (e.g., Russian drifting stations), and not from the integrated view presented here. The publication of this book is particularly timely given the large and rapid climate changes that are currently occurring in the Arctic. We anticipate that upper level undergraduate students will find it useful for Arctic-themed courses, that graduate students will find it a useful reference for their studies, and that professors and researchers will find it a useful summary of previous and ongoing research on Arctic ice shelves and ice islands.

We would like to acknowledge many people for their help and assistance in putting together this book. First, we would like to thank the patience and professionalism of all of the chapter authors and co-authors, and for the high quality manuscripts that they produced. Many anonymous reviewers provided invaluable feedback on individual chapters, for which we are very grateful, and we particularly wish to thank Nicole Couture for managing the peer-review for our own first-authored chapters. Our graduate students and others who have joined us during fieldwork on northern Ellesmere Island over the past 10 years continue to provide a source of inspiration, including: Canadian Rangers who participated in Operation Nunalivut in 2008, Andrew Hamilton, Sierra Pope, Colleen Mortimer, Adrienne White, Miriam Richer McCallum, Tyler de Jong, Nat Wilson, Jill Rajewicz, Kelly Graves, Sam Brenner, Abigail Dalton, Kevin Xu, Dorota Medrzycka and Greg Crocker. Finally, we thank our families for their patience and support while we spend long periods of time away from home on ice shelves and other parts of the cryosphere.

Ottawa, ON, Canada
December 2016

Luke Copland (https://cryospheric.org)
Derek Mueller (https://wirl.carleton.ca)

Contents

Contributors

Bea Alt Balanced Environments Associates, Carlsbad Springs, Ottawa, ON, Canada

William F. Althoff Geologist/Naval Historian, Whitehouse Station, New Jersey, USA

Dermot Antoniades Department of Geography and Centre for Northern Studies (Centre d'études nordiques), Université Laval, Québec, QC, Canada

Igor M. Belkin Graduate School of Oceanography, University of Rhode Island, Narragansett, RI, USA

Carsten Braun Geography and Regional Planning Department, Westfield State University, Westfield, MA, USA

Luke Copland Department of Geography, Environment and Geomatics, University of Ottawa, Ottawa, ON, Canada

Julian A. Dowdeswell Scott Polar Research Institute, University of Cambridge, Cambridge, UK

John H. England Department of Earth and Atmospheric Sciences, University of Alberta, Edmonton, AB, Canada

David J.A. Evans Department of Geography, Durham University, Durham, UK

Mark Fuglem C-CORE, St. John's, NL, Canada

Martin O. Jeffries Office of Naval Research, Arctic and Global Prediction Program, Arlington, VA, USA

Ian Jordaan Ian Jordaan and Associates Inc., St. John's, NL, Canada

Anne D. Jungblut Centre d'Études Nordiques (CEN), Institut de biologie intégrative et des systèmes (IBIS) and Département de Biologie, Laval University, Quebec City, QC, Canada

Life Sciences Department, The Natural History Museum, London, UK

Sergey A. Kessel Arctic and Antarctic Research Institute (retired), St. Petersburg, Russia

Thomas R. Lakeman Department of Earth and Atmospheric Sciences, University of Alberta, Edmonton, AB, Canada

Colleen Mortimer Department of Earth and Atmospheric Sciences, University of Alberta, Edmonton, AB, Canada

Derek Mueller Department of Geography and Environmental Studies, Carleton University, Ottawa, ON, Canada

Sierra Pope Department of Geography, Environment and Geomatics, University of Ottawa, Ottawa, ON, Canada

Niels Reeh National Space Institute, Technical University of Denmark, Lyngby, Denmark

Miriam Richer McCallum Department of Geography and Environmental Studies, Carleton University, Ottawa, ON, Canada

Wesley Van Wychen Department of Geography, Environment and Geomatics, University of Ottawa, Ottawa, ON, Canada

Warwick F. Vincent Centre d'Études Nordiques (CEN), Takuvik Joint International Laboratory and Département de Biologie, Laval University, Quebec City, QC, Canada

Adrienne White Department of Geography, Environment and Geomatics, University of Ottawa, Ottawa, ON, Canada

Part I
Distribution and Characteristics of Arctic Ice Shelves

Chapter 1
Arctic Ice Shelves: An Introduction

Julian A. Dowdeswell and Martin O. Jeffries

Abstract Ice shelves are relatively thick ice masses that are afloat but attached to coastal land rather than adrift. They form by the seaward extension of glaciers or ice sheets or by build up of multiyear landfast sea ice. They thicken further by surface accumulation of snow and superimposed ice and by accretion of ice from water beneath. Composite ice shelves are composed of sea ice and glacier ice. Glacier tongues are floating ice margins that are narrow relative to their length. Ice shelves comprise 55% or 18,000 km of the Antarctic coast. 'Classical' Antarctic ice shelves are fed from glaciers or ice streams and are dynamically part of the parent ice sheet; the largest, the Ross and Ronne, are 10^5 km^2 and hundreds of metres thick. Where they ground on isolated bedrock peaks, 'ice rises' are formed. Arctic ice shelves are restricted to several archipelagos fringing the Arctic Ocean and to a few Greenland fjords. The Ward Hunt Ice Shelf is the largest at about 400 km^2. Arctic and Antarctic ice shelves have expanded and contracted during the Holocene. The Ellesmere Ice Shelf developed about 5500 years ago in response to Holocene cooling. In the warmer Twentieth century, calving events have broken this continuous ice-shelf into several remnants. Floating glacier tongues of the Greenland Ice Sheet have also broken up recently. The entire Arctic Ocean may have been covered by a huge ice shelf during the coldest Late Cenozoic glacial periods. Large, often tabular icebergs calve from ice shelves. Ice islands are a form of tabular iceberg in the Arctic Ocean which have a characteristic undulating surface. Icebergs drift mainly under the influence of currents and Arctic Ocean ice islands have been used occasionally as research stations.

Keywords Ice shelf • Glacier • Sea ice • Sikussak • Surface accumulation • Bottom accretion • Iceberg • Ice Island • Ellesmere Island • Greenland • Severnaya Zemlya • Franz Josef Land

---

J.A. Dowdeswell (✉)
Scott Polar Research Institute, University of Cambridge, Cambridge, UK
e-mail: jd16@cam.ac.uk

M.O. Jeffries
Office of Naval Research, Arctic and Global Prediction Program, Arlington, VA, USA
e-mail: martin.jeffries@navy.mil

L. Copland, D. Mueller (eds.), *Arctic Ice Shelves and Ice Islands*, Springer Polar Sciences, DOI 10.1007/978-94-024-1101-0_1

1.1 Introduction

The cryosphere, all those parts of the Earth that are frozen, has a number of components: permafrost; snow; freshwater ice on lakes and rivers; sea ice and icebergs on the ocean; glaciers and ice sheets; and ice shelves. Among the ice categories, the thickness ranges from a few millimetres to metres (freshwater and sea ice), to hundreds to thousands of metres (glaciers and ice sheets). Ice shelves, ranging in thickness from tens to hundreds of metres, fall between the two extremes. Simply defined, an ice shelf is a relatively thick ice mass that is afloat on the ocean but attached to coastal land rather than free to drift under the influence of winds and currents, as happens to sea ice and icebergs.

Ice shelves are common at the seaward margins of the modern Antarctic Ice Sheet, where about 55% of the coast is made up of these floating ice features (Fig. 1.1) (e.g. Drewry et al. 1983; Griggs and Bamber 2011; Pritchard et al. 2012). The largest of the Antarctic ice shelves, the Ross and Ronne, are 10^5 km^2 in area and hundreds of metres thick. In the Arctic, ice shelves are much smaller and fewer than in the Antarctic. For example, the most extensive individual ice shelf in the Arctic, the Ward Hunt Ice Shelf, Nunavut, Canada, is, at about 224 km^2, three orders of magnitude smaller than the large Antarctic ice shelves. Geographically, the Arctic

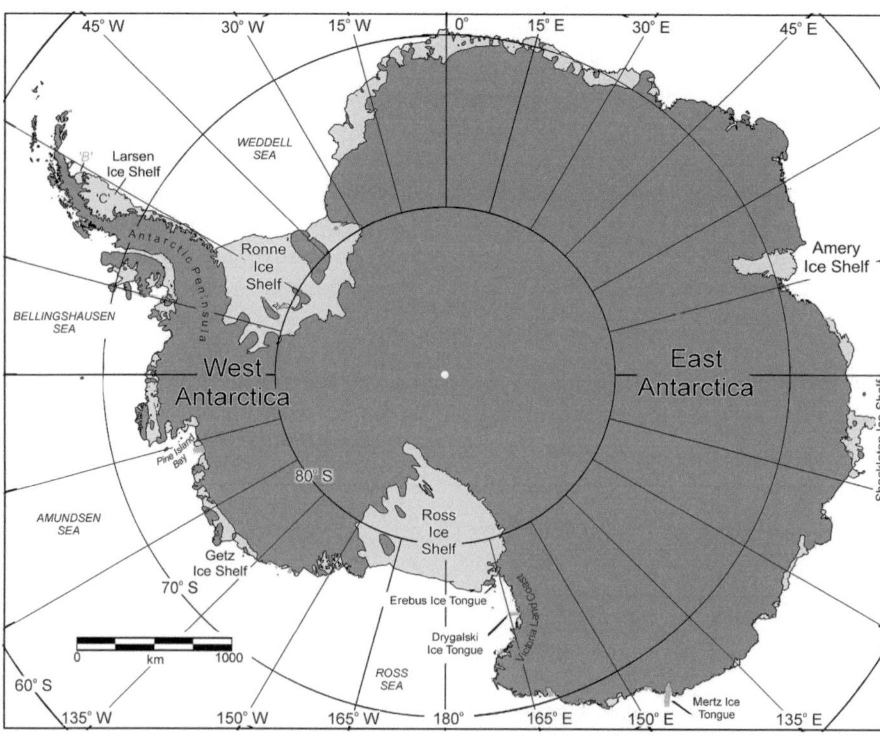

Fig. 1.1 Map of the Antarctic, with the locations of the major ice shelves shown

ice shelves are restricted to several of the high-Arctic archipelagos that fringe the Arctic Ocean (Dowdeswell 2017; Jeffries 2017) and to the fjords of Greenland, where a number of fast-flowing glaciers drain the Greenland Ice Sheet to the sea (Fig. 1.2) (Reeh 2017).

This book is concerned primarily with the Ellesmere Island ice shelves of the Canadian High Arctic, but it also describes Eurasian and Greenlandic ice shelves (Dowdeswell 2017; Reeh 2017). The aims of this chapter are to introduce the general characteristics and significance of modern Arctic ice shelves. We begin by developing a general definition of an Arctic ice shelf and describe the three main types of Arctic ice shelf and how they differ from 'classical' Antarctic ice shelves. The chapter continues with sections on the types and distribution of ice shelves, their physics and mass balance, ice shelves and environmental change, and the icebergs and ice islands that are produced by calving from them.

1.2 Defining an Arctic Ice Shelf

In the first paragraph we defined an ice shelf simply as a relatively thick ice mass that is afloat on the ocean but attached to the land (coast) rather than free to drift under the influence of winds and currents. This basic picture quickly becomes more complex when factors such as the morphology, thickness, mass balance, ice sources and composition of ice shelves are considered.

This complexity was recognized by Barkov (1985), who proposed a sixteen-category genetic classification for Antarctic ice shelves, each category being defined by the type of ice that formed the main mass at the time of initial ice-shelf formation, and the direction of subsequent mass transfer at the top and bottom surfaces. Vaughan (1998) proposed a simpler nine-category classification for both Antarctic and Arctic ice shelves, each category being defined by the dominant source of input (from glacier flow, *in situ* surface accumulation or basal accretion) and the dominant route for ablation (by iceberg calving, surface melting/sublimation or basal melting). According to Vaughan, there are well-documented contemporary examples of ice shelves for six of the nine possible categories. Thus, in the Canadian Arctic, the Ward Hunt Ice Shelf falls within Vaughan's Type E; that is, build up through basal accretion and mass loss mainly by surface melting/sublimation. The continuing disintegration of part of this ice shelf, however, would also put it in Vaughan's Type C, with iceberg calving being the dominant mechanism of mass loss. In the Eurasian Arctic, the Matusevich Ice Shelf on Severnaya Zemlya, which largely disintegrated in 2012 (Willis et al. 2015), would be Vaughan's Type A, with glacier input and calving loss (Williams and Dowdeswell 2001).

To define an *Arctic ice shelf*, we return to one of the earliest, and most simple, definitions of an *ice shelf* (Armstrong et al. 1966):

> A floating ice sheet of considerable thickness attached to the coast. Ice shelves are usually of great horizontal extent and have a level or undulating surface. They are nourished by accumulation of snow and often by seaward extension of land glaciers. Limited areas may be aground

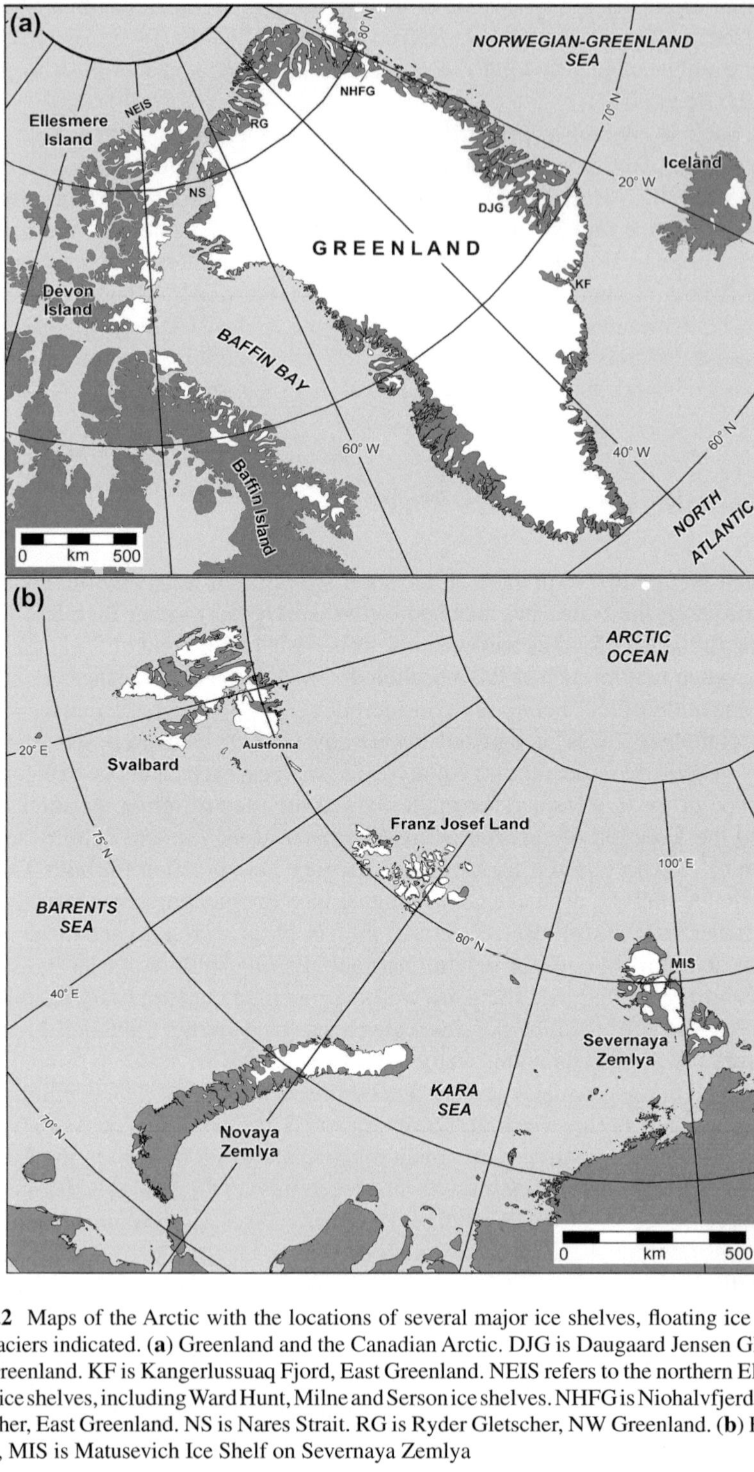

Fig. 1.2 Maps of the Arctic with the locations of several major ice shelves, floating ice tongues and glaciers indicated. (**a**) Greenland and the Canadian Arctic. DJG is Daugaard Jensen Gletscher, East Greenland. KF is Kangerlussuaq Fjord, East Greenland. NEIS refers to the northern Ellesmere Island ice shelves, including Ward Hunt, Milne and Serson ice shelves. NHFG is Niohalvfjerdsfjorden Gletscher, East Greenland. NS is Nares Strait. RG is Ryder Gletscher, NW Greenland. (**b**) Eurasian Arctic, MIS is Matusevich Ice Shelf on Severnaya Zemlya

However, there are also ice shelves that are nourished, or originate, from other sources. This is indeed the case in the Arctic, where sea ice is a significant component of some ice shelves (Jeffries 2017), as indicated by the definition of *fast ice* (Armstrong et al. 1966):

> Sea ice which remains fast along the coast, where it is attached to the shore, to an ice wall, to an ice front, or over shoals, or between grounded icebergs. Fast ice may extend a few m or several km from the shore. Fast ice may be more than a year old. When its surface level becomes higher than about 2 m above sea level, it is called an ice shelf

Combining the essential genetic elements of these two definitions, an ice shelf is an ice mass that forms by the seaward extension of glaciers or the formation of multiyear sea ice, or both, and thickens further by the accumulation of snow. However, from a genetic perspective, this definition remains incomplete, as it does not include ice accretion at the bottom surface, which is now known to be a relatively common process (e.g. Jenkins and Doake 1991; Jeffries 1992a; Wen et al. 2010).

We now have the essential genetic elements (glaciers, sea ice, snow accumulation, bottom accretion) for a definition of an Arctic ice shelf, but the morphology or appearance, particularly of the top surface, is also an important element, as recognized by the definition of an *ice island* (Armstrong et al. 1966):

> A form of tabular berg found in the Arctic Ocean, with a thickness of 30-50 m and from a few thousand square m to 500 square km in area. Ice islands often have an undulating surface, which gives them a ribbed appearance from the air.

The ribbed appearance of the original ice islands that were discovered in the late 1940s is inherited from their source; the ice shelves of northernmost Ellesmere Island (Koenig et al. 1952). While the Ellesmere ice shelves are generally considered to be the main source of ice islands, Higgins (1989) drew attention to the fact that large tabular icebergs with an undulating surface also calve from the long floating glacier tongues in the fjords of northernmost Greenland.

Finally, we note a minor contradiction between the minimum thickness of an ice shelf originating from sea ice (20 m, as inferred from a surface elevation of 2 m above sea level) and the minimum thickness of an ice island (30 m). Given that sea-ice formation is an essential element of the Ellesmere ice shelves, the source of the original ice islands, we choose 20 m as a minimum thickness for an Arctic ice shelf or, by implication, an ice island.

Now we are in a position to propose a general definition of an *Arctic ice shelf* that includes morphological and genetic elements, and thickness:

> An ice mass of considerable thickness (≥20 m) that is afloat on the ocean but attached to the coast. It is often of great horizontal extent (many km) and has, typically but not exclusively, a regularly undulating surface. An ice shelf can form by the seaward extension of a glacier or glaciers, or by the formation of multiyear sea ice, or both, and thicken further by the accumulation of snow at the top surface and the accretion of ice from water at the bottom surface.

In the next section, we begin with a description of the 'classical' type of ice shelf that is typically associated with Antarctica, and then describe the three types of Arctic ice shelf that are the subject of this book; each is sufficiently different from the 'classical' ice shelf to warrant separate description.

1.3 Types and Distribution of Ice Shelves

1.3.1 'Classical' Ice Shelf

The 'classical' ice shelves comprise most of the floating margins of the Antarctic Ice Sheet (Fig. 1.1). They extend up to hundreds of kilometres seaward of a 'grounding zone' where glaciers and ice sheets become decoupled from the bed and reach hydrostatic equilibrium with the underlying water. Often originating from many glaciers and/or ice streams that coalesce in very large embayments, these 'classical' ice shelves, such as the Ross, Ronne and Amery, are dynamically a part of the parent ice sheet. The largest Antarctic ice shelves are over a thousand metres thick at the grounding zone, thinning to roughly 200–300 m at their seaward fronts (Dowdeswell and Bamber 2007; Griggs and Bamber 2011; Fretwell et al. 2013), which can be several hundred kilometres long and a similar distance from the grounding zone. These ice shelves form extensive areas of low-gradient ice (Fretwell et al. 2013), with water hundreds of metres deep beneath them. Where their beds come into contact with isolated bedrock peaks, basal shear stress and ice-surface-gradient increase, forming domes known as 'ice rises'. Ice rises are easily identified on otherwise flat ice-shelf surfaces (Fig. 1.3a).

About 18,000 km of the seaward margin of the Antarctic Ice Sheet is afloat, with about 14,000 km of the coastline being made up of 'classical' ice shelves and a further 4000 km being outlet glaciers and ice streams with floating tongues (Fig. 1.1) (Drewry et al. 1983). The huge Ross, Ronne and Amery ice shelves extend up to 450 km from grounding zone to terminus and are about 300 m thick at their margins (Dowdeswell and Bamber 2007; Griggs and Bamber 2011). A large number of smaller fringing ice shelves, often only about 200 m thick at their margins and with flow lines of tens of kilometres, are located around both West and East Antarctica, and the Antarctic Peninsula (e.g. Getz, Shackleton, Larsen C ice shelves; Fig. 1.1).

In the Arctic, the Milne Ice Shelf of northernmost Ellesmere Island, confined within the steep-sided Milne Fiord, is composed of a number of small coalesced glacier tongues (Jeffries 1986, 2017). As such, it is probably the closest of all the Ellesmere ice shelves to being a classical ice shelf.

1.3.2 Glacier Tongue

By contrast with 'classical' ice shelves, a glacier tongue (sometimes also referred to as an ice tongue) is a floating ice margin that is narrow relative to its length (Hambrey 1994). Glacier tongues are usually fed from fast-flowing outlet glaciers or ice streams, which are constrained by valley sides or are bounded laterally by shear zones between fast- and slow-flowing ice, respectively. In either case, fast-flowing ice is confined to linear or curvilinear filaments within the ice sheet (Bamber et al. 2000) and the floating tongues represent the terminus regions of these ice masses.

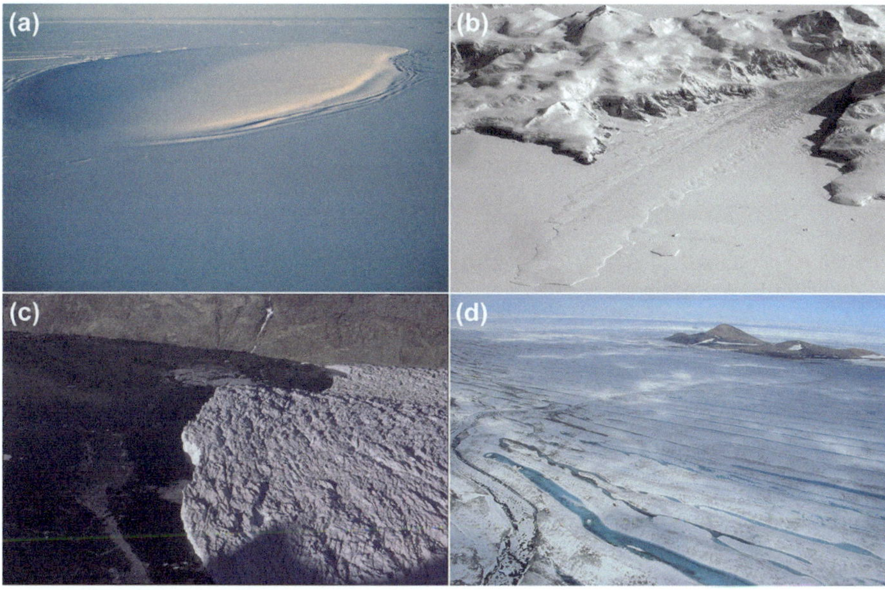

Fig. 1.3 Photographs of Antarctic and Arctic ice shelves. (**a**) Larsen C Ice Shelf and Gipps Ice Rise, eastern Antarctic Peninsula (Photo: C.W.M. Swithinbank). The ice rise is about 18 km long. (**b**) A floating glacier tongue, 2–4 km wide, embedded in sea ice immediately north of the larger Aviator Glacier Tongue, Victoria Land Coast, Antarctica (Photo: J.A. Dowdeswell). (**c**) The floating margin of Daugaard Jensen Gletscher, East Greenland (Photo: J.A. Dowdeswell). (**d**) The Ward Hunt Ice Shelf, northern Ellesmere Island, Arctic Canada (Photo: D.R. Mueller)

In the Antarctic, a number of glacier tongues extend well beyond constraining mountains and into the open ocean beyond (Fig. 1.3b); examples include the Erebus, Drygalski and Mertz glacier tongues (e.g. McIntyre 1985). Some of these ice tongues, usually less than 100 km long, are sources for among the thickest floating ice around Antarctica, with termini of 400 to 600 m in thickness (Fig. 1.3b) (Dowdeswell and Bamber 2007).

In the Arctic, glacier tongues are often confined to fjords, with the mountainous rock walls of a number of Greenland glacier tongues providing clear examples (Fig. 1.3c) (e.g. Higgins 1989, 1990; Mayer et al. 2000; Enderlin and Howat 2013). The very large (10^4 km^2 in area) interior basins of the Greenland Ice Sheet are drained by about fifteen fast-flowing outlet glaciers that have eroded deep valleys and troughs through the fringing mountains (e.g. Rignot and Kanagaratnam 2006; Reeh 2017). The margins of a number of these outlet glaciers form floating ice tongues in fjords which can exceed 1000 m in water depth (e.g. Nick et al. 2012). The ice tongues are often less than 10 km long, but can be as much as a few tens of kilometres in length, with a marginal thickness of about 400–600 m. The largest example is Niohalvfjerdsfjorden Gletscher, at 79°N in NE Greenland (Fig. 1.2a), which has a 60 km-long floating terminus (Mayer et al. 2000). The Greenland glacier tongues are described in more detail by Reeh (2017).

Finally, it should be noted that glacier tongues and tidewater glaciers are not synonymous. Whereas an ice tongue is afloat, tidewater glaciers are grounded; that is, resting entirely on the sea floor (but still able to calve icebergs). In principle, a tidewater glacier could become an ice tongue if thinning and/or sea-level rise made it sufficiently buoyant to float.

1.3.3 Sea-Ice Ice Shelf

Unlike classical ice shelves and glacier tongues, which are both composed mainly of glacier ice, some ice shelves owe their origin to sea-ice growth and remain composed predominantly of sea ice. Lemmen et al. (1988) coined the term 'sea-ice ice shelf' for these ice shelves. The best-known example of a sea-ice ice shelf, the 40–50 m thick and ~224 km^2 Ward Hunt Ice Shelf (Fig. 1.3d), is the largest remnant of the former Ellesmere Ice Shelf. Extending tens of kilometres offshore and hundreds of kilometres alongshore in the early years of the twentieth century (Koenig et al. 1952), the Ellesmere Ice Shelf was, in all likelihood, predominantly a sea-ice ice shelf. Sea-ice ice shelves form initially from multiyear fast ice (also known as multiyear landfast sea ice, MLSI; Jeffries et al. (1989)), often in sheltered inlets and embayments, but also along more exposed stretches of coast where multiyear pack ice provides protection from the influence of the open ocean. The multiyear landfast sea ice acts as a platform, or basement, for further thickening by both snow and superimposed ice accumulation at the top surface and accretion of ice from the water at the bottom surface. As indicated by the definition of fast ice in the previous section, sea-ice ice shelves are at least 20 m thick.

1.3.4 Composite Ice Shelf

A composite ice shelf is composed of both sea ice and glacier ice (Lemmen et al. 1988). The Serson Ice Shelf on Ellesmere Island was a good example of a composite ice shelf until summer 2008, when much of the sea ice unit and two glacier tongues calved and drifted away. The Serson Ice Shelf is now composed largely of just two remaining glacier tongues. The original glacier tongues that flowed offshore to form the glacier component of the ice shelf were clearly visible as 'filaments' of glacier ice (Jeffries 2017), as were the glacier tongues that formed part of Ice Island ARLIS II (Smith 1964), which calved from the Serson Ice Shelf in the mid-1950s (Jeffries 1992b).

In Greenland, the Inuit use the term 'sikussak' to describe the mélange of multiyear landfast sea ice and icebergs that is typically found in protected High Arctic fjords (Fig. 1.4). The many icebergs trapped in multiyear sea ice at the southern end of Yelverton Fiord, and in Disraeli Fiord to the south of the disintegrating Ward Hunt Ice Shelf, are examples of sikussak on Ellesmere Island. Extensive areas of

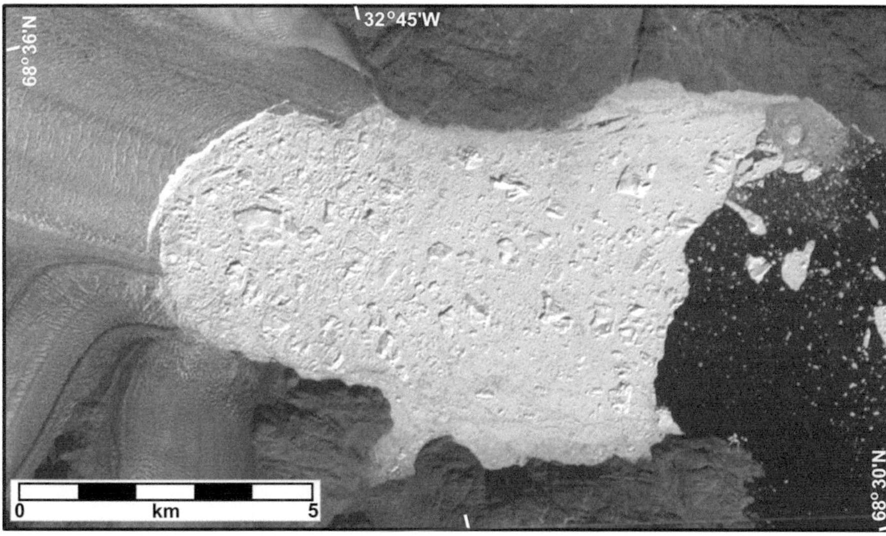

Fig. 1.4 Sikussak; a frozen ice mélange of multiyear sea ice and icebergs, Kangerlussuaq Fjord, East Greenland. Landsat ETM+ image, path 229 row 012, 16 August 2002

sikussak have also been reported in North and North-East Greenland (Koch 1945), including that in Sherard Osborne Fjord, into which Ryder Gletscher drains, where icebergs up to about 10 km in length have been observed trapped in the multiyear landfast sea ice (Higgins 1989). Recently, the term 'ice mélange' has been adopted by many workers to describe the mix of sea-ice floes and icebergs found in Greenland fjords with calving glaciers at their heads (e.g. Amundson et al. 2010; Cassotto et al. 2015).

In principle, continued thickening of sikussak, by accumulation of snow and ice at the top surface and accretion of ice at the bottom surface, could produce a form of composite ice shelf; the glacier component being icebergs that have calved from glaciers, rather than being glacier tongues that have flowed seaward from the ice sheet. More normally, sikussak or ice mélange breaks up at timescales that range from seasonal to decadal, with the release of trapped icebergs and a series of large multiyear ice floes (e.g. Reeh et al. 1999; Dowdeswell et al. 2000; Cassotto et al. 2015).

Sikussak-like ice is also found in the Eurasian Arctic, where calving outlet glaciers mix with multiyear landfast sea ice. In Matusevich Fjord, Severnaya Zemlya (Fig. 1.2b), for example, where sea ice has persisted for many years, it contains icebergs from outlet glaciers which enter the fjord from the surrounding ice caps (Williams and Dowdeswell 2001). However, the few ice shelves that do exist in the Eurasian Arctic, in Severnaya Zemlya and Franz Josef Land, are formed mainly from glacier ice rather than the build up of old sea ice (Dowdeswell 2017).

1.4 Physics and Mass Balance of Ice Shelves

Ice shelves are, for the most part, fully buoyant, in hydrostatic equilibrium with the underlying water (Robin 1979). Close to the grounding zone, at their lateral margins and at any pinning points where they ground on bedrock pinnacles as 'ice rises' (Fig. 1.3a), a part of their mass may be supported by the bed and side walls. Ice shelves have very low-gradient surfaces because, at the almost frictionless contact with the water beneath, the basal shear stress approaches zero. Ice shelves thin under their own weight by internal deformation, a process that is most rapid in unbounded ice shelves without side walls or pinning points where they touch the sea floor. Creep thinning is responsible, in part, for the systematic reduction in ice thickness with distance from the grounding zone. In an unbounded ice shelf, creep or deformation thinning is approximately proportional to the fourth power of ice thickness (Weertman 1957; Robin 1979; Thomas 1979).

The mass balance of ice shelves is the sum of gains and losses of mass to the system, usually measured over a balance year (from one late-summer to the next). An ice shelf is in equilibrium when these gains and losses are approximately equal. Classical ice shelves and ice tongues that are fed from a parent glacier or ice-sheet drainage basin gain mass in several ways; by advection of ice across the grounding zone flowing from an ice-sheet interior, by accumulation of snow on the ice shelf surface and by adfreezing of sea water at the bottom surface. Mass is lost by iceberg production, known as calving, by surface melting (and sometimes sublimation), and by basal melting at the ice-ocean interface.

At the base of large Antarctic ice shelves, basal melting generally occurs close to the grounding zone and towards the ice margin. Basal melting can reach more than 10 m year^{-1} near the grounding zone (e.g. Jenkins et al. 1997; Rignot et al. 2013). Dense (cold and saline) ocean water, often formed by sea-ice production beyond the ice-shelf edge, usually flows in towards the grounding zone along the sea floor, which typically slopes inshore due to ice-sheet loading and isostatic depression. The water becomes less dense as it mixes with meltwater close to the grounding line and, thus, flows up the ice-shelf basal boundary as it become more buoyant. As it rises, pressure decreases, the melting point is raised, and supercooled water may freeze to the ice-ocean interface (Jenkins and Doake 1991). This process can lead to the accretion of tens of metres of sea-water derived ice at the ice shelf base. Ice core and radar studies of Antarctic ice shelves have shown a three-part structure; glacier ice derived by flow from continental drainage basins is sandwiched between densified snow from accumulation on the ice-shelf surface, and sea-water derived ice frozen on at the floating ice-shelf base (Jenkins and Doake 1991; Fricker et al. 2001).

While their dimensions are certainly smaller, the Arctic ice shelves are similar, in many respects, to the Antarctic ice shelves in terms of mass balance and physics. Glaciers nourish part or all of some Arctic ice shelves, for example those of Greenland and the Russian Arctic. In the case of the Ellesmere ice shelves, there has been *in situ* surface accumulation of snow and superimposed ice, in addition to sea-ice thickening, although superimposed ice is currently being lost during a prolonged period of surface melting and negative surface mass balance in summer (Braun 2017; Jeffries 2017). The Arctic ice shelves also lose mass by iceberg (ice island) calving and bottom melting. Basal melting close to the grounding zones of Greenland ice tongues is often several metres to tens of metres per year (Rignot and Kanagaratnam 2006; Enderlin and Howat 2013). Unlike most of their Antarctic counterparts, however, surface melting is also important on Arctic ice shelves.

Bottom accretion is also known to occur on the Ellesmere ice shelves (Jeffries 2017). It is the nature of bottom accretion that perhaps distinguishes the Ellesmere ice shelves from their larger Antarctic counterparts; that is, bottom accretion commonly involves the freezing of fresh and brackish waters as well as seawater (Jeffries 2017). The primary source of fresh and brackish waters is drainage from epishelf lakes, which form behind the dam-like ice shelves. An epishelf lake is a tidal body of low-salinity water which, because of its lower density, "floats" on seawater. The epishelf lake in Disraeli Fiord to the south of the Ward Hunt Ice Shelf, and the role of the positive feedback that occurs between them in the mass balance of the ice shelf, are discussed in Jeffries (1991).

1.5 Ice Shelves and Environmental Change

In both the Arctic and the Antarctic there is evidence that ice shelves have expanded and contracted during the Holocene. The Ellesmere Ice Shelf appears to have developed about 5500 years ago as a response to cooling after the Early Holocene climatic optimum (Koerner and Fisher 1990; England et al. 2008; England et al. 2017). At the beginning of the Twentieth century, at the end of the centuries-long cold interval known as the 'Little Ice Age', the ice shelf extended unbroken for about 500 km with an area of 8900 km^2 along the coast of northern Ellesmere Island (Vincent et al. 2001). Since then, a series of calving events has broken the once single, continuous ice-shelf fringe into a number of small and isolated remnants (Mueller et al. 2017).

The floating tongues of several fast-flowing outlet glaciers of the Greenland Ice Sheet have also broken up in recent years; a consequence, in part at least, of the northward penetration of relatively warm Atlantic water into Greenland fjords, enhancing basal melting and thinning of the floating ice tongues (Holland et al. 2008; Christoffersen et al. 2011). Conversely, in periods of climatic cooling, such as the Little Ice Age or the Younger Dryas stadial, colder conditions would apply to more parts of the Arctic coastline and multiyear landfast sea ice and sikussak would be expected to spread southward, especially into fjords and protected inlets (Reeh et al. 1999; Dowdeswell et al. 2000). The presence of sikussak is also thought to help stabilize and protect floating glacier tongues from mass loss by iceberg calving (Reeh et al. 2001; Amundson et al. 2010; Cassotto et al. 2015).

At longer time scales, it has been suggested that the entire Arctic Ocean may have been covered by a huge ice shelf in one or more of the longest and coldest full-glacial periods in the Late Cenozoic (Mercer 1970). Sea-floor morphological features, interpreted by some to represent regions of ice-shelf grounding, have been observed on, for example, the Lomonosov Ridge and the Yermak Plateau in the Arctic Ocean (e.g. Vogt et al. 1994; Polyak et al. 2001; Jakobsson 2016). The dating of such events is still uncertain, but it appears that a major ice shelf may have occupied the whole Arctic Basin at about 660,000 years ago (e.g. Vogt et al. 1994; Flower 1997), and that ice shelves may also have developed around at least the fringes of the deep-water basin during Marine Isotope Stage 6 about 140–160,000 years ago (Jakobsson et al. 2010, 2014, 2016).

In the Antarctic, too, considerable variability in Holocene ice-shelf extent has been demonstrated. Although the relatively large Larsen B Ice Shelf appears not to have disintegrated in the Holocene prior to its collapse in 2002 (Domack et al. 2005), smaller ice shelves on the Antarctic Peninsula disappeared in the mid-Holocene and reformed about 2000 years ago (Pudsey and Evans 2001). More recent, dramatic retreat or disintegration of a number of ice shelves on the Antarctic Peninsula has been observed from satellite imagery over the past two decades and linked with climate warming (e.g. Doake and Vaughan 1991; Vaughan and Doake 1996; Scambos et al. 2003; Cook and Vaughan 2010). The rapid ice-shelf disintegration in the Antarctic Peninsula has been ascribed to surface meltwater penetration into crevasses, which then deepen to the ice-shelf base to cause break-up (e.g. Scambos et al. 2003; Banwell et al. 2013), although enhanced basal melting and thinning due to warming ocean waters has also been implicated (Shepherd et al. 2003). A mean summer air temperature of about 0°C is thought to represent an empirical upper climatic limit for the viability of Antarctic ice shelves (Robin and Adie 1964), and Vaughan and Doake (1996) make a similar suggestion concerning the −5°C mean annual isotherm. It has been proposed that, because of this empirical air temperature threshold, ice shelves may represent a particularly sensitive indicator of climate change (Mercer 1978; Vaughan and Doake 1996).

1.6 Icebergs and Ice Islands

Relatively large, often tabular icebergs calve from ice shelves (e.g. Dowdeswell et al. 1992; Dowdeswell and Bamber 2007). They are known as 'tabular' for their characteristically steep sides and low-gradient, regular surface, and are often kilometres to tens of kilometres in length (Fig. 1.5a, b). They contrast with the smaller and more irregular icebergs typically produced from grounded tidewater-glacier margins (Fig. 1.5c), where the relatively close spacing of marginal crevasses means that few large icebergs are produced (Dowdeswell and Forsberg 1992). The floating tongues at the margins of the outlet glaciers of the Greenland and Antarctic ice sheets produce the deepest-keeled icebergs at about 400–600 m (Dowdeswell et al. 1992; Dowdeswell and Bamber 2007). The Ross and Ronne ice shelves in Antarctica produce thinner icebergs, characteristically about 300 m thick, because

Fig. 1.5 Photographs of icebergs from the Antarctic and Arctic seas. (**a**) A tabular iceberg about 1 km long in Pine Island Bay, West Antarctica (Photo: J.A. Dowdeswell). (**b**) Heavily crevassed tabular iceberg (about 1 km long) calved from the floating tongue of the fast-flowing Daugaard Jensen Gletscher, East Greenland (Photo: J.A. Dowdeswell). (**c**) Relatively small icebergs (<100 m long) of irregular shape calved from a grounded tidewater glacier margin, Austfonna, eastern Svalbard (Photo: J.A. Dowdeswell). (**d**) Hobson's Choice Ice Island, calved from the Ward Hunt Ice Shelf, northern Ellesmere Island (Photo: M.O. Jeffries)

creep thinning and basal melting reduce ice-shelf thickness along flowlines that extend several hundred kilometres beyond the grounding zone.

The tabular icebergs, more commonly known as ice islands (Fig. 1.5d), that calve from the Ellesmere ice shelves have typically been up to 50 m thick (e.g. Copland et al. 2007; Jeffries 2017). Jeffries (1992a) defined an ice island as 'a tabular iceberg which has broken away or calved from an Arctic ice shelf. They have a gently undulating surface which gives them a ribbed appearance from the air'. A fuller description of the morphology and characteristics of the Ellesmere ice islands is provided by Jeffries (2017) and Van Wychen and Copland (2017). Although Higgins (1989) also used the term 'ice island' to describe the tabular icebergs that calve from the glacier tongues of North Greenland, many of the icebergs produced from fast-flowing outlet glaciers of the Greenland Ice Sheet have a surface dominated by heavy crevassing rather than being gently undulating (Fig. 1.5b); they are also an order of magnitude thicker than the Ellesmere ice islands (Dowdeswell et al. 1992). Thus, the term 'ice island' is usually restricted to icebergs derived from the Ellesmere ice shelves.

Once calved, and providing they do not become trapped by the shallow sills found in many fjords (Syvitski et al. 1987), icebergs drift in the ocean mainly under the influence of currents. Icebergs gradually deteriorate by fragmentation due to flexure in ocean-swell waves (Kristensen et al. 1982) and by melting as they float into warmer waters (e.g. Weeks and Campbell 1973; Enderlin and Hamilton 2014). In the Arctic Ocean, several ice islands are known to have circulated for many years in the Beaufort Gyre before becoming entrained in the Transpolar Drift and drifting south via Fram Strait into the North Atlantic Ocean along the east coast of Greenland (e.g., Jeffries 1992a; Van Wychen and Copland 2017). Icebergs calving from East Greenland outlet glaciers take a similar path. Ice islands also drift into the inter-island channels of the Canadian Arctic Archipelago (Koenig et al. 1952; Jeffries and Shaw 1993; Copland et al. 2007), and into Baffin Bay via Nares Strait (Nutt 1966), the channel separating Ellesmere Island from Greenland (Fig. 1.2a). Icebergs produced from north-west Greenland outlet glaciers follow a similar route, and those which calve directly into Baffin Bay from West Greenland can circle the bay anti-clockwise before reaching the waters off Labrador and Newfoundland (Robe 1980; Van Wychen and Copland 2017).

Because of their large size and relative stability, ice islands that have entered the Arctic Ocean have been used occasionally as platforms for drifting research stations by the USA, Canada and the Soviet Union/Russia (Althoff 2017; Belkin and Kessel 2017). In addition to supporting scientific investigations of the atmosphere, sea ice, ocean and seafloor, the ice islands themselves have revealed much about the ice shelves from which they calved (e.g. Jeffries 1992a, 2017).

1.7 Summary

The interface between ice sheets and the ocean can be made up of either floating ice shelves or grounded tidewater glaciers. Antarctic ice shelves, and some in the Arctic, are fed from fast-flowing ice streams and outlet glaciers advecting mass from interior ice-sheet drainage basins to their termini to offset mass loss through iceberg production and basal melting. The ice shelves of northern Ellesmere Island in the Canadian High Arctic, although much smaller than they were only 100 years ago, remain the largest in the circum-Arctic. They are distinguished by the role that multiyear landfast sea ice has often played in their expansion. Further bottom accretion of fresh and brackish ice, as well as sea ice, and the inflow of glacier tongues to form composite ice shelves of glacier ice and sea ice, means that the crystallographic and geochemical characteristics of the Ellesmere ice shelves differ from those ice shelves located at ice-sheet margins. Tabular icebergs (known as ice islands when they have calved from the Ellesmere ice shelves) are produced by calving from the margins of both glacier-fed and sea-ice ice shelves, but icebergs from the two sources can usually be differentiated by their internal structure and surface character (Fig. 1.5).

In defining and reviewing the different types of ice shelf and iceberg that occur in the Arctic and Antarctic, our aim has been to provide a general context for the more detailed chapters of this book (Copland and Mueller 2017) that focus, in particular, on the characteristics, history and human use of the ice shelves and ice islands found in northern Ellesmere Island and the Arctic Ocean to the north. The distribution and character of ice shelves and icebergs in the Canadian Arctic, Greenland and the archipelagos of the Eurasian Arctic are also considered briefly here, and in more detail in Jeffries (2017), Dowdeswell (2017) and Reeh (2017), to provide a wider geographical and glaciological perspective.

Acknowledgments Grants from the John Ellerman Foundation and the Arctic Environmental Program of ConocoPhillips supported JAD for parts of this work. MOJ contributed to this work while he was on leave from the University of Alaska Fairbanks and working on secondment from 2006 to 2010 at the National Science Foundation (NSF) under the terms of the Inter-Governmental Personnel Act; any opinion, findings, and conclusions or recommendations expressed in this material are his and do not necessarily reflect the views of either NSF or the Office of Naval Research. We thank Toby Benham for his work with the figures.

References

Althoff, W. (2017). The military importance and use of ice islands during the Cold War. In L. Copland & D. Mueller (Eds.), *Arctic ice shelves and ice islands* (p. 343–366). Dordrecht: Springer. doi:10.1007/978-94-024-1101-0_13.

Amundson, J. M., Fahnestock, M., Truffer, M., Brown, J., Lüthi, M. P., & Motyka, R. J. (2010). Ice mélange dynamics and implications for terminus stability, Jakobshavn Isbrae, Greenland. *Journal of Geophysical Research, 115*. doi:10.1029/2009JF001405.

Armstrong, T., Roberts, B. B., Swithinbank, C. W. M. (1966). *Illustrated glossary of snow and ice* (Special publication no. 4). Cambridge: Scott Polar Research Institute.

Bamber, J. L., Vaughan, D. G., & Joughin, I. (2000). Widespread complex flow in the interior of the Antarctic Ice Sheet. *Science, 287*, 1248–1250.

Banwell, A. F., MacAyeal, D., & Serglenko, O. V. (2013). Breakup of the Larsen B Ice Shelf triggered by chain reaction drainage of supraglacial lakes. *Geophysical Research Letters, 40*, 5872–5876.

Barkov, N. I. (1985). *Ice shelves of Antarctica* (Russian translations series 20). Rotterdam: Balkema.

Belkin, I., & Kessel, S. (2017). Russian drifting stations on Arctic ice islands. In L. Copland & D. Mueller (Eds.), *Arctic ice shelves and ice islands* (p. 367–393). Dordrecht: Springer. doi:10.1007/978-94-024-1101-0_14.

Braun, C. (2017). The surface mass balance of the Ward Hunt Ice Shelf and Ward Hunt Ice Rise, Ellesmere Island, Nunavut, Canada. In L. Copland & D. Mueller (Eds.), *Arctic ice shelves and ice islands* (p. 149–183). Dordrecht: Springer. doi:10.1007/978-94-024-1101-0_6.

Cassotto, R., Fahnestock, M., Amundson, J. M., Truffer, M., & Joughin, I. (2015). Seasonal and interannual variations in ice melange and its impact on terminus stability, Jakobshavn Isbrae, Greenland. *Journal of Glaciology, 61*, 76–87.

Christoffersen, P., Mugford, R., Heywood, K. J., Joughin, I., Dowdeswell, J. A., Syvitski, J. P. M., Luckman, A., & Benham, T. J. (2011). Warming of waters in an East Greenland fjord prior to glacier retreat: Mechanisms and connection to large-scale atmospheric conditions. *The Cryosphere, 5*, 701–714.

Cook, A. J., & Vaughan, D. G. (2010). Overview of areal changes of the ice shelves on the Antarctic Peninsula over the past 50 years. *The Cryosphere, 4*, 77–98.

Copland, L., & Mueller, D. (Eds.). (2017). *Arctic ice shelves and ice islands*. Dordrecht: Springer. doi:10.1007/978-94-024-1101-0.

Copland, L., Mueller, D. R., & Weir, L. (2007). Rapid loss of the Ayles Ice Shelf, Ellesmere Island, Canada. *Geophysical Research Letters, 34*, L21501. doi:10.1029/2007GL031809.

Doake, C. S. M., & Vaughan, D. G. (1991). Rapid disintegration of the Wordie Ice Shelf in response to atmospheric warming. *Nature, 350*, 328–330.

Domack, E., Duran, D., Leventer, A., Ishman, S., Doane, S., McCallum, S., Amblas, D., Ring, J., Gilbert, R., & Prentice, M. (2005). Stability of the Larsen B ice shelf on the Antarctic Peninsula during the Holocene epoch. *Nature, 436*, 681–685.

Dowdeswell, J. A. (2017). Eurasian Arctic ice shelves and tidewater ice margins. In L. Copland & D. Mueller (Eds.), *Arctic ice shelves and ice islands* (p. 55–74). Dordrecht: Springer. doi:10.1007/978-94-024-1101-0_3.

Dowdeswell, J. A., & Bamber, J. L. (2007). Keel depths of modern Antarctic icebergs and implications for sea-floor scouring in the geological record. *Marine Geology, 243*, 120–131.

Dowdeswell, J. A., & Forsberg, C. F. (1992). The size and frequency of icebergs and bergy bits from tidewater glaciers in Kongsfjorden, north-west Spitsbergen. *Polar Research, 11*, 81–91.

Dowdeswell, J. A., Whittington, R. J., & Hodgkins, R. (1992). The sizes, frequencies and freeboards of East Greenland icebergs observed using ship radar and sextant. *Journal of Geophysical Research, 97*, 3515–3528.

Dowdeswell, J. A., Whittington, R. J., Jennings, A. E., Andrews, J. T., Mackensen, A., & Marienfeld, P. (2000). An origin for laminated glacimarine sediments through sea-ice build-up and suppressed iceberg rafting. *Sedimentology, 47*, 557–576.

Drewry, D. J., Jordan, S. R., & Jankowski, E. J. (1983). Measured properties of the Antarctic Ice Sheet: Surface configuration, ice thickness, volume and bedrock characteristics. *Annals of Glaciology, 3*, 83–91.

Enderlin, E. M., & Hamilton, G. S. (2014). Estimates of iceberg submarine melting from high-resolution digital elevation models: Application to Sermilik Fjord, East Greenland. *Journal of Glaciology, 60*, 1084–1092.

Enderlin, E. M., & Howat, I. M. (2013). Submarine melt rate estimates for floating termini of Greenland outlet glaciers (2000–2010). *Journal of Glaciology, 59*, 67–75.

England, J. H., Lakeman, T. R., Lemmen, D. S., Bednarski, J. M., Stewart, T. G., & Evans, D. J. A. (2008). A millennial-scale record of Arctic Ocean sea ice variability and the demise of the Ellesmere Island ice shelves. *Geophysical Research Letters, 35*. doi:10.1029/2008GL034470.

England, J. H., Evans, D. J. A., & Lakeman, T. (2017). Holocene history of Arctic ice shelves. In L. Copland & D. Mueller (Eds.), *Arctic ice shelves and ice islands* (p. 185–205). Dordrecht: Springer. doi:10.1007/978-94-024-1101-0_7.

Flower, B. P. (1997). Overconsolidated section on the Yermak Plateau, Arctic Ocean: Ice sheet grounding prior to 660 ka? *Geology, 25*, 147–150.

Fretwell, P., Pritchard, H. D., Vaughan, D. G., Bamber, J. L., Barrand, N. E., Bell, R. & 54 others. (2013). Bedmap2: Improved ice bed, surface and thickness datasets for Antarctica. *The Cryosphere*, *7*, 375–393.

Fricker, H. A., Popov, S., Allison, I., & Young, N. (2001). Distribution of marine ice beneath the Amery Ice Shelf. *Geophysical Research Letters, 28*, 2241–2244.

Griggs, J. A., & Bamber, J. L. (2011). Antarctic ice-shelf thickness from satellite radar altimetry. *Journal of Glaciology, 57*, 485–498.

Hambrey, M. J. (1994). *Glacial environments*. London: UCL Press.

Higgins, A. K. (1989). North Greenland ice islands. *Polar Record, 25*, 207–212.

Higgins, A. K. (1990). North Greenland glacier velocities and calf ice production. *Polarforschung, 60*, 1–23.

Holland, D. M., Thomas, R. H., de Young, N., Ribergaard, M. H., & Lyberth, B. (2008). Acceleration of Jakobshavn Isbrae triggered by warm subsurface ocean waters. *Nature Geoscience, 1*, 659–664.

Jakobsson, M. (2016). Submarine glacial landform distribution in the central Arctic Ocean shelf-slope-basin system. In J. A. Dowdeswell, M. Canals, M. Jakobsson, E. K. Dowdeswell, & K. A. Hogan (Eds.), *Atlas of submarine glacial landforms* (Vol. 46, p. 469–476). London: Geological Society.

Jakobsson, M., Nilsson, J., O'Regan, M. A., Backman, J., Löwemark, L., Dowdeswell, J. A., Colleoni, F., Marcussen, C., Anderson, L., Bjork, G., Darby, D., Eriksson, B., Hanslik, D., Hell, B., Mayer, L., Polyak, L., Sellen, E., & Wallin, A. (2010). An Arctic Ocean ice shelf during MIS 6 constrained by new geophysical and geological data. *Quaternary Science Reviews, 29*, 3505–3517.

Jakobsson, M., Andreassen, K., Bjarnadottir, L. R., Dove, D., Dowdeswell, J. A., England, J. H., Funder, S., Hogan, K., Ingolfsson, O., Jennings, A., Larsen, N. K., Kirchner, N., Landvik, J. Y., Mayer, L., Mikkelsen, N., Moller, P., Niessen, F., Nilsson, J., O'Regan, M., Polyak, L., Norgaard-Pedersen, N., & Stein, R. (2014). Arctic Ocean glacial history. *Quaternary Science Reviews, 92*, 40–67.

Jakobsson, M., Nilsson, J., Anderson, L., Backman, J., Bjork, G., Cronin, T. M., Kirchner, N., Koshurnikov, A., Mayer, L., Noormets, R., O'Regan, M., Stranne, C., Ananiev, R., Barrientos Macho, N., Cherniykh, D., Coxall, H., Eriksson, B., Floden, T., Gemery, L., Gustafsson, O., Jerram, K., Johansson, C., Khortov, A., Mohammad, R., & Semiletov, I. (2016). Evidence of an ice shelf covering the central Arctic Ocean during the penultimate glaciation. *Nature Communications*. doi:10.1038/ncomms10365.

Jeffries, M. O. (1986). Glaciers and the morphology and structure of the Milne Ice Shelf, Ellesmere Island, N.W.T. *Arctic and Alpine Research, 18*, 397–405.

Jeffries, M. O. (1991). Perennial water stratification and the role of basal freshwater flow in the mass balance of the Ward Hunt Ice Shelf, Canadian High Arctic. In *Proceedings of the international conference on the role of the polar regions in global change, volume I, University of Alaska Fairbanks, Fairbanks, Alaska, 11–15 June 1990* (p. 332–337). Alaska: Geophysical Institute and Center for Global Change and Arctic System Research, University of Alaska Fairbanks.

Jeffries, M. O. (1992a). Arctic ice shelves and ice islands: Origin, growth and disintegration, physical characteristics, structural-stratigraphic variability, and dynamics. *Reviews of Geophysics, 30*, 245–267.

Jeffries, M. O. (1992b). The source and calving of ice island ARLIS-II. *Polar Record, 28*, 137–144.

Jeffries, M. O. (2017). The Ellesmere ice shelves, Nunavut, Canada. In L. Copland & D. Mueller (Eds.), *Arctic ice shelves and ice islands* (p. 23–54). Dordrecht: Springer. doi:10.1007/978-94-024-1101-0_2.

Jeffries, M. O., & Shaw, M. A. (1993). The drift of ice islands from the Arctic Ocean into the channels of the Canadian Arctic Archipelago: The history of Hobson's Choice Ice Island. *Polar Record, 29*(171), 305–312.

Jeffries, M. O., Krouse, H. R., Sackinger, W. M., & Serson, H. V. (1989). Stable isotope ($^{18}O/^{16}O$) tracing of fresh, brackish and sea ice in multiyear landfast sea ice, Ellesmere Island, Canada. *Journal of Glaciology, 35*, 9–16.

Jenkins, A., & Doake, C. S. M. (1991). Ice-ocean interaction on the Ronne Ice Shelf. *Journal of Geophysical Research, 96*, 791–813.

Jenkins, A., Vaughan, D. G., Jacobs, S. S., Hellmer, H. H., & Keys, J. R. (1997). Glaciological and oceanographic evidence of high melt rates beneath Pine Island Glacier, West Antarctica. *Journal of Glaciology, 43*, 114–121.

Koch, L. (1945). The East Greenland ice. *Meddelelser om Grønland, 130*, 374 pp.

Koenig, L. S., Greenaway, K. R., Dunbar, M., & Hattersley-Smith, G. (1952). Arctic ice islands. *Arctic, 5*, 67–103.

Koerner, R. M., & Fisher, D. A. (1990). A record of Holocene summer climate from a Canadian High-Arctic ice core. *Nature, 343*, 630–631.

Kristensen, M., Squire, V. A., & Moore, S. C. (1982). Tabular icebergs in ocean waves. *Nature, 297*, 669–671.

Lemmen, D. S., Evans, D. J. A., & England, J. (1988). Ice shelves of northern Ellesmere Island, NWT, Canadian landform examples. *Canadian Geographer, 32*, 363–367.

Mayer, C., Reeh, N., Jung-Rothenhausler, F., Huybrechts, P., & Oerter, H. (2000). The subglacial cavity and implied dynamics under Nioghalverdsfjorden Glacier, NE Greenland. *Geophysical Research Letters, 27*, 2289–2292.

McIntyre, N. F. (1985). The dynamics of ice-sheet outlets. *Journal of Glaciology, 31*, 99–107.

Mercer, J. H. (1970). A former ice sheet in the Arctic Ocean. *Palaeogeography, Palaeoclimatology and Palaeoecology, 8*, 19–27.

Mercer, J. H. (1978). West Antarctic ice sheet and CO_2 greenhouse effect: A threat of disaster. *Nature, 271*, 321–325.

Mueller, D., Copland, L., & Jeffries, M. (2017). Changes in Canadian Arctic ice shelf extent since 1906. In L. Copland & D. Mueller (Eds.), *Arctic ice shelves and ice islands* (p. 109–148). Dordrecht: Springer. doi:10.1007/978-94-024-1101-0_5.

Nick, F. M., Luckman, A., Vieli, A., van der Veen, C. J., van As, D., van de Wal, R. S. W., Pattyn, F., Hubbard, A. L., & Floricioiu, D. (2012). The response of Petermann Glacier, Greenland, to large calving events, and its future stability in the context of atmospheric and ocean warming. *Journal of Glaciology, 58*, 229–239.

Nutt, D. C. (1966). The drift of ice island WH-5. *Arctic, 19*, 244–262.

Polyak, L., Edwards, M., Coakley, B., & Jakobsson, M. (2001). Ice shelves in the Pleistocene Arctic Ocean inferred from glacigenic deep-sea bedforms. *Nature, 410*, 453–457.

Pritchard, H. D., Ligtenberg, S. R. M., Fricker, H. A., Vaughan, D. G., van den Broeke, M. R., & Padman, L. (2012). Antarctic ice-sheet loss driven by basal melting of ice shelves. *Nature, 484*, 502–505.

Pudsey, C. J., & Evans, J. (2001). First survey of Antarctic sub-ice shelf sediments reveals mid-Holocene ice shelf retreat. *Geology, 29*, 787–790.

Reeh, N. (2017). Greenland ice shelves and ice tongues. In L. Copland & D. Mueller (Eds.), *Arctic ice shelves and ice islands* (p. 75–106). Dordrecht: Springer. doi:10.1007/978-94-024-1101-0_4.

Reeh, N., Mayer, C., Miller, H., Thomsen, H. H., & Weidick, A. (1999). Present and past climate control on fjord glaciations in Greenland: Implications for IRD-deposition in the sea. *Geophysical Research Letters, 26*, 1039–1042.

Reeh, N., Thomsen, H. H., Higgins, A. K., & Weidick, A. (2001). Sea ice and the stability of north and northeast Greenland floating glaciers. *Annals of Glaciology, 33*, 474–480.

Rignot, E., & Kanagaratnam, P. (2006). Changes in velocity structure of the Greenland ice sheet. *Science, 311*, 986–990.

Rignot, E., Jacobs, S., Mouginot, J., & Scheuchl, B. (2013). Ice-shelf melting around Antarctica. *Science, 341*, 266–270.

Robe, R. Q. (1980). Iceberg drift and deterioration. In S. Colbeck (Ed.), *Dynamics of snow and ice masses* (p. 211–259). New York: Academic Press.

Robin, G. de Q. (1979). Formation, flow, and disintegration of ice shelves. *Journal of Glaciology, 24*(90), 259–271.

Robin, G. de Q., & Adie, R.J. (1964). The ice cover. In R. Priestley, R. J. Adie, & G. de Q. Robin (Eds.), *Antarctic research; a review of British scientific achievement in Antarctica* (p. 100–117). London: Butterworth.

Scambos, T., Hulbe, C., & Fahnestock, M. (2003). Climate-induced ice shelf disintegration in the Antarctic Peninsula. In E. Domack, A. Leventer, A. Burnett, R. A. Bindschadler, P. Convey, & M. Kirby (Eds.), *Antarctic Peninsula climate variability: Historical and paleoenvironmental perspectives* (pp. 79–92). Washington: American Geophysical Union.

Shepherd, A., Wingham, D., Payne, T., & Skvarca, P. (2003). Larsen ice shelf has progressively thinned. *Science, 302*, 856–859.

Smith, D. D. (1964). Ice lithologies and structure of ice island ARLIS-II. *Journal of Glaciology, 5*, 17–38.

Syvitski, J. P. M., Burrell, D. C., & Skei, J. (1987). *Fjords: Processes and products*. New York: Springer.

Thomas, R. H. (1979). Ice shelves: A review. *Journal of Glaciology, 24*, 273–286.

Van Wychen, W., & Copland, L. (2017). Ice island drift mechanisms in the Canadian High Arctic. In L. Copland & D. Mueller (Eds.), *Arctic ice shelves and ice islands* (p. 287–316). Dordrecht: Springer. doi:10.1007/978-94-024-1101-0_11.

Vaughan, D. G. (1998). A new classification scheme for ice shelves based on mechanisms of mass gain and loss. *Polar Record, 34*, 56–58.

Vaughan, D. G., & Doake, C. S. M. (1996). Recent atmospheric warming and retreat of ice shelves on the Antarctic Peninsula. *Nature, 379*, 328–331.

Vincent, W. F., Gibson, J. A. E., & Jeffries, M. O. (2001). Ice-shelf collapse, climate change, and habitat loss in the Canadian High Arctic. *Polar Record, 38*, 133–142.

Vogt, P. R., Crane, K., & Sundvor, E. (1994). Deep Pleistocene iceberg ploughmarks on the Yermak Plateau: Sidescan and 3.5 kHz evidence for thick calving ice fronts and a possible marine ice sheet in the Arctic Ocean. *Geology, 22*, 403–406.

Weeks, W. F., & Campbell, W. J. (1973). Icebergs as a fresh water source: An appraisal. *Journal of Glaciology, 12*, 207–233.

Weertman, J. (1957). Deformation of floating ice shelves. *Journal of Glaciology, 3*, 38–42.

Wen, J., Wang, Y., Wang, W., Jezek, K., Liu, H., & Allison, I. (2010). Basal melting and freezing under the Amery ice shelf, East Antarctica. *Journal of Glaciology, 56*, 81–90.

Williams, M., & Dowdeswell, J. A. (2001). Historical fluctuations of the Matusevich Ice Shelf, Severnaya Zemlya, Russian High Arctic. *Arctic, Antarctic and Alpine Research, 33*, 211–222.

Willis, M. J., Melkonian, A. K., & Pritchard, M. E. (2015). Outlet glacier response to the 2012 collapse of the Matusevich Ice Shelf, Severnaya Zemlya, Russian Arctic. *Journal of Geophysical Research, Earth Surface, 120*, 2040–2055.

Chapter 2
The Ellesmere Ice Shelves, Nunavut, Canada

Martin O. Jeffries

Abstract The Ellesmere ice shelves are located on the northernmost edge of Canada, the northwest coast of Ellesmere Island, facing the perennial pack ice of the Arctic Ocean. They are the only ice shelves in Canada and the most extensive in the entire Arctic. The special nature of the Ellesmere ice shelves extends to the essential role that sea ice has played in their initiation, maintenance and replacement. This chapter describes ice physical features, types and properties that inform the knowledge and understanding of the origin and development of the ice shelves. There are three main sections. The first is a brief history of the scientific investigation of the ice shelves. The second section describes features at the ice surface, including the characteristics and origin of the rolling topography, geological and biological materials, fractures and channels, and ice rises. The third section describes what is found below the ice surface, including the thickness and types of ice. These include marine ice (ancient and modern), formed by freezing of seawater and brackish water, and meteoric ice formed by surface accumulation of superimposed ice and lake ice (both originating as snow), and bottom accretion of freshwater.

Keywords Ice shelf • Ice island • Glacier • Multiyear landfast sea ice • Surface accumulation • Bottom accretion • Meteoric ice • Marine ice • Ice shelf basal topography • Ice shelf surface topography • Ellesmere Island

2.1 Introduction

The Ellesmere ice shelves are located along the northwest coast of Ellesmere Island, the northernmost coast of Canada, facing the perennial pack ice of the Arctic Ocean (Fig. 2.1). They are the only ice shelves in Canada and, despite a century of losses, they remain the most extensive in the entire Arctic. A number of outlet glaciers that flow to the sea from ice caps elsewhere on Ellesmere Island also meet the criteria for definition as ice shelves (Dowdeswell and Jeffries 2017). Those ice shelves are not discussed in this chapter and the reader who wants to learn more is referred to the

M.O. Jeffries (✉)
Office of Naval Research, Arctic and Global Prediction Program, Arlington, VA, USA
e-mail: martin.jeffries@navy.mil

L. Copland, D. Mueller (eds.), *Arctic Ice Shelves and Ice Islands*, Springer Polar Sciences, DOI 10.1007/978-94-024-1101-0_2

Fig. 2.1 Sub-scene of a RADARSAT-1 ScanSAR Wide-B image of the northwest coast of Ellesmere Island, Nunavut, Canada, showing the ice shelves and other features and locations mentioned in the text. This radiometrically corrected and geolocated (Albers Equal Area Projection) scene covers an area of 110 km by 325 km. The original image (20010307_r1_27873_swb_211) was acquired on 7 March 2001 by the Alaska Satellite Facility at the University of Alaska Fairbanks

literature, e.g., Short and Gray (2005), Williamson et al. (2008) and Mair et al. (2009).

The special nature of the Ellesmere ice shelves extends to the essential role that sea ice has played in their initiation, maintenance and replacement. Because of the importance of sea ice, Lemmen et al. (1988) coined the term 'sea ice-ice shelf' for those Ellesmere ice shelves, e.g., the Ward Hunt Ice Shelf, known to have originated as sea ice. Lemmen et al. (1988) also coined the self-explanatory term 'glacial-ice shelf' and the term 'composite-ice shelf' for one that is a combination of sea ice and glacier ice.

There is convincing historical evidence that in the early twentieth century there was a single ice shelf as much as 8900 km^2 in area and extending 500 km alongshore from Clements Markham Inlet to Nansen Sound (Koenig et al. 1952; Vincent et al. 2001). In 2001, all that remained of the former Ellesmere Ice Shelf were six smaller, remnant ice shelves in the inlets, fiords and bays that incise the coastline between Clements Markham Inlet and Phillips Inlet (Fig. 2.1). Beginning in 2002, ice shelf disintegration accelerated and by 2008 the number of major ice shelves was reduced to four: Ward Hunt, Milne, Peterson and Serson (Mueller et al. 2017). Of these four, the Ward Hunt Ice Shelf has continued to disintegrate at an alarming rate, with particularly heavy losses in summer 2010 (Mueller et al. 2017). Copland et al. (2017) discuss the factors that have contributed to recent ice shelf losses.

The scientific study of the Ellesmere ice shelves dates back to 1952 and almost all investigations have addressed physical characteristics and processes. This chapter describes ice physical features, types and properties that inform the knowledge and understanding of the origin and development of the ice shelves. It draws on the literature that describes observations and measurements of the ice shelves themselves and of the ice islands that have calved from them. It is not a comprehensive literature review. The reader who wishes to learn more is encouraged to read the references herein and previous overviews of ice shelf and ice island research (e.g.,

Jeffries 1987, 1992a, 2002), and consult the now-dated but still useful and only bibliographies of Ellesmere ice shelf and ice island publications (Ommanney 1982a, b).

Section 2.2 is a brief history of the scientific investigation of the ice shelves. Section 2.3 describes features at the ice surface, including the characteristics and origin of the rolling surface topography, geological and biological materials, fractures and channels, and ice rises. Section 2.4 describes what is found below the ice surface, including the thickness and types of ice. These include marine ice, ancient and modern, formed by freezing of seawater and brackish water, and meteoric ice formed by surface accumulation and bottom accretion. Meteoric ice, originating from atmospheric precipitation, is an alternative and preferred term for freshwater ice. In the case of the Ellesmere ice shelves, meteoric ice is derived primarily from snow and refrozen snow meltwater.

2.2 A Brief History of Science on the Ellesmere Ice Shelves

The history of the Ellesmere ice shelves, as represented by the people who have travelled over and studied the ice, and written about what they observed and learned, can be divided into five basic periods, one solely of exploration and four of scientific study. The first period, the exploration period, covers the British Arctic Expedition of 1875–76 (Parliamentary Paper v LVI 1877) and the Peary Arctic Expedition of 1905–06 (Peary 1907), when men travelled over the ice shelves hauling their own sledges and by dog-sled, respectively. Although they were dedicated almost exclusively to exploration, the expedition reports have some scientific value, particularly the descriptions of the ice surface topography from which the former extent of the ice shelves has been inferred (e.g., Koenig et al. 1952, Vincent et al. 2001).

After a hiatus of almost 50 years, the second period began with the first scientific expedition, in 1953 (Hattersley-Smith et al. 1955), and continued through the 1950s. During this time, Geoffrey Hattersley-Smith of the Defence Research Board (DRB) in Canada and Albert Crary of the Air Force Cambridge Research Laboratory (AFCRL) in the United States, were pioneers in the study of the ice shelves and Ice Island T-3, respectively. Ice islands are the tabular icebergs that calve from the Ellesmere ice shelves.

The third period covers the 1960s and early 1970s, when Hattersley-Smith continued DRB-sponsored studies, and the Arctic Institute of North America (AINA) managed AFCRL and U.S. Naval Engineering Laboratory contracts for ice shelf research (MacDonald 2005). During this time, John Lyons of Dartmouth College made important contributions to the knowledge of ice types, properties, stratigraphy and the origin of the Ward Hunt Ice Shelf, and the age of its ice rises (areas of the ice shelf grounded on the seafloor).

The fourth period was during the 1980s, when I had the good fortune to work on the ice shelves and on Hobson's Choice Ice Island, which calved from the east Ward Hunt Ice Shelf in 1982–83 (Jeffries and Serson 1983), with Harold Serson, who had previously worked for many years with Hattersley-Smith. Inspired by the work of

Lyons, and motivated by the interest and support of AINA, the oil industry and the Defence Research Establishment Pacific (Canada), Serson and I drilled many ice cores to learn more about ice types, properties, stratigraphy and origin of the ice shelves.

The fifth and final period covers the early years of the twenty-first century, when Luke Copland, Derek Mueller, Warwick Vincent and their students were active. Particularly noteworthy is the study of microscopic life on the ice surface by Mueller and Vincent. They were the first to look at the ice shelves from a biological perspective, the first to study what physical scientists had previously considered, at best, to be curious dirt accumulations on the ice surface. Mueller and Vincent showed instead that the "sediments" are part of a complex microbial ecosystem that flourishes in the extreme physical conditions on the ice, but which is now threatened as ice shelf disintegration continues and the loss of a rare habitat accelerates (Vincent et al. 2004; Mueller and Vincent 2006; Mueller et al. 2005, 2006; Jungblut et al. 2017).

2.3 At the Ice Surface

2.3.1 Top and Bottom Topography

Whether viewed from ground level, or from the air or from space, the ice shelves and the multiyear landfast sea ice (MLSI) attached to them have a distinctive topography (Fig. 2.2). In late winter, the low sun angle accentuates the rolling topography of mildly sinuous and sometimes bifurcated and elongated peaks and troughs on the ice shelves (Fig. 2.2a). In summer, the pooling of snow and ice meltwater in the troughs gives both the shelf ice and the MLSI a striped appearance (Fig. 2.2b). In most cases, the peaks and troughs are oriented parallel to the general trend of the outer coastline.

The rolling topography was first reported by the British Arctic Expedition of 1875–76 (Parliamentary Paper v LVI 1877). In spring 1876, a man-hauling sledge party encountered "ice rollers and ridges" on what we now know as the Ward Hunt Ice Shelf. Thirty years later, while travelling by dog-sled, Robert Peary referred to a "series of rolling swells" on the Ward Hunt Ice Shelf, and "long, prairie-like swells" which were a "constant and striking feature" of the "glacial fringe" along much of the coast (Peary 1907). In summer 1906, Peary noted that "the orography of the glacial fringe … was very strongly brought out by the streams and blue lakes which filled every depression and furrow" (Peary 1907). For a comprehensive summary of historical descriptions of the ice surface, the reader is referred to Koenig et al. (1952).

The first modern observations of the topography of the ice shelves were made during Royal Canadian Air Force aerial photographic missions in 1950 and U.S. Air Force reconnaissance missions in 1951 (Koenig et al. 1952). They confirmed that

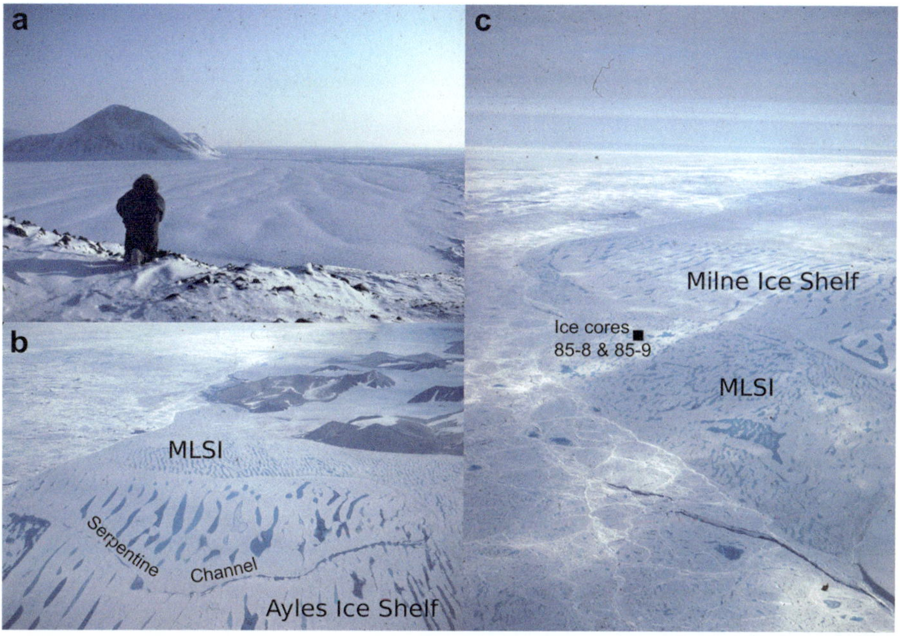

Fig. 2.2 (**a**) Looking west across the Markham Ice Shelf in April 1984; Harold Serson kneeling in the left foreground. (**b**) Looking east across the Ayles Ice Shelf in July 1984. (**c**) Looking northeast across the front of the Milne Ice Shelf in July 1984; the location of multiyear landfast sea ice (MLSI) cores 85-8 and 85-9 (Fig. 2.11) is marked. The Markham and Ayles ice shelves no longer exist; they disintegrated and drifted away as ice islands in summer 2005 (Copland et al. 2007) and summer 2008 (Mueller et al. 2017), respectively (Photographs by the author)

the ice shelves were the source of the Arctic Ocean ice islands, which bore the same distinctive topography that had been seen from the air as early as 1946. Since the early 1950s it has been common to refer to the undulations as "rolls" (e.g., Hattersley-Smith 1957).

The first scientific party to visit the Ward Hunt Ice Shelf, in 1954, completed a 19.2 km-long transit level survey to the west of Ward Hunt Island (Fig. 2.1). The mean wavelength (horizontal distance from the top of one peak to the next adjacent peak) of the rolls was 232 m (Table 2.1) and their mean height (vertical distance from a peak crest to the bottom of the next adjacent trough) was 2.1 m (Hattersley-Smith et al. 1955). No standard deviation value was given with the mean value, but the variable wavelength and height are evident in the topographic profile illustrated in Hattersley-Smith (1957: p. 37) and Crary (1958: p. 6).

Measurements on air photographs taken in 1984 provide data on the variability of the wavelength of the rolls on different ice shelves and the MLSI (Jeffries et al. 1990a). Those data (Table 2.1) illustrate quantitatively what was already known subjectively about the ice surface topography: (1) the mean wavelength on the MLSI is on the order of tens of metres while that on the ice shelves is on the order of hundreds of metres (Fig. 2.2b); (2) mean wavelength varies on any given ice

Table 2.1 Descriptive statistics for the wavelength of the undulating topography of multiyear landfast sea ice (MLSI), ice shelves and ice islands

Location	Minimum (m)	Maximum (m)	Mean ± 1 s.d. (m)
MLSI			
Moss Bay[a]	N.A.	N.A.	35.0, 40.0, 51.5
Doidge Bay	27	70	43 ± 8 (n = 85)
Markham Bay Re-entrant	40	120	71 ± 16 (89)
Ward Hunt Ice Shelf	25	105	55 ± 15 (63)
Ayles Fiord	45	121	79 ± 17 (138)
Milne Re-entrant	40	127	80 ± 21 (128)
Yelverton Bay	30	85	59 ± 21 (130)
Ice Shelves			
Ward Hunt Ice Shelf – East	87	355	212 ± 42 (84)
Ward Hunt Ice Shelf – South	110	398	225 ± 69 (111)
West Ward Hunt Ice Shelf[b]	N.A.	N.A.	232.0
Ward Hunt Ice Shelf – West[a]	110	382	258 ± 64 (101)
Ward Hunt Ice Shelf – West[b]	138	440	280 ± 53 (66)
Ayles Ice Shelf	108	311	211 ± 48 (80)
Milne Ice Shelf – Outer	135	450	330 ± 62 (86)
Ice Islands			
T-3[c]	230	260	245
SP-19[d]	200	300	210
SP-22[e]	N.A.	N.A.	220

Unless they are marked otherwise, most data are from Jeffries et al. (1990a) who also provided the number of measurements (*n*) made on air photographs

[a]Verrall and Todoeschuk (unpublished data)

[b]Hattersley-Smith et al. (1955)

[c]Crary (1960)

[d]Legen'kov and Chugui (1973)

[e]Grischenko and Simonov (1985)

shelf, e.g., Ward Hunt Ice Shelf, and on any given area of MLSI, e.g., Moss Bay; and (4) mean wavelength varies among areas of MLSI and among ice shelves, e.g., wavelength on the outer part of Milne Ice Shelf is significantly greater than on the Ward Hunt and Ayles ice shelves.

The massive calving of ice islands in 1961–62 from the Ward Hunt Ice Shelf (Hattersley-Smith 1963) carried away the northern end, ~50% of the total length, of the 1954 transit level survey profile. The survey traversed what became Ice Island WH-3, as named by Hattersley-Smith (1963). WH-3 was later renamed SP-18 and then SP-19 (Table 2.1) when the Soviet Union operated drifting research stations on it (Belkin and Kessel 2017). Not surprisingly, the few available data for ice island topography are similar to the ice shelf data, particularly those for the Ward Hunt and Ayles ice shelves (Table 2.1).

The topography data for the MLSI in Moss Bay (Table 2.1; Fig. 2.1) were obtained by ground-penetrating radar (GPR) in 1989 and 1990 (Verrall and Todoeschuk unpublished data). The value of these data lies not only in what they tell us about the surface of the ice, but also in what was learned about the bottom of the ice. Verrall and Todoeschuk found that the MLSI has a rolling bottom topography with the same wavelength as the top surface. While this is to be expected of ice that is afloat and in isostatic equilibrium, what was unexpected was that the bottom topography was not an exact mirror image of the top surface. Instead, Verrall and Todoeschuk found that the rolls on the bottom are offset horizontally from those at the top by one quarter wavelength. The rolling topography of the bottom of the MLSI affects its composition, i.e., the types of ice it contains, and their properties, as discussed in Sect. 2.4.2.2.

Hattersley-Smith (1957) expected, reasonably, that the surface topography of the ice shelves would be reciprocated at the bottom of the ice, i.e., it would also have a rolling topography that mirrors the top surface. However, airborne radio-echo sounding of the Milne Ice Shelf, for example, revealed that the surface topography had little or no expression at the bottom (Narod et al. 1988). Instead, they reported a very gently undulating bottom topography with a wavelength on the order of kilometres and a much lower relative relief than the top surface. The bottom of Ice Island T-3 was also essentially flat according to airborne radio-echo sounding and submarine upward-looking sonar data (Holdsworth 1987).

Narod et al. (1988) suggest that the negligible bottom topography indicates that the rolls at the top of the ice shelves are formed by surface processes, and that stresses resulting from incomplete isostatic compensation are too low to require substantial re-adjustment of the bottom topography. However, Fig. 3 in Mortimer et al. (2012), who used GPR to measure the thickness of the Milne Ice Shelf, shows a definite bottom topography with a wavelength and height similar to that at the top surface. Whether there is a horizontal offset between the top and bottom topography requires more than visual examination. The difference between what Narod et al. (1988) and Mortimer et al. (2012) observed at the bottom of the Milne Ice Shelf is probably explained by advances in radar technology and data processing, and suggests that it would be fruitful to further investigate the physics of the relationship between the top and bottom topography of the ice shelves and MLSI.

Narod et al. (1988) also reported what they interpreted as two crevasses at the bottom of the Milne Ice Shelf. The bottom crevasses appear to be related to fractures that are visible at the ice surface above, as described in Sect. 2.3.4. Mortimer et al. (2012) detected the same bottom crevasses in 2009 with GPR. The location of one of those crevasses, and the surface fracture above (both illustrated in Mortimer et al. (2012), Fig. 3), had remained unchanged since the early 1980s (Narod et al. 1988). Examination of that bottom crevasse/surface fracture pair in Fig. 3 in Mortimer et al. suggests that the height and shape of the bottom crevasse also remained largely unchanged over the course of more than 20 years.

2.3.2 *Origin of the Rolling Topography of the Ice Shelves and MLSI*

Any description of the surface rolls inevitably raises questions about their origin and evolution, which have been the subject of discussion since they were first observed (e.g., Hattersley-Smith 1957; Lister 1962; Holdsworth 1987). Lister (1962) classified the numerous explanations into genetic and superimposed categories. In the former, the rolls originate during the formation of an ice shelf. In the latter, the rolls are superimposed after ice shelf formation. Lister's classification was supplemented by Holdsworth (1987) and his review is summarized in Table 2.2.

Holdsworth (1987) favours a combination of the pack ice pressure and meltwater lake explanations (Table 2.2). Briefly, Holdsworth argues that the steady, gravity-induced creep expansion of the Ward Hunt Ice Shelf (Dorrer 1971; MacAyeal and Holdsworth 1986) can be briefly "paralyzed and even reversed" by occasional extreme pack ice pressure events. These events cause buckling instabilities and the development of the surface waveforms. Since pack ice pressure events occur only along limited sections of the ice shelf front at any one time, the resultant non-homogenous strain field can explain the bifurcating peaks (Fig. 2.2a).

Unlike Hattersley-Smith (1957), who attributed only a minor role to meltwater lakes and wind action in the development of the rolls, Holdsworth (1987) considered the pooling of meltwater in the troughs to be essential to the preservation of the deformation-induced rolls. Crary (1960) also strongly favoured the role of meltwater in the troughs. Observing that the rolls are generally oriented parallel to the direction of the prevailing winds and the shore, Crary suggested that winds blowing parallel to the coastline over an initially randomly distributed array of melt ponds

Table 2.2 Explanations for the origin and evolution of the rolls

Genetic	
1.	Thermal stress, in which rapid temperature increases cause compressive strains which lead to buckling
2.	Pack ice pressure, which compressses the ice at roughly right angles to the coastline, implying the development of a creep buckling instability in the ice
3.	Pressure ridge stacking, in which "hedges" of stacked ice alternate with sections of undeformed ice, and a surface waveform would subsequently evolve by snow drifting
4.	Glacier compression, in which the flow of glaciers causes folding of the ice
5.	Tidal action, in which the ice is assumed to be floating everywhere except where it is attached to the coast, and is deformed as the tide progresses into the coast
Superimposed	
6.	Snow dune formation parallel to the coastline and the direction of the prevailing winds, i.e., sief-type dunes
7.	Meltwater lake alignment, in which winds blowing parallel to the coastline over an initially randomly distributed array of melt ponds and lakes causes them to become elongated

After Holdsworth 1987

would, in time, cause them to become elongated as a result of a combination of solar heating of the water and resultant convection currents.

There is evidence that the rolls not only elongate under the influence of meltwater and wind, but they also migrate laterally, leading to a cyclic inversion of relief (Crary 1960; Smith 1961; Lister 1962; Jeffries et al. 1991). The occurrence of extensive meltwater lakes today (Figs. 2.2b and 2.4) and in the past is evidence of a negative surface mass balance. Whether a negative surface mass balance is due to increased air temperatures or a longer duration melt season, or both, has not been investigated. Braun et al. (2004) and Braun (2017) summarize the mass balance measurements that have been made on the Ward Hunt Ice Rise and the Ward Hunt Ice Shelf since 1959.

The effect of glaciers on ice shelf topography is most pronounced in the central unit of the Milne Ice Shelf. There, the rolls are particularly sinuous, due either to their alignment with former crevasses (Hattersley-Smith 1957) or glacier flow lines (Jeffries 1986). The troughs in this unit are also particularly deep and often steep-sided (Fig. 2.3), requiring circuitous travel following the maze of troughs as opposed to more direct routes across the less imposing and more symmetrical peaks elsewhere on this ice shelf and others.

The southern 'tongue' of the Ward Hunt Ice Shelf, which protruded into Disraeli Fiord (Figs. 2.1 and 2.6) until it disintegrated during the 2000s, had a more chaotic undulating topography than the regular, nearly linear rolls of the main body of the

Fig. 2.3 Looking west across the central unit of the Milne Ice Shelf. The snowmobile, sled and figure (Harold Serson) give some sense of the scale and severity of the topography and the size of the boulders (Photograph taken in May 1984 by the author)

Fig. 2.4 Aerial view looking southwest across the Milne Ice Shelf towards two glaciers that have advanced seaward into Milne Fiord. The 100 m thick shelf ice in this area is the thickest ice measured, by airborne radio-echo sounding, in the Ellesmere ice shelves. Note the linear lateral moraines on the ice shelf surface seaward of the small cirque glacier, and the bifurcations in the rolls (Photograph taken in July 1984 by the author)

ice shelf (D. R. Mueller personal communication), but similar to the topography of the central Milne Ice Shelf. Since the Ward Hunt tongue was not glacial in origin, the cause of its chaotic relief is not clear.

2.3.3 Geological and Biological Materials

The scattered rocks and boulders (Fig. 2.3) and lateral moraines (Fig. 2.4) on the Milne Ice Shelf are the most visible evidence of the contribution of glaciers to the ice shelf. The two glaciers that coalesce offshore to form the inner section of the Serson Ice Shelf (Fig. 2.5) have also transported a large amount of till that is exposed at the ice surface. It is evident from the widespread till at the surface of Ice Island ARLIS-II (Smith 1964), which calved from the Serson Ice Shelf in roughly 1955 (Jeffries 1992b), that the glacier section of the ice shelf was once much more extensive. On both the Milne and Serson ice shelves, the moraines are characterized by conical mounds of till that are as much as 10 m high (see Fig. 5 in Jeffries 1986).

There were also extensive zones of rocks on the surface of the former M'Clintock Ice Shelf. Since it is evident from aerial photographs that the ice shelf did not have

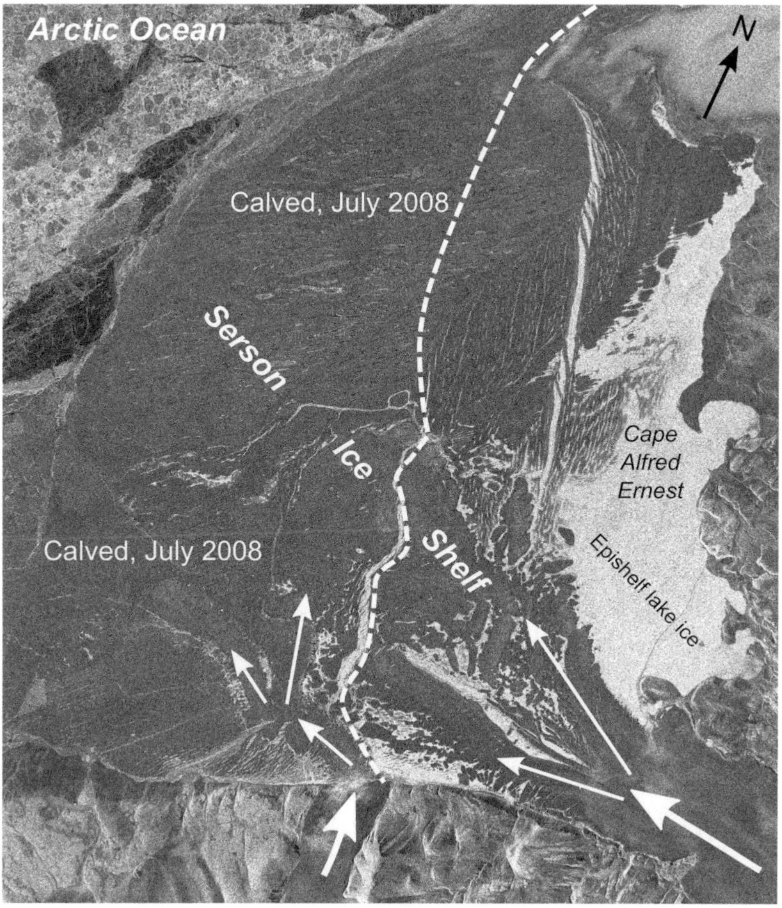

Fig. 2.5 Sub-scene of a RADARSAT-1 Fine Beam SAR image of the Serson Ice Shelf on 27 January 2007. This radiometrically corrected and geolocated (Albers Equal Area Projection) scene covers an area of 16.5 km by 19.5 km. The original image (20070129_r1_58658_fn1_208) was acquired by the Alaska Satellite Facility. Serson Ice Shelf is a composite ice shelf. The *dashed line* encloses the outer, and probably sea ice, section of the ice shelf that calved in July 2008. *Bold* and *fine arrows* indicate grounded glaciers and floating glacier tongues, respectively

a glacier component, it is possible that the rocks had either fallen from the cliffs above or had been exposed at the surface after being frozen in at the bottom of the ice, as proposed by Debenham (1954). Bottom freezing coupled with surface melting probably also accounts for the exposure of benthic sediments and fossil marine organisms at the surface of the ice shelves and ice islands. The organisms include silicious sponges and spicules, gastropods, pelecypods, bryozoa, annelids, echinoids, foraminifera, serpulid worm tubes and fish remains, including Arctic cod, *Arctogadus glacialis* (Crary 1960; Lyons and Mielke 1973; Smith 1964).

Lichen, moss, leaf, plant and tree, e.g., Arctic willow, *Salix arctica*, remains, and caribou antlers, were found at the surface of Ice Island T-3 (Crary et al. 1952; Polunin 1955, 1958). More recently, it has been shown that the ice shelves are a cryohabitat for an ecosystem of microbial communities, including microbial mats, that are particularly rich where aeolian and glacier-derived sediments occur in association with marine ice (Vincent et al. 2004; Mueller and Vincent 2006; Mueller et al. 2005, 2006). Microbial mats were particularly prevalent at the Markham Ice Shelf (Jungblut et al. 2017), which disintegrated completely in summer 2008.

2.3.4 Fractures and Channels in the Ice

The rolling topography is the dominant characteristic of the ice surface, but there are other noteworthy features, one typically linear or almost linear in appearance, the other serpentine or S-shaped. They cut across the grain of the rolls and are likely evidence of extreme events.

A serpentine feature appeared suddenly at the surface of the Ward Hunt Ice Shelf (Fig. 2.6) in summer 2002 (Mueller et al. 2003) and continued to widen in subsequent years (D. R. Mueller personal communication). Extending from the top to the bottom of the ice shelf, and from the southern margin to the northern, seaward margin, the serpentine feature is a channel that is now believed to have opened as early as 2000 and acted as a conduit for the catastrophic drainage of 4 km^3 of freshwater from the ice shelf-dammed epishelf lake in Disraeli Fiord (Mueller et al. 2003). A rehealed serpentine feature in the former Ayles Ice Shelf (Figs. 2.2b and 2.7)

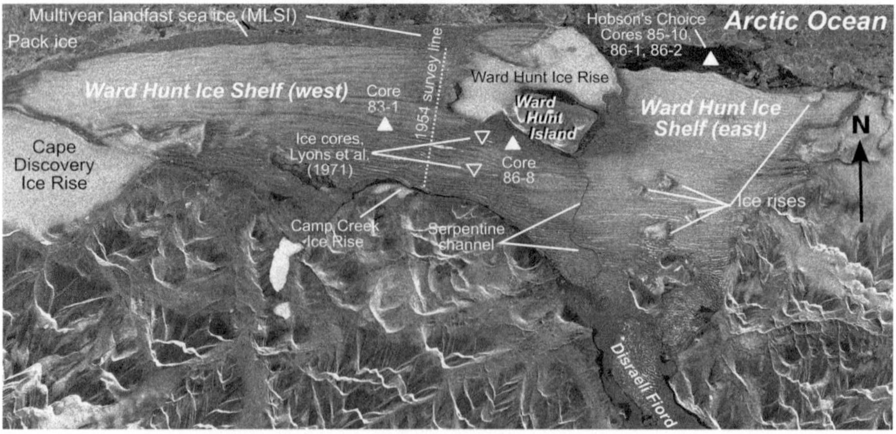

Fig. 2.6 Sub-scene of a RADARSAT-1 Standard Beam 1 SAR image of the Ward Hunt Ice Shelf on 14 February 2003. This radiometrically corrected and geolocated (Albers Equal Area Projection) scene covers an area of 19 km by 38 km. The original image (20030214_r1_38005_st1_215) was acquired by the Alaska Satellite Facility. The locations of ice cores, ice rises, the serpentine channel and other features and locations referred to in the text are marked

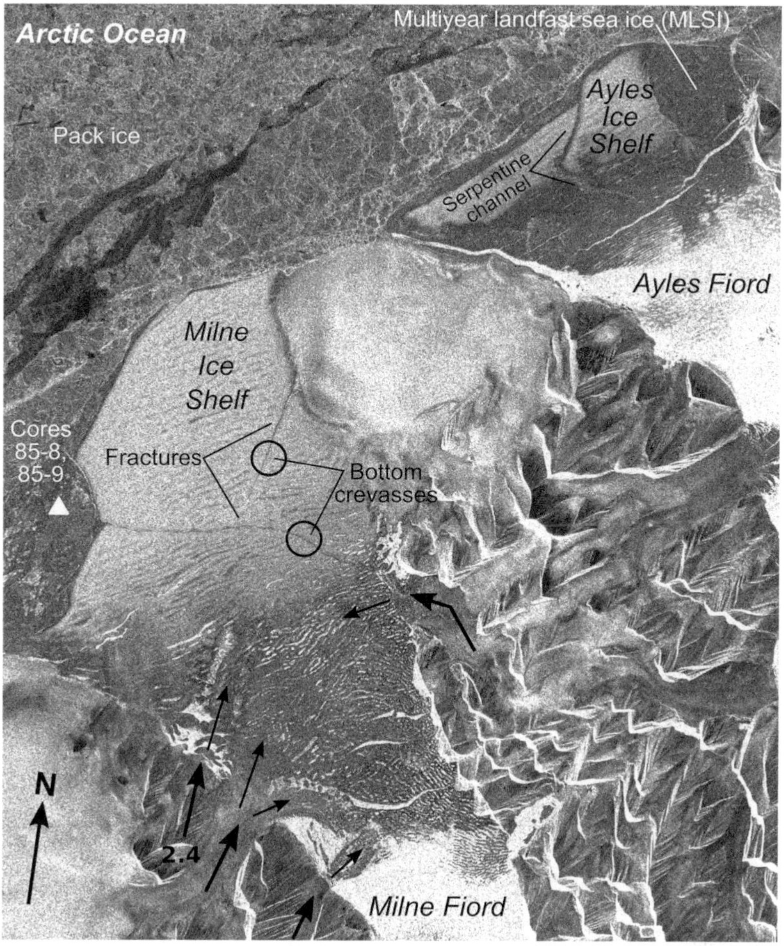

Fig. 2.7 Sub-scene of a RADARSAT-1 Standard Beam 1 SAR image of the Milne and Ayles ice shelves on 26 January 2005. This radiometrically corrected and geolocated (Albers Equal Area Projection) scene covers an area of 19 km by 24 km. The original image (20050126_r1_48181_st1_214) was acquired by the Alaska Satellite Facility. *Bold* and *fine arrows* indicate grounded glaciers and floating glacier tongues, respectively. The locations of multiyear landfast sea ice (MLSI) cores, and fractures and bottom crevasses in the outer Milne Ice Shelf, and a serpentine channel in the Ayles Ice Shelf, are also marked. The *bold arrow marked 2.4* points toward the camera in Fig. 2.4

suggests that a channel also developed in that ice shelf and could have acted as a conduit for the rapid drainage of an epishelf lake, if there was one, from Ayles Fiord. Epishelf lakes were common in this region (Veillette et al. 2008). In the case of the Ward Hunt Ice Shelf, the epishelf lake in Disraeli Fiord played an important role in the ice mass balance and maintaining the ice thickness via a feedback system composed of the ice shelf dam, the epishelf lake, and the outflow of epishelf lake water from which meteoric ice accreted at the bottom of the eastern ice shelf (Jeffries 1991a). The stratigraphy and origin of ice in the eastern Ward Hunt Ice Shelf is described in Sect. 2.4.2.3.

The opening of the serpentine channel in the Ward Hunt Ice Shelf was accompanied by extensive fracturing, with many fractures oriented parallel to the rolls and others cutting across the rolls. Since then, the southern and eastern sectors of the Ward Hunt Ice Shelf have become riddled with fractures that suggest a greatly weakened and vulnerable ice mass (D. R. Mueller personal communication). Two long features cutting across the rolls of the outer unit of the Milne Ice Shelf (Fig. 2.6) are believed to be ancient fractures (Jeffries 1986) that suggest some extreme event or events that the ice shelf survived long ago. It is likely that the bottom crevasses detected by airborne radio-echo sounding in the outer unit of Milne Ice Shelf (Narod et al. 1988) are basal manifestations of the former fractures (Fig. 2.7).

2.3.5 Ice Rises

The Ward Hunt Ice Shelf is the only Ellesmere ice shelf known to have ice rises, which indicate that the ice shelf is grounded on the seafloor at those locations (Fig. 2.6). Heat flow analysis indicates that the Ward Hunt Ice Rise grounded 250–350 years ago, while ice stratigraphy and dust layers in the Cape Discovery and Camp Creek ice rises suggest that they might be as much as 1600 years old (Lyons et al. 1972). The elevation of the ice rises is not known.

The only successful attempt to drill to the bottom of the ice rises occurred in 1960, when a 52.3 m long ice core was obtained from the Ward Hunt Ice Rise (Lyons and Ragle 1962; Ragle et al. 1964; Lyons et al. 1972). This revealed that it is sitting on till and that the lowermost 19 m of the ice rise is a layer of sea ice. The presence of sea ice in the ice rise is consistent with the sea-ice origin of the ice shelf, which began to form about 5500 years ago (England et al. 2008, 2017). The sea ice and other types of ice in the ice shelves are described in the next section.

2.4 Below the Ice Surface

2.4.1 Ice Thickness

The thickness of the ice shelves and MLSI has been determined by drilling (e.g., Jeffries et al. 1989, 1990a, 1991), GPR (Verrall and Todoeschuk unpublished data; Copland et al. 2007; Mortimer et al. 2012) and by airborne radio-echo sounding (Hattersley-Smith 1969; Hattersley-Smith et al. 1969; Narod et al. 1988). Drilling is labour-intensive and has yielded only a small number of widely dispersed data points (e.g., Jeffries et al. 1989, 1990a, 1991). Airborne radio-echo sounding has enjoyed mixed success due to navigation and technical difficulties, and signal loss in saline ice, which prevents the acquisition of bottom reflections, e.g., in the west Ward Hunt Ice Shelf (Hattersley-Smith 1969; Hattersley-Smith et al. 1969; Narod et al. 1988). GPR has been applied at four locations: MLSI in Moss Bay (Fig. 2.1) in 1989 and 1990 (Verrall and Todoeschuk unpublished data); the Ayles Ice Island, a proxy for the former Ayles Ice Shelf (Figs. 2.1 and 2.7), in 2007 (Copland et al. 2007); the Milne Ice Shelf in 2008 and 2009 (Mortimer et al. 2012); and the Petersen Ice Shelf in 2011 (White et al. 2015).

The thickest known shelf ice occurs in the Milne Ice Shelf. In the early 1980s it was as much as 90 m thick in the north-eastern section of the outer ice shelf and as much as 100 m thick in the western section of the central ice shelf (Narod et al. 1988: Fig. 6). The 100 m thick ice occurred where two glaciers have advanced seaward into Milne Fiord (Fig. 2.4). Since the Narod et al. (1988) investigation, the Milne Ice Shelf has thinned by an average of 8.1 m, with the greatest thinning having occurred in the central ice shelf (Mortimer et al. 2012). The latter also note that bottom melting has been a significant contributor to the thinning of the ice shelf, to which inflowing glaciers have also added less mass since the early 1980s.

The thickest MLSI measured along the northwest coast of Ellesmere Island, at the front of the Ward Hunt Ice Shelf, was 10 m beneath a peak at the top of a roll (Jeffries et al. 1989, 1990a). MLSI 12 m and 20 m thick was measured by drilling at Ice Island ARLIS-II (Smith 1964); it is discussed in Sect. 2.4.2.2.

In his assessment of the role of deformation in forming and maintaining the rolling topography, Holdsworth (1987) noted that any folding resulting from a buckling instability will cause the surface to have a characteristic wavelength (L) that is proportional to the ice thickness (H). This suggests that H can be estimated from L. Jeffries et al. (1990a) explored this possibility in the knowledge that the wavelength of the rolls on MLSI is significantly different from that of the rolls on the ice shelves (Fig. 2.2b, c, Table 1), and that the thickness of MLSI differs significantly from that of the ice shelves and ice islands. Now updated with additional data for MLSI in Moss Bay (Fig. 2.1) and four ice islands (Table 2.1), the original Jeffries et al. (1990a) data are plotted in Fig. 2.8. It shows a strong correlation between ice thickness and wavelength, which, as proposed by Jeffries et al. (1990a), offers a useful rule-of-thumb for approximating ice thickness when the wavelength of the rolls is known. Note that, for simplicity, only mean wavelength values are plotted in Fig. 2.8. The minimum and maximum wavelength values in Table 2.1 offer a means to estimate lower and upper bounds, respectively, for the thickness of shelf ice, ice islands and MLSI.

2.4.2 *Types of Ice*

2.4.2.1 Introduction

Ice cores have been the main source of information about the types of ice that occur in the ice shelves. Ice cores have been recovered from ice islands T-3, ARLIS-II, WH-3/SP-18/SP-19 and Hobson's Choice (a proxy for the east Ward Hunt Ice

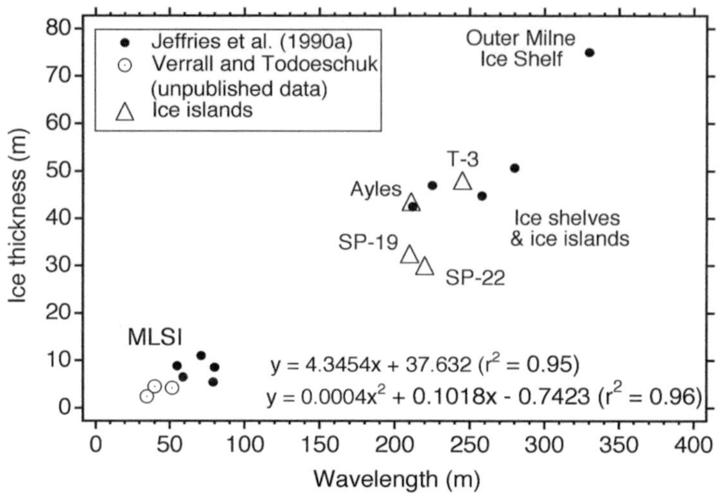

Fig. 2.8 Scatter plot of ice thickness and wavelength of the rolls on ice shelves, ice islands and MLSI, with linear and second-order polynomial regression equations. The MLSI data are those in the lower left corner only. The ice island data are from Crary (1958, 1960) for T-3, Legen'kov and Chugui (1973) and Grischenko and Simonov (1985) for SP-19 and SP-22, and Jeffries et al. (1990a) and Copland et al. (2007) for Ayles. All other data are from Jeffries et al. (1990a)

Shelf), from Ward Hunt Ice Shelf itself, and from the MLSI. The latter is a modern-day proxy for the earliest ice shelf formation thousands of years ago.

Ice core studies have revealed two basic types of ice: marine ice and meteoric ice. The second part of this section describes marine ice, which includes sea ice and brackish ice. The marine ice is also referred to as basement ice (Lyons et al. 1971), a term borrowed from geology to describe the ice on which snow accumulated and then became meteoric ice. The third part of this section describes meteoric ice, which includes superimposed ice and lake ice. The final part of the section describes the compressive strength of marine ice and meteoric ice.

2.4.2.2 Marine Ice

As noted in Sect. 2.3.5, a 19 m thick layer of sea ice was found at the base of the 52 m thick Ward Hunt Ice Rise (Lyons and Ragle 1962; Ragle et al. 1964; Lyons et al. 1972). The lower half of ice island SP-18/SP-19 (originally WH-3, which calved from the Ward Hunt Ice Shelf immediately northwest of the Ward Hunt Ice Rise (Hattersley-Smith 1963), was a 16 m thick layer of sea ice (Legen'kov 1973). The lowermost 17 m of the roughly 41 m thick Ice Island WH-4, the adjacent sister of WH-3/SP-18/SP-19, was primarily sea ice (Lyons and Ragle 1962; Ragle et al. 1964). In the central Ward Hunt Ice Shelf, between Ward Hunt Island and the mainland (Fig. 2.6), Lyons et al. (1971) reported a 20 m thick layer of sea ice, brackish ice layers at least 6 m thick, and interleaving of sea ice and brackish ice layers in the

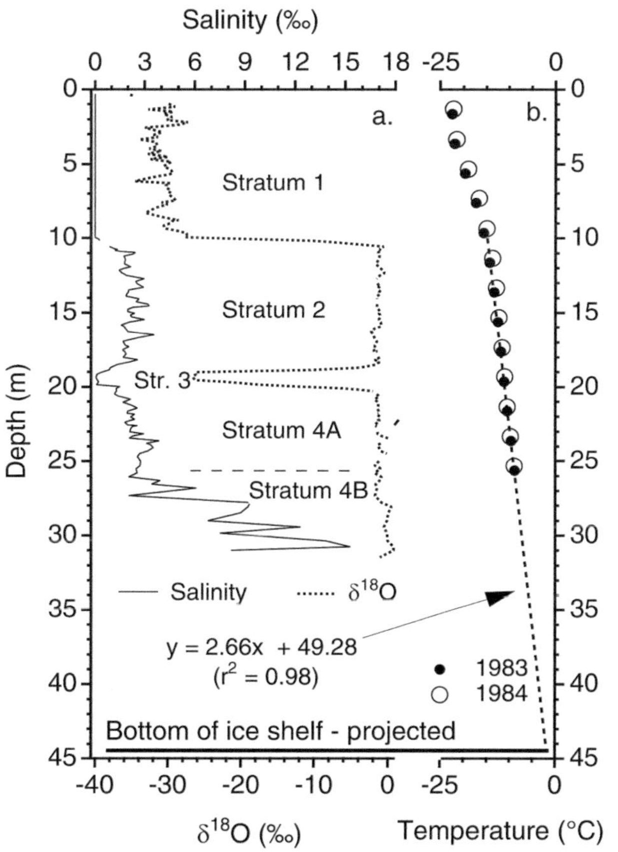

Fig. 2.9 (**a**) Salinity and $\delta^{18}O$ profiles in ice core 83-1, and (**b**) ice temperature profile at the same location. The ice core was recovered from the west Ward Hunt Ice Shelf at a location ~6 km from the area where Lyons et al. (1971) investigated brackish ice and sea ice (Fig. 2.6). The two sets of temperature data are offset vertically by 0.3 m, the amount of ablation that occurred in summer 1993. The regression line is for the 1983 temperature data only

shoreward parts of the ice shelf. It is these sea ice and brackish ice layers that Lyons et al. (1971) termed "basement ice".

The term "basement ice" was originally used to describe the ~25 m thick lower layer of Ice Island T-3. Marshall (1960) proposed that it was largely of glacial origin with some brine-soaked layers. On the other hand, with the exception of the lowermost 2 m thick layer of sea ice, Muguruma and Higuchi (1963) were uncertain of the origin of the basement ice in T-3. Lyons et al. (1971) concluded that it is brackish ice, since it shared many petrographic characteristics with the brackish basement ice in the Ward Hunt Ice Shelf.

Thick layers of marine ice, primarily sea ice, are illustrated in Fig. 2.9a, which shows salinity and $\delta^{18}O$ profiles in core 83-1 from the west Ward Hunt Ice Shelf (Fig. 2.6). The marine ice comprises a thin layer (1 m) of brackish ice (Stratum 3)

Table 2.3 Some physical properties of ancient sea ice in the Ward Hunt Ice Shelf

Core	Salinity (psu)	Brine volume (‰)	Density (kg m^{-3})	$\delta^{18}O$ (‰)
82-6[a]	2.1 ± 1.0 (29)	N.A.	903 ± 23 (16)	−0.1 ± 1.0 (27)
83-1, Stratum 2[a]	2.1 ± .0.5 (54)	9.9 ± 2.4 (23)	919 ± 16 (20)	−1.3 ± 0.4 (51)
83-1, Stratum 4A[a]	2.4 ± 0.6 (35)	13.2 ± 4.8 (18)	923 ± 15 (15)	−1.3 ± 0.4 (31)
83-1, Stratum 4B[a]	8.2 ± 7.6 (14)	N.A.	923 ± 15 (15)	−0.5 ± 1.6 (14)
83-9[a]	2.4 ± 0.9 (19)	N.A.	893 ± 36 (16)	0.0 ± 0.6 (20)

N.A. Not Available
[a]Jeffries (1985)

sandwiched between two thick layers of sea ice (Strata 2 and 4). Stratum 4 subdivides further into two layers, A & B. The salinity and $\delta^{18}O$ values of Strata 2, 4A and 4B, and of two much shorter cores from elsewhere on the Ward Hunt Ice Shelf (Fig. 2.1), are summarized in Table 2.3. The $\delta^{18}O$ values, together with the columnar crystal texture and cellular sub-structure of ice plates and brine layers (Jeffries 1985, 1991b), are unambiguous evidence that the ice grew from seawater. Stratum 1 (Fig. 2.9a), overlying the marine ice, is meteoric ice, which is discussed in Sect. 2.4.2.3.

Core 83-1 did not extend all the way to the bottom of the ice shelf (Jeffries 1991b). Drilling was abandoned because core recovery became increasingly difficult in ice with a high liquid brine content; hence the high salinity values in Stratum 4B. Within 68 hours of the cessation of coring, the borehole had filled with brine to a depth of 25.6 m, the boundary between Strata 4A and 4B (Fig. 2.9a). Jeffries (1991b) suggested that some of the brine in the borehole represented seawater that was being exchanged between the ice and the ocean below. It is also likely that some of the brine was local, having drained into the borehole from the marine ice immediately surrounding it.

Once the brine level had stabilized at a depth of 25.6 m in the borehole, thermocouples were installed and ice temperatures measured 26 days later (31 May 1983) and 379 days later (19 May 1994) (Fig. 2.9b). Extrapolating the 1983 linear ice temperature profile below 10 m to −1.8°C (freezing point of seawater assumed to flow below the ice shelf), the total ice thickness was estimated to be 44.5 m (Jeffries 1991b). Stratum 4 was therefore probably as much as 24.5 m thick (Fig. 2.9b), almost three times the thickness of Stratum 2 (9 m).

The thick sea ice layers noted in the first paragraph of this section and illustrated in Fig. 2.9a are evidence of the significant role that sea ice growth has played in the formation of the Ward Hunt and other ice shelves. Hence the term "sea ice-ice shelves" coined by Lemmen et al. (1988). Moreover, the crystal texture, including columnar crystals (Ragle et al. 1964: Fig. 10) and a characteristic cellular substructure of ice plates and brine cells (Jeffries 1985, 1991b), are immediately recognizable as those of undeformed congelation ice formed by Stefan growth due to heat conduction through the ice above. Such thick, undeformed sea ice is rare in the Arctic. On the basis of one-dimensional thermodynamic modeling, Walker and

Wadhams (1979) suggested that such thick ice growth is possible in locations where the oceanic heat flux is zero and there is 1 m of annual snowfall.

At the time core 83-1 was obtained, it was understood that the Ward Hunt Ice Shelf was ~3000 years old. It has since been shown that it is probably ~5500 years old (England et al. 2008, 2017). If, as Jeffries (1991b) suggested, Stratum 2 represents the original sea ice basement of the ice shelf, then it is truly ancient sea ice with remarkable properties for its age. Not only is the crystal texture well preserved, but the salinity and density values (Table 2.3) are similar to those of modern multiyear sea ice in the Arctic Ocean, while the brine volume values (Table 2.3) exceed those of modern multiyear ice at the same temperature (Jeffries 1991b). The preservation of the columnar crystal texture, cellular sub-structure and high brine content is attributed to the ice never having been exposed at the ice shelf surface and therefore not subject to brine flushing by meltwater, elevated summer temperatures and solar radiation (Jeffries 1991b).

As noted above, Stratum 3 (Fig. 2.9a) is brackish ice and can be described as interleaved with Strata 2 and 4. The interleaving of sea ice and brackish ice in Ward Hunt Ice Shelf is better illustrated by ice core 86-8 (Fig. 2.10a), which was obtained from the shoreward area (Fig. 2.6) where Lyons et al. (1971) first described the interleaving of the two ice types. Stratum 3, the brackish ice and sea ice of Lyons et al. (1971), and core 86-8 each have similar salinity and $\delta^{18}O$ values (Fig. 2.10b).

The salinity and $\delta^{18}O$ minima and maxima in core 86-8 are closely matched, resulting in a statistically significant correlation coefficient for the regression equation (Fig. 2.10b) and significantly different mean salinity and mean $\delta^{18}O$ values for each ice type (Fig. 2.10). The brackish ice and sea ice layers and their different properties reflect changes in the salinity of the water flowing below the ice shelf (Lyons et al. 1971; Jeffries and Sackinger 1989). Citing Crary (1960) on the age of the ice in these shoreward areas, Jeffries and Sackinger (1989) suggest that the interleaving of the two ice types and their different properties have been preserved for hundreds of years.

The interleaving of ancient brackish ice and sea ice observed in the Ward Hunt Ice Shelf (Fig. 2.10) is common in the MLSI that is attached to the ice shelves (Fig. 2.2b) (Jeffries and Krouse 1988; Jeffries et al. 1989). Salinity and $\delta^{18}O$ profiles in two cores, one from a peak (9.8 m thick) and the other from the adjacent trough (7.24 m thick), in MLSI at the front of the Milne Ice Shelf (Fig. 2.7) illustrate this phenomenon (Fig. 2.11).

Depending on whether zero or maximum isotopic fractionation during freezing is assumed, the MLSI in this region contains 0.4% freshwater ice (only below troughs), 42.3% brackish ice and 57.3% sea ice (zero fractionation), or 29.6% brackish ice and 70% sea ice (maximum fractionation) (Jeffries et al. 1989). As with the ancient ice, the modern brackish ice and sea ice have significantly different mean salinity values and significantly different mean $\delta^{18}O$ values (Table 2.4). The modern brackish ice and sea ice layers and their different properties represent changing under-ice water properties, with the ice below troughs tending to contain more brackish ice due to preferential pooling of brackish water in the inverted troughs at the bottom of the ice (Jeffries et al. 1989).

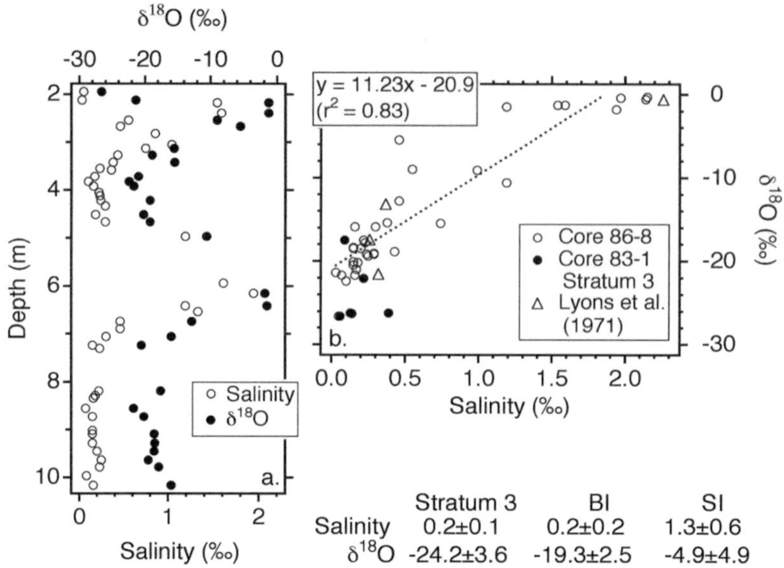

Fig. 2.10 (**a**) Salinity and $\delta^{18}O$ profiles, and (**b**) relationship between salinity and $\delta^{18}O$ in ice core 86-8 from central Ward Hunt Ice Shelf. Salinity and $\delta^{18}O$ values for brackish ice in Stratum 3, ice core 83-1 (Fig. 2.9) and from Lyons et al. (1971) are also shown in (**b**). Mean (±1 standard deviation) salinity and $\delta^{18}O$ values for brackish ice (BI) in Stratum 3 (core 83-1), and for brackish ice and sea ice (SI) in core 86-8, are tabulated at *lower right*

Ice Island ARLIS-II had a large area of "bluish stratified ice", that we now recognize as MLSI, attached to the main glacier ice body (Smith 1964). With the exception of some locally deformed areas up to 20 m thick, the MLSI was typically 12 m thick and strongly stratified with layers averaging 0.6–0.7 m thick. It was not known whether these were annual growth layers (Smith 1964), and there are insufficient salinity data and no isotopic data to determine whether they were composed of brackish ice and sea ice. The bulk salinity of the ARLIS-II MLSI (range 1.1–3.3‰, mean 1.9‰, n = 22) is similar to the ancient sea ice in Ward Hunt Ice Shelf and the modern Ellesmere MLSI.

Perhaps the most interesting characteristic of the ARLIS-II MLSI was the occurrence of preferred c-axis alignments. Smith (1964) was among the first to report such alignments in any sea ice, and it has since been documented in pack ice and landfast ice throughout the Arctic. It occurs as water flowing across the skeletal layer at the bottom of the ice causes the ice plates to align perpendicular to the flow direction (e.g., Weeks and Gow 1980). It is reasonable to assume that such alignments occur today in the Ellesmere MLSI.

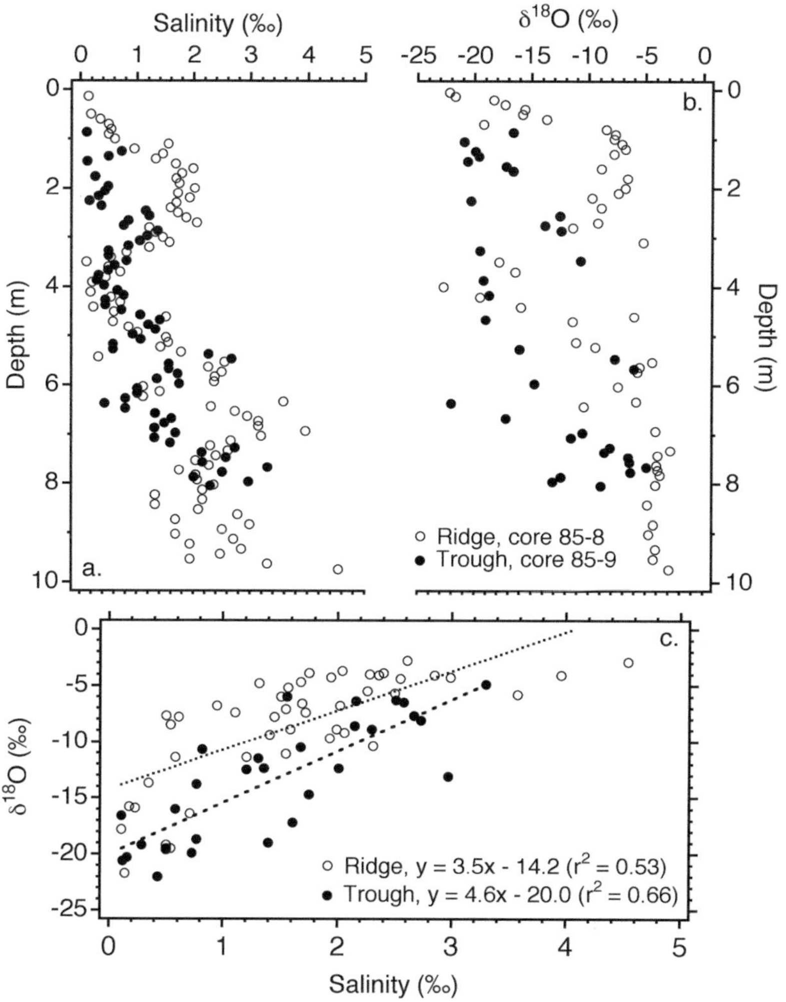

Fig. 2.11 (**a**) Salinity and (**b**) $\delta^{18}O$ profiles in adjacent peak (9.8 m thick) and trough (7.24 m thick) ice cores from MLSI at the front of the Milne Ice Shelf (Figs. 2.2c and 2.7). The profiles have been plotted to allow for the elevation difference between the peak and trough. The relationships between salinity and $\delta^{18}O$ in each core are shown in (**c**)

2.4.2.3 Meteoric Ice

In the previous section, Stratum 1 of core 83-1 (Fig. 2.9a) was described as meteoric ice, i.e., it originated as precipitation, which is evident in the negligible salinity and very negative $\delta^{18}O$ values. Stratum 1 is an example of the "iced-firn and interstratified lake ice" or "interlensed lake ice" described by a number of early investigators (Marshall 1955, 1960; Crary 1958; Lyons and Leavitt 1961). Ragle et al. (1964) used the term "glacial ice" to describe the iced-firn, but made it clear that this was

Table 2.4 Mean (±1 standard deviation) salinity and $\delta^{18}O$ values for brackish ice and sea ice in modern MLSI

	Salinity (psu)	$\delta^{18}O$ (‰)	n
Case 1			
Brackish ice	0.66 ± 0.53	−19.9 ± 1.9	111
Sea ice	1.88 ± 1.35	−6.5 ± 3.1	264
Case 2			
Brackish ice	0.75 ± 0.56	−18.1 ± 2.8	153
Sea ice	2.03 ± 1.46	−5.2 ± 2.6	222

Source: Jeffries et al. (1989)
Case 1 assumes no isotopic fractionation ($\alpha = 1.0$) during freezing
Case 2 assumes maximum isotopic fractionation ($\alpha = 1.0003$) during freezing

not to imply that it was ice that had flowed from the land. Rather, it had accumulated *in situ* on the original sea ice basement. For a description of true glacier ice in an Ellesmere ice shelf, the reader is referred to Smith (1964) who investigated the structure of the large body of glacier ice in ice island ARLIS-II. In modern glaciological terminology "iced-firn" is superimposed ice. This section, then, describes superimposed ice and lake ice. A third meteoric ice category, freshwater ice accreted at the base of the shelf ice, is also described.

Ice core 85-10 (Fig. 2.12a) represents the entire thickness of Hobson's Choice Ice Island/east Ward Hunt Ice Shelf. It is composed entirely of meteoric ice, in sharp contrast to core 83-1 (Fig. 2.9a) and the large amount of marine ice observed in the west Ward Hunt Ice Shelf. The differences in ice stratigraphy and ice type between the two parts of the ice shelf are discussed elsewhere (Jeffries et al. 1988a, 1991). The meteoric ice in core 85-10 sub-divides further into a 37 m thick layer that accumulated at the surface, and a 5 m thick basal layer that accreted from freshwater flowing below the ice shelf (Jeffries et al. 1988a, 1991).

Poplin and Ralston (1992) reported a mean density of 897 ± 2 kg m^{-3} and both granular and columnar crystal textures in the basal layer. Like the thicker ice above, the basal layer has a very low level of dissolved impurities (the mean specific electrolytic conductivity (SEC) of 8 μS cm^{-1} is roughly equivalent to a salinity of 0.005‰). High tritium values in this layer indicate that accretion occurred after 1952 from meteoric water flowing below the ice shelf from the epishelf lake in Disraeli Fiord (Jeffries et al. 1988a, 1991).

The very thick upper layer of Hobson's Choice is primarily superimposed ice with a polygonal granular texture (Fig. 2.13a–d) and many air inclusions. The grain size is primarily <10 mm (Fig. 2.12b) and the crystals have horizontal c-axes (Jeffries et al. 1991) defined by a single maximum perpendicular to the ice surface in all ice samples (Barrette and Sinha 1996). A variety of other textural characteristics of the superimposed ice are reported in Barrette and Sinha (1996). Embedded in the superimposed ice are a number of thin layers with grain sizes >10 mm (Fig. 2.12b). These are primarily the lake ice that previous investigators described as "interstratified" and "interlensed" with the superimposed ice ("iced-firn"). A few lake ice layers had vertical c-axes, but most had horizontal c-axes with large-grained polygonal granular textures (Fig. 2.13e, f) (Jeffries et al. 1991).

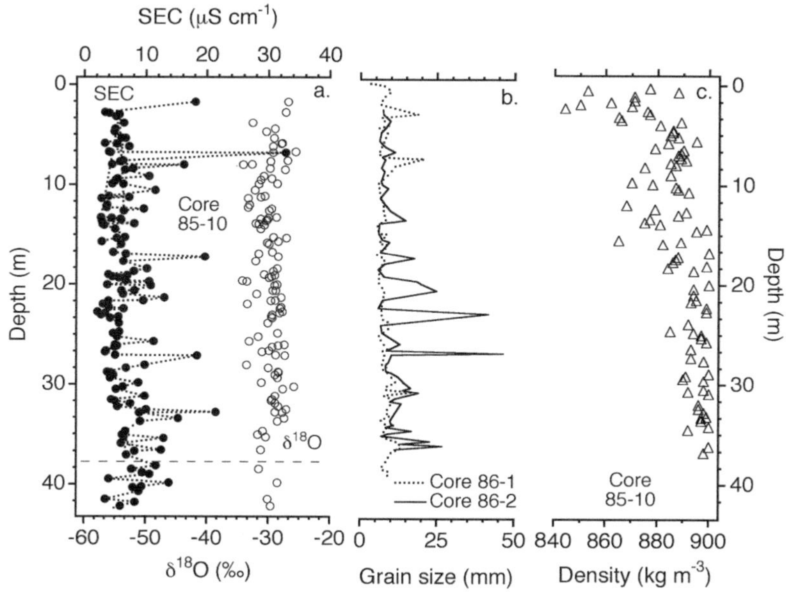

Fig. 2.12 Ice core profiles from Hobson's Choice Ice Island: (**a**) specific electrolytic conductivity (SEC) and $\delta^{18}O$ in core 85-10; (**b**) grain size in cores 86-1 and 86-2; (**c**) density in core 85-10. The horizontal *dashed line* in (**a**) is the boundary between meteoric ice that accumulated from snow at the top surface and meteoric ice that accreted from freshwater flowing at the bottom surface. Cores 85-10 and 86-1 were obtained from a peak. Core 86-2 was obtained from a trough. The profiles are plotted to allow for the elevation differences among them. The data in (**a**) and (**b**) are from Jeffries et al. (1991). The data in (**c**) are from Poplin and Ralston (1992)

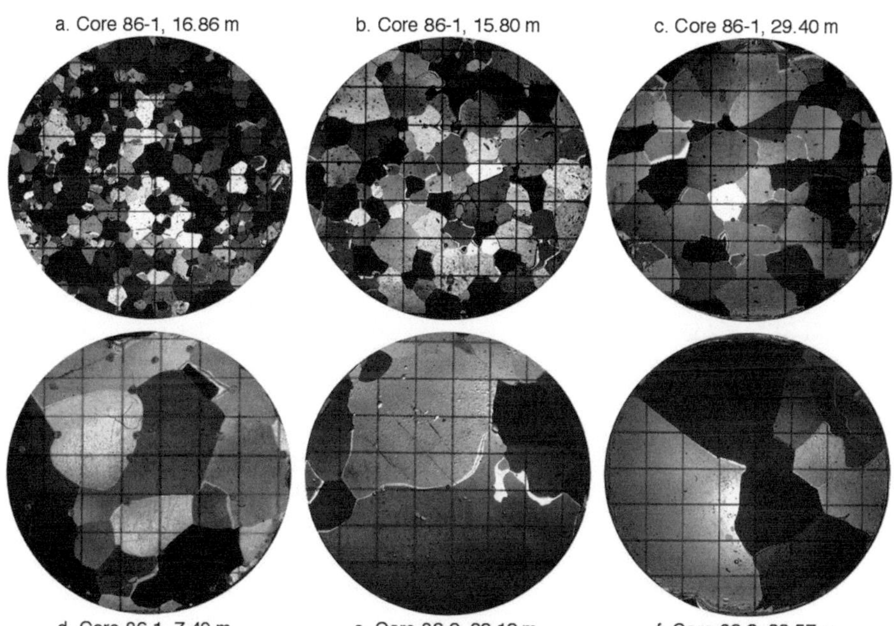

Fig. 2.13 Crystal texture of superimposed ice in Hobson's Choice Ice Island (Photographs by the author)

Table 2.5 Some physical properties of modern lake ice in troughs

Core and location	Specific electrolytic conductivity (μS cm^{-1})	Density (kg m^{-3})	δ^{18}O (‰)
82–3, WHIS-West[a]	105.3 ± 207 (12)	914 ± 5 (12)	−23.1 ± 3.5 (13)
84–6, WHIS-West[a]	59.6 ± 124.1 (7)	N.A.	−21.9 ± 5.3 (7)
83–6, Milne Ice Shelf-Outer[a]	13.9 ± 17.8 (9)	864 ± 12 (4)	−26.8 ± 3.7 (9)

N.A. Not Available
[a]Jeffries (1985)

Table 2.6 Mean density of the mixed superimposed ice and lake ice layer in Hobson's Choice Ice Island

Core	85-10[a]	86-1[b]	86-2[b]	87-1[b]
Density (kg m^{-3})	888 ± 21 n = 27	870 ± 21 n = 156	874 ± 13 n = 129	894 ± 12 n = 77

[a]Poplin and Ralston (1992)
[b]Jeffries et al. (1991)

Ice density measurements in the mixed superimposed ice and lake ice layer of Hobson's Choice have not discriminated between the two ice types. The mean density of this layer is summarized in Table 2.6. A characteristic feature of the ice density observed in three cores is an increase in density with depth (Fig. 2.12c) reported by Jeffries et al. (1991) and Poplin and Ralston (1992). It is possible that the depth-dependent density increase is caused by the same process that Barrette and Sinha (1996) proposed to explain the single c-axis orientation maxima. That is, compression by the mass of ice above, which, as Barrette and Sinha (1996) suggest, implies that there was once a much greater thickness of superimposed ice that has since thinned by surface melting.

The lake ice described above is very old, on the order of hundreds to a few thousand years old. Much younger lake ice forms each winter when the lakes in troughs (Fig. 2.2b) freeze. Some properties of this ice are summarized in Table 2.5. Both the δ^{18}O and SEC values are higher than those in Stratum 1, core 83-1 (Fig. 2.9a) and core 85-10 (Fig. 2.12a), for example. The isotopic difference has been attributed to evaporation in summer and the mixing of summer precipitation with meltwater from the previous winter's precipitation (Jeffries 1985). The difference in SEC values has been attributed to the dissolution of wind-blown dust and the addition of meltwater from sea ice exposed at the ice shelf surface (Jeffries 1985). Summertime evaporation might also be a factor, as it would concentrate dissolved salts in the remaining water. Mueller and Vincent (2006) reported SEC values of the same order of magnitude and higher in their study of the lakes as a habitat for micro-organisms.

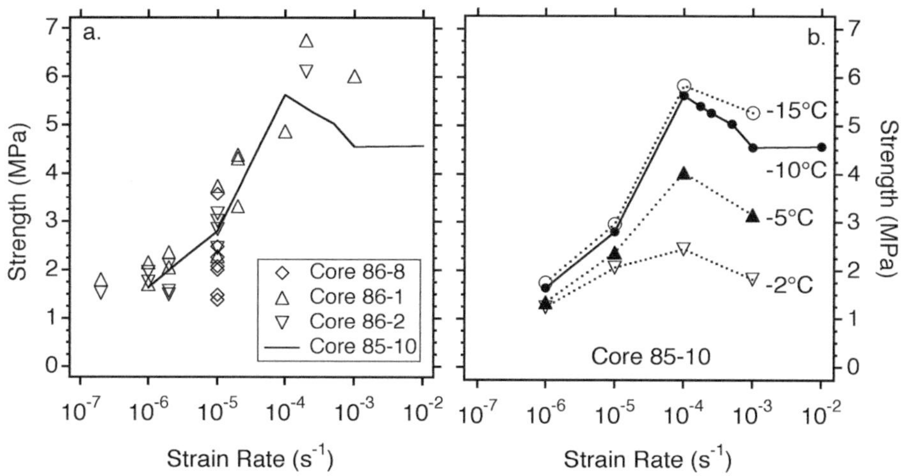

Fig. 2.14 Compressive strength of marine ice (core 86-8, Ward Hunt Ice Shelf) and meteoric ice (cores 85-10, 86-1, 86-2, Hobson's Choice Ice Island) as a function of strain rate and measurement temperature: (**a**) test temperature of −10°C only; (**b**) four different test temperatures

2.4.2.4 Compressive Strength of Marine Ice and Meteoric Ice

The strength of the ice is of interest to the Arctic offshore oil and gas industry. Poplin and Ralston (1992) note that even a small ice island fragment would represent an "ice load concern" if it collided with an offshore oil/gas production facility in the southern Beaufort Sea. Consequently, there have been a few investigations of ice mechanical properties, particularly of the compressive strength of meteoric ice from Hobson's Choice Ice Island and marine ice from Ward Hunt Ice Shelf. The results are summarized in Fig. 2.14.

There are insufficient data to give meaningful mean values for the brackish ice and sea ice in core 86-8 (Fig. 2.10). Combined, however, the two marine ice types have a mean compressive strength of 2.2 ± 0.7 MPa (n = 7, range 1.38–3.57 MPa) at a test temperature of −10°C and strain rate of 10^{-5} s^{-1} (Fig. 2.14a) (Jeffries et al. 1988b). The 2.2 ± 0.7 MPa value is slightly lower than the compressive strength of the meteoric ice in Hobson's Choice Ice Island at the same test temperature and strain rate: 2.9 ± 0.5 MPa (n = 6) for cores 86-1 and 86-2 combined (Jeffries et al. 1990b) and 2.8 ± 0.6 MPa (n = 35) for core 85–10 (Poplin and Ralston 1992). This is consistent with previous studies of the strength of other types of marine and meteoric ice (e.g., Schulson 2003). Figure 2.14b, which shows the temperature dependence of compressive strength in Hobson's Choice, is, in effect, a simulation of the change in strength that would be expected between the colder upper layers and warmer lower layers of the ice island. The temperature-dependence of ice strength coupled with the decrease in ice density with increasing depth (Fig. 2.12c) explains the observed decrease in ice strength with increasing depth (Poplin and Ralston

1992). No clear relationship between ice strength and grain size or texture was observed in Hobson's Choice (Poplin and Ralston 1992).

2.5 Conclusion

Though few in number and small in area, and continuing to shrink, the Ellesmere ice shelves are, nevertheless, significant ice masses. As noted in the Introduction, they are the only ice shelves in Canada and the most extensive in the entire Arctic. Unusual among ice shelves anywhere, they owe their origin and subsequent development primarily to marine ice formation by freezing of seawater and brackish water. There is also evidence of meteoric ice accretion at the bottom surface, but most meteoric ice has accumulated as superimposed ice and lake ice at the top surface. While superimposed ice is a term associated with glaciers, most of the superimposed ice (and lake ice) found in the Ellesmere ice shelves has accumulated *in situ* on the marine ice basement. The seaward flow of glaciers has played a minor role in the growth and maintenance of the Ellesmere ice shelves. But, ironically, as the ice shelves have declined in number and area, the proportion of the remaining ice that has originated from glaciers has increased.

In all likelihood, the marine, or basement, ice that remains in the Ellesmere ice shelves, most of it in the western and southern sections of the Ward Hunt Ice Shelf, contains the thickest and oldest sea ice and brackish ice in the entire Arctic. The salinity, brine volume and isotopic composition of the marine ice have been documented, but there is still much to learn, particularly about the fabric of the ice, e.g., c-axis orientation and crystal size, shape and sub-structure (ice plate width/brine layer spacing). Such information would be of local interest, e.g., a history of the direction of water flow below the shelf ice, and would also contribute to the broader knowledge and understanding of marine ice physical properties and processes, and perhaps of other porous and permeable materials. The marine ice might also preserve a record of microbial activity in the water below the ice, and whether the micro-organisms remained viable once they were entrained in the ice. The microbial ecosystem at the ice surface has now been described in some detail, but the microbial ecosystem of the marine ice, if there is one, remains to be investigated.

Besides the ancient marine ice, the other distinguishing feature of the Ellesmere ice shelves is the rolling topography of the surface. The cause of the rolls remains uncertain and finding an unambiguous explanation is of more than local and academic interest; an explanation for the rolls is likely to provide insights into climate change-related ice island calving and ice shelf disintegration. For if, as Holdsworth (1987) concluded, pack ice pressure and buckling instability explain the rolls, the long-term persistence of the ice shelves might also be explained by confinement due to the pack ice. The age, thickness and annual minimum extent of the Arctic Ocean pack ice have decreased significantly in recent years (e.g., Maslanik et al. 2007; Giles et al. 2008; Perovich et al. 2014). If the pack ice continues its decline, and there is a growing consensus that it will, it is reasonable to assume that the Ellesmere ice shelves will respond accordingly and continue to disintegrate.

Continued disintegration of the ice shelves will create more ice islands, which pose a potential risk to offshore oil/gas production and vessel navigation. To date, ice strength studies have focussed primarily on compressive strength and its dependence on strain rate. Much remains to be learned about the compressive strength, tensile strength and flexural strength of the ancient marine and meteoric ice in the ice shelves, and the modern MLSI attached to the ice shelves, and their dependence on a variety of factors, e.g., strain rate, ice fabric (grain size and orientation, and sub-structure) and salinity. In view of the vulnerability of the ice shelves, and the great age of the ice, such comprehensive and systematic ice strength studies would have significant practical and academic value.

The Ward Hunt Ice Shelf, where the eastern and southern sections are already riddled with fractures, is probably the most vulnerable to continued pack ice and climate change. The Milne, Peterson and Serson ice shelves might be less vulnerable in the near-term because they derive a large proportion of their mass from the seaward advance of glaciers. However, in the long-term their survival will depend on the mass balance of their grounded source areas, and the ability of the floating sections to withstand changing ocean and atmospheric forces resulting from further pack ice and MLSI decline. Indeed, since this chapter was originally written, it has been reported that the area of the Peterson Ice Shelf decreased by about 60% since 2005, most likely due to the massive loss of protective MLSI from Yelverton Bay and exposure to more open water and less pack ice pressure during subsequent, record warm summers (Pope et al. 2012; White et al. 2015).

Acknowledgements This chapter was written while I was on leave from the University of Alaska Fairbanks and working on secondment from 2006 to 2010 at the National Science Foundation (NSF), Office of Polar Programs, Division of Arctic Sciences, under the terms of the Inter-Governmental Personnel Act. Any opinion, findings, and conclusions or recommendations expressed in this material are mine and do not necessarily reflect the views of either NSF or the Office of Naval Research. I am very grateful to Ron Verrall (1941–2014, formerly of the Defence Research Establishment Pacific and Defence Research and Development Canada) for allowing me to use unpublished data on multiyear landfast sea ice topography and thickness in Moss Bay. The SAR images were provided by the Alaska Satellite Facility using data credits made available by NASA. I thank Derek Mueller and Luke Copland for inviting me to write this chapter, and Derek Mueller and two anonymous reviewers for their thoughtful comments and suggestions.

References

Barrette, P. D, Sinha, N. K. (1996). Crystallographic characterization of a core from Ward Hunt Ice Shelf, Canada. In K. C. Agarwal (Ed.), *Proceedings of the international symposium on snow and related manifestations, 26–28 September 1994, Snow and Avalanche Study Establishment, Manali* (p. 114–124).

Belkin, I., & Kessel, S. (2017). Russian drifting stations on Arctic ice islands. In L. Copland & D. Mueller (Eds.), *Arctic ice shelves and ice islands* (p. 367–393). Dordrecht: Springer. doi:10.1007/978-94-024-1101-0_14.

Braun, C. (2017). The surface mass balance of the Ward Hunt Ice Shelf and Ward Hunt Ice Rise, Ellesmere Island, Nunavut, Canada. In L. Copland & D. Mueller (Eds.), *Arctic ice shelves and ice islands* (p. 149–183). Dordrecht: Springer. doi:10.1007/978-94-024-1101-0_6.

Braun, C., Hardy, D. R., Bradley, R. C., & Sahanatien, V. (2004). Surface mass balance of the Ward Hunt Ice Rise and Ward Hunt Ice Shelf, Ellesmere Island, Nunavut, Canada. *Journal of Geophysical Research, 109*, D22110. doi:10.1029/2004JD004560.

Copland, L., Mueller, D. M., & Weir, L. (2007). Rapid loss of the Ayles Ice Shelf, Ellesmere Island, Canada. *Geophysical Research Letters, 34*, L21501. doi:10.1029/2007GL031809.

Copland, L., Mortimer, C., White, A., Richer McCallum, M., & Mueller, D. (2017). Factors contributing to recent Arctic ice shelf losses. In L. Copland & D. Mueller (Eds.), *Arctic ice shelves and ice islands* (pp. 263–285). Dordrecht: Springer. doi:10.1007/978-94-024-1101-0_10.

Crary, A. P. (1958). Arctic ice island and ice shelf studies. Part I. *Arctic, 11*(1), 3–42. doi:10.14430/arctic3731.

Crary, A. P. (1960). Arctic ice island and ice shelf studies, Part II. *Arctic, 13*(1), 32–50. doi:10.14430/arctic3687.

Crary, A. P., Cotell, R. D., & Saxton, T. F. (1952). Preliminary report on scientific work on "Fletcher's Ice Island", T-3. *Arctic, 5*(4), 211–223. doi:10.14430/arctic3913.

Debenham, F. (1954). The ice islands of the Arctic: A hypothesis. *Geographical Review, 44*(4), 495–507. doi:10.2307/212156.

Dorrer, E. (1971). Movement of the Ward Hunt Ice Shelf. *Journal of Glaciology, 10*(59), 211–225. doi:10.3198/1971JoG10-59-211-225.

Dowdeswell, J. A., & Jeffries, M. O. (2017). Arctic ice shelves: An introduction. In L. Copland & D. Mueller (Eds.), *Arctic ice shelves and ice islands* (p. 3–21). Dordrecht: Springer. doi:10.1007/978-94-024-1101-0_1.

England, J. H., Lakeman, T. R., Lemmen, D. S., Bednarski, J. M., Stewart, T. G., & Evans, D. J. A. (2008). A millennial-scale record of Arctic ocean sea ice variability and the demise of the Ellesmere Island ice shelves. *Geophysical Research Letters, 35*, L19502. doi:10.1029/2008GL034470.

England, J. H., Evans, D. A., & Lakeman, T. R. (2017). Holocene history of Arctic ice shelves. In L. Copland & D. Mueller (Eds.), *Arctic ice shelves and ice islands* (p. 185–205). Dordrecht: Springer. doi:10.1007/978-94-024-1101-0_7.

Giles, K. A., Laxon, S. W., & Ridout, A. L. (2008). Circumpolar thinning of Arctic sea ice following the 2007 record ice extent minimum. *Geophysical Research Letters, 35*, L22502. doi:10.1029/2008GL035710.

Grischenko, V. D., & Simonov, I. M. (1985). Morphological and structural peculiarities of the SP-22 drifting ice island. *Problemy Arktiki i Antarktiki, 59*, 60–68.

Hattersley-Smith, G. (1957). The rolls on the Ellesmere Ice Shelf. *Arctic, 10*(1), 32–44. doi:10.14430/arctic3753.

Hattersley-Smith, G. (1963). The Ward Hunt Ice Shelf: Recent changes at the ice front. *Journal of Glaciology, 4*(34), 415–424. doi:10.3198/1963JoG4-34-415-424.

Hattersley-Smith, G. (1969). Results of radio echo sounding in northern Ellesmere Island, 1966. *The Geographical Journal, 135*(4), 553–557. doi:10.2307/1795101.

Hattersley-Smith, G., Crary, A. P., & Christie, R. L. (1955). Northern Ellesmere Island, 1953 and 1954. *Arctic, 8*(1), 3–46. doi:10.14430/arctic3802.

Hattersley-Smith, G., Fuzesy, A., Evans, S. (1969). *Glacier depths in northern Ellesmere Island: Airborne radio echo sounding in 1966*. DREO Technical Note 69–6, Defence Research Establishment Ottawa, Defence Research Board, Department of National Defence Canada.

Holdsworth, G. (1987). The surface waveforms on the Ellesmere Island ice shelves and ice islands. In *Workshop on extreme ice features, technical memorandum 141 (NRCC 28003), National Research Council of Canada* (p. 385–403).

Jeffries, M. O. (1985). *Physical, chemical and isotopic investigations of Ward Hunt Ice Shelf and Milne Ice Shelf, Ellesmere Island, NWT*. Ph.D. thesis, University of Calgary, Calgary, pp. 310.

Jeffries, M. O. (1986). Glaciers and the morphology and structure of the Milne Ice Shelf, Ellesmere Island, N.W.T. *Arctic and Alpine Research, 18*(4), 397–405. doi:10.2307/1551089.

Jeffries, M. O. (1987). The growth, structure and disintegration of Arctic ice shelves. *Polar Record, 23*(147), 631–649. doi:10.1017/S0032247400008342.

Jeffries, M. O. (1991a). Perennial water stratification and the role of basal freshwater flow in the mass balance of the Ward Hunt Ice Shelf, Canadian High Arctic. In *Proceedings of the International Conference on the Role of the Polar Regions in Global Change, Volume I, University of Alaska Fairbanks, Fairbanks, Alaska, 11–15 June 1990*. Geophysical Institute and Center for Global Change and Arctic System Research, University of Alaska Fairbanks, pp. 332–337.

Jeffries, M. O. (1991b). Massive, ancient sea-ice strata and preserved physical-structural characteristics in the Ward Hunt Ice Shelf. *Annals of Glaciology, 15*, 125–131.

Jeffries, M. O. (1992a). Arctic ice shelves and ice islands: Origin, growth and disintegration, physical characteristics, structural-stratigraphic variability, and dynamics. *Reviews of Geophysics, 30*(3), 245–267. doi:10.1029/92RG00956.

Jeffries, M. O. (1992b). The source and calving of ice island ARLIS-II. *Polar Record, 28*(165), 137–144. doi:10.1017/S0032247400013437.

Jeffries, M. O. (2002). Ellesmere Island ice shelves and ice islands. In R. S. Williams Jr. & J. G. Ferrigno (Eds.), *Satellite image atlas of glaciers of the world: Glaciers of North America. U. S. Geological Survey Professional Paper 1386-J-1* (p. J147–J164). Reston: United States Geological Survey.

Jeffries, M. O., & Krouse, H. R. (1988). Salinity and isotope analysis of some multi-year landfast sea-ice cores, northern Ellesmere Island, Canada. *Annals of Glaciology, 10*, 63–67. doi:10.3198/1988AoG10-63-67.

Jeffries, M. O., & Sackinger, W. M. (1989). Somc measurements and observations of very old sea ice and brackish ice, Ward Hunt Ice Shelf, N.W.T. *Atmosphere-Ocean, 27*(3), 553–564. doi:10.1080/07055900.1989.9649352.

Jeffries, M. O., & Serson, H. V. (1983). Recent changes at the front of Ward Hunt Ice Shelf. *Arctic, 36*(3), 289–290. doi:10.14430/arctic2278.

Jeffries, M. O., Sackinger, W. M., Krouse, H. R., & Serson, H. V. (1988a). Water circulation and ice accretion beneath Ward Hunt Ice Shelf (northern Ellesmere Island, Canada) deduced from salinity and isotope analysis of ice cores. *Annals of Glaciology, 10*, 68–72.

Jeffries, M. O., Sackinger, W. M., Frederking, R. M. W., & Timco, G.W. (1988b, August 23–27). Initial mechanical and physical-structural property measurements of old sea and brackish ice from Ward Hunt Ice Shelf, Canada. In *Proceedings of the 9th International Symposium on Ice* (Vol. 1, p. 177–187). Sapporo: International Association for Hydraulic Research.

Jeffries, M. O., Krouse, H. R., Sackinger., W. M., & Serson, H. V. (1989). Stable isotope ($^{18}O/^{16}O$) tracing of fresh, brackish and sea ice in multiyear landfast sea ice, Ellesmere Island, Canada. *Journal of Glaciology, 35*(119), 9–16. doi:10.3198/1989JoG35-119-9-16.

Jeffries, M. O., Krouse, H. R., Sackinger, W. M., & Serson, H. V. (1990a). Surface topography, thickness and ice core studies of multiyear landfast sea ice and Ward Hunt Ice Shelf, northern Ellesmere Island, N.W.T. In C. R. Harington (Ed.), *Canada's missing dimension: Science and history in the Canadian Arctic Islands* (Vol. 1, p. 229–254). Ottawa: Canadian Museum of Nature.

Jeffries, M. O., Sinha, N. K., & Sackinger, W. M. (1990b, August 20–23). Deformation of natural ice island ice under constant strain rate uniaxial compression. In *Proceedings of the 10th International Symposium on Ice* (Vol. 1, p. 238–251). Espoo: International Association for Hydraulic Research.

Jeffries, M. O., Serson, H. V., Krouse, H. R., & Sackinger, W. M. (1991). Ice physical properties, structural characteristics and stratigraphy in Hobson's Choice Ice Island and implications for the growth history of East Ward Hunt Ice Shelf, Canadian High Arctic. *Journal of Glaciology, 37*(126), 247–260. doi:10.3198/1991JoG37-126-247-260.

Jungblut, A. D., Mueller, D., & Vincent, W. F. (2017). Arctic ice shelf ecosystems. In L. Copland & D. Mueller (Eds.), *Arctic ice shelves and ice islands* (p. 227–260). Dordrecht: Springer. doi:10.1007/978-94-024-1101-0_9.

Koenig, L. S., Greenaway, K. R., Dunbar, M., & Hattersley-Smith, G. (1952). Arctic ice islands. *Arctic, 5*(2), 67. doi:10.14430/arctic3901.

Legen'kov, A. P. (1973). Thermal stresses and deformations of the drifting ice island station North Pole 19. *Oceanology, 13*(6), 804–808.

Legen'kov, A. P., & Chugui, I. V. (1973). Results of morphometric measurements of the ice island of the Severnyi Polyus-19 drifting station. *Problemy Arktiki i Antarktiki, 42*, 44–48.

Lemmen, D. S., Evans, D. J. A., & England, J. (1988). Ice shelves of northern Ellesmere Island, NWT, Canadian landform examples-10. *Canadian Geographer, 32*(4), 363–367.

Lister, H. (1962). *Heat and mass balance at the surface of the Ward Hunt Ice Shelf* (p. 54). Scientific Report 1, Research Paper 19. Montreal: Arctic Institute of North America.

Lyons, J. B., & Leavitt, F.G. (1961). *Structural and stratigraphic studies on the Ward Hunt Ice Shelf.* Final report on Contract AF 19(604)-6188, Bedford: Geophysics Research Directorate, U.S. Air Force Cambridge Research Laboratories.

Lyons, J. B., & Mielke, J. E. (1973). Holocene history of a portion of northernmost Ellesmere Island. *Arctic, 26*(4), 314 . doi:10.14430/arctic2930.323.

Lyons, J. B., & Ragle, R. H. (1962, September 10–18). Thermal history and growth of the Ward Hunt Ice Shelf. In *I Colloque d'Obergurgl* (p. 88–97), International Union of Geodesy and Geophysics-International Association of Hydrological Sciences.

Lyons, J. B., Savin, S. M., & Tamburi, A. J. (1971). Basement ice, Ward Hunt Ice Shelf, Ellesmere Island, Canada. *Journal of Glaciology, 10*(58), 93–100. doi:10.3198/1971JoG10-58-85-92.

Lyons, J. B., Ragle, R. H., & Tamburi, A. J. (1972). Growth and grounding of the Ellesmere Island ice rises. *Journal of Glaciology, 11*(61), 43–52.

MacAyeal, D. R., & Holdsworth, G. (1986). An investigation of low-stress ice rheology on the Ward Hunt Ice Shelf. *Journal of Geophysical Research, 91*(B6), 6347–6358. doi:10.1029/JB091iB06p06347.

MacDonald, R. (2005). Challenges and accomplishments: A celebration of the Arctic Institute of North America. *Arctic, 58*(4), 440–451.

Mair, D., Burgess, D., Sharp, M., Dowdeswell, J. A., Benham, T., Marshall, S., & Cawkwell, F. (2009). Mass balance of the Prince of Wales Icefield, Ellesmere Island, Nunavut, Canada. *Journal of Geophysical Research, 114*(F2), F001082. doi:10.1029/2008JF001082.

Marshall, E. W. (1955). Structural and stratigraphic studies of the northern Ellesmere Ice Shelf. *Arctic, 8*(2), 109–114. doi:10.14430/arctic3810.

Marshall, E. W. (1960). Structure and stratigraphy of T-3 and the Ellesmere Ice Shelf. In V. C. Bushnell (Ed.), *Scientific studies at Fletcher's ice island T-3 (1952–1955)* (Geophysical research paper 6, Volume 3, p. 45–57). Geophysics Research Directorate, Air Force Cambridge Research Laboratory.

Maslanik, J. A., Fowler, C., Stroeve, J., Drobot, S., Zwally, J., Yi, D., & Emery, W. (2007). A younger, thinner Arctic ice cover: Increased potential for rapid, extensive sea-ice loss. *Geophysical Research Letters, 34*, L24501. doi:10.1029/2007GL032043.

Mortimer, C. A., Copland, L., & Mueller, D. R. (2012). Volume and area changes of the Milne Ice Shelf, Ellesmere Island, Nunavut, Canada, since 1950. *Journal of Geophysical Research, 117*, F04011. doi:10.1029/2011JF002074.

Mueller, D. R., & Vincent, W. F. (2006). Microbial habitat dynamics and ablation control on the Ward Hunt Ice Shelf. *Hydrocarbon Processing, 20*(4), 856–876. doi:10.1002/hyp.6113.

Mueller, D. R., Vincent, W. F., & Jeffries, M. O. (2003). Break-up of the largest Arctic ice shelf and associated loss of an epishelf lake. *Geophysical Research Letters, 30*(20), 2031. doi:10.1029/2003GL017931.

Mueller, D. R., Vincent, W. F., Bonilla, S., & Laurion, I. (2005). Extremotrophs, extremophiles and broadband pigmentation strategies in a high arctic ice shelf ecosystem. *FEMS Microbiology Ecology, 53*(1), 73–87. doi:10.1016/j.femsec.2004.11.001.

Mueller, D. R., Vincent, W. F., & Jeffries, M. O. (2006). Environmental gradients, fragmented habitats and microbiota of a northern ice shelf cryoecosystem, Ellesmere Island, Canada. *Arctic, Antarctic, and Alpine Research, 38*(4), 593–607. doi:10.1657/1523-0430.

Mueller, D., Copland, L., & Jeffries, M. O. (2017). Changes in Canadian Arctic ice shelf extent since 1906. In L. Copland & D. Mueller (Eds.), *Arctic ice shelves and ice islands* (p. 109–148). Dordrecht: Springer. doi:10.1007/978-94-024-1101-0_5.

Muguruma, J., & Higuchi, K. (1963). Glaciological studies on ice island T-3. *Journal of Glaciology, 4*(36), 709–730. doi:10.3198/1963JoG4-36-709-730.

Narod, B. B., Clarke, G. K. C., & Prager, B. T. (1988). Airborne UHF radar sounding of glaciers and ice shelves, northern Ellesmere Island, Arctic Canada. *Canadian Journal of Earth Sciences, 25*(1), 95–105. doi:10.1139/e88-010.

Ommanney, C. S. L. (1982a). *Bibliography of Canadian Glaciology, 1982 – Bibliography No. 2, Ellesmere Island glaciers and ice shelves* (Paper No. 20, Environment Canada). Saskatoon: National Hydrology Research Centre.

Ommanney, C. S. L. (1982b). *Bibliography of Canadian Glaciology, 1982 – Bibliography No. 3, Ice Islands of the Arctic Ocean* (Environment Canada, Paper No. 21). Saskatoon: National Hydrology Research Centre.

Parliamentary Paper v LVI. (1877). *Journals and proceedings of the Arctic expedition of 1875–76 under the command of captain Sir George Nares, R.N., K.C.B.* London.

Peary, R. E. (1907). *Nearest the Pole: A narrative of the polar expedition of the Peary Polar Club in the S.S. Roosevelt, 1905–1906*. London: Hutchinson.

Perovich, D., Gerland, S., Hendricks, S., Meier, W., Nicolaus, M., & Tschudi, M. (2014). *Sea ice. Arctic report card: Update for 2014: Tracking recent environmental changes*. http://www.arctic.noaa.gov/reportcard/sea_ice.html

Polunin, N. V. (1955). Long distance plant dispersal in the north polar regions. *Nature, 176*(4470), 22–24. doi:10.1093/aobpla/plv036.

Polunin, N. V. (1958). The botany of ice island T-3. *Journal of Ecology, 46*(2), 323–347. doi:10.2307/2257399.

Pope, S., Copland, L., & Mueller, D. (2012). Loss of multiyear landfast sea ice from Yelverton Bay, Ellesmere Island, Nunavut, Canada. *Arctic, Antarctic, and Alpine Research, 44*(2), 210–221. doi:10.1657/1938-4246-44.2.210.

Poplin, J. P., & Ralston, T. D. (1992). Physical and mechanical properties of Hobson's Choice Ice Island cores. *Cold Regions Science and Technology, 20*(2), 207–223. doi:10.1016/0165-232X(92)90018-P.

Ragle, R. H., Blair, R. G., & Persson, L. E. (1964). Ice core studies of Ward Hunt Ice Shelf, 1960. *Journal of Glaciology, 5*(37), 39–59. doi:10.3198/1964JoG5-37-39-59.

Rothrock, D. A., Yu, Y., & Maykut, G. A. (1999). Thinning of the Arctic sea-ice cover. *Geophysical Research Letters, 26*(23), 3469–3472. doi:10.1029/1999GL010863.

Schulson, E. M. (2003). Brittle fracture of ice. *Engineering Fracture Mechanics, 68*(17–18), 1839–1887. doi:10.1016/S0013-7944(01)00037-6.

Short, N. H., & Gray, A. L. (2005). Glacier dynamics in the Canadian High Arctic from RADARSAT-1 speckle tracking. *Canadian Journal of Remote Sensing, 31*(3), 225–239. doi:10.5589/m05-010.

Smith, D. D. (1961). Sequential development of surface morphology on Fletcher's Ice Island, T-3. In G. O. Rausch (Ed.), *Geology of the Arctic* (Vol. 2, p. 896–914). Canada: University of Toronto Press.

Smith, D. D. (1964). Ice lithologies and structure of ice island ARLIS-II. *Journal of Glaciology, 5*(37), 17–38.

Veillette, J., Mueller, D. R., Antoniades, D., & Vincent, W. F. (2008). Arctic epishelf lakes as sentinel ecosystems: Past, present and future. *Journal of Geophysical Research, 113*, G04014. doi:10.1029/2008JG000730.

Vincent, W. F., Gibson, J. A. E., & Jeffries, M. O. (2001). Ice-shelf collapse, climate change, and habitat loss in the Canadian High Arctic. *Polar Record, 38*(201), 133–142. doi:10.1017/S0032247400026954.

Vincent, W. F., Mueller, D. R., & Bonilla, S. (2004). Ecosystems on ice: The microbial ecology of Markham Ice Shelf in the high Arctic. *Cryobiology, 48*(2), 103–112. doi:10.1016/j.cryobiol.2004.01.006.

Walker, E. R., & Wadhams, P. (1979). Thick sea-ice floes. *Arctic, 32*(2), 140–147. doi:10.14430/arctic2612.

Weeks, W. F., & Gow, A. J. (1980). Crystal alignments in the fast ice of Arctic Alaska. *Journal of Geophysical Research, 85*(C2), 1137–11146. doi:10.1029/JC085iC02p01137.

White, A., Copland, L., Mueller, D., & Wychen, W. V. (2015). Assessment of historical changes (1959–2012) and the causes of recent break-ups of the Petersen Ice Shelf, Nunavut, Canada. *Annals of Glaciology, 56*(69), 65–76. doi:10.3189/2015AoG69A687.

Williamson, S. N., Sharp, M. J., Dowdeswell, J. A., & Benham, T. J. (2008). Iceberg calving rates from northern Ellesmere Island ice caps, Canadian Arctic, 1999–2003. *Journal of Glaciology, 54*(186), 391–400. doi:10.3189/002214308785837048.

Chapter 3
Eurasian Arctic Ice Shelves and Tidewater Ice Margins

Julian A. Dowdeswell

Abstract Despite the presence of about 4000 km of marine-terminating glaciers and ice caps in the Eurasian Arctic, there are few floating ice shelves. Neither are there extensive areas of multi-year shorefast sea ice which might thicken into composite ice shelves themselves. The archipelagos of Severnaya Zemlya and Franz Josef Land contain some ice shelves in addition to grounded tidewater ice fronts. The largest Eurasian Arctic ice shelf was the Matusevich Ice Shelf, Severnaya Zemlya, at about 240 km^2 with a drainage basin of about 1100 km^2; this ice shelf largely broke up in 2012. In Franz Josef Land, a number of ice caps have smooth and very low surface gradient seaward margins, covering over 300 km^2 or 2% of the total area of the ice caps in the archipelago. These low-gradient areas are located mainly in relatively protected embayments and produce large tabular icebergs of up to several kilometres in length. Whether individual areas are floating in hydrostatic equilibrium or are simply close to buoyancy, they provide the major modern source of tabular icebergs to the Barents Sea. Svalbard has about 860 km of coastal ice cliffs, but almost none of the ice margin appears to be afloat. There may be short periods, during the active phase of the surge cycle, where marine margins become afloat. Neither is there evidence that the margins of the marine-terminating glaciers on Novaya Zemlya are floating. Twenty-five to fifty percent of the bed of the three largest ice caps in the Eurasian Arctic lies below sea level. Thus, in a warming Arctic, the ice margin would eventually retreat onto land, curtailing mass loss by iceberg production and providing a break on rapid ice-cap disintegration through calving.

Keywords Ice shelves • Tidewater glaciers • Eurasian Arctic • Icebergs • Sea ice

J.A. Dowdeswell (✉)
Scott Polar Research Institute, University of Cambridge, Cambridge, UK
e-mail: jd16@cam.ac.uk

L. Copland, D. Mueller (eds.), *Arctic Ice Shelves and Ice Islands*, Springer Polar Sciences, DOI 10.1007/978-94-024-1101-0_3

3.1 Introduction

Glaciers and ice caps with extensive marine margins cover an area of about 90,000 km^2 in the Eurasian Arctic archipelagos (Fig. 3.1; Dowdeswell 1995; Dowdeswell and Hagen 2004). In western Eurasia, Svalbard has an ice cover of about 36,500 km^2, representing about 60% of its land area (Hagen et al. 1993). Austfonna, in north-eastern Svalbard, is the largest ice cap in the whole Eurasian Arctic at about 8000 km^2 (Dowdeswell 1986; Dowdeswell et al. 2008). East of Svalbard, the Russian Arctic archipelagos of Franz Josef Land, Severnaya Zemlya and Novaya Zemlya hold about 52,000 km^2 of ice (Moholdt et al. 2012a), with glacierized areas of 85%, 50% and 30%, respectively (Dowdeswell et al. 2010). In the Russian Far East, ice totalling less than 100 km^2 is present in the DeLong archipelago east of the New Siberian Islands (Verkulich et al. 1992).

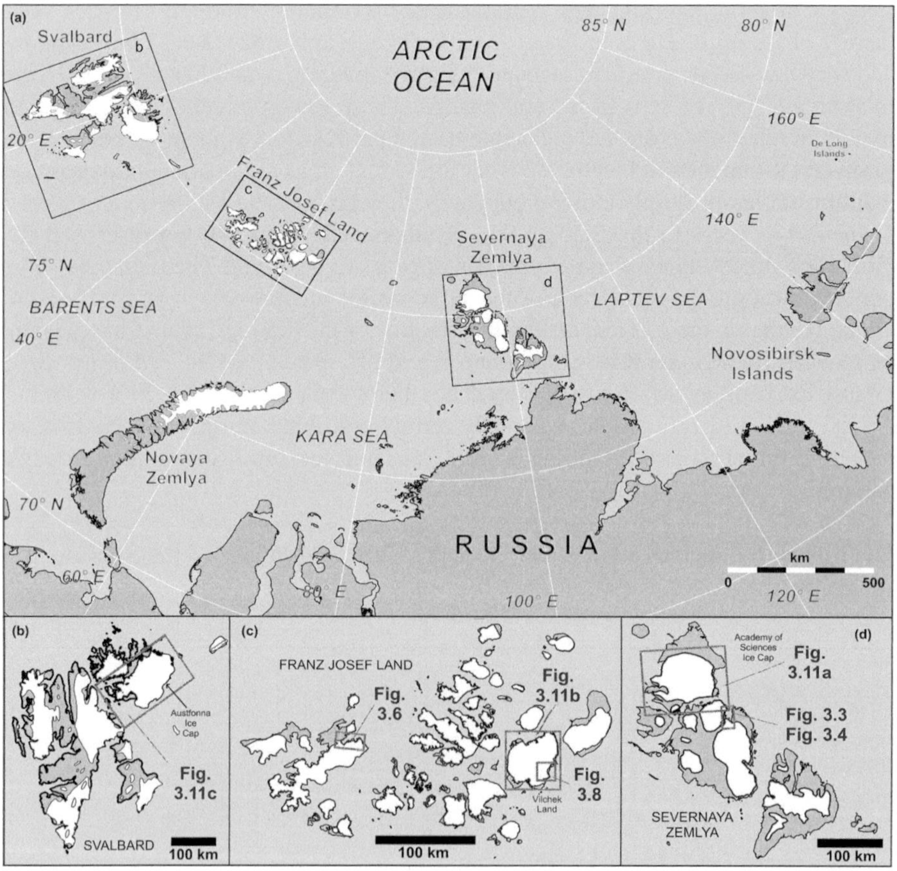

Fig. 3.1 (**a**) Map of the Eurasian Arctic showing the locations of the island archipelagos and the distribution of ice caps and glaciers. (**b**) Svalbard. (**c**) Franz Josef Land. (**d**) Severnaya Zemlya. The locations of several subsequent figures are shown (numbered on parts **b**, **c** and **d**)

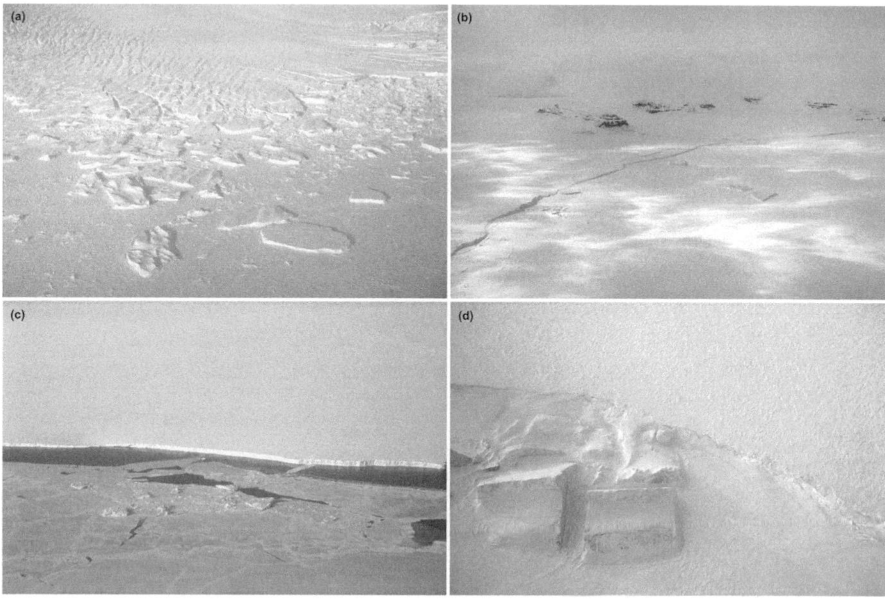

Fig. 3.2 Oblique aerial photographs of marine ice margins and icebergs calved from Eurasian Arctic ice caps. (**a**) Large numbers of tabular icebergs produced from a fast-flowing and probably floating ice stream on the east side of Academy of Science Ice Cap, Severnaya Zemlya, May 1998 (Dowdeswell et al. 2002). (**b**) Tabular iceberg about 0.5 km long embedded in shorefast sea ice in Franz Josef Land, May 1994. Note the new icebergs being produced at the calving margin of a low-profile ice cap. (**c**) Icebergs of irregular shape calved from the grounded margin of the Austfonna ice cap in eastern Svalbard in May 1983. The ice cliffs are about 30 m high. (**d**) An almost vertical view of the calving margin of Nergibreen in eastern Spitsbergen, May 1983. Several icebergs of almost 0.5 km across have been calved and are trapped in winter shorefast sea ice (Photographs: J.A. Dowdeswell)

Given the relatively high proportions of ice-covered land in the Eurasian Arctic islands, ice reaches the sea along much of this coastline (e.g. Fig. 3.2). Overall, there are approximately 4000 km of coastal ice cliffs in the Eurasian Arctic (Sharov 2005). There are about 860 km of marine ice cliffs in Svalbard (Blaszczyk et al. 2009), 2500 km in Franz Josef Land, 490 km in Severnaya Zemlya and 200 km in Novaya Zemlya (Sharov 2005). Comparisons between recent satellite datasets and maps produced from aerial photographs acquired in the 1950s suggest that the length of this ice-ocean interface has declined by about 8% (Sharov 2005), presumably reflecting the general retreat and thinning of glacier and ice-cap margins over this period (Dowdeswell and Hagen 2004; Carr et al. 2014; Melkonian et al. 2016).

The character of the margins of glaciers and ice caps in the Eurasian Arctic is a fundamental control on their mass balance (Hagen and Reeh 2004). For glaciers and ice caps that terminate on land, mass loss is almost entirely through surface melting and runoff. Where ice reaches the sea, the calving of icebergs and the melting of vertical ice cliffs provide additional mechanisms of mass loss (e.g. Weeks and Campbell 1973; Rignot et al. 2010). If a marine ice margin is afloat, a further increment of mass loss takes place by melting at the floating underside of the ice shelf. Rates of basal ice-shelf melting can reach tens of metres per year (e.g. Rignot and Kanagaratnam 2006; Enderlin and Howat 2013). The sensitivity of ice caps and glaciers to projected climate warming is also linked to whether they are marine- or land-terminating. Eurasian Arctic ice masses whose coastal margins are likely to retreat onto land will decay more slowly than those marine-based ice masses where the retreating margin remains a marine one and, thus, mass loss by iceberg production and melting of the ice-ocean interface continues (e.g. Dowdeswell et al. 2008). In this contribution, the glaciers and ice caps of the Eurasian Arctic are described and discussed in the context of their marine margins and, in particular, the presence or otherwise of floating ice shelves.

3.2 Identification and Terminology for Floating and Grounded Ice

The terminology used here to describe the margins of glaciers and ice caps ending in marine waters is as follows. When the term 'ice shelf' is used in the context of marine-terminating glacier ice, the margin is observed or interpreted to be floating. The term 'ice tongue' has also been used by some researchers to describe floating glacier ice that is constrained by valley walls, with a floating portion that is long relative to its width (Hambrey 1994). When the term 'tidewater' is applied in this contribution, it implies that the marine ice margin is observed or interpreted to be grounded below sea level. When neither term is used in the context of marine ice cliffs, there is no implication that the margin is either grounded or afloat; it simply ends in the sea and it is not known whether or not the margin is grounded.

3.3 Floating Ice Shelves in the Eurasian Arctic

Despite the presence of about 4000 km of marine-terminating glaciers and ice caps in the Eurasian Arctic (Sharov 2005), very few areas of floating ice shelves fed by flow from glaciers and ice caps have been reported (Fig. 3.2). Neither do there appear to be extensive and thick areas of multi-year shorefast sea ice, similar to those making up the so-called ice mélange or sikussak of Greenland fjords (e.g. Reeh et al. 1999; Dowdeswell et al. 2000; Amundson et al. 2010; Reeh 2017), or significant parts of the northern Ellesmere Island ice shelves in Canada (e.g. Jeffries 2017). Shorefast sea ice seldom lingers through the Eurasian Arctic summer.

The Russian Arctic archipelagos of Severnaya Zemlya and Franz Josef Land appear to contain some ice shelves in addition to grounded tidewater ice fronts (Fig. 3.2a, b), whereas the ice-ocean interface in Svalbard and Novaya Zemlya consists of grounded ice alone (Fig. 3.2c). The marine margins of the glaciers and ice caps in the four major archipelagos of the Eurasian Arctic are now described in turn, with particular reference to the presence of floating ice.

3.3.1 Severnaya Zemlya

The best-known ice shelf in the Eurasian Arctic is the Matusevich Ice Shelf, located in Severnaya Zemlya (Figs. 3.1d, 3.3, 3.4) (Williams and Dowdeswell 2001). In fact, ice in two areas of the archipelago is reported to be floating, according to the Russian Glacier Inventory (Govorukha et al. 1980). The more extensive of these ice shelves was the Matusevich Ice Shelf, which occupied the Matusevich Fjord on the northeast coast of October Revolution Island, centered on 79° 54'N, 98° 10'E. An ASTER image of the ice shelf is shown in Fig. 3.3. The fjord is almost 50 km long, with a maximum width of 12 km. The Matusevich Ice Shelf had an area of approximately 240 km^2 of floating ice in 1955 (Govorukha et al. 1980), with a surface slope of <0.2° near its margin (Fig. 3.5). Ice from a total of eight outlet glaciers from five source regions drains into Matusevich Fjord from the Rusanova and Karpinsky ice caps (Fig. 3.3). The total area of ice feeding the ice shelf was about 1100 km^2 in 1994 (Williams and Dowdeswell 2001). Terminus fluctuations of the Matusevich Ice Shelf appear to be cyclical, with a weak periodicity of about 30 years (Fig. 3.4f) (Williams and Dowdeswell 2001). This cyclicity is interpreted to be a result of the nature of iceberg calving from the ice shelf. Iceberg production is by the occasional breaking off of large tabular icebergs, of up to several kilometres in length (Fig. 3.4).

Between 10 August and 7 September 2012, the Matusevich Shelf largely broke up (Willis et al. 2015). This has resulted in the rapid thinning and speed-up of the outlet glaciers of the adjacent ice caps that fed the ice shelf. It is not yet clear whether the ice shelf may begin to grow again in the protected inlet beyond the present glacier margins (Fig. 3.3), as it has done after previous fluctuations in area over the twentieth century. This will probably depend on future atmospheric and ocean temperatures in the wider Russian Arctic (e.g. Walsh 2014).

A second ice shelf in Severnaya Zemlya, identified as ice mass No. 7 in the Russian Glacier Inventory (Govorukha et al. 1980), has a much smaller floating area of about 11 km^2 at the northeastern marine margin of the Academy of Sciences Ice Cap on Komsomolets Island (Fig. 3.1d). Many large tabular icebergs, of hundreds of metres in length, have been imaged calving from this floating ice margin (Fig. 3.2a; Moholdt et al. 2012b). Low gradient ice-surface profiles derived from 1950s Russian aerial photographs (Dowdeswell et al. 1994), together with the occurrence of a number of areas of open seawater bounding the larger Matusevich Ice Shelf (Zinger and Koryakin 1965), and radio-echo sounding data from Glacier No. 7 (Govorukha et al. 1980), have been used to confirm that these ice masses are afloat (Fig. 3.5).

Fig. 3.3 ASTER image of the Matusevich Ice Shelf acquired on 25 June 2005. Areas of floating ice shelf are shown are outlined in *red* in the inset. The ice shelf is fed from ice caps on Komsomolets and October Revolution islands in the Severnaya Zemlya archipelago (located in Fig. 3.1d). This ice shelf is the largest expanse of floating glacier ice in the Eurasian Arctic (Williams and Dowdeswell 2001). It is possible that some of the brash ice and sea ice in the areas between the icebergs and components of the floating ice shelf remain for more than a single year, forming so-called sikussak similar to that found in some Greenland fjords

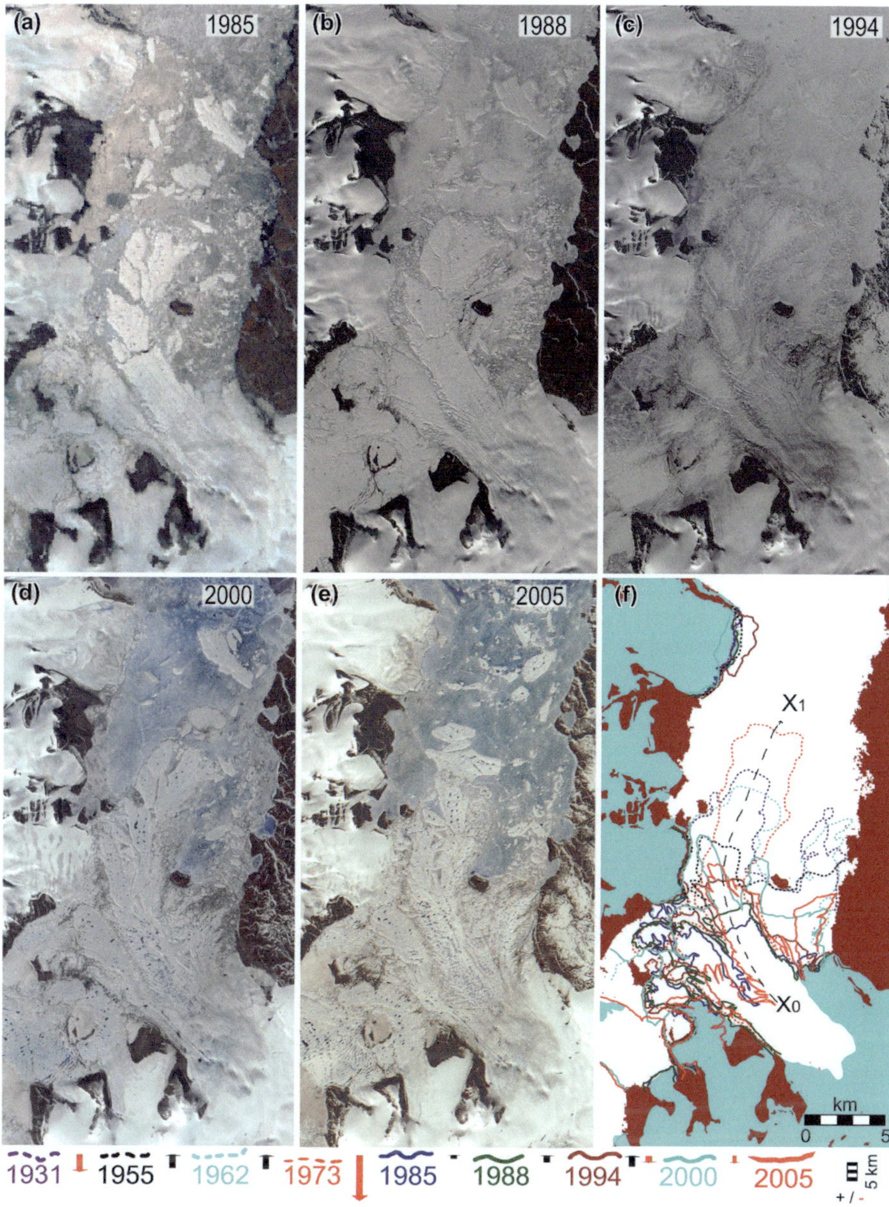

Fig. 3.4 Five satellite images of the Matusevich Ice Shelf, and adjacent icebergs derived from it, located between Komsomolets and October Revolution islands in the Russian archipelago of Severnaya Zemlya (located in Fig. 3.1d). (**a**) 12 August 1985 (Landsat TM image, Path/Row 170/002). (**b**) 26 August 1988 (Landsat TM image, Path/Row 164/003). (**c**) 20 June 1994 (Landsat TM image, Path/Row 168/002). (**d**) 9 July 2000 (Landsat ETM+ image, Path/Row 165/002). (**e**) 25 June 2005 (ASTER image). (**f**) The positions of the ice-shelf margin mapped for 1931, 1955, 1962, 1973, 1985, 1988, 1994, 2000 and 2005 (bare land is *brown* and grounded ice caps are *blue*). Changes in margin position were measured along transect X_0 to X_1. *Arrows* at the *bottom* of the diagram show the magnitude of advance (*black arrow*) or retreat (*red arrow*) between each period

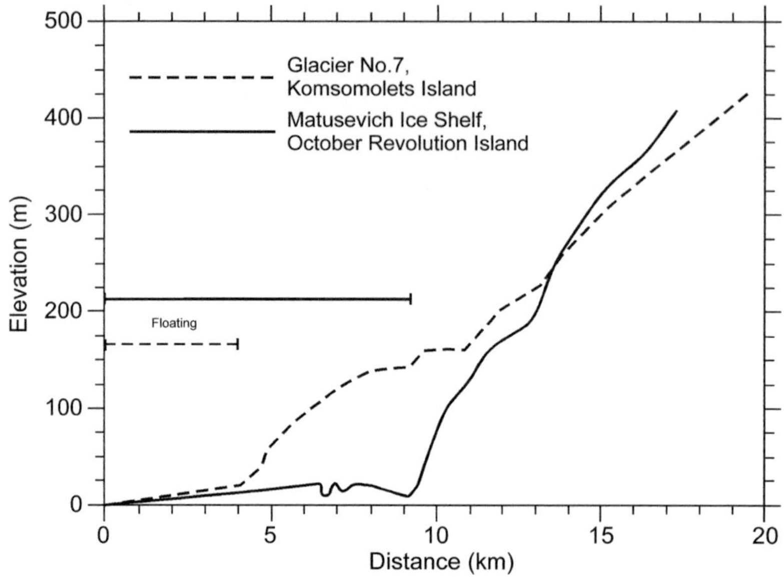

Fig. 3.5 Ice surface profiles from two ice shelves and their parent drainage basins on Severnaya Zemlya, derived from analysis of Russian aerial photographs from the 1950s. The Matusevich Ice Shelf is imaged in Fig. 3.3, and the smaller ice shelf on Komsomolets Island is fed by an eastern outlet of the Academy of Sciences Ice Cap (Fig. 3.2a) (Modified from Dowdeswell et al. 1994, Copyright Regents of the University of Colorado)

3.3.2 Franz Josef Land

There is some controversy in the early Russian literature as to the occurrence of ice shelves in Franz Josef Land. Spizharskiy (1936), Shumskiy (1949), and Govorukha (1968) each suggest that some small ice shelves may be present, whereas Grosswald et al. (1973) argue that none exist. Little direct evidence is offered in support of either case, although Govorukha (1968) states specifically that an ice shelf is present in the northeast of George Land (Figs. 3.6, 3.7).

While examining digital Landsat Thematic Mapper (TM) and Multispectral Scanner (MSS) satellite imagery of Franz Josef Land, a number of ice caps with smooth and apparently very low surface gradients at their seaward margins were observed (Figs. 3.6 and 3.7) (Dowdeswell et al. 1994). These areas were often associated with the production of relatively large tabular icebergs (Fig. 3.2b). The areas are dynamically part of the parent ice mass, and have a marked break of slope at their inner margins. Most, although not all, of the low gradient margins are located in relatively protected embayments (Fig. 3.6), rather than along sections of open coastline, and often have relatively deep water offshore. These flat features account

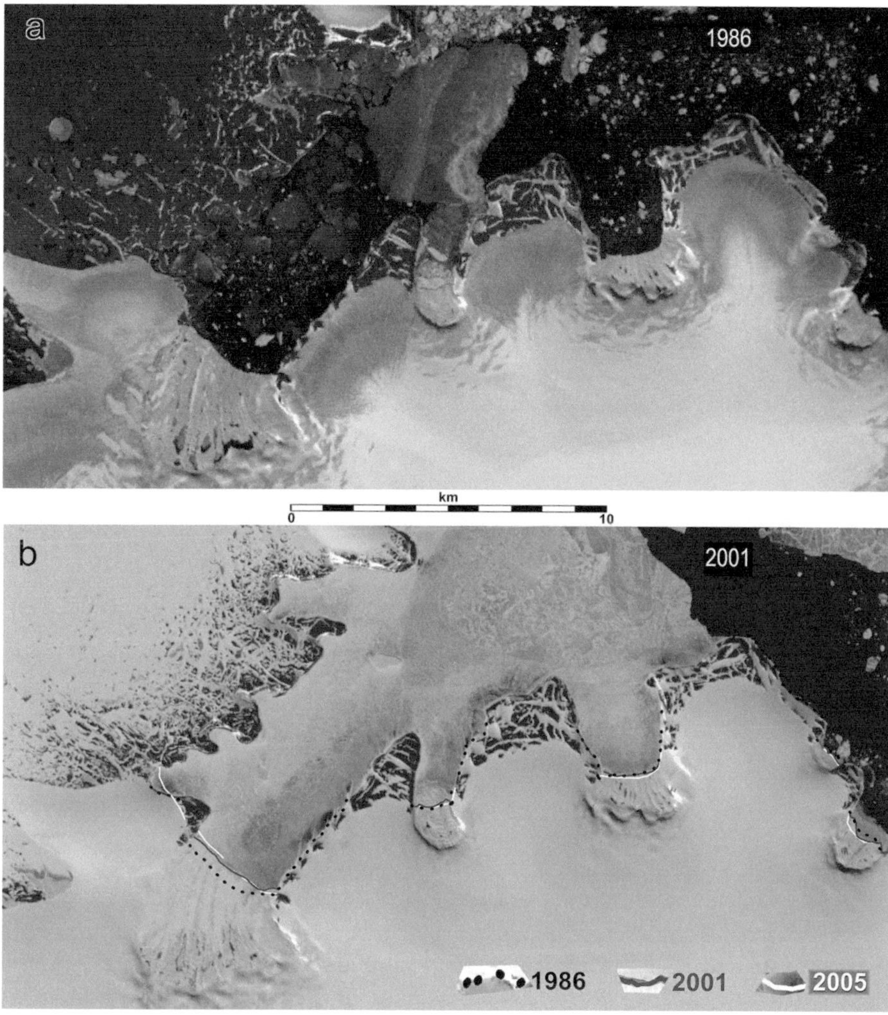

Fig. 3.6 Two satellite images of several areas of possible floating ice shelves in Geographer's Bay, George Land, in the Franz Josef Land archipelago (located in Figs. 3.1c and 3.7). (**a**) 25 July 1986 (Landsat TM image, Path/Row 199/002). (**b**) 17 June 2001 (Landsat ETM+ image, Path/Row 198/002). The ice margins for 1986, 2001 and 2005 are shown

for 315 km^2 or about 2% of the total area of the ice caps in the archipelago and the largest of these low gradient areas is 45 km^2 (Fig. 3.7). They make up a total of 175 km or about 7% of the ice-ocean interface from which iceberg calving takes place in the archipelago. Ice surface profiles, derived from analysis of vertical aerial photographs, show slopes of 0.5° on these features, as compared with 3.5–5° on other ice caps in Franz Josef Land (Dowdeswell et al. 2010).

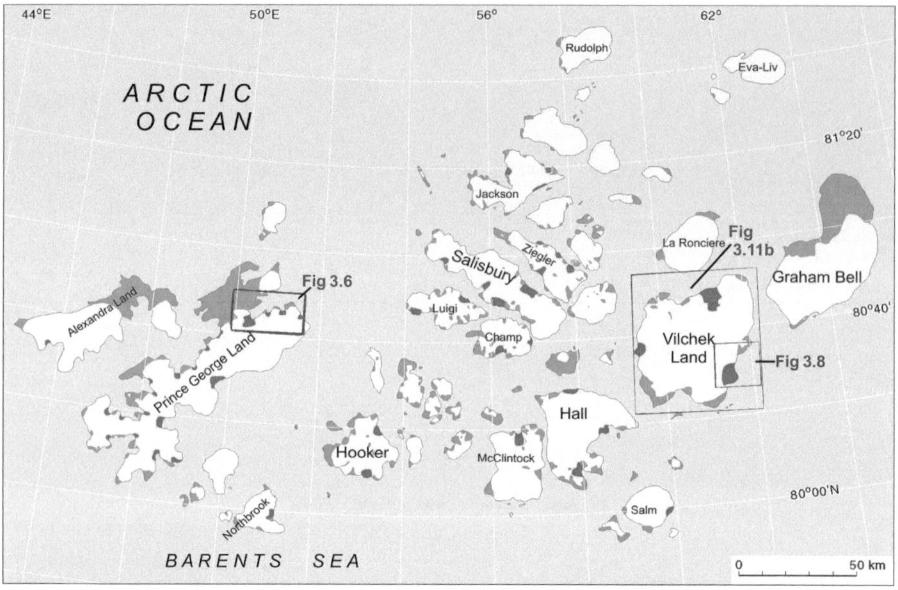

Fig. 3.7 Map of possible floating or nearly buoyant ice in the Franz Josef Land archipelago, shown by dark shading. *Lighter shading* is bare land (Modified from Dowdeswell et al. 1994, Copyright Regents of the University of Colorado)

Two areas of Franz Josef Land with ice inferred to be at or close to full buoyancy are shown in Figs. 3.6 and 3.8. First, the particularly flat areas of ice in five embayments in the deep-water Geographer's Bay on George Land appear likely candidates to be close to hydrostatic equilibrium. Their form is also very like that of the known ice shelves that fill numerous embayments along the coast of the Antarctic Peninsula (Fox and Vaughan 2005). Interestingly, the positions of the margins of these features have remained little altered over the period since 1985, and the largest of the possible shelves has undergone an advance of about 0.5 km over the last 20 years (Fig. 3.6). Secondly, an area of about 30 km^2 at the margin of Zneminity Glacier on Vilczek Land also has a very low-profile surface (Fig. 3.8). This ice margin has retreated by up to 4 km over the past 45 years and continues to produce tabular icebergs up to about 2 km in length.

At least some of the smooth, low gradient features mapped from around the margins of Franz Josef Land ice caps are likely to be floating ice shelves (Fig. 3.7) (Dowdeswell et al. 1994). They have similar ice surface gradients to the known ice shelves on Severnaya Zemlya (Fig. 3.5). There is no requirement for deep water to occur beneath these features, but only that they become buoyant over a significant part of their base. Glacier thinning, due to reduced mass balance since the termination of the Little Ice Age (Dowdeswell 1995; Dowdeswell and Hagen 2004),

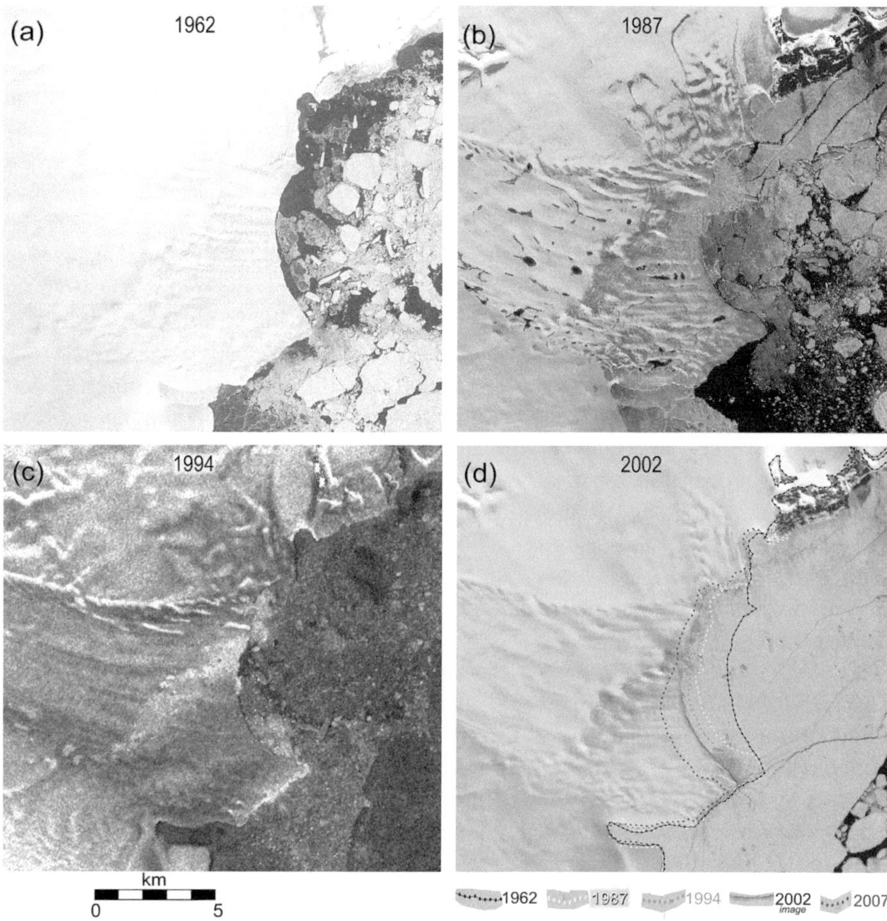

Fig. 3.8 Four satellite images of Znamenity Glacier and the icebergs offshore of Vilczek Land in the Franz Josef Land archipelago. The area is located in Figs. 3.1c and 3.7. (**a**) 18 September 1962 (Declassified Satellite Imagery, DSI). (**b**) 8 August 1987 (Landsat TM image, Path/Row 196/001). (**c**) 22 February 1994 (ERS-1 SAR image). (**d**) 4 June 2002 (Landsat ETM+ image, Path/Row 198/001). Ice margins for 1962, 1987, 1994 2002 and 2007 are shown in **d**, indicating a pattern of consistent ice-margin retreat

may have contributed to the presence of these features. However, an origin for some of these low gradient margins by deformation of an unlithified substrate cannot be ruled out. Irrespective of the basal boundary conditions beneath these features, and of whether individual areas are floating in hydrostatic equilibrium or are merely approaching buoyancy, they provide one of the major modern sources of tabular icebergs to both the eastern Barents Sea and to the Eurasian Basin of the Arctic Ocean (Fig. 3.2b) (Voevodin 1972; Dowdeswell et al. 1994).

3.3.3 Svalbard

Glaciers and ice caps reach the sea along about 860 km of the coastline of the islands of Svalbard (Blaszczyk et al. 2009), but almost none of these marine margins appear to be floating (Dowdeswell 1989) (Fig. 3.2c). This conclusion is reached through several lines of evidence. First, examination of the ice-surface long profiles of many glaciers and ice-cap drainage basins in the archipelago shows none of the very low slopes characteristic of floating ice with a near-zero basal shear stress. These surface-slope data are derived from both airborne radar investigations at megahertz frequencies and the analysis of existing maps (e.g. Dowdeswell 1986). Secondly, basal radar reflections from the marginal parts of many Svalbard ice caps do not show, either qualitatively or quantitatively, the very bright or high-power basal reflectors typical of the ice-water interface at the base of floating ice (e.g. Dowdeswell 1986, 1989; Dowdeswell and Bamber 1995). Thirdly, the icebergs produced from the marine margins of Svalbard glaciers and ice caps are usually small and irregular, rather than the sometimes kilometre-long tabular icebergs characteristic of ice that is floating or close to full buoyancy (Dowdeswell 1989). Fourthly, comparison of offshore bathymetry and radar-derived marginal ice thickness, where both are available, shows values that are similar within the limits of the errors in each method.

There may be short periods, during the active phase of a surge cycle as some outlet glaciers advance into deepening water, when marine margins may become floating prior to subsequent retreat through iceberg calving. Photographic evidence from the margins of Negribreen in eastern Spitsbergen shows the presence of tabular icebergs during the early 1980s (Fig. 3.2d), which may be indicative of short term floatation. However, such phenomena are likely to be transient and unrepresentative of the generally grounded tidewater margins of Svalbard ice masses. In addition, there are few measurements of the transient ice thickness or surface slope during the active phase of surging (e.g. Liestøl 1973).

3.3.4 Novaya Zemlya

Similarly to Svalbard, there is little evidence that the margins of the marine-terminating glaciers on Novaya Zemlya are floating. Most glaciers are retreating and thinning at present and it is noted that retreat into deepening water may induce short-term floatation and rapid calving at the terminus (Moholdt et al. 2012a; Carr et al. 2014). Several of the larger outlet glaciers are known to be of surge-type (Dowdeswell and Williams 1997; Grant et al. 2009). In such cases, short-term floatation during surge advance into deep water cannot be ruled out, but the marine margins of the ice caps on Novaya Zemlya appear typically to be grounded tidewater glaciers.

3.4 Icebergs Derived from Eurasian Arctic Ice Masses

Floating and grounded ice margins in the Eurasian Arctic tend to produce icebergs of rather different dimensions (Figs. 3.2, 3.9). Icebergs produced from floating ice shelves are characteristically tabular in plan, with a surface that is initially flat, reflecting the often smooth and low-elevation ice shelves from which they are calved (Fig. 3.9a). Willis et al. (2015) report tabular icebergs with a calculated thickness of about 120 m produced during the 2012 breakup of Matusevich Ice Shelf in Severnaya Zemlya.

The size-frequency distribution of tabular icebergs in the Barents Sea, produced from the margins of Franz Josef Land ice caps, is shown in Fig. 3.10b (Voevodin 1972). The largest tabular icebergs are several kilometres in diameter. Icebergs of similar size have been observed close to the margin of Znemenity Glacier, one of the largest single drainage basins in the archipelago at about 400 km^2, where the very low ice-surface slope suggests that the margin is at or close to full buoyancy (Dowdeswell et al. 1994). Icebergs of up to about a kilometre in maximum length can be seen in several of the satellite images comprising Fig. 3.8. Other reports and analyses of icebergs derived from the Russian Arctic islands include Zubin et al. (2005) and Kubyshkin et al. (2006).

By contrast, icebergs derived from tidewater ice margins, grounded below sea level, are usually much smaller and more irregular in shape than those produced from ice shelves (Fig. 3.9b). Their morphology is often restricted by the spacing of crevasses close to the margin, where the ice is usually in longitudinal tension (Hodgkins and Dowdeswell 1994). Individual icebergs are usually tens to about one hundred metres in diameter when calved. The size-frequency distribution of icebergs derived from the fast-flowing glacier, Kongsvegen, in north-western Svalbard is shown in Fig. 3.10a (Dowdeswell and Forsberg 1992). In this fjord, icebergs are usually irregular in shape and less than about 20 m in diameter, and are seldom greater than 100 m across.

3.5 Retreating Marine Ice-Cap Margins and Their Sensitivity

The greatest volumes of ice in the Eurasian Arctic are held in a number of large ice caps, each of several 1000 km^2 in area; the largest are Austfonna in eastern Svalbard (area 8000 km^2, volume 2500 km^3), Academy of Sciences Ice Cap in Severnaya Zemlya (area 5500 km^2, volume 2200 km^3) and the Vilczek Land Ice Cap in Franz Josef Land (area 1840 km^2, volume 340 km^3). At present, many of these ice caps have extensive marine margins, although ice shelves are rare or absent. Mass is lost, therefore, by both surface melting and iceberg production, with iceberg loss accounting for about 30–40% of total mass loss (Dowdeswell et al. 2002, 2008). Most of these ice caps are undergoing retreat and thinning (e.g. Moholdt et al. 2012a).

Fig. 3.9 Images of icebergs calved from floating and grounding ice margins in the Eurasian Arctic (Photographs: J.A. Dowdeswell). (**a**) Tabular icebergs from the very flat marine margin of a glacier on Franz Josef Land, May 1994. The low-gradient ice-surface profile indicates that the margin is at or close to full buoyancy. Note the fracture suggesting that a long but rather narrow iceberg of almost a kilometre in maximum dimension is about to be calved. (**b**) The crevassed margin of a small grounded tidewater glacier on George Land in Franz Josef Land, May 1994. Icebergs of irregular shape and maximum dimension about 100 m across are about to be calved from the ice margin, with release taking place when the winter sea-ice cover in the photograph breaks up

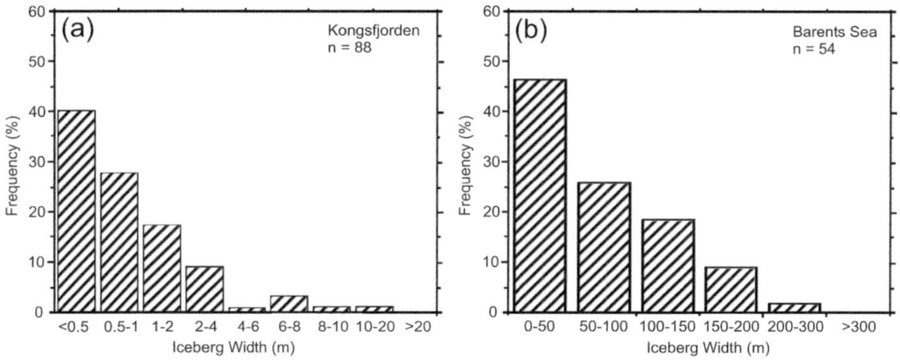

Fig. 3.10 Size-frequency distributions of the width of icebergs calved from grounded and floating ice margins, respectively, in the Eurasian Arctic. (**a**) Icebergs of irregular shape in Kongsfjorden, north-west Svalbard (*Source*: Dowdeswell and Forsberg 1992). (**b**) Tabular icebergs observed in the Barents Sea (*Source*: Voevodin 1972)

The sensitivity of these ice caps to atmospheric and ocean warming over the coming century, where mean annual temperature rise is predicted to be 3–6°C or even more for the Arctic depending on which predictive scenario is used (IPCC 2013), will be influenced in part by the proportions of their beds that are located below present sea level and whether or not they are floating. For tidewater ice masses in general, the mechanisms and pattern of retreat are likely to be complex, involving both thinning and buoyancy effects linked to bed geometry; retreat is unlikely to be a simple linear function of temperature change. In Greenland, the penetration of warmer waters into a number of fjords in the past few years has resulted in enhanced basal melting, thinning and breakup of several floating ice tongues at the margins of fast-flowing outlet glaciers (e.g. Holland et al. 2008; Nick et al. 2009; Christoffersen et al. 2011). Although little is known about changing water-mass temperatures around the Russian Arctic islands, a similar mechanism could in principle lead to the rapid thinning and breakup of floating ice shelves on Franz Josef Land and the Matusevich Ice Shelf on Severnaya Zemlya. In the longer term, the retreat of tidewater ice margins onto land will mean that iceberg production and marine melting cease. Further mass loss would, therefore, be by surface melting and runoff alone, providing a natural constraint on the rate of subsequent decay. A similar argument on sensitivity to warming would, of course, apply to the large marine-terminating outlet glaciers draining, for example, the upland icefields of Spitsbergen and Novaya Zemlya; ice-penetrating radar data exist for only some of these glaciers.

The bed topography of the three largest ice caps in the Eurasian Arctic, derived from ice-penetrating radar studies, shows that significant areas of each ice cap lie below sea level (Fig. 3.11). The elevation of the bed of Austfonna is shown in Fig. 3.11c. About 28% of the bed lies below present sea level (Dowdeswell et al. 2008). The marginal ice thickness often exceeds 100 m and the minimum ice thickness is about 85 m. About 50% of the bed of the Academy of Sciences Ice Cap is below present sea level (Fig. 3.11a), and long profiles of the ice cap show that parts

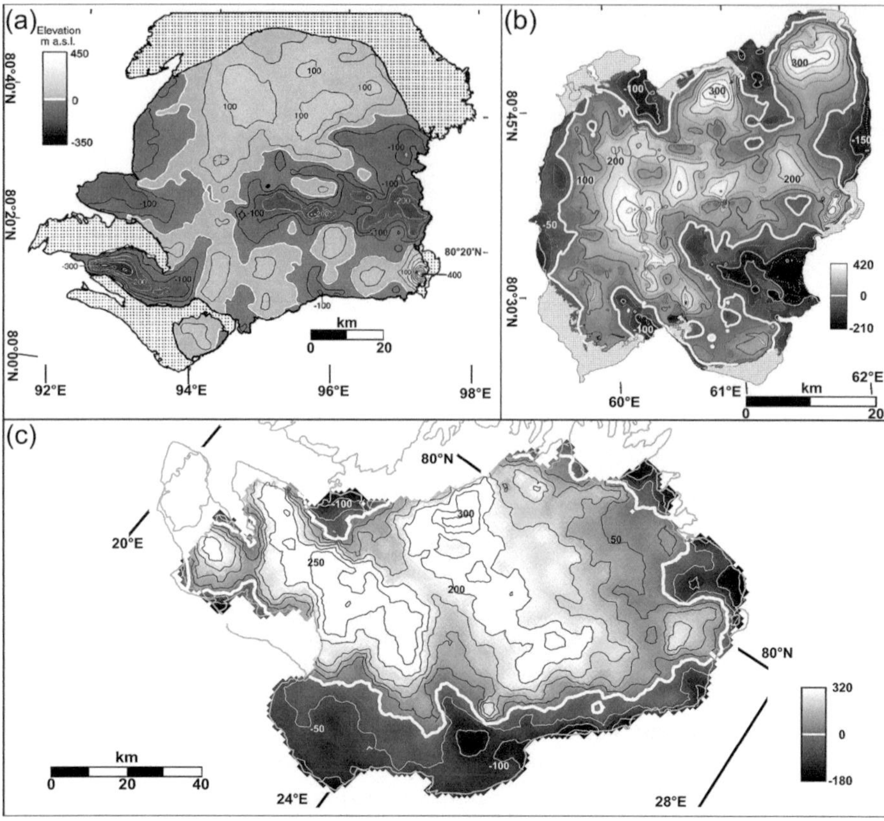

Fig. 3.11 Bed topography of three large ice caps in the Eurasian Arctic, showing the proportions above and below present sea level. (**a**) Academy of Sciences Ice Cap, Severnaya Zemlya, located in Fig. 3.1d (Data from Dowdeswell et al. 2002). (**b**) Vilczek Land Ice Cap, Franz Josef Land, located in Figs. 3.1c and 3.7. (**c**) Austfonna, Nordaustlandet, Svalbard, located in Fig. 3.1b (Data from Dowdeswell 1986; Dowdeswell et al. 2008)

of its bed lie below sea level over 40 km inland of the ice margin (Dowdeswell et al. 2002). The minimum bed-elevation is −207 m on the eastern side of the ice cap, and about −317 m in the south-west. The northern part of the ice cap is, by contrast, underlain by a relatively smooth topography that is above modern sea level. The ice cap on Vilczek Land has about 25% of its bed below present sea level and a maximum below water bed-elevation of about −200 m (Fig. 3.11b).

In each of these ice caps, the lowest elevation areas of the bed are close to the present ice margins. Thus, ice will usually retreat into shallower water, making the likelihood of the development of floating ice margins and rapid iceberg production low, although any atmospheric warming will still enhance the rate of mass loss due to surface melting. This bed-topographic situation contrasts with that in much of West Antarctica, where the ice-sheet bed becomes deeper with distance inland due to long-term isostatic loading by ice several kilometres in thickness (e.g. Anderson 1999).

Taking the case of the largest ice cap in the Eurasian Arctic, Austfonna (Fig. 3.11c), its overall calving loss has been calculated at between about 1.3 and 2.5 km^3 $year^{-1}$ (Dowdeswell et al. 2008; Moholdt et al. 2010). Given that Austfonna has a volume of about 2500 km^3, a simple calculation demonstrates that, if this rate of loss continued, the ice cap could disappear within about 1000 years. However, this is a pessimistic view even in a warming Arctic, since the bed elevation of Austfonna shows that less than 30% of its area lies below sea level (Fig. 3.11c). Thus, retreat of the ice margin onto land would immediately halt the component of mass loss through iceberg calving and, hence, slow the rate of decay; warming would, nonetheless, continue to enhance the rate of mass loss due to surface melting. Much of bed topography beneath the Academy of Sciences and Vilczek Land ice caps also lies close to or above sea level (Fig. 3.11a, b), suggesting that there is a natural break on the rapid collapse of these ice caps through floatation and rapid iceberg production.

3.6 Conclusions

- Despite the presence of about 4000 km of marine-terminating glaciers and ice caps in the Eurasian Arctic (Fig. 3.1) (Sharov 2005), there are few areas of floating ice shelves. Neither are there extensive areas of multi-year shorefast sea ice, which might slow iceberg calving and, indeed, thicken into composite ice shelves themselves.
- The Russian Arctic archipelagos of Severnaya Zemlya and Franz Josef Land appear to contain some ice shelves in addition to grounded tidewater ice fronts (Fig. 3.2a, b). The largest ice shelf was the Eurasian Arctic is the Matusevich Ice Shelf, Severnaya Zemlya, with an area of about 240 km^2, fed from drainage basins totaling about 1100 km^2 (Figs. 3.3, 3.4) (Williams and Dowdeswell 2001). This ice shelf largely broke up in 2012 (Willis et al. 2015)
- In Franz Josef Land, a number of ice caps have smooth and very low surface gradients at their seaward margins, covering over 300 km^2 or about 2% of the total area of the ice caps in the archipelago (Fig. 3.7) (Dowdeswell et al. 1994). They are located mainly in relatively protected embayments (Fig. 3.6), and produce large tabular icebergs (Fig. 3.9a). Whether individual areas are floating in hydrostatic equilibrium or are merely approaching buoyancy, they provide the major modern source of tabular icebergs to the Barents Sea (Fig. 3.10b).
- Although Svalbard has about 860 km of coastal ice cliffs, almost none of the ice margin appears to be afloat (Dowdeswell 1989) (Fig. 3.2c). There may be short periods, during the active phase of the surge cycle, where marine margins become afloat prior to retreat through iceberg calving. Neither is there evidence that the margins of the marine-terminating glaciers on Novaya Zemlya are floating.
- 25–50% of the bed of the three largest ice caps in the Eurasian Arctic lies below sea level (Fig. 3.11). Thus, in a warming Arctic, the ice margin would eventually retreat onto land, curtailing mass loss by iceberg production and providing a break on rapid ice-cap disintegration through calving (Dowdeswell et al. 2008).

Acknowledgments Grants from the John Ellerman Foundation and the Arctic Environmental Program of ConocoPhillips supported parts of this work. Airborne radar campaigns to measure ice thickness in the Eurasian Arctic archipelagos were funded by a series of grants from the UK Natural Environment Research Council. Toby Benham, Evelyn Dowdeswell, Andrey Glazovsky, Jon Ove Hagen, Yuri Macheret and Martin Sharp are thanked for their helpful comments on the paper.

References

Amundson, J. M., Fahnestock, M., Truffer, M., Brown, J., Lüthi, M. P., & Motyka, R. J. (2010). Ice mélange dynamics and implications for terminus stability, Jakobshavn Isbrae, Greenland. *Journal of Geophysical Research, 115*. doi:10.1029/2009JF001405.

Anderson, J. B. (1999). *Antarctic marine geology*. Cambridge: Cambridge University Press.

Blaszczyk, M., Jania, J. A., & Hagen, J. O. (2009). Tidewater glaciers of Svalbard: Recent changes and estimates of calving fluxes. *Polish Polar Research, 30*, 85–141.

Carr, J. R., Stokes, C., & Vieli, A. (2014). Recent retreat of major outlet glaciers on Novaya Zemlya, Russian Arctic, influenced by fjord geometry and sea-ice conditions. *Journal of Glaciology, 60*, 155–170.

Christoffersen, P., Mugford, R., Heywood, K. J., Joughin, I., Dowdeswell, J. A., Syvitski, J. P. M., Luckman, A., & Benham, T. J. (2011). Warming of waters in an East Greenland fjord prior to glacier retreat: Mechanisms and connection to large-scale atmospheric conditions. *The Cryosphere, 5*, 701–714.

Dowdeswell, J. A. (1986). Drainage-basin characteristics of Nordaustlandet ice caps, Svalbard. *Journal of Glaciology, 32*, 31–38.

Dowdeswell, J. A. (1989). On the nature of Svalbard icebergs. *Journal of Glaciology, 35*, 224–234.

Dowdeswell, J. A. (1995). Glaciers in the high Arctic and recent environmental change. *Philosophical Transactions of the Royal Society, Series A, 352*, 321–334.

Dowdeswell, J. A., & Bamber, J. L. (1995). On the glaciology of Edgeøya and Barentsøya, Svalbard. *Polar Research, 14*, 105–122.

Dowdeswell, J. A., & Forsberg, C. F. (1992). The size and frequency of icebergs and bergy bits from tidewater glaciers in Kongsfjorden, north-west Spitsbergen. *Polar Research, 11*, 81–91.

Dowdeswell, J. A., & Hagen, J. O. (2004). Arctic glaciers and ice caps. In J. L. Bamber & A. J. Payne (Eds.), *Mass balance of the cryosphere* (p. 527–557). Cambridge: Cambridge University Press.

Dowdeswell, J. A., & Williams, M. (1997). Surge-type glaciers in the Russian high Arctic identified from digital satellite imagery. *Journal of Glaciology, 43*, 489–494.

Dowdeswell, J. A., Gorman, M. R., Glazovsky, A. F., & Macheret, Y. Y. (1994). Evidence for floating ice shelves in Franz Josef Land, Russian High Arctic. *Arctic and Alpine Research, 26*, 86–92.

Dowdeswell, J. A., Whittington, R. J., Jennings, A. E., Andrews, J. T., Mackensen, A., & Marienfeld, P. (2000). An origin for laminated glacimarine sediments through sea-ice build-up and suppressed iceberg rafting. *Sedimentology, 47*, 557–576.

Dowdeswell, J. A., Bassford, R. P., Gorman, M. R., Williams, M., Glazovsky, A. F., Macheret, Y. Y., Shepherd, A. P., Vasilenko, Y. V., Savatyuguin, L. M., Hubberten, H.-W., & Miller, H. (2002). Form and flow of the Academy of Sciences ice cap, Severnaya Zemlya, Russian High Arctic. *Journal of Geophysical Research, 107*. doi:10.1029/2000/JB000129.

Dowdeswell, J. A., Benham, T. J., Strozzi, T., & Hagen, J. O. (2008). Iceberg calving flux and mass balance of the Austfonna ice cap on Nordaustlandet, Svalbard. *Journal of Geophysical Research, 113*, F03022. doi:10.1029/2007JF000905.

Dowdeswell, J. A., Dowdeswell, E. K., Williams, M., & Glazovsky, A. F. (2010). The glaciology of the Russian High Arctic from Landsat imagery. *U.S. Geological Survey Professional Paper, 1386-F*, 94–125.

Enderlin, E. M., & Howat, I. M. (2013). Submarine melt rate estimates for floating termini of Greenland outlet glaciers (2000–2010). *Annals of Glaciology, 59*, 67–75.

Fox, A. J., & Vaughan, D. G. (2005). The retreat of Jones Ice Shelf, Antarctic Peninsula. *Journal of Glaciology, 51*, 555–560.

Govorukha, L. S. (1968). The present state of ice cap islands in the Soviet Union. *Polar Geography and Geology, 12*, 312–316.

Govorukha, L. S., Semenov, I. V., Popova, N. M., Shamont'yeva L. A., & Bazheva, V. Ya. (1980) Chast' 1. Severnaya Zemlya. In Katalog lednikov SSSR. Tom 16. Leningrad, Gidrometeoizdat, p. 5–49.

Grant, K. L., Stokes, C. R., & Evans, I. S. (2009). Identification and characteristics of surge-type glaciers on Novaya Zemlya, Russian Arctic. *Journal of Glaciology, 55*, 960–972.

Grosswald, M., Krenke, A. N., Vinogradov, O. N., Markin, V. A., Psariova, T. V., Razumeiko, N. G., & Sukhodrovsky, V. L. (1973). *Glaciers of Franz Josef Land: Results of research under the programme of the International Geophysical Year*. Moscow: Nauka.

Hagen, J. O., Liestøl, O., Roland, E., & Jørgensen, T. (1993). *Glacier atlas of Svalbard and Jan Mayen*. Oslo: Norsk Polarinstitutt.

Hagen, J. O., & Reeh, N. (2004). In situ measurement techniques: Land ice. In J. L. Bamber & A. J. Payne (Eds.), *Mass balance of the cryosphere* (p. 12–42). Cambridge: Cambridge University Press.

Hambrey, M. J. (1994). *Glacial environments*. London: UCL Press.

Hodgkins, R., & Dowdeswell, J. A. (1994). Tectonic processes in Svalbard tidewater glacier surges: Evidence from structural glaciology. *Journal of Glaciology, 40*, 553–560.

Holland, D. M., Thomas, R. H., de Young, N., Ribergaard, M. H., & Lyberth, B. (2008). Acceleration of Jakoshavn Isbrae triggered by warm subsurface ocean waters. *Nature Geoscience, 1*, 659–664.

IPCC. (2013). Summary for policymakers. In T. F. Stocker, D. Qin, G.-K. Plattner, M. Tignor, S. K. Allen, J. Boschung, A. Nauels, Y. Xia, V. Bex, & P. M. Midgley (Eds.), *Climate change 2013: The physical science basis. Contribution of Working Group I to the fifth assessment report of the Intergovernmental Panel on Climate Change*. Cambridge/New York: Cambridge University Press.

Jeffries, M. O. (2017). The Ellesmere ice shelves, Nunavut, Canada. In L. Copland & D. Mueller (Eds.), *Arctic ice shelves and ice islands* (p. 23–54). Dordrecht: Springer. doi:10.1007/978-94-024-1101-0_2.

Kubyshkin, N. V., Buzin, I. V., Glazovsky, A. F., & Skutin, A. A. (2006). Determination of the area of generation of big icebergs in the Barents Sea – temperature distribution analysis. In *Proceedings of the Sixteenth International Offshore and Polar Engineering Conference, San Francisco*, May 28-June 2, 2006 (p. 634–638).

Liestøl, O. (1973). Glaciological work in 1971. Norsk Polarinstitutt Årbok 1971. Oslo.

Melkonian, A. K., Willis, M. J., Pritchard, M. E., & Stewart, A. J. (2016). Recent changes in glacier velocities and thinning at Novaya Zemlya. *Remote Sensing of Environment, 174*, 244–257.

Moholdt, G., Hagen, J. O., Eiken, T., & Schuler, T. V. (2010). Geometric changes and mass balance of the Austfonna ice cap, Svalbard. *The Cryosphere, 4*, 21–34.

Moholdt, G., Wouters, B., & Gardner, A. S. (2012a). Recent mass changes of glaciers in the Russian High Arctic. *Geophysical Research Letters, 39*. doi:10.1029/2012GL051466.

Moholdt, G., Heid, T., Benham, T., & Dowdeswell, J. A. (2012b). Dynamic instability of marine glacier basins of Academy of Sciences Ice Cap, Russian high Arctic. *Annals of Glaciology, 53*, 193–201.

Nick, F. M., Vieli, A., Howat, I. M., & Joughin, I. (2009). Large-scale changes in Greenland outlet glacier dynamics triggered at the terminus. *Nature Geoscience, 2*, 110–114.

Reeh, N. (2017). Greenland ice shelves and ice tongues. In L. Copland & D. Mueller (Eds.), *Arctic ice shelves and ice islands* (p. 75–106). Dordrecht: Springer. doi:10.1007/978-94-024-1101-0_4.

Reeh, N., Mayer, C., Miller, H., Thomsen, H. H., & Weidick, A. (1999). Present and past climate control on fjord glaciations in Greenland: Implications for IRD-deposition in the sea. *Geophysical Research Letters, 26*, 1039–1042.

Rignot, E., & Kanagaratnam, P. (2006). Changes in velocity structure of the Greenland Ice Sheet. *Science, 311*, 986–990.

Rignot, E., Koppes, M., & Velicogna, I. (2010). Rapid submarine melting of the calving faces of West Greenland glaciers. *Nature Geoscience, 3*, 187–191.

Sharov, A. I. (2005). Studying changes of ice coasts in the European Arctic. *Geo-Marine Letters, 25*(2), 153–166.

Shumskiy, P. A. (1949). Modern glaciation of the Soviet Arctic. *Trudy Arkticheskogo Instituta, 111*, 11–39.

Spizharskiy, T. N. (1936). Glaciation of Franz Josef land. *Trudy Arkticheskogo Instituta, 36*, 5–37.

Verkulich, S. R., Krasanov, A. G., & Anisimov, M. A. (1992). The present state of, and trends displayed by, the glaciers of Bennett Island in the past 40 years. *Polar Geography and Geology, 16*, 51–57.

Voevodin, V. A. (1972). Dimension of icebergs in the region of Franz-Josef Land and Spitsbergen. *Problemy Arktiki i Antarktiki, 39*, 138–140.

Walsh, J. E. (2014). Intensified warming of the Arctic: Causes and impacts on middle latitudes. *Global and Planetary Change, 117*, 52–63.

Weeks, W. F., & Campbell, W. J. (1973). Icebergs as a fresh water source: An appraisal. *Journal of Glaciology, 12*, 207–233.

Williams, M., & Dowdeswell, J. A. (2001). Historical fluctuations of the Matusevich Ice Shelf, Severnaya Zemlya, Russian High Arctic. *Arctic, Antarctic and Alpine Research, 33*, 211–222.

Willis, M. J., Melkonian, A. K., & Pritchard, M. E. (2015). Outlet glacier response to the 2012 collapse of the Matusevich Ice Shelf, Severnaya Zemlya, Russian Arctic. *Journal of Geophysical Research, Earth Surface, 120*, 2040–2055.

Zinger, Y. M., & Koryakin, V. S. (1965). Yest li schelfove lednike na Severnoy Zemle [are there ice shelves on Severnaya Zemlya?] (Russian). *Materialy Glyatsiolicheskikh Issledovaniy, 11*, 250–253.

Zubin, G. K., Naumov, A. K., & Skutin, Y. A. (2005). Icebergs of the western sector of the Russian Arctic. In *Proceedings of the 18th International Conference on Port and Ocean Engineering under Arctic Conditions (POAC '05)* (Vol. 2, p. 565–574, ISSN 2077-7841).

Chapter 4
Greenland Ice Shelves and Ice Tongues

Niels Reeh

Abstract This chapter focuses on a review of the glaciers on north and northeast Greenland that terminate in fiords with long glacier tongues and floating, ice-shelf-like margins. There is some debate as to whether these glacier tongues can be classified as a traditional ice shelf, so the relevant literature and physical properties are reviewed. There exists a difference between: (1) Floating glaciers in northern Greenland (>77°N) which experience bottom melting as their dominant ablation mechanism and calve relatively thin, but large (km-sized) tabular icebergs ('ice islands'), and (2) Grounded glaciers further south (<77°N), where iceberg calving provides the dominant ablation mechanism. The relatively smaller iceberg discharge in northern Greenland is closely related to the occurrence of extended floating glacier sections, allowing bottom melting estimated at up to 10 m year^{-1} for locations such as Petermann Glacier. A case study is described of the physical characteristics and historical changes of Nioghalvfjerdsfjorden Glacier, NE Greenland, based on field and remote sensing studies.

Keywords Greenland • Ice tongue • Ice shelf • Outlet glacier • Sea ice • Icebergs • Ice Island

4.1 Introduction

The large outlet glaciers in north and northeast Greenland form extended, relatively thin, floating ice tongues which often fill fjords cut into the mountainous coastal landscape. The terminal parts of the floating glaciers disintegrate into kilometre sized, relatively thin tabular icebergs ("ice islands") (e.g. Ahnert 1963), that during long periods are hindered from drifting away by semi-permanent fast-ice (Koch 1928; Higgins 1988). Bottom melting, averaging several metres of ice per year, is the dominant ablation mechanism (Reeh 1994; Thomsen et al. 1999; Rignot et al.

Niels Reeh passed away on May 26, 2009. This chapter is based on an original manuscript by Niels Reeh, modified for clarity by Luke Copland

N. Reeh (✉)
National Space Institute, Technical University of Denmark,
Oersteds Plads, Dk-2800 Kgs, Lyngby, Denmark

L. Copland, D. Mueller (eds.), *Arctic Ice Shelves and Ice Islands*, Springer Polar Sciences, DOI 10.1007/978-94-024-1101-0_4

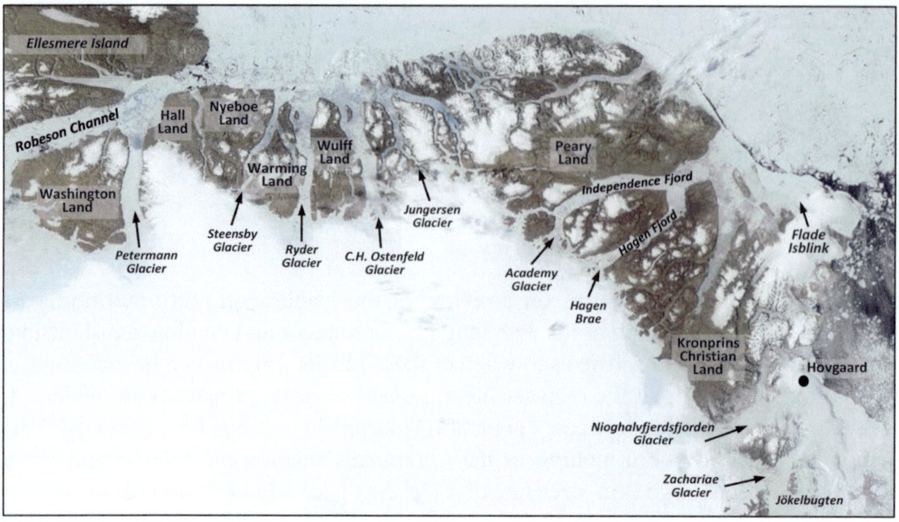

Fig. 4.1 MODIS satellite image (July 3, 2002) of north and northeast Greenland showing main locations described in the text (Image credit: Jacques Descloitres, MODIS Land Rapid Response Team, NASA/GSFC)

1997, 2001). This is in sharp contrast to the conditions in southeast and west Greenland where the large outlet glaciers terminate with calving fronts that are several hundred metres thick at or near their grounding-lines. The dominant mass loss from these glaciers is by calving of irregularly shaped icebergs with maximum dimensions that seldom exceed 1 km.

In this review the emphasis will be on the north and northeast Greenland glaciers that terminate with extended, floating ice-shelf-like margins. Other marine terminating outlet glaciers in this and other parts of Greenland will only be occasionally discussed. In general, the term floating ice tongue or glacier tongue will be used for the glaciers dealt with, a term that, in an Antarctic context is most often used for narrow floating glaciers protruding out from the coast into open sea. A few such glacier tongues exist in north and northernmost East Greenland, for example as seaward extensions of Nioghalvfjerdsfjorden and Zachariae Glaciers (Fig. 4.1). The Ellesmere Island type ice shelves attached to the coast, but not nourished by a supply of glacier ice, probably don't occur in present day Greenland, although a couple of glaciers in northernmost Peary Land have been suggested to belong to this category (Higgins 1991).

4.2 Observations of North and Northeast Greenland Floating Glaciers

4.2.1 Early Twentieth Century Exploration History

A first description of a floating glacier in north Greenland was given by Koch and Wegener (1912, pp. 7–19) based on their observations during the Danmark-Expedition to the northeast coast of Greenland in 1906–1908. The apparent absence of icebergs in front of the floating glaciers in Jökelbugten and Nioghalvfjerdsfjorden led these authors to suggest that the mass loss from the glaciers terminating in these waters was mainly due to melting by air flow and solar radiation at the surface and, probably more importantly, melting from beneath due to inflow of warm, salty water from the adjacent sea (a hypothesis that was confirmed by observations almost 90 years later).

During the Second Thule Expedition (1916–1918) and the Danish Bicentenary Jubilee Expedition (1920–1923), surveys were undertaken of the glaciers in north Greenland from Melville Bay in northwest Greenland to Peary Land in northernmost East Greenland including the glaciers on the north coast of Greenland. The report published by Koch (1928) included descriptions of many of the floating outlet glaciers along the Greenland Arctic Ocean coast such as the Petermann, Steensby, Ryder, C.H. Ostenfeld, Jungersen and Academy glaciers (Figs. 4.1, 4.2). Koch (1928) also summarizes glaciological observations made by earlier expeditions. In general the descriptions are based on observations made from a distance; either from the sea ice when passing the fjords occupied by the glaciers, from attempts at penetrating into the fjords by dog-sledge or on skis (which were often stopped by vertical cliffs of icebergs or glaciers), or from the ice sheet when crossing the drainage basins of the respective glaciers. Koch (1928) concluded that Petermann, Ryder, Jungersen and Academy glaciers had floating tongues. As the most important conditions for developments of floating glacier tongues in north Greenland, Koch (1928) mentions the small surface slope in the lower portions of the glaciers so that '*generally no crevasses occur at the boundary between the fixed and the floating parts of the glacier*', and that the '*floating glacier tongues are found in fjords with (semi)-permanent sea ice which presses against the glacier fronts*'.

Using data collected during field summer observations in 1953 and 1956–1960, aerial photographs from 1947, and earlier descriptions, Davies and Krinsley (1962) published an overview of the regimen of the Greenland Ice Sheet margin in north Greenland. Besides the glaciers with floating tongues described by Koch (1928), Davies and Krinsley (1962) also mention that the outlet glaciers from the ice sheet draining into Hagen Fjord and Spalteglacier, a northern branch of Nioghalvfjerdsfjorden Glacier (Fig. 4.1) have floating tongues, whereas the floating tongue of Academy Glacier reported by Koch (1928) had broken up into tabular icebergs. Davies and Krinsley (1962) concluded that '*during the last 50 years, the glacier margin in North Greenland from Kap York on the west coast north through Peary Land and south to Nioghalvfjerdsfjorden on the east coast had retreated generally 2 to 8 km*'. These general descriptions of north Greenland glaciers, however, give little quantitative information on glacier thickness and velocity.

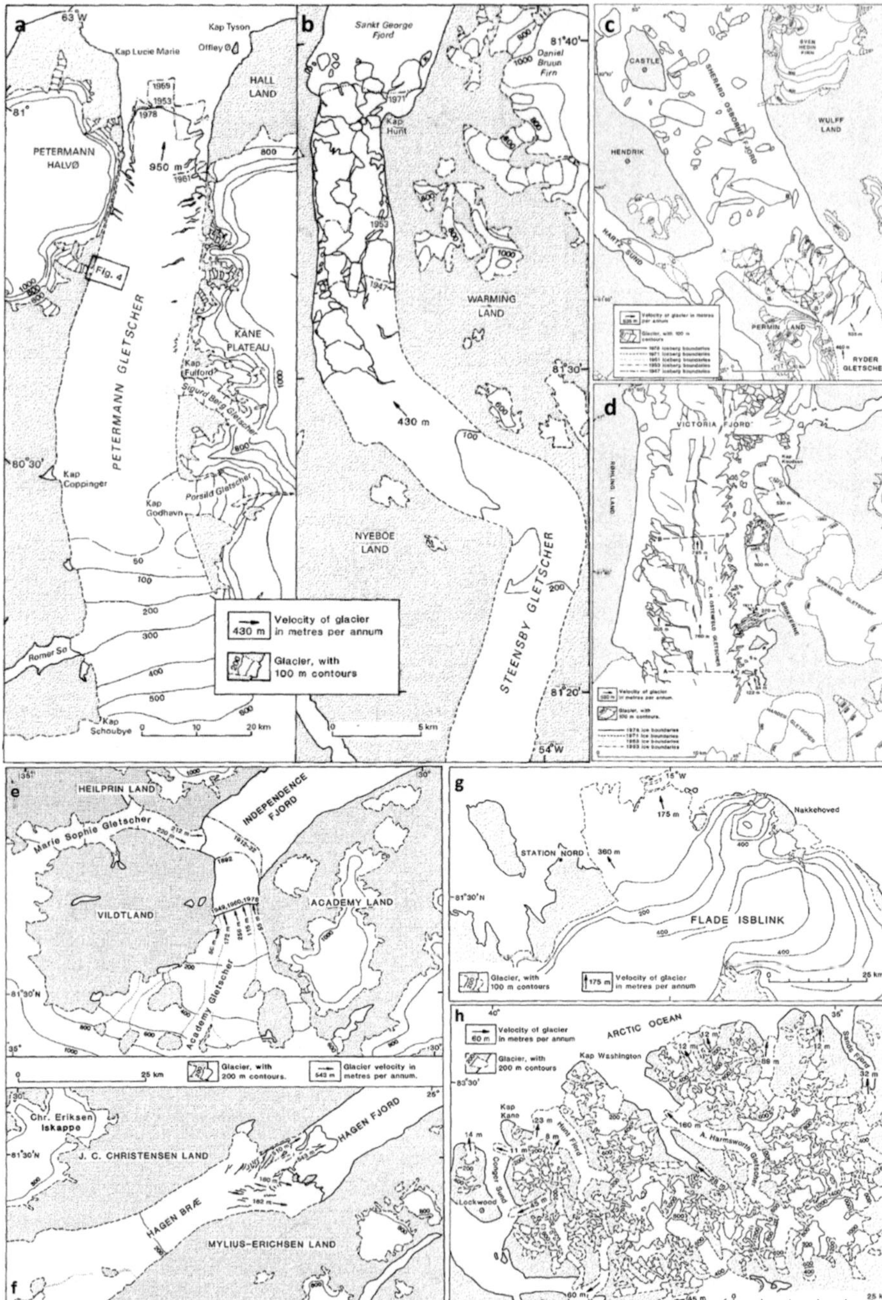

Fig. 4.2 Maps of north and northeast Greenland floating glacier tongues: (**a**) Petermann Glacier; (**b**) Steensby Glacier; (**c**) Ryder Glacier; (**d**) C.H. Ostenfeld Glacier; (**e**) Marie Sofie Glacier and Academy Glacier; (**f**) Hagen Glacier; (**g**) Flade Isblink; (**h**) Kape Kane, Northern Peary Land (From Higgins (1991), reproduced with permission of Polarforschung)

4.2.2 Repeat Aerial Photography

Vertical aerial photography of north Greenland was flown by the Danish Geodetic Institute in the period 1959–1963, and repeated again in 1978. The airborne programs included photographs of all the glaciers along the Arctic Ocean coast in north Greenland. A study of glacier velocities and calf ice production of the glaciers in the region extending from Petermann Glacier in the west to Flade Isblink in the east (Fig. 4.1) was performed by Higgins (1988, 1991). The study included all large outlet glaciers from the north Greenland Ice Sheet with floating sections: the Petermann, Steensby, Ryder, C.H. Ostenfeld, Jungersen, and Hagen glaciers (Figs. 4.1, 4.2). A few minor glaciers, among these the surging Brikkerne Glacier (Higgins and Weidick 1990) and a glacier which extends northwards from Flade Isblink as a 20 km long floating ice tongue, were also included in the study.

Photogrammetric measurements of the surface altitude of the different glaciers were made using the 1978 vertical aerial photographs. Contrary to the conclusion mentioned above by Davies and Krinsley (1962), Higgins (1988) argues that his study of vertical aerial photographs demonstrates that for floating glacier tongues the position of the glacier terminus is not a reliable indicator of advance and retreat, because the break-up of the floating glacier tongue in rare summers when the fast ice melts completely may give the impression of a sudden retreat unrelated to changes in mass balance. However, the nearly 30 years of photographic coverage permitted calculation of average velocities on floating segments of the glaciers, where these had preserved distinctive patterns of meandering streams and meltwater pools recognizable on photographs taken decades apart. A transverse profile was measured across each of the floating glacier tongues close to its calving front, to obtain the average altitude of the glacier surface above sea level. Knowing the width of the glacier, and assuming that 7/8ths of the volume of a floating ice mass is below sea level, the calf-ice production could then be calculated (Table 4.1).

Higgins (1991) also describes two slow moving glaciers on both sides of Kap Kane in extreme northwest Peary Land (Fig. 4.1), giving '*rise to floating sheets of glacier ice with the undulating surface features characteristic of ice shelves*'. According to Higgins (1991), '*the Kap Kane ice shelves are the only features in North Greenland which compare closely with the better known Ellesmere Island ice shelves*'. However, an aerial photograph (Weidick 1995, his Fig. 40) seems to indicate that the Kap Kane ice shelves are more like equivalents of Antarctic type floating ice tongues.

4.2.3 Satellite Remote Sensing

In the last few decades, space- and airborne remote sensing techniques have increasingly provided data that has revolutionized studies of ice dynamics and mass balance of glaciers and ice sheets. Examples include SAR interferometry (InSAR) for

Table 4.1 Calving flux from north and northeast Greenland floating glaciers draining the Greenland Ice Sheet

Glacier/fjord	Width (km)	Velocity (m year^{-1})		Surface elevation (m)		Glacier thickness (m)	Calving rate (km^3 year^{-1})	
		Range	Average	Range	Average		Individual	Glacier total
Petermann (west)	1.3		855	2–7	4.0	32	0.03	
Petermann (central)	12.0	932–988	950	4–6	5.0	40	0.46	0.59
Petermann (east)	2.2		890	8–10	9.0	72	0.10	
Steensby/Sankt George (west)	0.4		420		13.0	104	0.02	
Steensby/Sankt George (central)	3.7	410–435	430	18–25	22.6	181	0.29	0.32
Steensby/Sankt George (east)	0.4		410		7.0	56	0.01	
Ryder/Sherard Osborne (main)	8.2	460–535	500	18–23	20.0	160	0.66	0.70
Ryder/Sherard Osborne (east)	0.8		500	13–15	14.0	112	0.04	
C.H. Ostenfeld/Victoria	8.0	760–805	795	8–14	11.6	93		0.54
Jungersen/Nordenskiöld	2.5		350	12–19	15.0	120		0.10
Hagen (main)	7.0	510–543	540	10–14	11.9	95	0.36	0.47
Hagen (south)	4.5	180–182	180	14–23	17.0	136	0.11	
Flade Isblink (west)	25.0		360	n.a.	n.a.	n.a.		n.a.
Flade Isblink (east)			175					
Nioghalvfjerdsfjorden (main)	18.0	35–160	n.a.	n.a.	n.a.	n.a.		n.a.
Nioghalvfjerdsfjorden (north)	n.a.		220					
Zachariœ/Jökelbugten	20.0	220–280	n.a.	n.a.	n.a.	n.a.		n.a.
Total								**2.72**[a]

Data from Higgins (1988, 1991)

[a]According to Higgins (1991, Table 1), the total calving flux from north Greenland glaciers is 3.44 km^3 year^{-1}. This number includes the calving flux from non-floating glaciers, a surging glacier and three minor outlet glaciers from the Greenland Ice Sheet

simultaneous determination of glacier topography and surface velocity (Joughin et al. 1996a, b; Mohr et al. 1998), radar- or laser altimetry for measuring accurate surface elevations (Davis et al. 1998; Krabill et al. 1995), and visible, thermal infrared and passive microwave data for determining the distribution of glaciers (e.g., Weidick 1995), estimating surface temperature and albedo (Steffen et al. 1993) and estimating accumulation rates (Zwally and Giovinetto 1995).

Weidick (1995) used Landsat imagery as the basis for a discussion of the glaciological features of the Greenland Ice Sheet and other Greenland outlet glaciers, local ice caps and valley glaciers. Because of orbital constraints, the northernmost part of Greenland (> ~81.5°N) is not covered by Landsat images. However, Weidick's discussion is also illustrated by several oblique and vertical aerial photographs, thus providing a complete combined satellite imagery and aerial photographic coverage of the glaciers of Greenland including the floating glaciers of north and northeast Greenland.

A comprehensive study of surface velocity, topography, and grounding line position of the major outlet glaciers draining the northern sector of the Greenland Ice Sheet using data collected by remote sensing methods was presented by Rignot et al. (2001). This study was limited to the large outlet glaciers in the region extending from Harald Moltke Glacier near Thule Air Base to Storstrømmen Glacier in northeast Greenland (Fig. 4.1). The InSAR data of northern Greenland were collected in the winters of 1992 and 1995–1996. Multiple pairs of InSAR data were combined in a differential manner to separate ice sheet topography from ice motion (Mohr et al. 1998). On floating ice, the InSAR measurements contain a mixture of tidal deformation and ice creep deformation. To remove the tidal signal, the method described by Rignot et al. (2001) was applied using a quadruple difference interferogram. The ice thickness was measured by the airborne NASA/University of Kansas ice sounding radar (Chuah et al. 1996) with a 10 m precision. Most of these measurements were made in an along-flow direction, with only few transverse tracks completed along the grounding line. Grounding line ice thicknesses were therefore estimated from the ice-shelf elevation above mean sea level obtained from a digital elevation model (Ekholm 1996). A comparison of the estimated thicknesses with those measured with ice sounding radar showed large (±30%) inter-glacier variations. Rignot et al. (2001) estimate the uncertainty of the resulting ice thicknesses to be comparable to that of the ice radar thickness measurement. Laser altimetry data (10 cm vertical precision) were employed to validate the estimation of floating glacier thickness and to detect the line of hydrostatic equilibrium (grounding line).

4.2.4 Field Observations

From 1996 to 1998 a field program was carried out on Nioghalvfjerdsfjorden Glacier (Fig. 4.1), making it the best studied floating glacier tongue in Greenland. A detailed description of the results of this study is given in Sect. 4.5.

4.3 Mass Balance

Using Higgins' calving rate estimates, a compilation of snow accumulation data from the Greenland Ice Sheet (Ohmura and Reeh 1991), and a Greenland melt-rate model (Reeh 1991), were used to set up the mass budget for the North Greenland Ice Sheet sector. An apparent imbalance of this ice-sheet sector corresponding to an increase of surface elevation of 0.12 m year^{-1} was noted by Reeh (1994). However, Reeh (1994) pointed out that mass balance would be obtained if significant bottom melting occurred beneath the North Greenland floating outlet glaciers as already suggested by Koch and Wegener (1912) and Higgins (1991). The importance of this process was ultimately documented by a study of Petermann Glacier using InSAR derived velocity data and ice-radar measured ice thickness data (Rignot 1996). The bottom melt rate, averaged over the 70 km long and 20 km wide floating glacier tongue, was found to be about 10 m ice year^{-1} (Rignot 1996). The study was later extended to all floating outlet glacier tongues in north and northernmost East Greenland (Rignot et al. 2001) indicating that bottom melting from these glaciers totals about 35 km^3 ice year^{-1} (Tables 4.2, 4.3). This equates to ~5% of the total annual mass loss from the Greenland Ice Sheet (Reeh 1999).

The study by Rignot et al. (2001) concluded that the total discharge at the grounding line of the studied North Greenland glaciers is about 40 km^3 ice year^{-1} (Table 4.2), and that the mass imbalance is negative, but small. Rignot et al. (2001) remark that this would suggest a North Greenland Ice Sheet sector close to mass equilibrium. However, they argue that several of the studied glaciers exhibit the characteristics of surge-type glaciers, such as Brikkerne Glacier (Higgins and Weidick 1990), Storstrømmen (Reeh et al. 1994), Ryder Glacier (Joughin et al. 1996b), and Hagen Glacier (Rignot et al. 2001). The derived ice discharge of these glaciers may therefore have been underestimated compared to their ice discharge averaged over one surge cycle. Consequently, it is more likely that the North Greenland Ice Sheet sector showed a negative mass balance in the study period (1992–1996). Rignot et al. (2001) note that this finding is consistent with the general retreat of the North Greenland glacier margins reported by Davies and Krinsley (1962) for the period ~1910–1960, and the general thinning and retreat at the grounding lines of the floating glaciers documented by their 2001 study.

4.4 Climate Control on Formation of Floating Glacier Tongues

As outlined above, there is a difference between: (1) Floating glaciers in northern Greenland (>77°N) which experience bottom melting as their dominant ablation mechanism and calve relatively thin, but large (km-sized) tabular icebergs, and (2) Grounded glaciers further south (<77°N) where iceberg calving provides the dominant ablation mechanism (Table 4.3, Fig. 4.3). It appears that iceberg calving in

Table 4.2 North and northeast Greenland floating glaciers. Ice thickness, velocity, ice flux, and mass balance at the grounding line. Calving flux, area and average basal melt rate assuming steady state of floating sections, and grounding line migration between 1992 and 1996

Glacier/Fjord	Thickness (m)	Velocity (m $year^{-1}$)	Ice flux (km^3 $year^{-1}$)	Calving rate (km^3 $year^{-1}$)	Area of floating section (km^2)	Average basal melt rate (m $year^{-1}$)	Grounding line migration (m)
Petermann	630	1160	11.67 ± 0.2	0.59	1304	−6	−450
Steensby/Sankt George	477	329	0.51	0.32	36	−5	n.a.
Ryder/Sherard Osborne	620	540	2.29 ± 0.06	0.70	245	−8	−4200
C.H. Ostenfeld/Victoria	600	810	2.27 ± 0.1	0.54	137	−11	n.a.
Hagen	200	61	0.07 ± 0.02	0.36	188	n.a.	−390
Nioghalvfjerdsfjorden	650	1300	14.19 ± 0.02	0.90	1874	−5	−450 to −650
Zachariœ/Jökelbugten	550	1100	10.81 ± 0.2	1.10	1065	−8	n.a.
Storstrømmen/Borg Fjord	530	15	0.03 ± 0.005	0.90	82	n.a.	+658 to −1395
Total			**41.33 ± 0.4**	**5.05**	**4921**		

Data from Rignot et al. (2001)

Table 4.3 Area and mass balance terms for sectors and selected outlet glacier basins of the Greenland ice sheet (from Reeh 2004). Area, accumulation (AC) and surface melting (SM) are estimated by using the Greenland mass balance model described by Reeh (1991). Bottom melting (BM) is estimated based on Thomsen et al. (1999) and Rignot et al. (2001). Iceberg calving (IC) is determined on the assumption that each ice-sheet sector is in balance, i.e. IC = AC-SM-BM

Location	Area (km^2)	Accumulation (km^3 $year^{-1}$)	Surface melting (km^3 $year^{-1}$)	Surface melting (%)	Bottom melting (km^3 $year^{-1}$)	Bottom melting (%)	Iceberg calving (km^3 $year^{-1}$)	Iceberg calving (%)
Total ice sheet	1,707,400	602	304	50	35	6	263	44
North of 77°N	499,950	94	55	59	35	37	4	4
South of 77°N	1,207,450	508	249	49	~0	0	259	51
Daugaard-Jensen Glacier	48,500	10.8	0.8	7.4	~0	0	10.0	92.6
Storstrømmen	54,300	9.8	8.5	86.7	~0	0	1.3	13.3
Nioghalvfjerdsfjorden Glacier	120,300	16.2	3.0	18.5	13.0	81.2	0.2	0.3

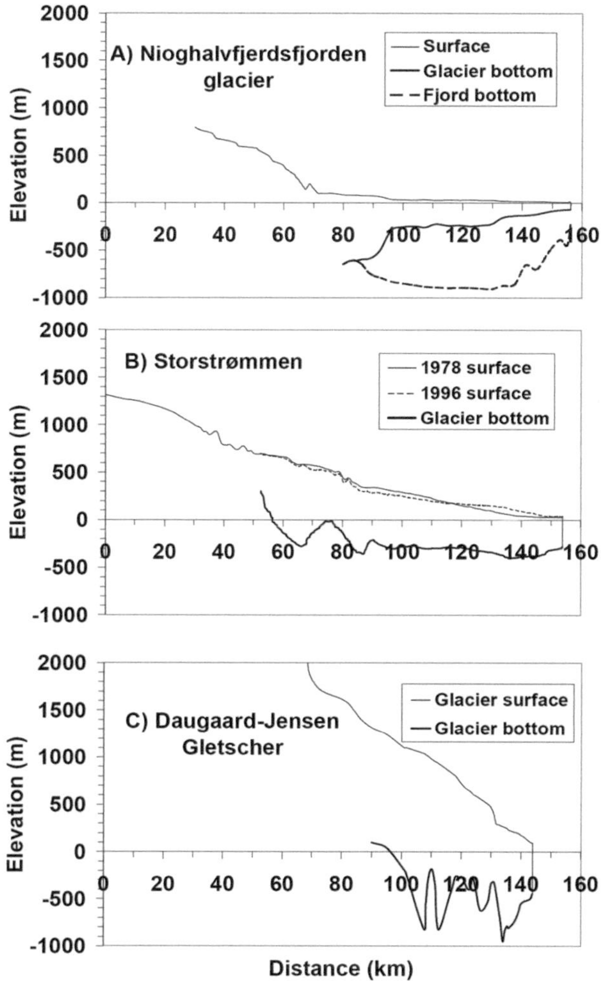

Fig. 4.3 Elevation profiles along the central flow lines of three northeast Greenland outlet glaciers illustrating the transition from north to south from floating to grounding glacier tongues: (**a**) Nioghalvfjerdsfjorden Glacier (79°30′N); (**b**) Storstrømmen (77°15′N); (**c**) Daugaard-Jensen Gletscher, Scoresby Sund (71°45′N) (Modified from Funder et al. (1998), reproduced with permission of Elsevier and Quaternary Science Reviews)

southern Greenland accounts for ~50% of the mass loss in contrast to only ~4% in northern Greenland. The smaller iceberg discharge in northern Greenland is closely related to the occurrence of extended floating glacier sections, allowing massive bottom melting. The total ice flux across the grounding line of north Greenland floating glaciers is estimated to be ~40 km^3 year^{-1} (Rignot et al. 2001). Augmented by an ice flux of ~4 km^3 year^{-1} from calving of grounded marine glacier fronts in this region, this mass loss amounts to ~50% of the accumulation (94 km^3 year^{-1}) falling on the northern Greenland sector. The remaining ~50% of the accumulation is removed by melting at the surface.

These calculations indicate that the percentage of the total mass input discharged via outlet glaciers is actually the same in northern and southern Greenland (Table 4.3). However, the dominant ablation process responsible for this mass loss is different in the north (i.e., bottom melting) versus the south (i.e., iceberg calving). It should also be remembered that the total ice covered area in northern Greenland is less than half that in southern Greenland.

4.4.1 Present Climate Conditions in East Greenland

The difference between the dominant ablation mechanism of southern and northern Greenland outlet glaciers is likely to be climatically controlled (Funder et al. 1998; Reeh et al. 1999a, 2001). As mentioned in Sect. 4.4.1, glaciers forming extended floating sections only occur in regions with semi-permanent fast sea ice. The growth and decay of fast ice depends primarily on winter cold and summer warmth, respectively, as well as on sea water temperature and salinity. Annual snowfall is also important: a thick snow cover, for example, will reduce ice growth because of its insulating effect.

In East Greenland there are large gradients of temperature and precipitation from south to north (Figs. 4.4, 4.5). From Fig. 4.4, a mean July temperature decrease of 0.4°C per degree latitude is derived, which is only about half the 0.7°C decrease per

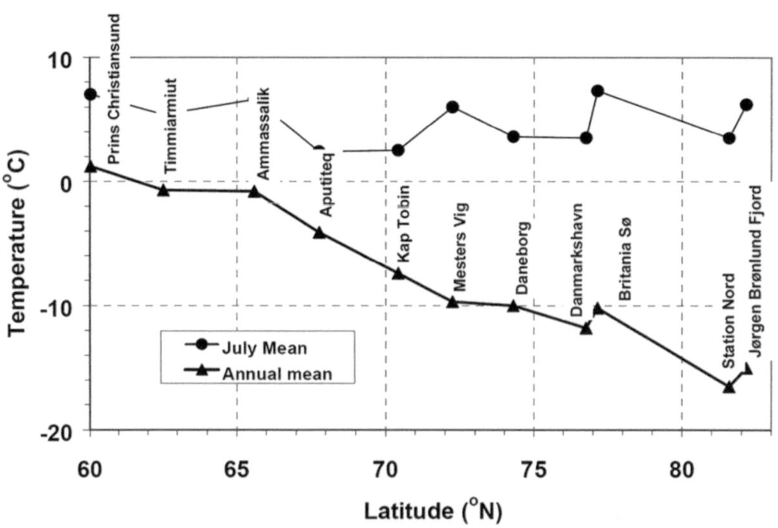

Fig. 4.4 Mean annual and mean July air temperatures measured at meteorological stations in East Greenland, reduced to the standard 1951–1960 period (Ohmura 1987). The stations are generally located on the coast at elevations between 12 and 76 m a.s.l. except Britannia Sø and Jørgen Brønlund Fjord, which are inland stations located at 229 m and 5 m a.s.l., respectively (Modified from Funder et al. (1998), reproduced with permission of Elsevier and Quaternary Science Reviews)

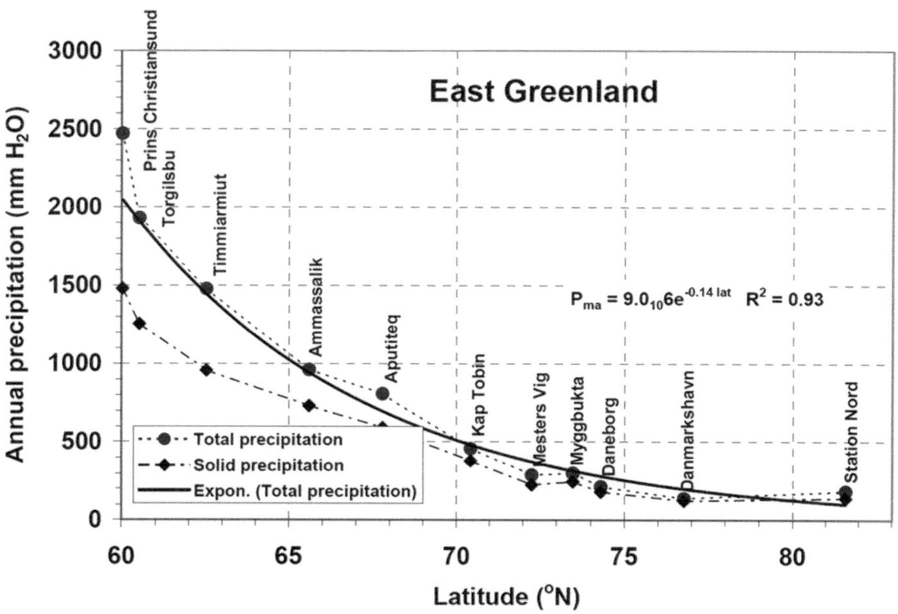

Fig. 4.5 Mean annual total and solid (snow) precipitation measured at meteorological stations in east Greenland. The *thick line* represents a least squares exponential fit to the total precipitation (Modified from Funder et al. (1998), reproduced with permission of Elsevier and Quaternary Science Reviews)

degree latitude of the mean annual temperature. In terms of total annual precipitation near the coast, values approximately half for each 5° latitude shift to the north. Figures 4.4 and 4.5 therefore illustrate the increasing potential for inter-annual survival of fast ice from south to north in Greenland favoured by both lower temperature (particularly winter temperature) and less snow cover in the north than in the south.

4.4.2 *Fast Ice Index*

A simple fast ice index (not accounting for changes of sea water temperature and salinity, and for the insulating effect of the snow cover) can be expressed as (Assur 1956; Reeh et al. 1999a):

$$FI = \sqrt{\frac{DG-}{DG+}} \tag{4.1}$$

where DG− and DG+ are annually summed air temperature degree days below and above the freezing temperature of seawater. The larger the FI, the more likely the survival of fast ice in protected fjords and bays from 1 year to another. Using

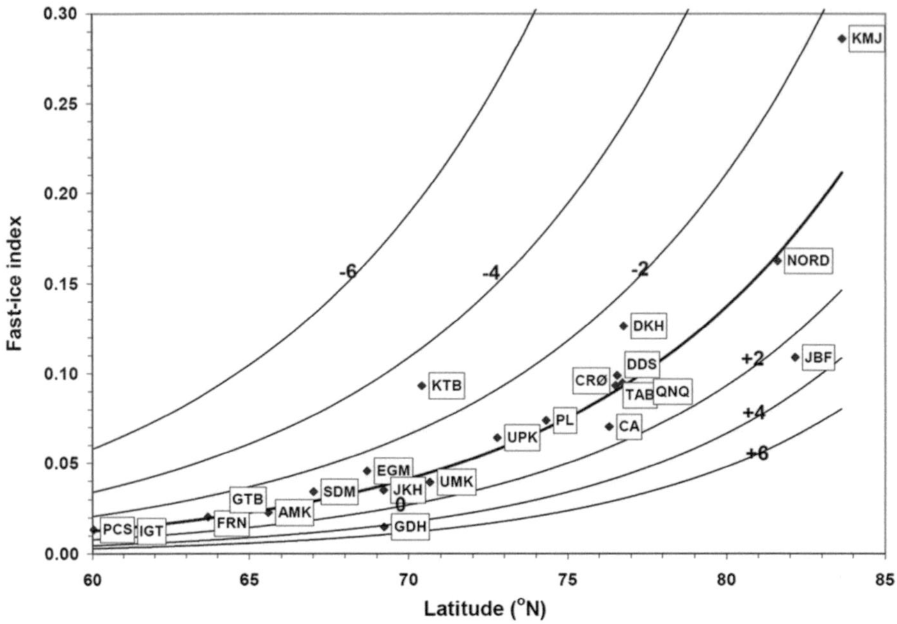

Fig. 4.6 Fast ice index for West Greenland calculated by means of a degree-day model (Reeh 1991), using temperature data reduced to the standard period of 1951–1960 (Ohmura 1987). The *thick line* represents an exponential least squares fit to the data, while the *thin lines* show exponential least squares fits calculated by raising/lowering the mean annual temperature by 2°C, 4°C and 6°C (*PCS* Prins Christian Sund, *IGT* Ivigtut, *FRB* Frederikshåb, *GTB* Godthåb, *AMK* Ammassalik, *SDM* Søndre Strømfjord, *EGM* Egedesminde, *JKH* Jakobshavn, *GDH* Godhavn, *UMK* Uummanak, *KTB* Kap Tobin, *UPK* Upernavik, *PL* Peary Lodge, *CA* Cap Athol, *TAB* Thule Air Force Base, *CRØ* Carey Øer, *DDS* Dundas, *QNQ* Qanaq, *DKH* Danmarkshavn, *JFB* Jørgen Brønlund Fjord, *Nord* Station Nord, *KMJ* Kap Morris Jesup; Modified from Reeh et al. (1999a), reproduced with permission of the American Geophysical Union)

data from west Greenland, it appears that a FI value of ~0.1 corresponds to 77°N, the latitude that presently separates regions with semi-permanent fast-ice from regions without semi-permanent fast ice (Fig. 4.6). This suggests that the regional potential for formation of floating glaciers can be characterized by FI >0.1. The exponential FI-latitude relationships suggest a higher sensitivity to climate change of fast ice survival from year to year in the north than in the south.

4.5 Nioghalvfjerdsfjorden Glacier: A Case Study

From 1996 to 1998, a field program was carried out on Nioghalvfjerdsfjorden Glacier (Fig. 4.1) as a collaboration between the Geological Survey of Denmark and Greenland (GEUS), the Danish Polar Center (DPC), the Alfred Wegener Institute for Polar and Marine Research (AWI) and the Danish Center for Remote Sensing

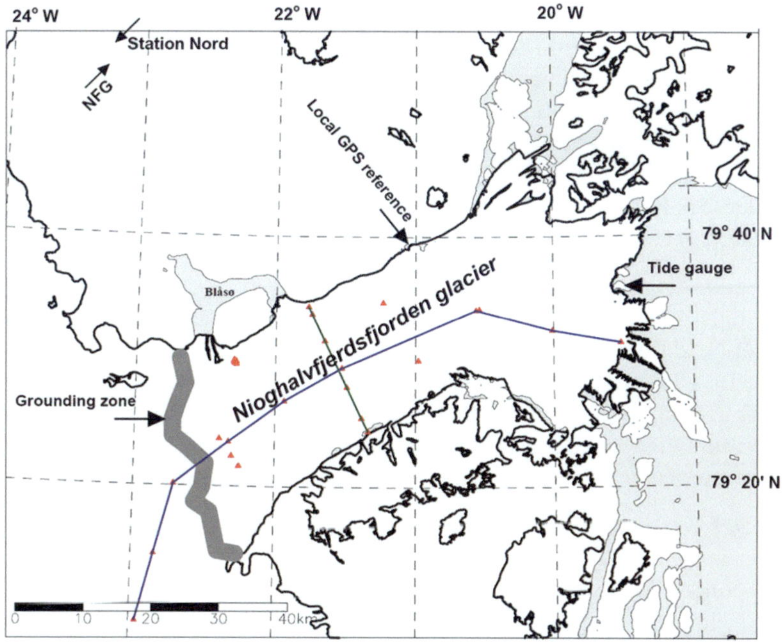

Fig. 4.7 Location map of Nioghalvfjerdsfjorden Glacier. Position of GPS velocity stakes measured in 1996, 1997 and 1998 are shown as *red triangles*. *Blue line* shows central stake line. *Green line* shows position of transverse velocity profile. Locations of tide gauge, local GPS reference site and grounding zone are indicated

(DCRS, now part of the National Space Center, DTU) (Thomsen et al. 1999). The study, aiming at assessing past glacier changes, mass balance, dynamics, and stability of the floating glacier, comprised observations of surface climate and mass balance, bottom melting, surface elevation, ice thickness, seismic sounding of the subglacial cavity under the glacier, glacier velocity, tidal movement, and proglacial and subglacial bathymetry and CTD measurements. The results of the studies of past glacier changes were published by Reeh et al. (1999b, 2001), tidal movement by Reeh et al. (2000, 2003), and bathymetry and CTD measurements by Mayer et al. (2000).

4.5.1 *Geographic Setting, Surface Morphology and Subglacial Cavity*

Nioghalvfjerdsfjorden Glacier (79°30′N, 22°W) is an 80 km long and 20 km wide outlet glacier which flows northeast from the Greenland Ice Sheet (Fig. 4.7). From a relatively smooth interior ice sheet region, the ice flows through a relatively steep, chaotic, 10–20 km long and 500 m high transition section into a ~60 km long floating section. The grounding region is near the western branch of the ice-dammed lake Blåsø (Fig. 4.7). The floating part of the glacier constitutes an extremely flat ice plain, with a characteristic meltwater drainage pattern of numerous small and large supraglacial rivers and shallow lakes. The main ice front of Nioghalvfjerdsfjorden Glacier is divided into five sections by four islands, of which the three southernmost are overrun by glacier ice forming ice rises up to 40 m high. A northern branch of the glacier (Spalte Glacier), 8 km wide, drains into the fjord Dijmphna Sund, west of Hovgaard Ø (Fig. 4.7).

Seismic Sounding of Subglacial Cavity The geometry of the subglacial cavity under Nioghalvfjerdsfjorden Glacier was derived from 98 seismic depth soundings (Mayer et al. 2000). The maximum depth of the trough reaches >900 m below sea level (b.s.l.). Close to the main glacier terminus the sea bottom rises within 20 km from 800 m b.s.l. to 200 m b.s.l. (and in some parts even higher), forming several small ice rumples. In contrast, the water depth at the divergence into Dijmphna Sund is close to 600 m.

Norske Øer Ice Barrier (NØIB) The bathymetry outside Nioghalvfjerdsfjorden Glacier is dominated by the Belgica Bank with water depths of 100–200 m (Fig. 4.8). It is surrounded by a 200–500 m deep trough system, the Belgica Trough in the south, the Norske Trough running parallel to the shore and the Westwind Trough to the north. The main terminus of the glacier is enclosed in a semi-permanent fast-ice barrier which protects the ice margin and prevents calving. The fast-ice barrier covers an extensive area of the sea between latitudes 78°–80°N and extends eastwards beyond the line of islands that include the Norske Øer (Fig. 4.8). The characteristics, extent and variation of this fast-ice barrier were described by Budéus and Schneider (1995) and Schneider and Budéus (1995, 1997). They refer to the feature as the Norske Øer Ice Barrier (NØIB). The NØIB extends 75–150 km from the coast, forming the southern limit of the Northeast Water (NEW) polynya. Significant interannual variations in size and horizontal extent of NØIB occur (Schneider and Budéus 1995).

A striking feature on Nioghalvfjerdsfjorden Glacier is the existence of successions of km long, up to 30 m high ice ridges close to the lateral margins of the floating part of the glacier, tentatively named "Midgårdsorme" (after the giant monster-snake that, according to Nordic mythology, winds all the way around the World). Inspection of aerial photographs and satellite images has revealed a widespread occurrence of similar ice ridges on other floating glaciers in North and East Greenland. In the creation area of the ridges the ice transport is apparently not parallel to the lateral grounding line. This leads to a shear/compressive deformation of the ice column, when the floating ice is pushed back onto firm ground at a narrow angle. The visible expression on the surface is a narrow, in the order of tens of metres high, heavily serrated ridge, which is oriented in the direction of the lateral grounding line. Obviously the creation of the ridges is not continuous, as several generations exist along the downstream margin of the glacier. The mechanics of the ridge-creation as well as its temporal development are not yet fully explained.

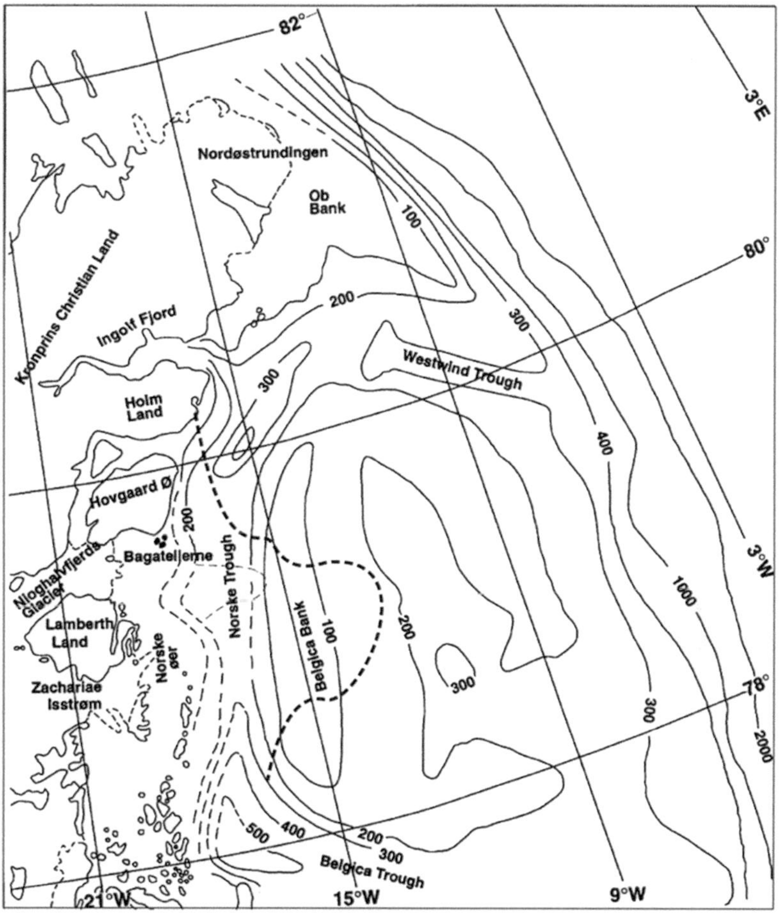

Fig. 4.8 Nioghalvfjerdsfjorden and surrounding areas. Bathymetry of the North-East Greenland Shelf as reported by Bourke et al. (1987). The late summer minimum and maximum fast-ice extent during the period are delineated with *thin* and *thick dashed lines*, respectively (From Reeh et al. (2001), reprinted from the Annals of Glaciology with permission of the International Glaciological Society)

4.5.2 Data Collection and Processing

For a detailed description of the data collection and processing on Nioghalvfjerdsfjorden Glacier the reader is referred to an unpublished report (Reeh et al. 2008). Here a brief account will be given.

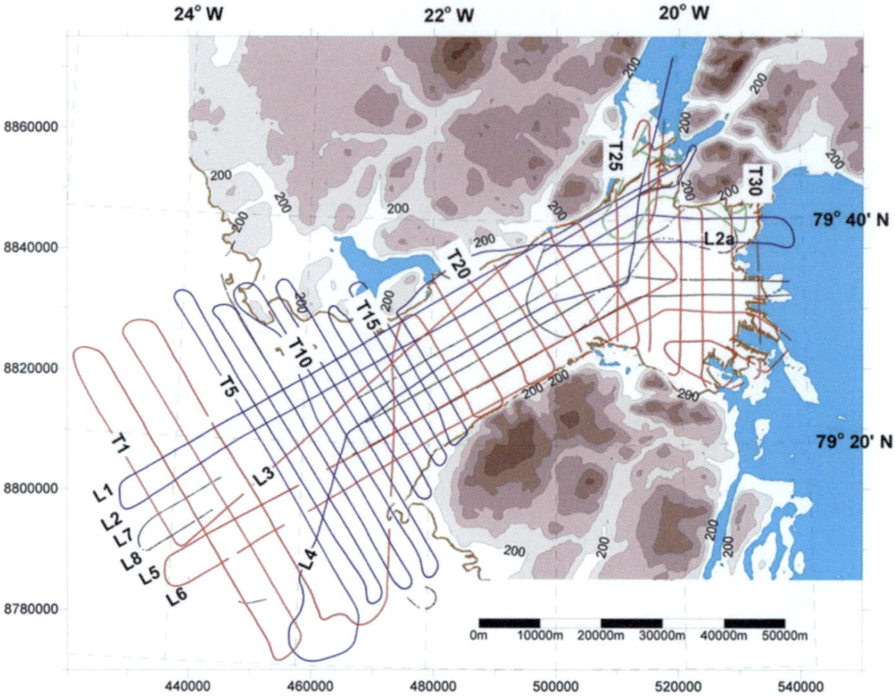

Fig. 4.9 Map showing ice radar and laser altimeter flight tracks acquired on July 10, 1998 (*black line*), July 11, 1998 (*green line*) and July 14, 1998 (*blue* and *red lines*) on Nioghalvfjerdsfjorden Glacier. Laser altimeter data was only acquired on one of the July 14 flights (*blue line*). Coordinates are UTM zone 27. Longitudinal and transverse profile lines are labelled L1–L8 and T1–T30, respectively. Missing profile line sections indicate missing bottom radar reflections

4.5.2.1 Airborne Laser-Altimeter and Ice Radar Measurements

In July 1998, about 3000 km of high quality surface and ice thickness profiles of the glacier were acquired by means of airborne ice-radar and laser-altimeter surveys from a Greenlandair Twin Otter aircraft (Fig. 4.9). An Optech 401 SX infrared pulsed laser altimeter was used for measuring the aircraft-ice surface distance, with positioning based on differential GPS receivers supplemented with an inertial measurement unit (IMU). An upgraded version of the DTU 60 MHz radar from 1969 was used for the ice sounding measurements (Christensen et al. 1970, 2000).

The airborne survey was accomplished in four flights in July 1998 (Fig. 4.9). The horizontal spacing between flight tracks was ~5 km on the relatively uniform floating part of the glacier, but was ~2.5 km in the grounding zone. Unfortunately, the laser altimeter was only operational during one flight (on July 14, 1998), whereas good coverage of radar recordings of the glacier bottom were obtained on all flight missions.

From the measurements taken on July 14, 1998 (when both accurate ice surface elevations h and ice thickness measurements H are available), the ratio h/H for the floating sections of the glacier can be derived as 0.113 ± 0.001, corresponding to a glacier ice/sea water density ratio of $\rho_i/\rho_w = 0.887 \pm 0.001$. These values agree reasonably well with the independently derived values of h/H = 0.108 and $\rho_i/\rho_w = 0.892$ given by Olesen et al. (1998). We take the average values as the best estimates (i.e. h/H = 0.110 and $\rho_i/\rho_w = 0.890$). For the floating sections of the glacier for which laser altimeter data are not available, the surface elevation values h were thus derived by multiplying the measured ice thickness H by 0.110 (Figs. 4.10, 4.11).

The cross profiles of the glacier surface and bed reveal a rough relief of the glacier bottom upstream of the grounding zone, with depth amplitudes of several hundred metres over distances of a few kilometres (Fig. 4.11, profiles T09 - T11). From the grounding line near T11 the thin/thick ice sections continue for several kilometres in the down-glacier direction as troughs cut up into the glacier base and as ridges extending downwards from the average bottom. However, the relief is gradually smoothed and eventually vanishes some tens of kilometres down-glacier from the grounding zone near profile T22.

The derived surface elevation, ice thickness, and glacier-bottom maps are displayed in Fig. 4.12a–c, respectively. A contour map showing the ratio *h/H* is shown in Fig. 4.12d; the contour line corresponding to *h/H* = 0.110 is used to roughly distinguish between grounded and floating sectors of the glacier.

4.5.2.2 Surface Velocity Measurements

In situ GPS A network of 13 stakes was established on Nioghalvfjerdsfjorden Glacier in 1996 for measurements of mass balance, ice velocity and deformation. The network extends from the front of the glacier tongue near sea level and follows the centre-line upstream to an elevation of ~530 m a.s.l. Ice motion was determined several times by differential GPS measurements in 1996 and 1997 relative to a reference point on a rock outcrop at the northern margin of the glacier (Fig. 4.7). The data were processed to give annual or summer values of magnitude and direction of horizontal ice motion (Fig. 4.13a). In 1997 a transverse velocity profile was established halfway down the main channel (Fig. 4.13b), together with three velocity stations in the grounding zone at the head of the glacier. In 1998 a stake network for strain and velocity measurement was established across an ice ridge close to northern margin of the glacier with a view to studying tidal deformations and the mechanics of ridge formation (Mayer et al. 2000). Positions of all GPS velocity points are shown in Fig. 4.7.

Aerial Photography Photogrammetric mapping of the Nioghalvfjerdsfjorden Glacier was based on vertical aerial photographs from 29 July 1963 and 2 August 1978 (Higgins, personal communication). A map of mean surface velocities over the 15 year period was produced by feature tracking of elements in the surface drainage system, such as shorelines of lakes and tight meanders in meltwater streams, and distinct features of the glacier front (Fig. 4.14a).

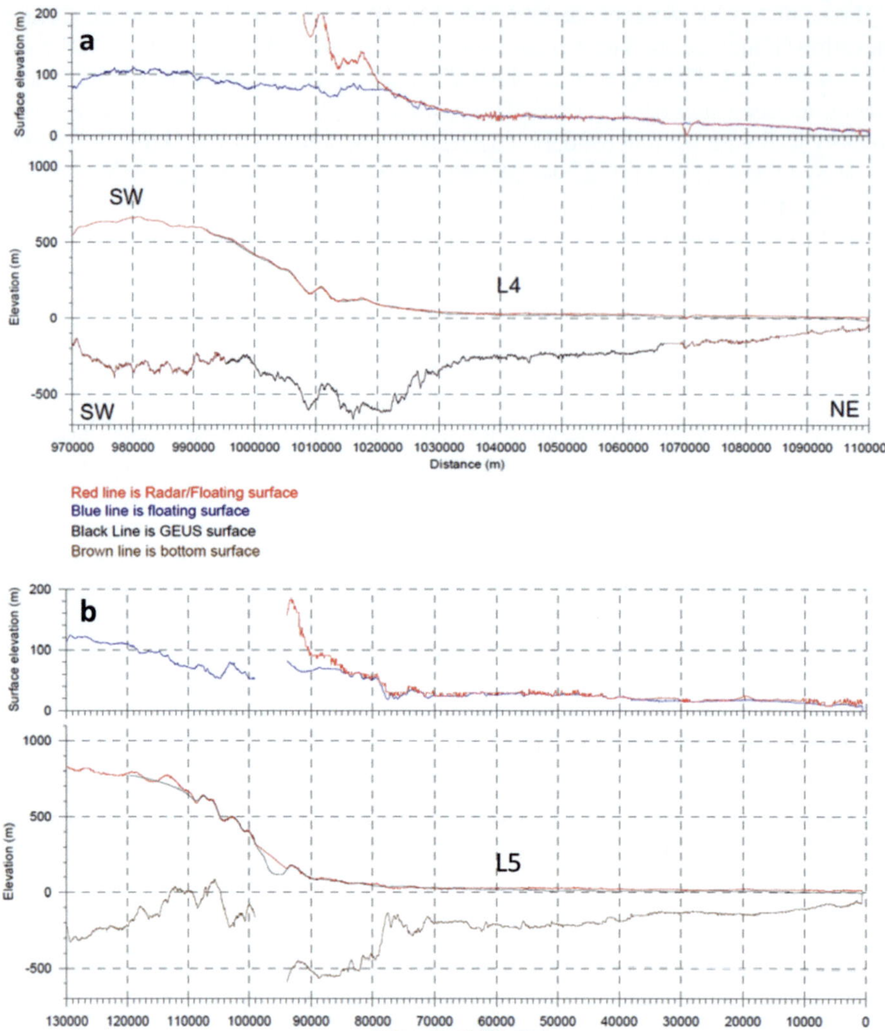

Fig. 4.10 Longitudinal ice surface and *bottom* surface profiles for Nioghalvfjerdsfjorden Glacier along: (**a**) flight line L4; (**b**) flight line L5 (see Fig. 4.9 for location). The *blue lines* are the "floating surface" profiles calculated as d_iH where d_i is the ratio of ice density to sea water density, and H is ice thickness. Locations where the *blue* and *red lines* intersect indicate floating sections, whereas grounded sections are characterized by an ice surface above the "floating surface"

InSAR Measurement Horizontal velocities were derived from Synthetic Aperture Radar Interferometry (InSAR) analysis of multiple pairs of ERS-1/2 tandem mode images collected in winter 1995–1996 (Rignot 1996; Rignot et al. 2001). SAR data were also acquired in 2004, with Fig. 4.14b showing a composite map mainly based on the 1995–1996 data supplemented by the 2004 data in the slow-moving, grounded sectors of the glacier.

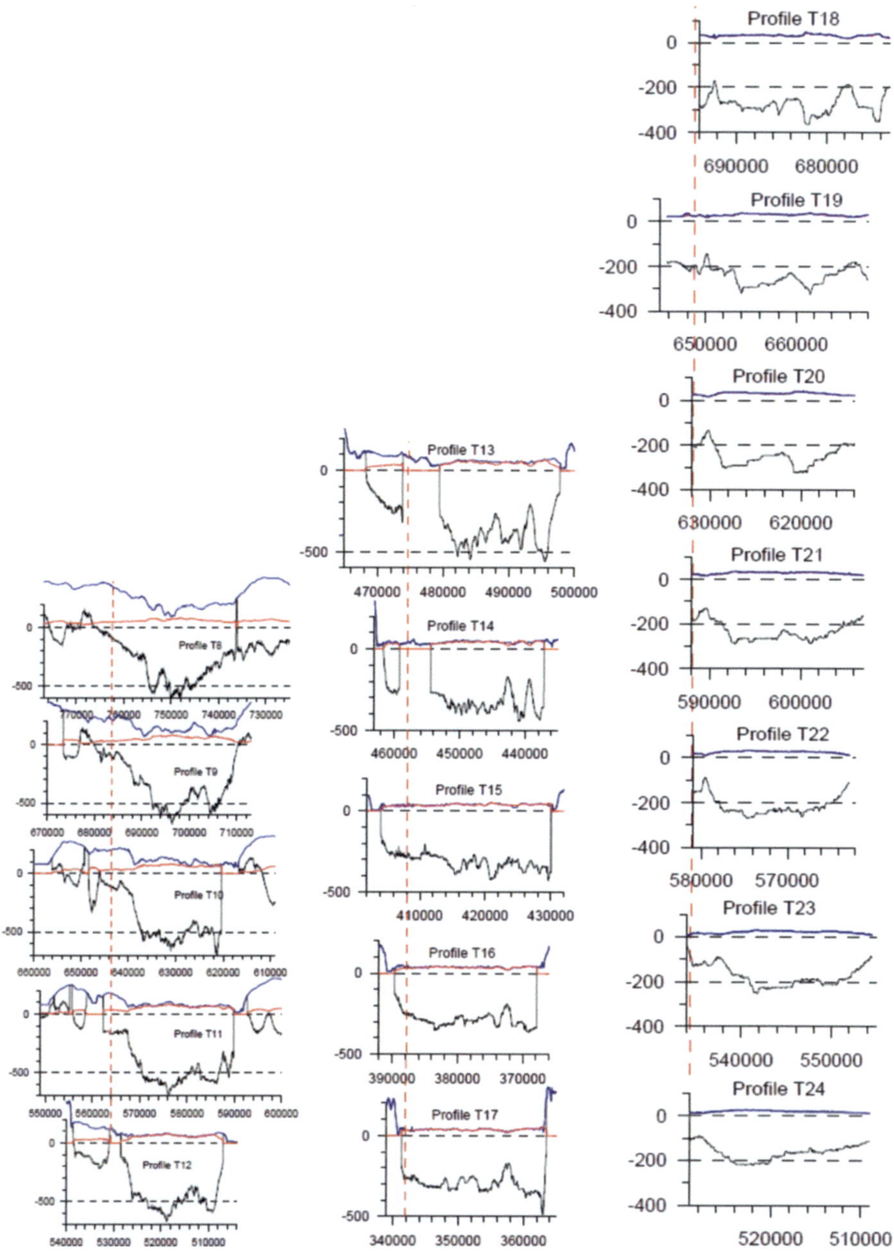

Fig. 4.11 Transverse ice surface (*blue lines*) and glacier bottom (*black lines*) profiles for Nioghalvfjerdsfjorden Glacier along flight lines T8– T24 (see Fig. 4.9 for location). The *red lines* are "floating surface" profiles calculated as $d_i H$ where d_i is the ratio of ice density to sea water density, and H is ice thickness. Locations where the *blue* and *red lines* intersect indicate floating sections, whereas grounded sections are characterized by an ice surface above the "floating surface". The *red dashed line* indicates the intersection of the cross profile with longitudinal profile line L1 shown in Fig. 4.9

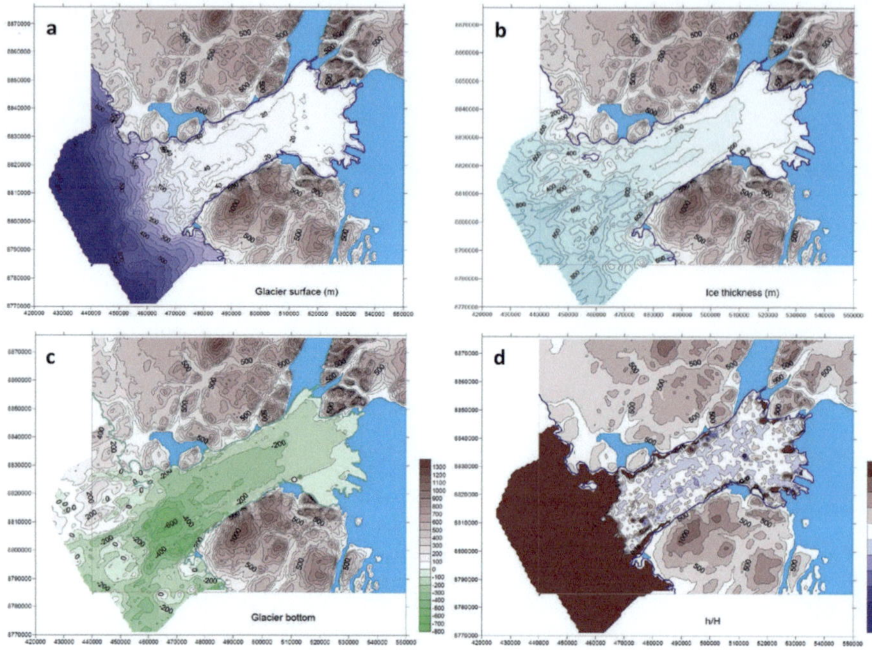

Fig. 4.12 (**a**) Surface elevation map of Nioghalvfjerdsfjorden Glacier and surroundings; (**b**) Ice thickness map; (**c**) Map of bottom of Nioghalvfjerdsfjorden Glacier, (**d**) "Floating ratio" h/H. Coordinates refer to UTM zone 27

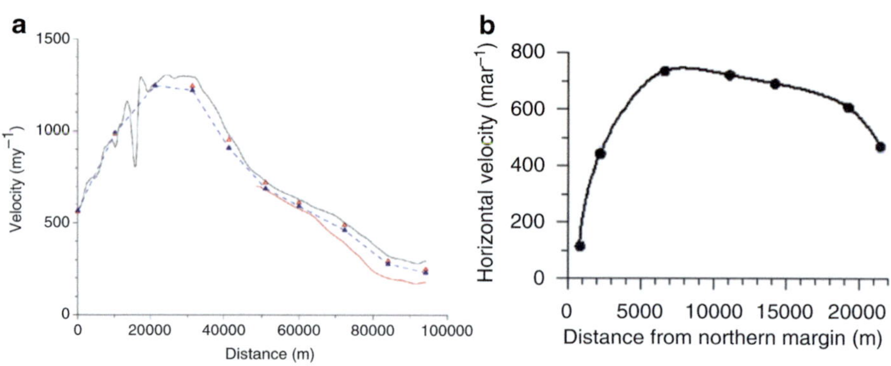

Fig. 4.13 (**a**) Horizontal surface velocities along the central stake line on Nioghalvfjerdsfjorden Glacier (see Fig. 4.7 for location). Points marked with *blue* and *red triangles* represent mean annual velocities (August 1996–August 1997) and summer velocities (mid July 1996 - mid August 1996), respectively, derived from repeated GPS measurements. The *red curve* represents 15-year mean velocities based on repeated aerial photography from 29 July 1963 and 2 August 1978. The *black curve* is based on InSAR measurements in December 1996. (**b**) Transverse velocity profile measured in summer 1997 (see Fig. 4.7 for location)

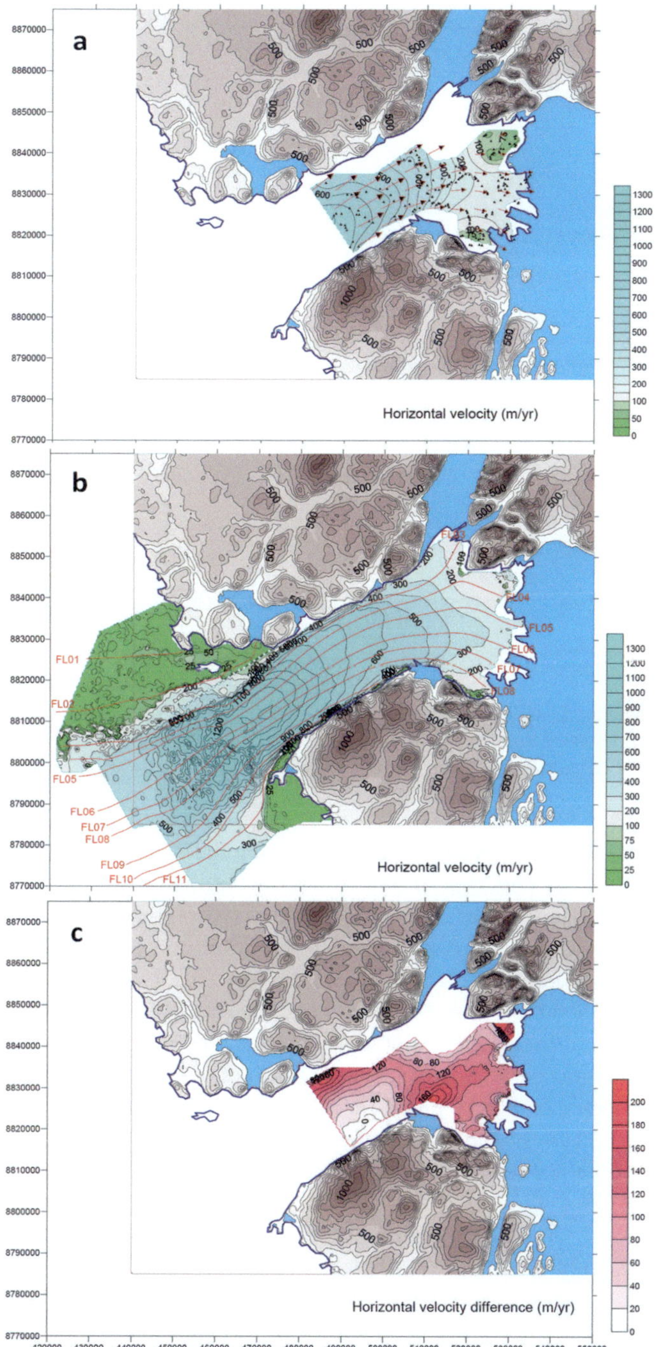

Fig. 4.14 (**a**) Horizontal surface velocities of Nioghalvfjerdsfjorden Glacier averaged over a 15-year period derived by comparing recognizable surface features in two sets of aerial photographs from 1963 and 1978, respectively. *Red arrows* indicate ice flow directions. *Black dots* indicate measurement points. (**b**) Surface velocities derived from InSAR measurements in 1996 and 2004. *Red curves* indicate the course of ice flow lines numbered FL01 to FL11 derived from the InSAR velocity data. (**c**) Difference (m year^{-1}) between the InSAR derived 1995–1996 winter velocities and the 1963–1978 average velocities derived from repeated aerial photography. Coordinates refer to UTM zone 27

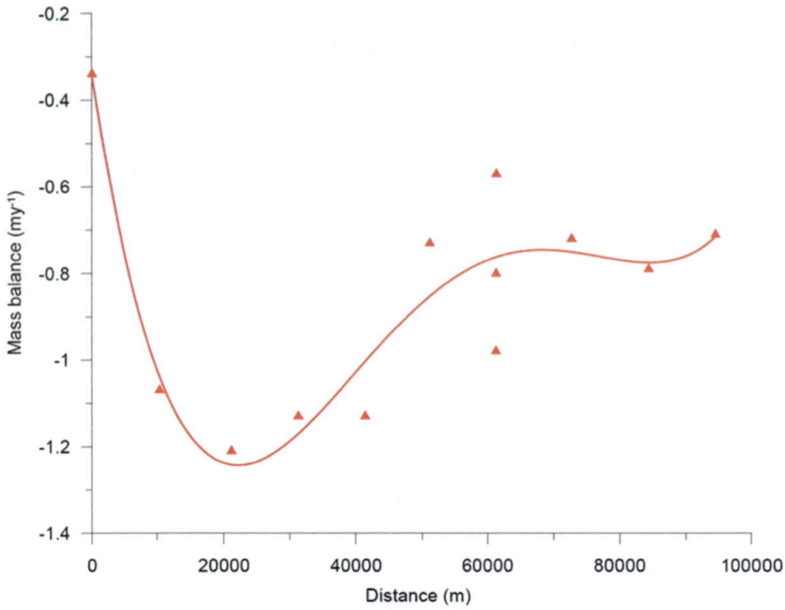

Fig. 4.15 Annual mass balance (August 1996–August 1997) along the central stake line on Nioghalvfjerdsfjorden Glacier (see Fig. 4.7 for location). Points marked with *red dots* are measured values. *Red curve* is a fourth degree polynomial fit

Comparison of Velocity Data From comparison between the different datasets described above it appears that summer velocities in the floating section of the glacier are a few percent higher than mean annual velocities (Fig. 4.13a). However, there is no indication of a seasonal velocity variation in the grounded sector of the glacier. As to inter-annual variations, it appears that the 1996–1997 annual velocities were higher than the 1963–1978 average. The rate of increase varies from ~2% some 40 km behind the glacier front to ~30% near the front (Fig. 4.14c). The 1996 data shows relatively large amplitude velocity fluctuations over a short distance at the grounded section (reflecting perturbations induced by flow over the undulating glacier bottom), as compared to the much lower amplitude velocity variations in the floating part of the glacier.

4.5.2.3 Mass Balance

The surface balance along the centre line of Nioghalvfjerdsfjorden Glacier was measured between August 1996 and August 1997 (Fig. 4.15). The surface balance varies between −1.2 and −0.35 m year^{-1}, with the maximum ablation rate (minimum mass balance) occurring at km 20. It appears that this location coincides with the grounding zone of the glacier (i.e., at the transition from the steeply sloping ice fall to the almost horizontal floating section of the glacier).

The average flux divergence over the 1640 km^2 grounded sector of the glacier is calculated as −0.85 m year^{-1}. This is in good agreement with the average measured ablation rate (−0.93 m year^{-1}) in this part of the glacier, and can therefore be most likely ascribed to surface ablation. The average flux divergence over the 1710 km^2 floating sector is −7.3 m year^{-1}, most of which must be ascribed to bottom melting. The total mass losses from the grounded and floating sectors of the glacier are 1.4 and 12.5 km^3 year^{-1}, respectively. The latter number compares well with the findings of Rignot et al. (2001).

4.5.2.4 Variations in Glacier Front Position and Interaction with Adjacent Fast Sea Ice

The Nioghalvfjerdsfjorden Glacier project also comprised collection and compilation of data to document short- and long-term variations in glacier-front position and extent of the adjacent fast sea ice. Variations on millennial and century time scales were derived from glacial geological information (Bennike and Weidick 1999). Descriptions and sketch maps in expedition reports give information about ice margin positions in the first half of the twentieth century (Koch and Wegener 1912, pp. 7–19; Mikkelsen 1922; Knuth 1942; Koch 1945). However, many of the descriptions stress the difficulty in distinguishing the low floating glacier front from the surrounding sea ice, especially during winter and spring visits when the surface is covered with snow. The information on glacier front positions can therefore only be regarded as a guideline.

The first photographic documentation of the Nioghalvfjerdsfjorden Glacier is from oblique aerial photographs taken in 1950 and 1951. Vertical aerial photographs are also available from 1962/63 and 1978. Subsequent ice front positions can be documented in cloud-free Landsat TM scenes from 1986, 1990 and 1993. Furthermore, a number of NOAA AVHRR and ERS-2 scenes have been used to document the development of the sea ice conditions in the vicinity of Nioghalvfjerdsfjorden.

The Holocene glaciation history of Nioghalvfjerdsfjorden and its surroundings has been documented by means of radiocarbon dated samples of shells, driftwood, plant remains and bones collected along a transect from Blåsø in the inland area (Fig. 4.7) to the outer coast, and by observations of the marine limit. The data indicate that the eastern margin of the Greenland ice sheet was still located on the offshore continental shelf in the early Holocene ~9.7 cal. ka BP (thousand calibrated 14C years before 1950; Bennike and Weidick 1999). Between ~7.7 and 4.5 cal. ka BP, the front of Nioghalvfjerdsfjorden Glacier retreated to a position at least as far west as Blåsø. Dates of driftwood and whale bones are restricted to the period 7.0 to 5.4 cal. ka BP, and indicate open water in the fjord at least during this interval of the middle Holocene. The mollusc fauna and plant remains indicate that temperatures, particularly in the summer, were a few degrees higher in the middle Holocene than at present (Bennike and Weidick 1999). According to Fig. 4.6, a temperature of ~2°C higher than present is sufficient to reduce the FI for North Greenland to values below the critical level for maintaining semi-permanent fast ice. Such a temperature increase is therefore likely to have caused disintegration of the floating sections of Nioghalvfjerdsfjorden Glacier, as documented by Bennike and Weidick (2001). The consequences for the interpretation of the record of ice rafted debris (IRD) on the east Greenland continental shelf of such Holocene disintegration of floating glaciers in northeast Greenland are discussed by Reeh et al. (1999a).

Evidence that can throw light on the variation in extent of other North and Northeast Greenland floating glaciers during the Holocene is scarce. Dated biogenic material of mid-Holocene age (Weidick et al. 1996) indicate the existence of a "Storstrømmen Sound", separating Germania Land from mainland Greenland in this period. Also, one shell sample collected between 5 and 10 km behind the present front of Steensby Glacier (81°30′N, 54°W) dates at 4.7 cal. ka BP (Kelly and Bennike 1992), showing that this glacier also experienced a substantial frontal retreat in the mid-Holocene. Most likely, during the Holocene Climatic Optimum, these glaciers, and probably all north Greenland Ice Sheet outlet glaciers presently terminating with floating ice tongues, were 'Daugaard-Jensen type' glaciers producing icebergs by grounding line calving (Fig. 4.3). Since ~4 ka BP, Nioghalvfjerdsfjorden has been permanently occupied by a floating glacier. This neo-glaciation probably culminated around AD 1900, after which a slight thinning and recession has occurred.

Observations from 1906/07 (Koch and Wegener 1912), 1910 (Mikkelsen 1922) and 1939 (Knuth 1942) indicate that the margin of Nioghalvfjerdsfjorden Glacier was located close to the Bagatellerne Islands (Fig. 4.8). Furthermore, Koch (1945) documented the existence of the Norske Øer Ice Barrier (NØIB) in 1933. Later positions of the main glacier front are based on oblique aerial photographs from 1950 and 1951. They show an ice margin with long glacier tongues stretching all the way to Bagatellerne. Vertical aerial photographs from 1963 show that major calving events occurred between 1951 and 1963, probably due to break-up of the protecting NØIB, which resulted in a 25 km retreat of the ice tongues. Although the 1951 and 1963 photographs do not document the easternmost extent of the fast ice cover, both sets of photographs show a glacier front confined in extended sea ice. This suggests that the NØIB ice barrier was present in both years. Support for the interpretation that the ice barrier broke up sometime in the intervening period is found in a letter of November 18, 1960 from L. Koch to A.E. Porsild (Wadhams 1981). According to Koch, the fast ice barrier, which was formerly attached to the glacier margin, broke away in the 1950s. Aerial photographs taken in 1978 and Landsat imagery from 1990 also show the presence of NØIB. The characteristic saw-toothed ice tongues advanced slowly in this period, pushing eastward through the fast ice barrier (Fig. 4.16). This advance took place without any major changes in the ice tongue configuration, and icebergs in front of the glacier were slowly pushed seaward without being rotated. The same pattern of advance continued until 1997, as documented by field observations in that year.

The ice margin in Dijmphna Sund has not shown a similar advance. The more frequent disappearance of the fjord ice in Dijmphna Sund has apparently resulted in frequent calving events there.

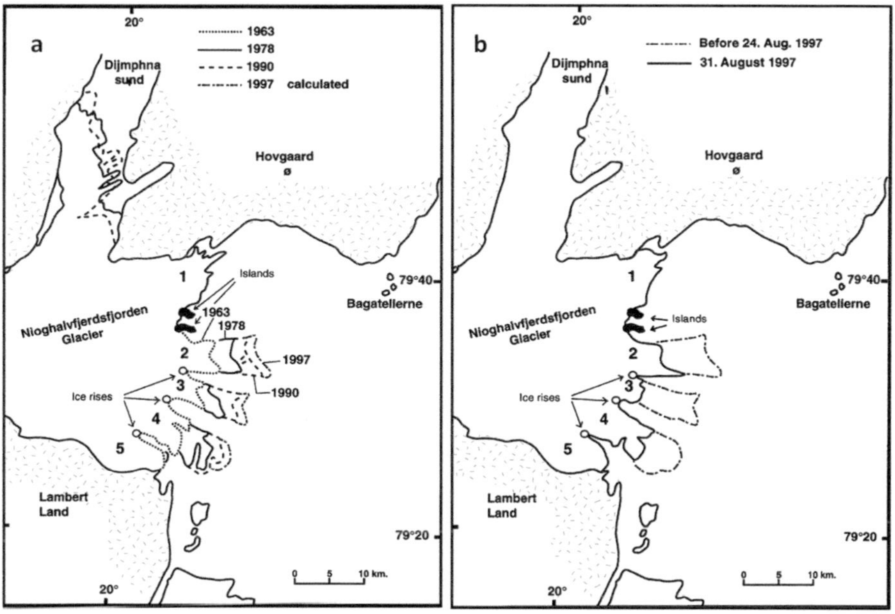

Fig. 4.16 (**a**) Ice margin variations of Nioghalvfjerdsfjorden Glacier from 1963 to 1997. The data sources are aerial photographs from 1963 and 1978 and Landsat data from 1990. The 1997 position is projected from the mapped ice velocities at the ice margin. (**b**) The main ice margin of the Nioghalvfjerdsfjorden Glacier before and after a major calving activity triggered by the break-up of the protecting Norske Øer Ice Barrier between 24 and 26 August 1997 (From Reeh et al. (2001), reprinted from the Annals of Glaciology with permission of the International Glaciological Society)

NOAA AVHRR data document that the NØIB broke up again between August 24 and 26 1997 (Fig. 4.16), apparently after having existed since at least 1963. On September 3 1997 open water was observed from the Northeast Water polynya, all the way south to the margin of Nioghalvfjerdsfjorden Glacier (Fig. 4.16). The break-up of the fast ice barrier (NØIB) triggered a major calving activity from the glacier (as documented by an ERS-2 image from August 31, 1997), which removed a glacier area of ~90 km^2. With an estimated ice thickness of 80 m (Reeh et al. 1999b) the calved-off ice volume amounts to ~7.4 km^3. Compared to a total ice flux of 0.2 km^3 year^{-1} from this part of the ice margin (Reeh et al. 1999a), the August 1997 calving event therefore corresponds to ~35 years of ice flux at the main ice front. This suggests that no major calving or ice-front break-up occurred during this 35 year period. It is worth noting that this period without major calving activity coincides with the time when the NØIB was most likely permanently present.

Recent NOAA AVHRR and Landsat 7 images reveal that the NØIB has once again re-formed after the sudden break-up in August 1997. Field observations in the area in 1998 show that the large icebergs released from the glacier front at the break-up in 1997 are now trapped within the fast ice of the re-formed NØIB.

The course of events described above strongly suggests that the infrequent occurrences of fast ice break-up and break-up of the ice tongues from Nioghalvfjerdsfjorden Glacier are interrelated. The stability of the floating ice margin seems to be dependent on the presence of a protecting fast ice cover in front of the glacier. This also seems to apply to other north Greenland floating glacier tongues, as documented by Higgins (1989, 1991) on the basis of aerial photographs taken at decadal intervals. Other mechanisms by which a semi-permanent fast-ice cover could inhibit calving of a floating glacier front are discussed by Reeh et al. (2001), who conclude that the main cause is probably the process by which the fast ice keeps the partly disintegrated glacier front from completely splitting up and preventing the fragments from drifting away as ice islands.

4.6 Response to Climate Warming

As discussed above, there is strong evidence that the change in fjord-glaciation characteristics in Greenland from grounding-line calving glaciers in the south to extended floating glaciers subject to intensive bottom melting in the north is controlled by climate. Temperatures in North Greenland only a few degrees warmer than present would most likely cause seasonal break-up of the near-shore sea-ice in contrast to break-up at intervals of several decades under present climate conditions (Reeh et al. 1999a). Under such warmer conditions, the floating glaciers will likely disintegrate.

The co-existence of floating glaciers and semi-permanent fast ice does not in itself prove a causal relationship between their occurrences. Both phenomena could very well owe their existence to the present climate conditions in north and north-east Greenland. A simple melt-rate model (Reeh 1991) shows that, at sea level, the melt rate at the upper surface of glaciers in North Greenland will increase by ~2.5 m year^{-1} if summer temperatures increase by 2°C. Using measured ice velocities on Nioghalvfjerdsfjorden Glacier, a simple calculation shows that it takes on the order of 100 years for glacier ice to travel ~60 km from the grounding zone to the glacier terminus. After 100 years with summer temperatures 2°C warmer than now, the frontal region of the glacier would have thinned by ~250 m due to increased surface melting. As large sections of Nioghalvfjerdsfjorden Glacier (and other North Greenland floating glaciers) are <250 m thick, increased surface melting would in such a scenario result in the disappearance of large areas of the present floating glaciers in North Greenland over <100 years, assuming constant precipitation, unchanged basal melt rates and no dynamic effect from the increase in surface slope. Reduced contact with pinning points such as ice rises, and reduced support along the side margins would further enhance the rate of deterioration by promoting disintegration of the floating glacier into large ice-islands.

Although the floating glaciers in North Greenland will thus respond to climate change independent of changes of the fast-ice cover, it is also clear that the occurrence of floating glaciers and fast ice is in some respect interdependent. Disintegration of the glacier ice tongues into large ice islands only occurs during the infrequent fast ice break-up events. On the other hand, the existence of the fast ice cover probably depends on the large volume of fresh water produced by melting at the bottom of the floating glaciers. The total fresh water volume produced by bottom melting from North Greenland floating glaciers is estimated at ~35 km^3 $year^{-1}$. The meltwater moves up the gradient at the underside of the floating glaciers and spreads beneath the fast ice cover in front of the glacier terminus. This layer of cold, fresh and relatively low-density meltwater stabilizes the sea-ice cover by preventing its contact with the warmer, more salty and denser sea water beneath (Mayer et al. 2000), thus reducing melting of the sea ice.

In a climate a few °C warmer than at present, the direct meltwater input to the ocean produced by bottom melting of North and Northeast Greenland floating glaciers (~35 km^3 $year^{-1}$) is likely to decrease to near zero if these ice tongues disappear. This could have an impact on local sea ice growth, and hence influence deep-water formation and the extent of polynyas. In addition, a significant increase in the iceberg flux along the coast of East Greenland is to be expected, particularly north of Scoresby Sound (~70°N), where, presently, drifting icebergs are seldom encountered. At present, the total ice flux across the grounding lines of North Greenland floating glaciers is ~50 km^3 $year^{-1}$ (Rignot et al. 1997). Instead of being removed by bottom melting, as happens today, most of this volume would eventually be released as icebergs by grounding line calving in a warmer climate. The icebergs, no longer trapped by semi-permanent sea ice, would join the large-scale oceanic circulation and be transported southward, as both bathymetry and the near-cost sea currents enable access of the icebergs to the East Greenland Polar Current. The increased iceberg flux would mean increased hazards for iceberg encounters, particularly if there is future offshore oil construction on the East Greenland continental shelf.

References

Ahnert, F. (1963). The terminal disintegration of Steensby Gletscher, North Greenland. *Journal of Glaciology, 4*(35), 537–545.

Assur, A. (1956). *Airfields on floating ice sheets for routine and emergency operations* (Sipre Report, 36). Wilmette: Snow Ice and Permafrost Research Establishment, Corps of Engineers, U.S. Army.

Bennike, O., & Weidick, A. (1999). Observations on the Quaternary geology around Nioghalvfjerdsfjorden, eastern North Greenland. *Geology of Greenland Survey Bulletin, 183*, 57–60.

Bennike, O., & Weidick, A. (2001). Late Quaternary history around Nioghalvfjerdsfjorden and Jøkelbugten, North-East Greenland. *Boreas, 30*, 205–227.

Bourke, R. H., Newton, J. L., Paquette, R. G., & Tunnicliffe, M. D. (1987). Circulation and water masses of the East Greenland Shelf. *Journal of Geophysical Research, 92*(C7), 6729–6740.

Budéus, G., & Schneider, W. (1995). On the hydrography of the Northeast Water Polynya. *Journal of Geophysical Research, 100*(C3), 4287–4299.

Christensen, E. L., Gundestrup, N., Nilsson, E., & Gudmandsen, P. (1970, January). *Radio glaciology - 60 MHz radar*. Electromagnetics Institute, Technical University of Denmark, R 77, pp. 82.

Christensen, E. L., Reeh, N., Forsberg, R., Jørgensen, J. H., Skov, N., & Woelders, K. (2000). A low-cost glacier-mapping system. *Journal of Glaciology, 46*(154), 531–537.

Chuah, T. S., Gogineni, S. P., Allen, C., & Wohletz, B. (1996). *Radar thickness measurements over the northern part of the Greenland Ice Sheet* (Radar Systems and Remote Sensing Laboratory Technical Report 10470-3). University of Kansas, Lawrence, Kansas.

Davies, W. E., & Krinsley, D. B. (1962). The recent regime of the ice cap margin in North Greenland. *International Association of Hydrological Sciences, 58*, 119–130.

Davis, C. H., Kluever, C. A., & Haines, B. J. (1998). Elevation change of the Southern Greenland Ice Sheet. *Science, 279*, 2086–2088.

Ekholm, S. (1996). A full coverage, high-resolution, topographic model of Greenland computed from a variety of digital elevation data. *Journal of Geophysical Research, 101*(B10), 21961–21972.

Funder, S., Hjort, C., Landvik, J. Y., Nam, S., Reeh, N., & Stein, R. (1998). History of a stable ice margin - East Greenland during the middle and upper Pleistocene. *Quaternary Science Reviews, 17*, 77–123.

Higgins, A. K. (1988). Glacier velocities in North and North-East Greenland. *Rapport Grønlands Geologiske Undersøgelse, 140*, 102–105.

Higgins, A. K. (1989). North Greenland ice islands. *Polar Record, 25*(154), 207–212.

Higgins, A. K. (1991). North Greenland glacier velocities and calf ice production. *Polarforschung, 60*(1), 1–23.

Higgins, A. K., & Weidick, A. (1990). The world's northernmost surging glacier? *Zeitschrift für Gletscherkunde und Glazialgeologie, 24*, 111–123.

Joughin, I., Kwok, R., & Fahnestock, M. (1996a). Estimation of ice-sheet motion using satellite radar interferometry: Method and error analysis with application to Humboldt Glacier, Greenland. *Journal of Glaciology, 42*(142), 564–575.

Joughin, I., Tulaczyk, S., Fahnestock, M., & Kwok, R. (1996b). A mini-surge on the Ryder Glacier, Greenland, observed by satellite radar interferometry. *Science, 274*(5285), 228–230.

Kelly, M., & Bennike, O. (1992). Quaternary geology of western and central North Greenland. *Rapp. Grønlands geol. Unders., 153*, 1–34.

Knuth, E. (1942). Dansk Nordøstgrønlands Ekspedition 1938-39. Report on the expedition and subsequent work at the Mørkefjord Station. *Meddelelser om Grønland, 126*(1), 1–159.

Koch, L. (1928). Contributions to the glaciology of North Greenland. *Meddelelser om Grønland, 45*, 181–464.

Koch, L. (1945). The East Greenland ice. *Meddelelser om Grønland, 130*(3), 1–373.

Koch, I. P., & Wegener, A. (1912). Die glaziologischen Beobachtungen der Danmark-Expedition. *Meddelelser om Grønland, 46*(1), 1–77.

Krabill, W., Thomas, R., Jezek, K., Kuivinen, K., & Minazade, S. (1995). Greenland ice thickness changes measured by laser altimetry. *Geophysical Research Letters, 22*(17), 2341–2344.

Mayer, C., Reeh, N., Jung-Rothenhäusler, F., Huybrechts, P., & Oerter, H. (2000). The subglacial cavity and implied dynamics under Nioghalvfjerdsfjorden Glacier, NE-Greenland. *Geophysical Research Letters, 27*(15), 2289–2292.

Mikkelsen, E. (1922). Alabama-Expeditionen. Til Grønlands Nordøstkyst 1909-1912. *Meddelelser om Grønland, 52*, 1–142.

Mohr, J. J., Reeh, N., & Madsen, S. N. (1998). Three dimensional glacial flow and surface elevations measured with radar interferometry. *Nature, 391*, 273–276.

Ohmura, A. (1987). New temperature distribution maps for Greenland. *Zeitschrift für Gletscherkunde und Glazialgeologie, 23*(1), 1–45.

Ohmura, A., & Reeh, N. (1991). New precipitation and accumulation maps for Greenland. *Journal of Glaciology, 37*(125), 140–148.

Olesen, O. B., Thomsen, H. H., Reeh, N., & Bøggild, C. E. (1998). Attempts at measuring bottom melting at Nioghalvfjerdsfjorden Glacier. In J. A. Dowdeswell, E. K. Dowdeswell & J. O. Hagen (Eds.), *International Arctic Science Committee (IASC), Arctic Glaciers Working Group Meeting and Workshop on Arctic Glaciers Mass Balance held at Gregynog, Wales* 29–30 January 1998. Centre for Glaciology Report 98–01. Prifysgol Cymru Aberystwyth, The University of Wales, May 1998. 5-page abstract.

Reeh, N. (1991). Parameterization of melt rate and surface temperature on the Greenland Ice Sheet. *Polarforschung, 59*(3), 113–128.

Reeh, N. (1994). Calving from Greenland glaciers: Observations, balance estimates of calving rates, calving laws. In N. Reeh (Ed.), *Report on the workshop on the calving rate of West Greenland glaciers in response to climate change* (p. 85–102). Copenhagen: Danish Polar Center.

Reeh, N. (1999). Mass balance of the Greenland Ice Sheet: Can modern observation methods reduce the uncertainty? In *Proceedings from the workshop on methods of mass balance measurements and modelling, Tarfala Sweden*, August 10–12, 1998. *Geografiska Annaler, 81A*(4), 735–742.

Reeh, N. (2004). Holocene climate and fjord glaciation in Northeast Greenland: Implications for IRD-deposition in the North Atlantic. *Sedimentary Geology, 165*(3–4), 333–342.

Reeh, N., Bøggild, C. E., & Oerter, H. (1994). Surge of Storstrømmen, a large outlet glacier from the Greenland Ice Sheet. *Rapport. Grønlands Geologiske Undersøgelse, 162*, 201–209.

Reeh, N., Mayer, C., Miller, H., Thomsen, H. H., & Weidick, A. (1999a). Climate control on fjord glaciations in Greenland: Implications for IRD-deposition in the sea. *Geophysical Research Letters, 26*(8), 1039–1042.

Reeh, N., Thomsen, H. H., Higgins, A. K., Weidick, A., & Starzer, W. (1999b). Stability conditions of North-East Greenland floating ice margins. In *Climate change and sea level: final report of work undertaken for the Commission of the European Communities under contract No. ENV4-CT95–0124*, 1st March 1996–28th February 1999. Report No. 9 from the Danish Polar Center.

Reeh, N., Mayer, C., Olesen, O. B., Christensen, E. L., & Thomsen, H. H. (2000). Tidal movement of Nioghalvfjerdsfjorden Glacier, Northeast Greenland: Observations and modelling. *Annals of Glaciology, 31*, 111–117.

Reeh, N., Thomsen, H. H., Higgins, A. K., & Weidick, A. (2001). Sea ice and the stability of North and Northeast Greenland floating glaciers. *Annals of Glaciology, 33*, 474–480.

Reeh, N., Christensen, E. L., Mayer, C., & Olesen, O. B. (2003). Tidal bending of glaciers: A linear visco-elastic approach. *Annals of Glaciology, 37*, 83–89.

Reeh, N., Christensen, E. L., Thomsen, H. H., Forsberg, R., & Rignot, E. (2008). *Nioghalvfjerdsfjorden Glacier, northeast Greenland: report on data collection, processing, and analysis*. Lyngby, Denmark: National Space Institute, Technical University of Denmark.

Rignot, E. (1996). Tidal motion, ice velocity and melt rate of Petermann Gletscher, Greenland, measured from radar interferometry. *Journal of Glaciology, 42*(142), 476–485.

Rignot, E., Gogineni, S. P., Krabill, W. B., & Ekholm, S. (1997). North and Northeast Greenland ice discharge from satellite radar interferometry. *Science, 276*(5314), 934–937.

Rignot, E., Gogineni, S. P., Joughin, I., & Krabill, W. B. (2001). Contribution to the glaciology of northern Greenland from satellite radar interferometry. *Journal of Geophysical Research Atmospheres, 106*(D24), 34007–34019.

Schneider, W., & Budéus, G. (1995). On the generation of the Northeast Water Polynya. *Journal of Geophysical Research Oceans, 100*(C3), 4269–4286.

Schneider, W., & Budéus, G. (1997). A note on Norske Ø Ice Barrier (Northeast Greenland), viewed by Landsat 5 TM. *Journal of Marine Systems, 10*(1–4), 99–106.

Steffen, K., Abdalati, W., & Stroeve, J. (1993). Climate sensitivity studies of the Greenland Ice Sheet using satellite AVHRR, SMMR, SSM/I and in situ data. *Meteorology and Atmospheric Physics, 51*, 239–258.

Thomsen, H. H., Reeh, N., Olesen, O. B., Starzer, W., & Bøggild, C. E. (1999). Bottom melting, surface mass balance and dynamics of floating North-East Greenland ice tongues. In *Climate change and sea level. Final report of work undertaken for the Commission of the European Communities under contract No. ENV4-CT95–0124*. Report No. 8 from the Geological Survey of Denmark and Greenland.

Wadhams, P. (1981). The ice cover in the Greenland and Norwegian seas. *Reviews of Geophysics and Space Physics, 19*(3), 345–393.

Weidick, A. (1995). Greenland. In R. S. Williams & J. C. Ferrigno (Eds.), *Satellite image atlas of glaciers of the world* (U.S. Geological Survey Professional Paper 1386C). C1–C105. Denver, Colorado, USA: U.S. Geological Survey

Weidick, A., Andreasen, C., Oerter, H., & Reeh, N. (1996). Neoglacial glacier changes around Storstrømmen, North-East Greenland. *Polarforschung, 64*(3), 95–108.

Zwally, H. J., & Giovinetto, M. B. (1995). Accumulation in Antarctica and Greenland derived from passive-microwave data: A comparison with contoured compilations. *Annals of Glaciology, 21*, 123–130.

Part II
Physical Processes and Historical Changes of Arctic Ice Shelves

Chapter 5
Changes in Canadian Arctic Ice Shelf Extent Since 1906

Derek Mueller, Luke Copland, and Martin O. Jeffries

Abstract The ice shelves along the northern coast of Ellesmere Island have been in a state of decline since at least the early twentieth century. Available data derived from explorers' journals, aerial photographs and satellite imagery have been compiled into a single geospatial database of ice shelf and glacier ice tongue extent over 13 observation periods between 1906 and 2015. During this time there was a loss of 8,061 km^2 (94%) in ice shelf area. The vast majority of this loss occurred via episodic calving, in particular during the first six decades of the twentieth century. More recently, between 1998 and 2015, 515 km^2 of shelf ice calved. Some ice shelves also thinned *in situ,* transitioning to thinner and weaker ice types that can no longer be considered ice shelf, although the timing of this shift is difficult to constrain with the methods used here. Some ice shelves composed partly of ice tongues (glacier or composite ice shelves) also disintegrated to the point where the ice tongues were isolated, representing a loss of ice shelf extent. Our digitization methods are typically repeatable to within 3%, and generally agree with past determinations of extent. The break-up of these massive features is an ongoing phenomenon, and it is hoped that the comprehensive dataset presented here will provide a basis for comparison of future changes in this region.

Keywords Ice shelf • Ice tongue • Break-up • Calving • Climate change • Change detection • Remote sensing • Geographic Information System (GIS) • Arctic

D. Mueller (✉)
Department of Geography and Environmental Studies, Carleton University, Ottawa, ON, Canada
e-mail: derek.mueller@carleton.ca

L. Copland
Department of Geography, Environment and Geomatics, University of Ottawa, Ottawa, ON, Canada
e-mail: luke.copland@uottawa.ca

M.O. Jeffries
Office of Naval Research, Arctic and Global Prediction Program, Arlington, VA, USA
e-mail: martin.jeffries@navy.mil

L. Copland, D. Mueller (eds.), *Arctic Ice Shelves and Ice Islands*, Springer Polar Sciences, DOI 10.1007/978-94-024-1101-0_5

5.1 Introduction

Historical losses of the Ellesmere Island ice shelves in Nunavut, Canada, have been documented in many previous articles and reports. Most of these are concerned with a particular calving event, or events, between sporadic field or remote sensing observations (Hattersley-Smith 1963, 1967; Jeffries 1982, 1986a, 1992b; Jeffries and Serson 1983; Jeffries and Sackinger 1991; Vincent et al. 2001, 2009, 2011; Mueller et al. 2003; Copland et al. 2007). Some have focused on the changes that have occurred to a particular ice shelf over time (Copland et al. 2007; Pope et al. 2012; White et al. 2015), while others have described the extent of all the Ellesmere ice shelves at once (Mueller et al. 2006). In addition, some articles have calculated the past extent of all ice shelves and compared them to various points in history (Vincent et al. 2001) or examined the extent lost during calving events (Jeffries 1987, 1992a). Despite this work, there has not yet been a comprehensive and rigorous analysis of the change in all the Ellesmere Island ice shelf extents over the entire record of available satellite remote sensing, aerial photograph and anecdotal data.

The ice shelves along the northern coast of Ellesmere Island were formed by a combination of multiyear landfast sea ice accretion and glacier input followed by thickening via snow accumulation (Dowdeswell and Jeffries 2017). The ice shelves are thought to be several thousand years old (Antoniades 2017; England et al. 2017) and are characterized by an undulating surface (Jeffries 2017). Multiyear landfast sea ice may also have these surface features, but with a closer spacing, which is thought to relate the thinner nature of this less-developed ice type. Ice shelves were known to occupy the mouths of embayments and fiords along the northern coast of Ellesmere Island. This often leaves an area of relatively thin freshwater ice between the head of the fiord and the ice shelf. This situation, known as an epishelf lake, is caused by the impoundment of a fresh meltwater layer by the ice shelf that lies atop denser seawater beneath (Veillette et al. 2008; Jungblut et al. 2017). Over the last decade there has been a dramatic loss of ice shelves, multiyear sea ice and epishelf lakes (Mueller et al. 2003; Copland et al. 2007; Veillette et al. 2008; Pope et al. 2012, 2017; White et al. 2015).

The purpose of this chapter is to review available information on the extent of the Ellesmere ice shelves and to refine the timing of calving events between 1906 and 2015. We lay out a series of methodological principles for quantifying ice shelf extent and change, and use geospatial data and Geographic Information System (GIS) techniques to accomplish this. This chapter also discusses issues related to the available data and their interpretation. The underlying data compiled for this project include maps generated from explorers' journals, maps created from air photographs, as well as optical and radar remote sensing imagery of various resolutions and polarizations. The temporal resolution of observations increases over time and increasingly over the past two decades it has been possible to obtain excellent satellite coverage of the northern coast of Ellesmere Island on a regular basis. The limitations of each method and the challenges of combining them are evaluated.

This chapter provides a description of temporal changes in ice shelf extent, but does not attempt a detailed explanation of their causes as this is covered by other chapters in this book (e.g., Copland et al. 2017). The geospatial data that were produced from the analyses conducted for this chapter are published separately (Mueller et al. 2017) and the metadata is accessible online (http://polardata.ca/12721), so that others can study and improve on them, but the underlying imagery remain with the copyright holder or, in some cases, on separate online repositories.

5.2 Methods

The collection of data proceeded as follows: each data source (Sect. 5.2.1; Table 5.1) was converted to an electronic format by scanning, if it was not already available as a digital image. Data sources (Sect. 5.2.1) were georeferenced or a correction to the existing georeferencing was applied (Sect. 5.2.2). Digitizing of each feature followed a protocol (Sect. 5.2.3), and data were cleaned and verified (Sect. 5.2.4). Following data entry, data display and analysis were conducted.

Table 5.1 Data sources by observation year and season

Year	Season	Sources	Data type	Notes
1906	Summer	Vincent et al. (2001), Bushnell (1956)	Anecdotal/survey	
1959	Summer	NTS maps; Aerial photos	Aerial photo/map	
1963	Summer	Corona	Optical satellite	Snow-covered, difficult to discern ice types
1988	Summer	SPOT-1	Optical satellite	Missing east of central Ward Hunt Ice Shelf
1992	Winter	ERS-1	SAR satellite	Missing east of western Ward Hunt Ice Shelf
1998	Winter	RADARSAT-1	SAR satellite	Hard to see fragments in Ayles Fiord
2003	Winter	RADARSAT-1	SAR satellite	
2006	Winter	RADARSAT-1	SAR satellite	
2009	Winter	RADARSAT-1	SAR satellite	
2011	Winter	RADARSAT-2	SAR satellite	
2012	Winter	RADARSAT-2	SAR satellite	
2013	Winter	RADARSAT-2	SAR satellite	
2015	Winter	RADARSAT-2	SAR satellite	

5.2.1 Data Sources

Four main data types were used to examine ice shelf extent along the northern coast of Ellesmere Island: anecdotal reports and survey data; aerial photographs and published maps; optical satellite remote sensing and synthetic aperture radar (SAR) satellite remote sensing. These vary substantially with respect to accuracy, as well as spatial and temporal coverage. As described below, we used each of the data types in succession and prioritized observation years that had the greatest spatial coverage. This meant that some datasets were not incorporated into the analysis due to restricted spatial coverage (e.g., aerial photographs from 1974 and 1987, and airborne real aperture radar from the 1980s) or, in later years, because of the availability of imagery with a high temporal frequency that did not show significant changes between scenes. When possible, imagery with the highest spatial resolution and accuracy was prioritized and observation years were selected to bracket major calving events.

5.2.1.1 Anecdotal/Survey

The first documented visit to the 'Ellesmere Ice Shelf' (unofficial name) was in 1876 by a sledge party led by Lieutenant Pelham Aldrich as part of the British Arctic Expedition (Aldrich 1877; for selected quotes see Copland et al. 2017). Aldrich described travelling over ice shelf ice, but a subsequent visit by Commander Robert Peary in 1906 yielded more precise descriptions of the ice type and he explored further along the coast than his predecessor (Peary 1907; for selected quotes see Jeffries 2017). Therefore, anecdotal evidence in Peary's journal was used to map the extent of the ice shelf from Cape Hecla to the northern end of Axel Heiberg Island (Fig. 5.1) by Vincent et al. (2001). This estimation assumed that the northern edge of the ice shelf went from headland to headland and that shelf ice filled the outer portion of bays and fiords along the coast. From Cape Richards to Point Moss (adjacent to ice shelves that bear these names; Fig. 5.1), the northern limit of the ice shelf was determined by the position of ocean depth soundings by Peary's compatriot, Ross G. Marvin. He travelled along the edge of the ice shelf and his sightings of mountains along the coast were used to fix his position by triangulation (Bushnell 1956). This information was also incorporated by Vincent et al. (2001), and Fig. 2 in Bushnell (1956) was used as a data source for the ice shelf front in this study.

5.2.1.2 Aerial Photographs/Maps

In 1959 and 1960, the first complete vertical (nadir) aerial photography of the region was conducted by the Royal Canadian Air Force and topographic maps were published based on photogrammetry of this data source. Scans of the printed 1:250,000

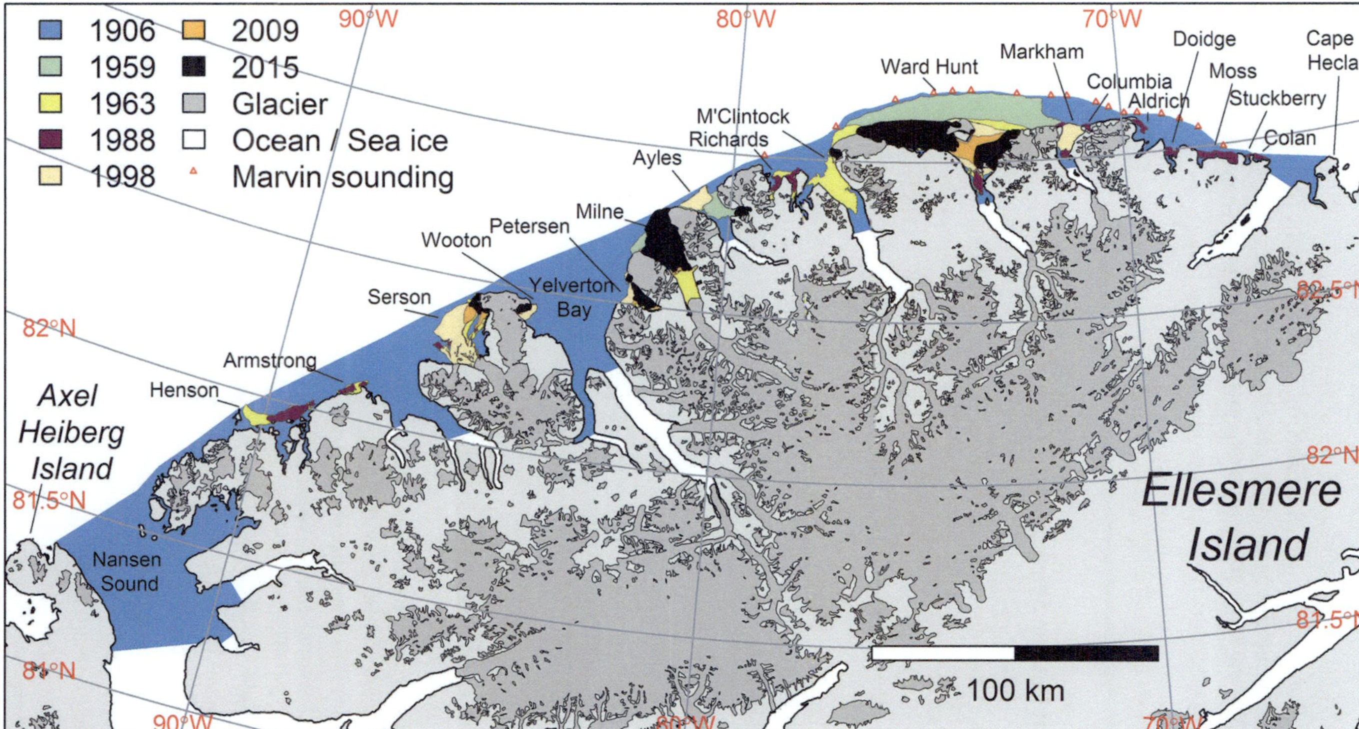

Fig. 5.1 Ice shelf extent for selected years from 1906 to 2015 in the study area. The ice shelves from the mid-twentieth century are labeled from west to east: Henson*, Armstrong*, Serson†‡, Wooton§, Petersen§, Milne†, Ayles†, Richards*, M'Clintock†, Ward Hunt†, Markham§, Columbia*, Aldrich*, Doidge*, Moss*, Stuckberry* and Colan* ice shelves (†official name; ‡formerly-known as Alfred Ernest Ice Shelf; §named unofficially in previous literature; *named unofficially for the first time herein). The colours indicate the year when each extent was last observed, but note that the extent from more recent years cover the extent of previous years. The 1906 extent (all *non-grey* and *white regions*) is the Ellesmere Ice Shelf. A portion of the northern edge of this large ice shelf was surveyed by Ross Marvin (*red triangles*; Bushnell 1956). The most recent extent is represented by *black*

scale National Topographic System (NTS) map sheets (120F, 340E, 340F, 560D) were used in this study, since the ice shelf feature type is not available in the published digital versions of these maps. One of these maps (340F) was erroneously constructed with some earlier aerial (trimetrogon) photographs that were acquired at select locations along the coast. As a consequence, the extent of the Serson Ice Shelf on the map reflects its extent in 1950 (Jeffries 1992b). We created a mosaic from the aerial photographs from the National Air Photo Library (Ottawa, Canada) that should have been used to create the map. This layer supplemented the map to ensure that the 1959 ice shelf extent was captured as accurately as possible.

5.2.1.3 Optical Remote Sensing

We obtained a declassified image of the northern coast of Ellesmere Island taken in 1963 by the U.S. reconnaissance satellite Corona. The Corona system involved taking photographs from low Earth orbit with 70 mm film and then releasing the undeveloped negatives to re-enter the Earth's atmosphere in a capsule for airborne retrieval (Ruffner 1995). The photographs were scanned at high resolution (90 m pixel spacing) and released to the public in 2004. A series of ten Level 2A, panchromatic SPOT 1 (Satellite Pour l'Observation de la Terre) images, covering the northern coast of Ellesmere Island in 1988, was obtained in 2014 from SPOT Image Corporation.

5.2.1.4 Synthetic Aperture Radar

From 1992 to 2015, we relied on spaceborne synthetic aperture radar (SAR) imagery. SAR is routinely used for ice discrimination and offers several advantages over optical imagery. SAR can operate through the polar night, can penetrate clouds and snow cover and is available on polar-orbiting satellites, which offer excellent acquisition possibilities (e.g., Flett 2003; De Abreu et al. 2011). Once SAR imagery was available from 1992 onwards, this image type was used exclusively to avoid mixing image types and any associated biases which can arise due to ice type discrimination. Numerous scenes were provided to us via data grants, through government collaboration and public repositories, and via organizations such as the Canadian Ice Service and Canadian Space Agency. SAR imagery from mid-winter typically provides the best contrast between ice types due to the absence of liquid water, so use of this imagery was preferred (Mueller et al. 2006; De Abreu et al. 2011). Following melt onset, the SAR backscatter signal is reduced due to absorption of microwave energy by liquid water in the overlying snow cover or on the ice surface itself, which makes it more difficult to distinguish ice types (Onstott and Shuchman 2004; Shokr and Sinha 2015). A collection of available SAR images was searched to identify candidate images that covered the ice shelves within the study area during winter (January to April). Years with complete or near complete coverage of all the ice shelves were prioritized, as were images with the highest possible spatial resolution and the most image bands.

SAR data included imagery from C-band (frequency: ~5.4 GHz, wavelength ~5.6 cm) satellites ERS-1 (European Remote-Sensing Satellite-1), RADARSAT-1 and RADARSAT-2 at a variety of spatial resolutions and polarizations (Table 5.2). The SAR scenes were provided at a smaller pixel spacing than their spatial resolution due to oversampling (Table 5.2). ERS-1 transmitted radar waves with vertical polarization and its receive antenna was likewise sensitive to vertically polarized backscatter (VV polarization). RADARSAT-1 transmitted and received in horizontal polarization (HH), whereas RADARSAT-2 is capable of transmitting and receiving in both H and V and can produce separate bands including co-polarized SAR (HH or VV), cross-polarized SAR (HV or VH) or quad-polarized SAR (HH, VV, HV and VH). Single band SAR images were analysed in grey scale, dual band images were false-coloured (Red: HH, Green and Blue: HV) prior to analysis, and quad-polarized images were converted to a Pauli Decomposition (Red: HH − VV, Green: (HV + VH)/2, Blue: HH + VV), which can be interpreted as the contribution made by the ice targets to single-bounce, volume and double-bounce scattering mechanisms, respectively (Lee 2009).

5.2.2 *Georeferencing and Projections*

To minimize distortion over the study area we used an Albers equal area projection based on the WGS-84 ellipsoid with standard parallel latitudes at 82°N and 83°N and an origin at 82°N, 75°W (http://spatialreference.org/ref/sr-org/7968/). A vector coastline layer was derived from the water body layer provided in the 1:250,000 National Topographic Data Base (NTDB) dataset for NTS map sheets 120F, 340E, 340F, 560D published in May 2009 (with a stated positional accuracy of 58–68 m). The NTDB data set was created from the original NTS map sheets (based on air photographs from 1959) and updated with Landsat and RADARSAT-1 imagery until 2004. This coastline served as a fixed reference throughout the project. Each satellite image was examined at a scale of 1:40,000 against this coastline to assess the quality of the original georeferencing. If the image deviated from the coastline, it was translated by an x- and y-offset to visually optimize a fit. This method was adequate to correct the georeferenced satellite imagery because the ice features are essentially at sea level and orthorectification issues can be ignored. In addition, shifting of the scenes meant that there was no need to further resample the imagery.

Depending on the incidence angle of the SAR images, there were substantial local effects from shadow or layover (an increased amount of backscatter that is a local artifact in front of steep slopes). Following image translation, deviation from the coast in areas where this could be assessed with confidence was at most 150 m (typically in the low resolution ScanSAR imagery; representing 2–3 pixels). For data acquired before the 1980s, a more sophisticated georeferencing method was employed, which used several control points along the coastline vector layer described above and an affine or polynomial transformation. For example, sections

Table 5.2 Data sources used to digitize ice shelves, ice tongues and fragments. Where more than one image/map was available on a single date, the number of images used is indicated by a superscript after the beam/mode

Platform	Date/time	Beam/mode/frame	Resolution (m)	Source
Survey	Summer 1906	NA	NA	Bushnell (1956)
Anecdotal	Summer 1906	NA	500[a]	Vincent et al. (2001)
Map	1959-08-18	Map[4]	50[a]	NTS
Aerial photo	1959-07-29	A16760-90	2	NAPL
	1959-08-18	A16760-91	2	NAPL
	1959-08-18	A16760-92	2	NAPL
	1959-08-18	A16841-19	2	NAPL
	1959-08-18	A16841-21	2	NAPL
Corona	1963-08-29	Corona	110	USGS
SPOT-1	1987-08-12	Beam 1[5]	10	SPOT Image Corp.
ERS-1	1992-01-24	Standard	27	ASF
	1992-01-29	Standard	27	ASF
	1992-03-16	Standard	27	ASF
	1992-03-23	Standard	27	ASF
Radarsat-1	1998-01-13	ScanSAR	100	ASF
	1998-04-19	ScanSAR	100	ASF
	2003-01-07	Standard-1	27	ASF
	2003-01-11	ScanSAR	100	ASF
	2006-01-13	Standard-1	27	ASF
	2006-01-14	Standard-1	27	ASF
	2006-01-15	Standard-1	27	ASF
	2006-01-16	Standard-1	27	ASF
Radarsat-2	2009-01-04	ScanSAR	160	ASF
	2009-04-26	ScanSAR	160	ASF
	2011-01-01	ScanSAR	160	SOAR-E
	2011-01-03	ScanSAR	160	SOAR-E
	2011-02-28	Fine 21f	10	SOAR-E
	2011-03-19	Fine 2n	10	SOAR-E
	2011-03-23	Fine 5n	10	SOAR-E
	2011-04-25	Ultra-Fine 78	3	SOAR-E
	2012-02-03	Ultra-Fine 3w[2]	3	SOAR-E
	2012-02-05	ScanSAR	160	CIS
	2012-02-08	Ultra-Fine 10w[2]	3	SOAR-E
	2012-02-19	ScanSAR	160	CIS
	2012-03-22	Ultra-Fine 3w[2]	3	SOAR-E
	2012-04-17	Fine Quad 19w	8	SOAR-E
	2012-04-18	Fine Quad 21w	8	SOAR-E
	2013-03-12	ScanSAR	160	CIS
	2013-03-16	ScanSAR	160	CIS
	2013-04-27	Fine Quad 13w[2]	8	CIS

(continued)

Table 5.2 (continued)

Platform	Date/time	Beam/mode/frame	Resolution (m)	Source
	2013-04-29	Fine Quad 10w^{2}	8	CIS
	2013-04-30	Fine Quad 11w	8	CIS
	2015-03-22	ScanSAR	160	CIS
	2015-03-23	Fine Quad 5w	8	CIS
	2015-03-23	Fine Quad 16w	8	CIS
	2015-03-24	Fine Quad 9w	8	CIS
	2015-03-26	Fine Quad 10w^{2}	8	CIS
	2015-03-27	Fine Quad 15w^{2}	8	CIS

Sources include: National Topographic Service of Canada (NTS), National Air Photo Library (NAPL), United States Geological Survey (USGS), Alaska Satellite Facility (ASF), Canadian Space Agency's Science and Operational Applications Research – Education (SOAR-E) program and the Canadian Ice Service (CIS)

[a]These resolutions were set arbitrarily to provide a comparable error estimate to other data sources. See text for further details

of the 1:250,000 scale map sheets were georeferenced to a positional root mean square error (RMSE) of 64 m or better and the aerial photographs used for the Serson Ice Shelf had an RMSE of 71 m. The Corona image was georeferenced with 26 control points to yield a RMSE of 1331 m. Ross Marvin's soundings and the ice shelf front in 1906 from Bushnell (1956) had an RMSE of 670 m after correction using 31 control points.

5.2.3 *Ice Feature Identification*

The study area was defined as the region covered by ice shelves in 1906 as determined by Vincent et al. (2001; Fig. 5.1). For each observation year listed in Table 5.1, the extent of ice features was digitized offshore of the coastline vector layer. Identification and delineation of the ice types was performed at a scale of 1:25,000 using the visual cues provided by the data at hand, reports from the literature, personal communication with other experts and field knowledge. The following ice features were identified for each image date since they represent the primary forms of thick floating ice in the study area and are reasonably easy to distinguish:

Ice Shelf Ice shelves are defined as thick (≥20 m) and extensive landfast ice features (Dowdeswell and Jeffries 2017). Note that the thickness criterion cannot be directly evaluated with the imagery we used, although the undulating surface morphology common to all Ellesmere ice shelves (Fig. 5.2; Hattersley-Smith 1957; WMO 1970; Jeffries 2002) and field observations (if available) made them readily distinguishable from other ice types (e.g., glacier ice, multiyear landfast sea ice, first year sea ice and freshwater ice). Since we are concerned with an accounting of the changes in ice shelf extent over time we do not set a lower limit on the extent of an ice shelf.

Fig. 5.2 Recent oblique photographs of selected ice features. (**a**) The easternmost portion of the Ward Hunt Ice Shelf south of Ward Hunt Island and ice rise. The distance to the coast of Ward Hunt Island is approximately 5 km. (18 July 2015, courtesy of A Culley); (**b**) Ward Hunt East Ice Shelf looking to the east. The distance from the ice shelf edge to the coast is approximately 8 km (18 July

We do not distinguish between glacial, sea-ice and composite ice shelves, the three main types of Ellesmere ice shelves defined by Lemmen et al. (1988). Grounded sections of ice shelves, known as ice rises, were excluded from the ice shelf extent. These were delineated using the coastline vector layer described above. This represents a 'best guess' of the true grounding line and in some cases the coastline was modified based on calving front retreat (see below).

Ice Tongue This is defined as the floating extension of a single valley glacier over the ocean. Ice tongues were considered distinct from ice shelves if they simply abutted against them. If they were incorporated into ice shelves and provided a source for them, this ice was considered to be ice shelf (e.g., northern margins of Petersen Ice Shelf; White et al. 2015). Glacier ice typically has characteristic surface flow features (e.g., medial moraines) and typically high-freeboard edges that are apparent in the imagery we analyzed. Only ice tongues within the study area that were ~>2 km wide and had (at least at one point in time) an apparently extensive floating section were digitized and our analysis focuses on ice tongues that were at one time associated with ice shelves.

Ice Shelf/Tongue Fragments These are relatively small (typically <1 km^2) free-floating ice features that broke away from ice shelves and icc tongues but remained within the bays and fiords that contain parent ice masses. These are essentially ice islands that have not drifted away from the study area. Their provenance is not always apparent and was not tracked.

Several principles were applied to maintain a consistent definition of the ice types and their outlines between data sources:

(a) The coverage of each ice type was mutually exclusive, so that no two ice types occupy the same space at the same time.
(b) Ice features were considered to be distinct when it was clear that they were separated by a substantial distance (at least 50 m) from each other following calving. Break-up (*in situ* fracturing) of ice features was not recorded explicitly due to the lack of sufficient remotely sensed data. When ice shelves disintegrated into more than one ice shelf, the largest portion retained the original name, smaller ice shelves that did not already have a name were named after their position (e.g., Ward Hunt East Ice Shelf). Non-landfast pieces that calved from ice shelves and ice tongues were considered to be ice shelf/tongue fragments.

Fig. 5.2 (continued) 2015, courtesy of WF Vincent); (**c**) The western side of the Petersen Ice Shelf (29 June 2012); (**d**) The inner portion of Petersen Bay showing the Petersen Ice Shelf in the background and fragments from the Petersen South Glacier ice tongue in the foreground. The island in the centre of the bay is 1.5 km long (29 June 2012); (**e**) Wooton (foreground) and Wooton East (behind) ice shelves with Yelverton Bay in the background. For scale, the undulating ice shelf ice in the photo is approximately 3 km along the line of sight (29 June 2012); (**f**) The Serson Glacier ice tongue with fragments in the foreground. The ice tongue is 7 km long (13 July 2015); (**g**) Panorama of the Milne Ice Shelf looking east from the west side of the fiord. The tributary glacier in the foreground merges with the Central Unit of the ice shelf. The fiord is 6.5 km wide at this point (17 July 2014); (**h**) Panorama of the Milne Glacier ice tongue looking west near the grounding line. The fiord is 5 km wide at this point and the detached portion of the ice tongue is 14.8 km long. (16 July 2014)

(c) In observation years when there was no available imagery or when ice type delineation was not possible for a given ice feature (e.g., a snow-covered optical image or poor quality SAR image), the feature was assumed to persist intact. In this case, the extent digitized from the previous observation was copied to the observation year in question. If, however, there was no previous observation or there was evidence that a given calving event occurred prior to or after the missing observation year, then the extent digitized from the closest available observation year was copied to the observation year with missing data. In 1906, due to lack of evidence to the contrary, all ice tongues were assumed to be fully merged with the ice shelf.
(d) The image used to digitize each feature was recorded in the metadata. In the case where two images were used in a digitization, due to incomplete coverage or for supplementary ice interpretation, the reference images were recorded in order of importance (i.e., the image that covered most of the feature or, in the case of a change in extent, the imagery that indicated that change, when part of the feature was missing).

The 2009 coastline vector consistently deviated from the coastline observed in imagery in a few places. This was due to errors introduced in the generation of the original map (and the NTDB data set that the map was based on), as well as the loss of ice rise and glacier ice after 1959. In these cases, it was necessary to delineate ice types on the landward side of the 2009 coastline. In the case of undefined coastlines such as at the calving front of ice tongues, the minimum extent of the ice cover over the study period was taken as the coastline.

5.2.4 Quality Control

Following digitization, the data were checked to make sure that no ice types overlapped, that each ice type was properly identified, and that no slivers or gaps were found in the coverage. The area of each polygon was calculated and each ice shelf and ice tongue was assigned an object identification number that was coded to reveal the relationship between ice features as they calved into different pieces.

The error associated with each observation was determined in two ways. First by considering the resolution of the image used (Ghilani 2000):

$$\sigma_{area} = D\sigma_D \sqrt{2} \tag{5.1}$$

where the uncertainty (σ_{area}) is computed from the equivalent-area square (D), taken as the square root of the area of the polygon, multiplied by the uncertainty in the image (σ_D) or half the spatial resolution. This method has been used by other glaciologists (e.g., Hoffman et al. 2007; Crawford 2013) but it does not account for errors related to delineating ice types or operator error (Crawford 2013; c.f. White and Copland 2015), which were quantified by using four test images (SPOT-1, RADARSAT-1 ScanSAR, RADARSAT-1 Standard Beam and RADARSAT-2 Fine Quad) containing ice shelves ranging in area from 8.8 to 73 km^2. These ice shelves were digitized five times on separate occasions to obtain a coefficient of variation that served as a proxy for digitization precision (Paul et al. 2013).

The error estimation using the first method (Ghilani 2000) yielded a relative error of 0.07% for Ward Hunt East Ice Shelf in 2015 using a RADARSAT-2 Fine Quad Beam image, 0.24% for M'Clintock Ice Shelf in 1988 using a SPOT-1 image, 0.29% for Markham Ice Shelf in 2006 using a RADARSAT-1 Standard Beam image, and 1.1% for the same ice shelf but at ScanSAR resolution. The repeat digitization method gave coefficients of variation of 0.04, 0.59, 1.29 and 2.8%, respectively, for these ice shelves. The last image, a ScanSAR RADARSAT-1 image of the Markham Ice Shelf in 2006, had an extent that was on average 6.3% larger than the Standard Beam image, which is a significant difference ($p = 0.001$). This indicates that lower resolution data yield less accurate results but, in this case, the digitization is repeatable within 3%. We found that the relative error derived from the Ghilani (2000) equation was related to the coefficient of variation given by repeat digitization (Adjusted R^2: 0.91, $p = 0.03$). Therefore, to estimate the repeatability error (typically the largest error), we multiplied the result of Eq. 5.1 by a factor of 2.54 and added an offset of 0.1%, which yielded an RMSE of 0.26%. For the purposes of generating an error estimate, the 'resolution' of the anecdotal information in 1906 was arbitrarily set at 500 m and the resolution of the scanned maps from 1959 was set at 50 m. All other image resolutions were either the pixel size (optical imagery) or published worst-case resolution for SAR imagery (Table 5.2).

5.3 Results

5.3.1 Overview

Over 2000 ice shelf, ice tongue and ice fragment polygons were digitized over the thirteen observation periods. The original contiguous 'Ellesmere Ice Shelf' in 1906, as recreated from Vincent et al. (2001), was 8597 km^2 in area (Figs. 5.1 and 5.3a; Table 5.3). In 1959 there were 15 ice shelves with a total area of 2168 km^2. Then, by 2015 ice shelf extent had decreased by 75% to 535 km^2 (Table 5.3) distributed among 13 ice shelves (see Fig. 5.2 for current photographs). As ice shelves calved and ice islands moved away from the coast, some feeder glaciers that once merged with ice shelf ice were reclassified from ice shelf to ice tongue (Fig. 5.3b; Table 5.3). This partly offset the loss of ice tongues due to calving. In 1959, there were eight ice tongues with a total area of 94 km^2. In 2015, there were nine ice tongues with a total area of 23 km^2 (Table 5.3).

The break-up of ice shelves and glaciers resulted in the production of fragments and ice islands. Most were advected from the study area by a combination of pack ice drift, wind and currents (Sackinger et al. 1985; Jeffries 1992a; Jeffries and Shaw 1993), yet many fragments remained in the fiords and bays. They totalled 9 km^2 in area in 1959 and 80 km^2 (mostly in Milne Fiord) in 2015, but with a wide variability

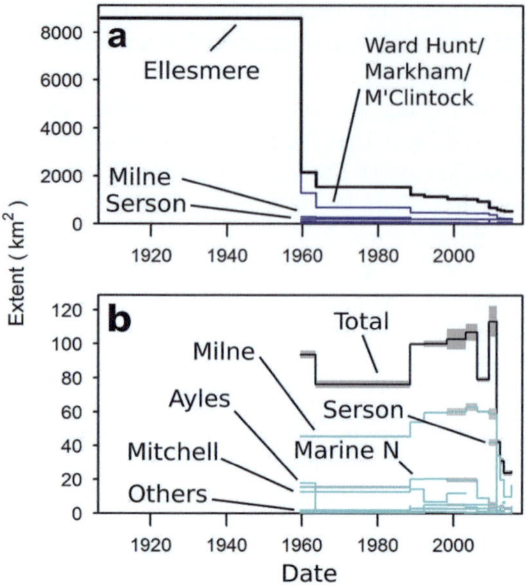

Fig. 5.3 Total and individual extents for (**a**) ice shelves and (**b**) ice tongues. The *black line* is the total extent, the thinner coloured lines indicate the extent of individual ice shelves (*dark blue*) or ice tongues (*cyan*). *Grey shading* is the uncertainty in the extent. At this scale, only the large ice shelves are visible. Prominent ice tongues include the Milne, Ayles, Richards North and Fanshawe glaciers (in order of extent). Note that changes in extent are registered only at each of the observation years and do not reflect precise timing of calving events. For example, much of the large 'Ellesmere Ice Shelf' likely calved away in the 1930s and 40s, producing large ice islands that were discovered in the 1940s (Belkin and Kessel 2017; Dowdeswell and Jeffries 2017)

Table 5.3 Extent of ice types by year for the northern coast of Ellesmere Island

	Ice shelves		Ice tongues		Fragments	
Year	Area (km^2)	Error (km^2)	Area (km^2)	Error (km^2)	Area (km^2)	Error (km^2)
1906	8597	94.9	–	–	–	–
1959	2168	15.3	94	2.3	9	0.8
1963	1549	20.2	76	2.1	13	1.2
1988	1222	6.4	100	0.7	58	1.6
1992	1143	7.1	100	1.9	49	3.5
1998	1050	18.1	103	6.2	27	3.2
2003	1037	12.6	107	5.2	27	1.7
2006	947	5.1	79	1.5	20	0.7
2009	680	19.6	113	9.2	1	0.6
2011	605	2.5	42	1.5	97	4.9
2012	557	3.2	31	1.8	85	1.1
2013	537	3.4	24	1.7	79	1.7
2015	535	2.2	25	0.2	80	1.3

between years (Table 5.3). The disintegration of ice shelves and ice tongues into fragments as well as their thinning is described below. The following sections give details on each ice shelf that was present in 1959 in order of extent.

5.3.2 M'Clintock/Ward Hunt/Markham Ice Shelf

In 1959, the largest remnant of the former Ellesmere Ice Shelf was the contiguous M'Clintock/Ward Hunt/Markham Ice Shelf. Its change in extent since then is documented in Figs. 5.4a and 5.5 as well as Table 5.4. This ice shelf disintegrated over the next several years, losing 598.7 km^2 in a calving event which created five large ice islands between August 1961 and April 1962, and caused the separation of Ward Hunt/M'Clintock and Markham ice shelves (Hattersley-Smith 1963). The M'Clintock Ice Shelf broke up between August 1963 (Corona imagery, this study) and April 1966 (Hattersley-Smith 1967) and became detached from the Ward Hunt Ice Shelf at that time or soon after (Serson 1983). Between spring 1980 and 1982 a 40 km^2 loss was estimated near Discovery Ice Rise (Jeffries 1982).

The Ward Hunt Ice Shelf shrank again between June 1982 and April 1983, causing the calving of another ~40 km^2 of ice islands, including Hobson's Choice Ice Island (Jeffries and Serson 1983; Hobson 1989). No further calving was reported until 2002 when the Ward Hunt Ice Shelf calved ~6 km^2 of ice, and extensive fracturing resulted in drainage of the epishelf lake in Disraeli Fiord (Mueller et al. 2003). In February 2008, extensive cracks were found over large areas of the eastern part of the Ward Hunt Ice Shelf during an airborne survey and calving occurred that summer (Mueller et al. 2008; Derksen et al. 2012). This isolated two small parts of the ice shelf (Ward Hunt Northwest and North) on the northwest and east side of the Ward Hunt Ice Rise. In August 2010, there was a further calving from the Ward Hunt Ice Shelf to the south and east of Ward Hunt Island that created Ward Hunt East Ice Shelf and left large ice fragments between the two ice shelves (Vincent et al. 2011). This central area disintegrated in 2011 and 2012 along with portions of the Ward Hunt Ice Shelf (Figs. 5.2a, 5.4a and 5.5a).

There is little information about Markham Ice Shelf from 1963 until 1998 since, unlike most of the other ice shelves, no imagery was available in 1988 and 1992. From SAR imagery in the late 1990s it appeared that the ice shelf grew laterally to fill most of the width of the fiord (Figs. 5.4e and 5.5a; Table 5.4). However, the southern edge of the ice shelf moved north over that period. This may have been due to a calving event since some standard beam RADARSAT-1 imagery from 2000 shows what appears to be an ice fragment in the fiord. The southern and western edge of the ice shelf was poorly resolved, particularly in ScanSAR imagery, and variation in ice shelf size may be the result of image interpretation issues. A 2.4 km^2 calving event along the northern edge of the ice shelf was evident after comparing imagery from 1998 and 2003 (Fig. 5.5a). An examination of ancillary SAR imagery indicated that this calving occurred in 2000. On August 6, 2008, the ice shelf broke up into two main pieces and calved away completely. This event was captured by Moderate Resolution Imaging Spectroradiometer (MODIS) imagery (Mueller et al. 2008).

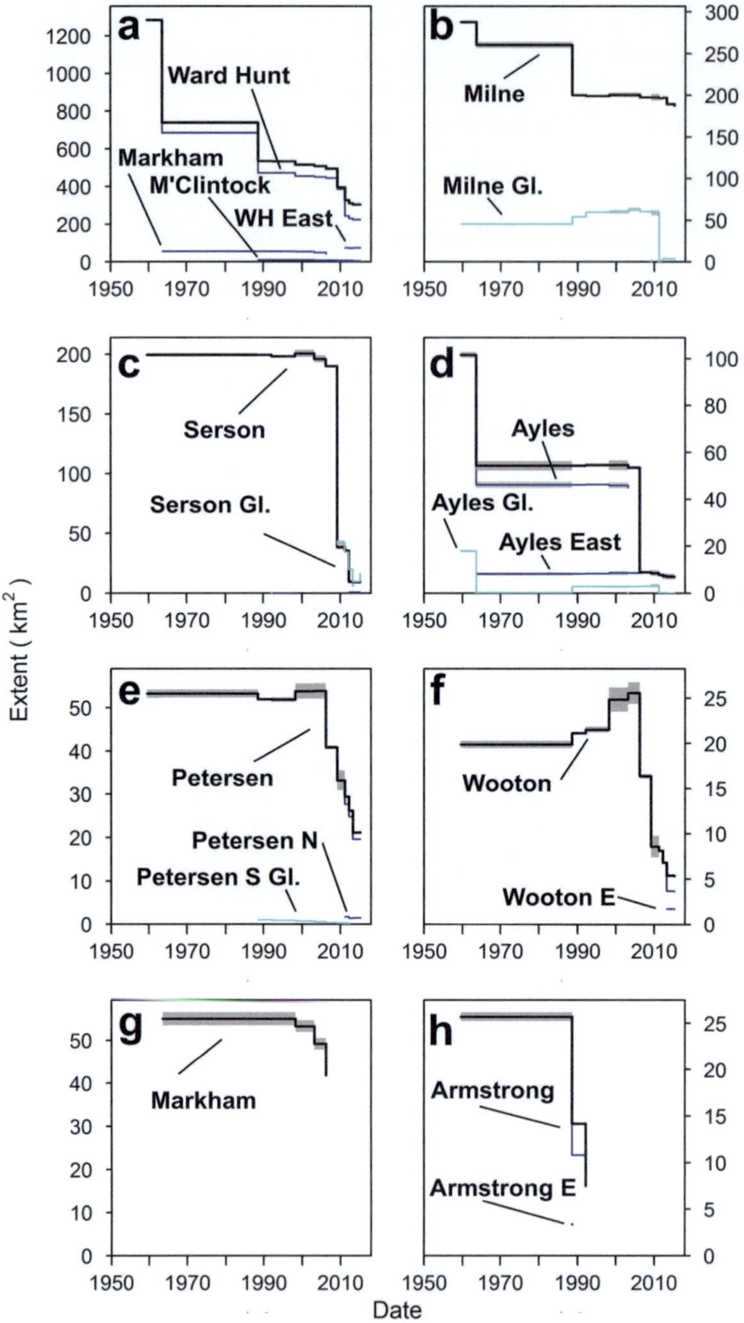

Fig. 5.4 Extents for the following ice shelves and associated ice tongues: (**a**) M'Clintock/Ward Hunt/Markham, (**b**) Milne, (**c**) Serson, (**d**) Ayles, (**e**) Petersen, (**f**) Wooton, (**g**) Markham, and (**h**) Armstrong. The *thick black line* is the total extent, the *thinner coloured lines* indicate the extent of individual ice shelves (*dark blue*) and ice tongues (*cyan*) that were created in the disintegration process. *Grey shading* is the uncertainty in the extent. Note that changes in extent are registered only at each of the observation years and do not reflect precise timing of calving events

5.3.3 *Milne Ice Shelf*

In 1959, Milne Fiord was nearly completely covered in thick ice shelf ice from the ice tongue at the head of the fiord to the northern edge of the Milne Ice Shelf at the mouth of the fiord (Fig. 5.6a). A 27 km^2 calving event at the northern edge of the Milne Ice Shelf (Jeffries 1986a) occurred between 1959 and 1963 as evidenced by the Corona image used in this study (Figs. 5.4b and 5.6a; Table 5.4). Following that calving, little change occurred at the seaward (northern) edge of the ice shelf, but the landward (southern) side changed substantially (Fig. 5.6a).

Jeffries (1986b) defined three regions on the Milne Ice Shelf: the Outer Unit (northernmost third), the Central Unit and the Inner Unit (which abutted against the floating terminus of the Milne Glacier). In 1984, Jeffries (1985) measured the thickness of ice in the Inner Unit to be 3.19 m, which corroborated a thickness of <10 m determined by airborne radio-echo sounding (Prager 1983; Narod et al. 1988). This suggests that the Inner Unit was no longer shelf ice at that time (Mortimer et al. 2012). The area occupied by the Inner Unit (from the 1980s on) is now considered to be an epishelf lake (Mueller et al. 2006; Veillette et al. 2008; Mortimer et al. 2012) with an ice thickness of <2 m measured annually by the authors (Mueller and Copland) between 2008 and 2015. The Central Unit of the Milne Ice Shelf (Fig. 5.2g) also showed regions of smooth, featureless ice which were interpreted as 'internal' epishelf lakes starting in 1988 (Fig. 5.6a). These regions of lake ice became even more apparent in SAR imagery owing to their bright radar return (c.f., Jeffries 2002; Veillette et al. 2008; White et al. 2015; Fig. 5.6b), and grew over time in spite of their variable appearance in imagery at times.

Between 2008 (*in situ* observation; Mortimer et al. 2012) and 2009, a straight crack formed across the Central Unit of the ice shelf. This was followed by calving at the southern edge of the ice shelf in late August 2012, when the northwestern third of the epishelf lake ice broke up. There was also extensive break-up within the ice shelf; the southeastern portion of the ice shelf disintegrated in place leaving many small fragments that were not captured by the methods outlined in this study since they did not drift apart appreciably. During 2012–13, a 3 km^2 portion of ice shelf also calved away near Cape Evans (Figs. 5.4b and 5.6a).

The Milne Glacier tongue also changed considerably over the period 1959–2015 (Figs. 5.4b and 5.6a; Table 5.4). In 1959 the ice tongue was flanked by epishelf lakes on both sides. These increased in size and the glacier advanced (Jeffries 1984; Mortimer 2011) between then and 1988. The ice tongue advanced even further during 1988–2009 and the margins became more deeply and sharply incised by transverse rifts over time. Between 2009 and 2011 the ice tongue calved into the fiord and started to break apart. As of 2015 it was not very cohesive, although it was digitized as several large polygons as most of these pieces are still held in place by epishelf lake ice (Fig. 5.2h).

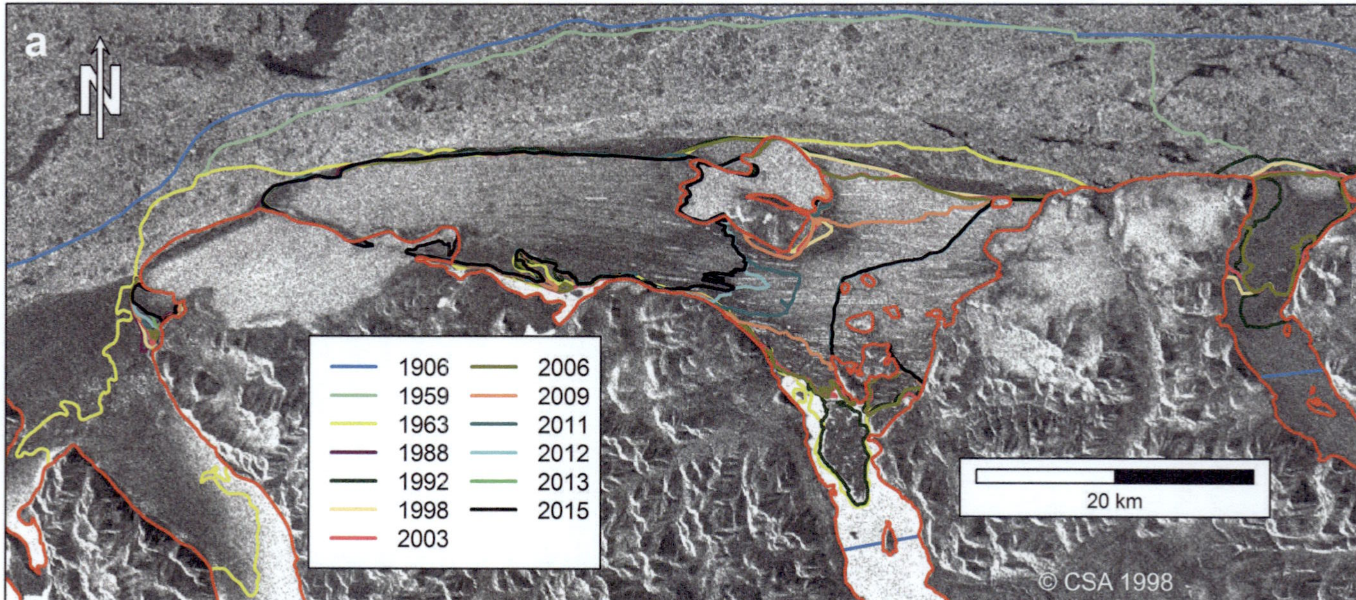

Fig. 5.5 The extent of the M'Clintock/Ward Hunt/Markham Ice Shelf (**a**) from 1959 to 2015 and (**b**) in 2015. In (**a**) ice shelf extents are denoted by *coloured lines*; note that the extent in more recent years may cover the extent of previous years. Background images are (**a**) a RADARSAT-1 ScanSAR HH image acquired on January 13, 1998 and (**b**) a RADARSAT-2 ScanSAR image acquired on March 22, 2015 (*Red*: HH, *Green* and *Blue*: HV) and two RADARSAT-2 Fine Beam Wide images from March 26, 2015 (*Red*: HH – VV, *Green*: (HV + VH)/2, *Blue*: HH + VV), which can be interpreted as the contribution made by the ice targets to single-bounce, volume and double-bounce scattering mechanisms, respectively)

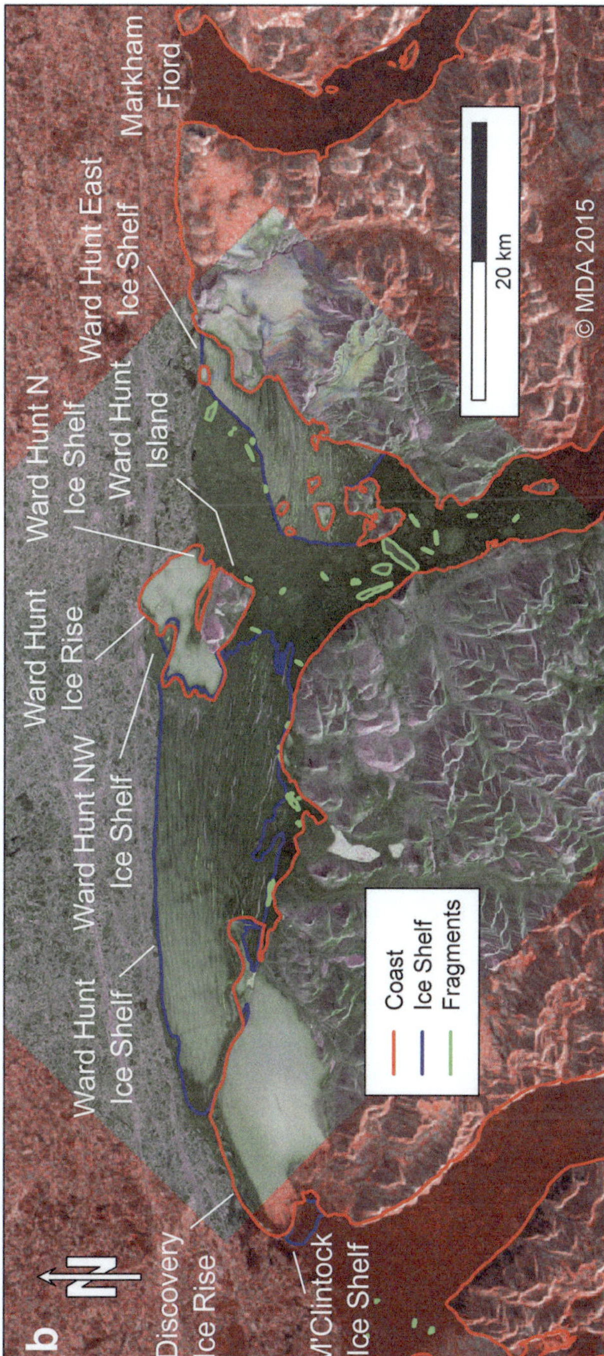

Fig. 5.5 (continued)

Table 5.4 Extent of ice shelves and ice tongues (km^2) by observation year for the northern coast of Ellesmere Island

Ice feature		1959	1963	1988	1992	1998	2003	2006	2009	2011	2012	2013	2015
Ice shelves													
	Ward Hunt	1282.8	684.1	470.8	470.6	454.0	450.3	444.0	385.3	244.1	230.1	224.0	224.3
	Markham	–	55.0	55.0	55.0	53.3	49.2	41.8	–	–	–	–	–
	M'Clintock	–	–	8.8	8.7	8.6	7.8	7.7	7.0	6.5	6.2	5.4	5.1
	Ward Hunt Northwest	–	–	–	–	–	–	–	2.0	2.1	2.0	2.0	2.0
	Ward Hunt East	–	–	–	–	–	–	–	–	73.9	72.1	73.2	73.2
	Ward Hunt North	–	–	–	–	–	–	–	–	0.9	0.2	0.2	0.2
	Milne	287.6	260.2	200.2	198.9	200.6	200.4	197.4	197.4	196.3	196.9	189.1	187.4
	Serson	199.5	199.5	199.5	198.1	200.5	196.0	190.0	38.5	35.5	9.1	8.8	8.9
	Serson East	–	–	–	–	–	–	–	–	–	0.7	0.7	0.8
	Ayles	101.5	46.1	46.1	46.2	45.9	44.8	–	–	–	–	–	–
	Ayles East		8.2	8.2	8.3	8.6	8.6	8.9	8.6	7.9	7.3	7.0	6.9
	Henson	84.1	83.9	48.4	48.2	–	–	–	–	–	–	–	–
	Petersen	53.2	53.2	51.9	51.8	53.8	53.8	40.9	33.2	27.6	24.7	19.6	19.6
	Petersen North	–	–	–	–	–	–	–	–	1.8	1.4	1.5	1.6
	Richards	42.9	42.9	15.6	15.6	–	–	–	–	–	–	–	–
	Richards West			12.1	12.3	–	–	–	–	–	–	–	–
	Moss	35.8	35.8	35.8		–	–	–	–	–	–	–	–
	Armstrong	25.7	25.7	10.8	7.5	–	–	–	–	–	–	–	–

	Armstrong East	–	–	3.4	–	–	–	–	–	–	–	–	–
	Wooton	19.9	19.9	21.1	21.5	24.8	25.6	16.3	8.6	8.1	6.8	3.7	3.6
	Wooton East	–	–	–	–	–	–	–	–	–	–	1.7	1.7
	Aldrich	11.4	11.4	11.4	–	–	–	–	–	–	–	–	–
	Doidge	10.6	10.6	10.6	–	–	–	–	–	–	–	–	–
	Colan	6.5	6.5	6.5	–	–	–	–	–	–	–	–	–
	Columbia	4.3	4.3	4.3	–	–	–	–	–	–	–	–	–
	Stuckberry	1.9	1.9	1.9	–	–	–	–	–	–	–	–	–
	Total	2167.7	1549.1	1222.3	1142.8	1050.2	1036.6	947.0	680.4	604.7	557.5	537.0	535.2
Ice tongues													
	Milne Glacier	45.4	45.4	51.5	57.6	58.6	61.2	58.1	59.2	0.4	1.8	1.9	2.9
	Ayles Glacier	17.8	0.4	2.8	2.8	2.8	2.9	3.0	3.2	0.3	0.3	0.2	0.4
	Mitchell Glacier	12.5	12.5	14.2	8.4	11.8	11.8	0.0	0.0	0.0	0.3	0.2	0.3
	Marine Glacier North S	10.4	10.0	13.7	13.9	13.3	13.3	3.5	0.1	0.4	0.3	0.1	0.1
	Marine Glacier North N	5.0	5.4	6.4	6.4	6.3	6.3	5.5	1.6	0.9	1.6	0.3	0.3
	Richards Glacier North	1.8	1.8	2.2	2.2	2.2	2.3	2.3	2.5	2.0	2.0	0.9	0.5
	Fanshawe Glacier	0.6	0.6	2.9	2.9	2.8	3.7	1.0	0.1	0.1	0.3	0.2	0.3

(continued)

Table 5.4 (continued)

Ice feature		1959	1963	1988	1992	1998	2003	2006	2009	2011	2012	2013	2015
	Richards Glacier South	0.2	0.2	2.6	2.8	2.8	2.8	2.8	2.8	2.1	2.1	2.1	2.0
	Petersen South Glacier	–	–	1.0	0.9	0.8	0.6	0.3	0.3	0.2	0.1	–	–
	Serson Glacier	–	–	–	–	–	–	–	42.0	34.1	19.8	16.4	16.6
	Total	93.7	76.2	99.8	100.1	102.9	107.0	79.1	113.5	42.3	31.1	23.8	24.9

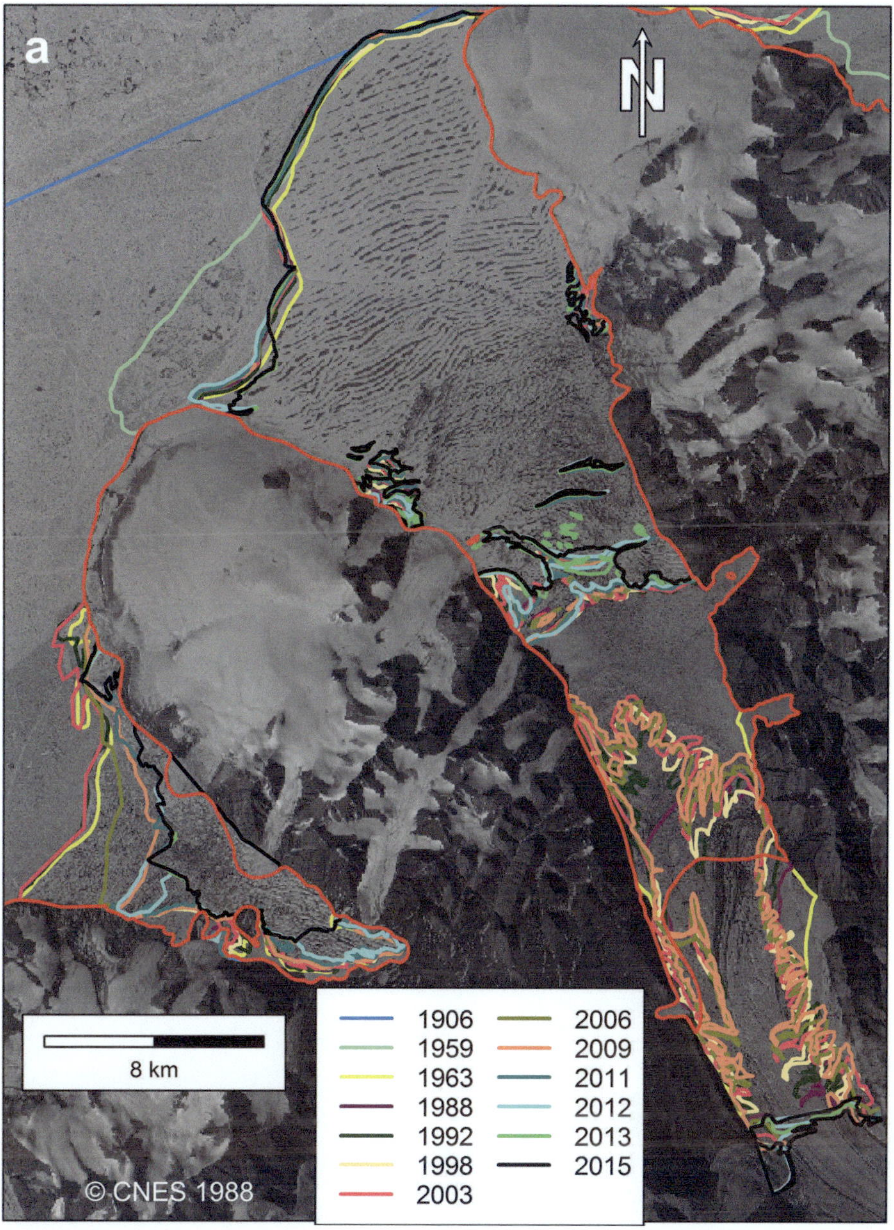

Fig. 5.6 The extent of the Milne and Petersen ice shelves and the Milne Glacier Ice Tongue (**a**) from 1959 to 2015 and (**b**) in 2015. In (**a**) ice shelf and ice tongue extents are denoted by *coloured lines*; note that the extent in more recent years may cover the extent of previous years. Background images are (**a**) SPOT-1 true colour multispectral image acquired on August 8, 1988 and (**b**) two RADARSAT-2 Fine Beam Wide images acquired on March 27, 2015 (*Red*: HH – VV, *Green*: (HV + VH)/2, *Blue*: HH + VV), which can be interpreted as the contribution made by the ice targets to single-bounce, volume and double-bounce scattering mechanisms, respectively)

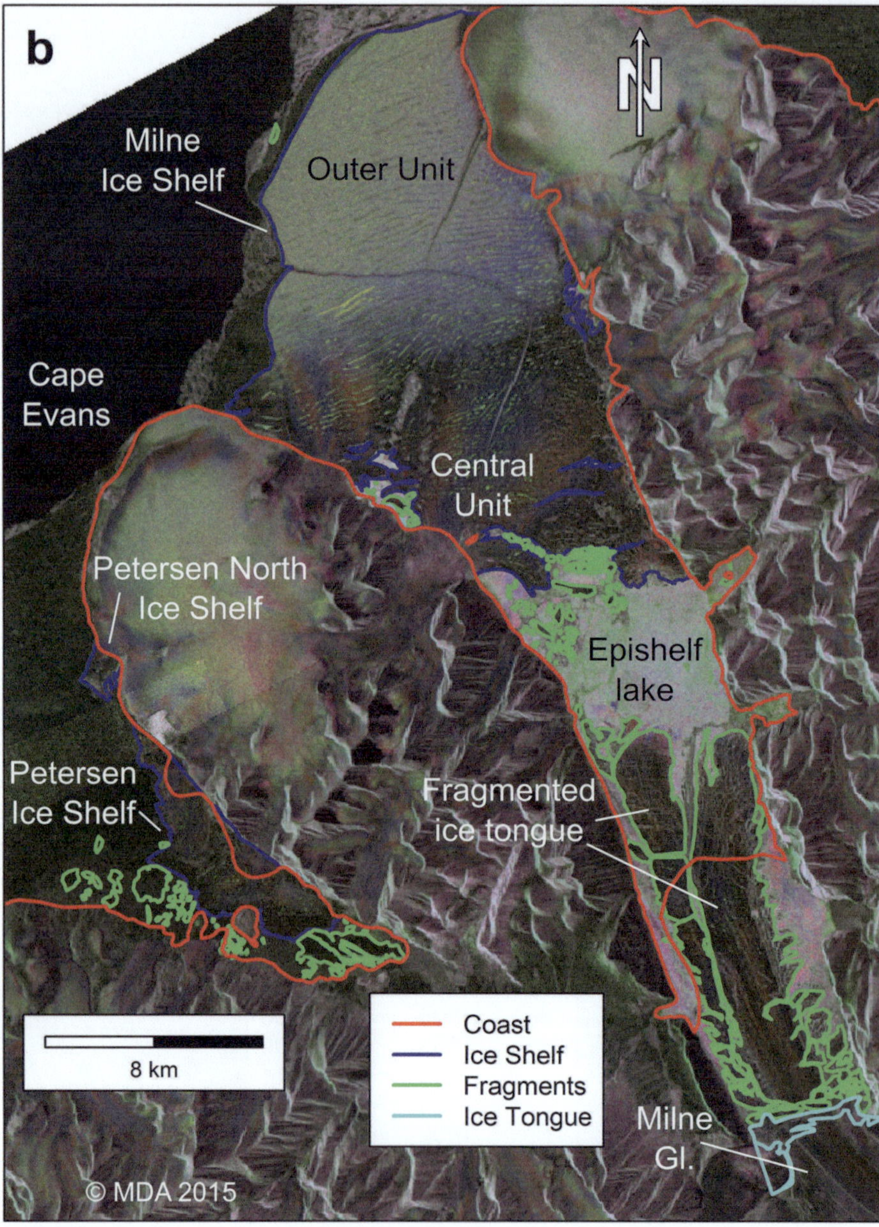

Fig. 5.6 (continued)

5.3.4 *Serson Ice Shelf*

The Serson Ice Shelf was relatively stable from the 1960s until summer 2008, when most of it calved away (Figs. 5.4c and 5.7a; Table 5.4; Mueller et al. 2008). The 1959 map, that was created with data from 1950, suggests a 12 km^2 greater extent than the polygon we corrected using the aerial photographs from 1959. Much of that discrepancy can be explained by the calving of the ice island ARLIS-II (3 × 6 km) in 1955 (Jeffries 1992b). The Serson Ice Shelf is considered to be a composite of a glacial ice shelf and sea ice ice shelf (Lemmen et al. 1988) and influx of the Serson Glacier (unofficial name) from the southeast appears to have generally compensated for some minor peripheral losses until 2006. There was also continuing development of internal epishelf lakes from at least 1998, particularly in the glacial ice shelf section as well as at the interface between the glacial and sea ice shelf sections (in 2006). This foreshadowed the breaking apart of the two sections of the ice shelf in the summer of 2008 (Mueller et al. 2008).

Following this breakup event, the floating tongue of the Serson Glacier that was, until 2008, an integral part of the ice shelf, was reclassified as ice tongue (Fig. 5.2f). The ice tongue of the glacier to the west of the Serson Glacier calved back, presumably to a grounding line (Fig. 5.7a). The 2008 ice shelf loss of 151 km^2 (Figs. 5.4c and 5.7a; Table 5.4) can therefore be partitioned into loss of ice extent (110 km^2) and reclassification of former ice shelf to ice tongue (41 km^2). The ice shelf was further diminished by 26.5 km^2 between 2011 and 2012 and a small (<1 km^2) portion (Serson East) was isolated. The Serson Glacier tongue also lost mass due to calving which was only partially offset by advance during 2008–2015 (Figs. 5.4c and 5.7a; Table 5.4).

5.3.5 *Ayles Ice Shelf*

The Ayles Ice Shelf, with an area of 101 km^2 in 1959, filled the outer portion of Ayles Fiord past the terminus of a large ice tongue that extended from the glacier at the west side of the fiord (Figs. 5.4d and 5.8a; Table 5.4). Jeffries (1986a) reported that between 1959 and 1974 a 15 km^2 ice island calved from the Ayles Ice Shelf, and the ice shelf itself detached from the coast and moved seaward 5 km. The Corona image along with aerial photographs (Hattersley-Smith 1967) reveal that these changes actually occurred between 1962 and 1963. At that time, the ice tongue also disintegrated, leaving numerous ice fragments in the fiord. A small portion of the Ayles Ice Shelf remained attached to two coalesced ice tongues in the eastern arm of the fiord. This comprised another smaller ice shelf that was ignored in previous research (e.g., Jeffries 1992a; Mueller et al. 2006). The large portion of the Ayles Ice Shelf (45 km^2) that was displaced seaward in the early 1960s remained landfast at the mouth of the fiord until August 2005, when it broke away completely, together with the multiyear landfast sea ice surrounding it (Copland et al. 2007). The Ayles

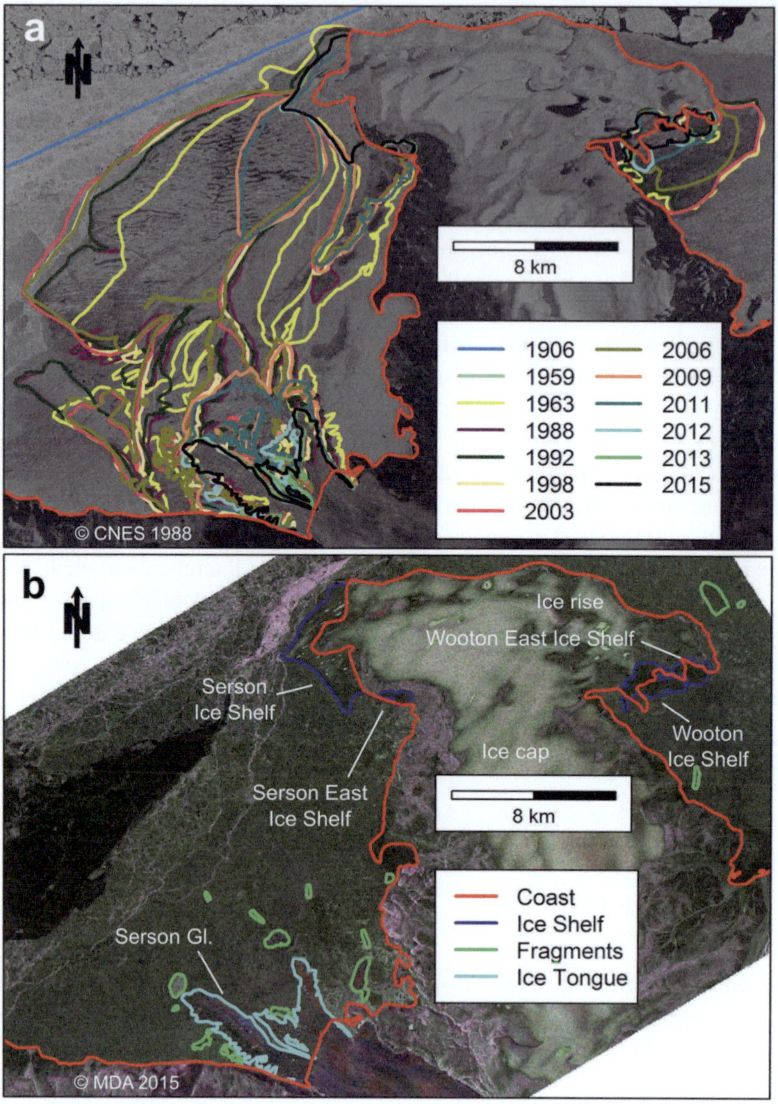

Fig. 5.7 The extent of the Serson and Wooton ice shelves and the Serson Glacier ice tongue (**a**) from 1959 to 2015 and (**b**) in 2015. In (**a**) ice shelf and ice tongue extents are denoted by *coloured lines*; note that the extent in more recent years may cover the extent of previous years. Background images are (**a**) SPOT-1 true colour multispectral image acquired on August 8, 1988 and (**b**) a RADARSAT-2 Fine Beam Wide images acquired on March 24, 2015 (*Red*: HH – VV, *Green*: (HV + VH)/2, *Blue*: HH + VV), which can be interpreted as the contribution made by the ice targets to single-bounce, volume and double-bounce scattering mechanisms, respectively)

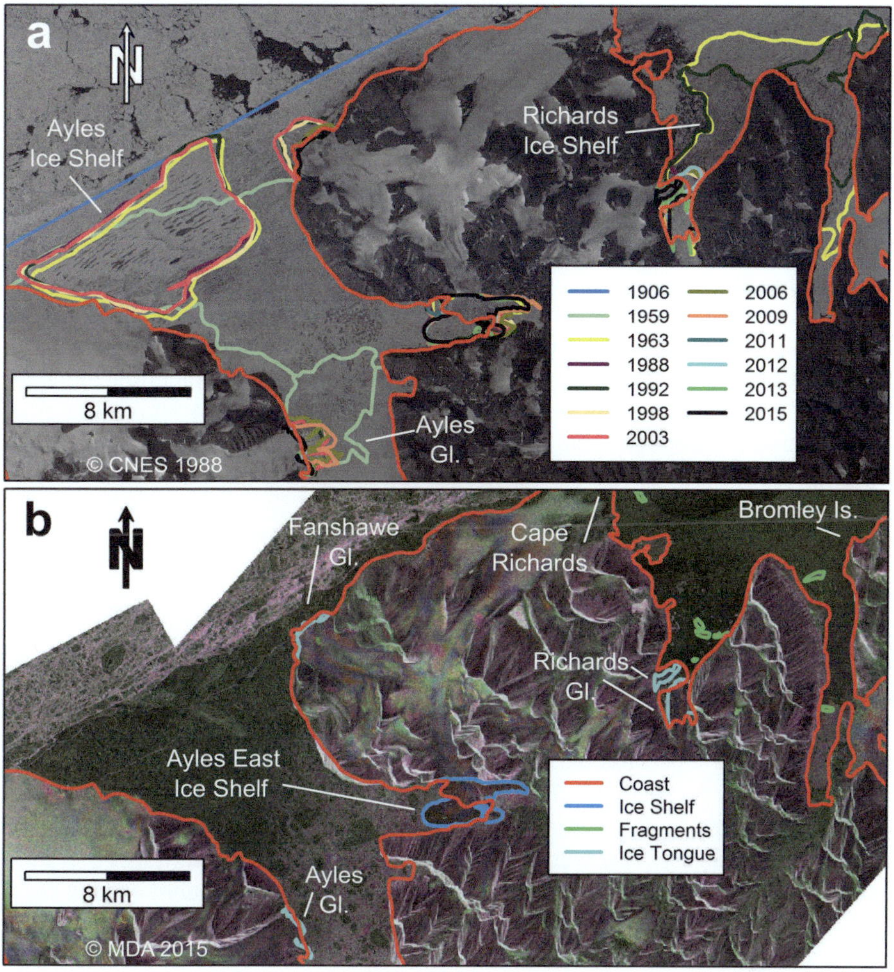

Fig. 5.8 The extent of the Ayles and Richards ice shelves and the Ayles, Fanshawe and Richards glacier ice tongues (**a**) from 1959 to 2015 and (**b**) in 2015. In (**a**) ice shelf and ice tongue extents are denoted by *coloured lines*; note that the extent in more recent years may cover the extent of previous years. Background images are (**a**) SPOT-1 true colour multispectral image acquired on August 8, 1988 and (**b**) a RADARSAT-2 Fine Beam Wide image acquired on March 23, 2015 (*Red*: HH – VV, *Green*: (HV + VH)/2, *Blue*: HH + VV), which can be interpreted as the contribution made by the ice targets to single-bounce, volume and double-bounce scattering mechanisms, respectively)

East Ice Shelf was unaffected by this calving and retained approximately the same extent from 1988 until 2011, before disintegrating slightly (1 km^2) between 2011 and 2015 (Fig. 5.4d). The Ayles Glacier advanced between 1959 and 1988, and again slightly yet continuously between 1998 and 2005, before calving right back to its grounding line in 2012 (Fig. 5.8a; Table 5.4).

5.3.6 Petersen Ice Shelf

The extent of the Petersen Ice Shelf was generally stable until the 59–79 year-old multiyear landfast sea ice in Yelverton Bay calved in summer 2005 (Copland et al. 2007; Pope et al. 2012; White et al. 2015), although there was some fluctuation in area with slight growth in the late 1990s (Figs. 5.4e and 5.6a; Table 5.4). After the August 2005 losses there was a marked decline in the ice shelf extent from 2005 to 2012 (White et al. 2015), and continued loss from 2013 to 2015 (Fig. 5.4e). Today, Petersen Ice Shelf consists of two portions, a small ice shelf to the north that separated from the main section between 2009 and 2012 (Fig. 5.6a). The second, main section has since calved to the west into Yelverton Bay (Fig. 5.2c) as well as at the head of Petersen Bay where long ice fragments were formed by fracturing along the east-west trending meltwater lake troughs (Fig. 5.2d). In 2015, most of these fragments remained trapped near the head of the bay behind the main ice shelf (Fig. 5.6b).

5.3.7 Wooton Ice Shelf

The Wooton Ice Shelf is located in Yelverton Bay to the east of Serson Ice Shelf (Fig. 5.7). Changes in this ice shelf (1959 to July 2009) were documented by Pope et al. (2012). In this study, the Wooton Ice Shelf was found to have grown in size from 1988 to 2003 because the ice rise to its west took on an ice shelf-like appearance at its margins (Fig. 5.4f). This included the appearance of linear lakes in 1988 and a marked drop in backscatter to a level consistent with the remainder of the ice shelf in SAR imagery. This was followed by a large calving event (a loss of 9 km^2) between 2003 and 2006 (Pope et al. 2012) as well as another smaller calving event (1.3 km^2) between 2011 and 2012, and yet another in the following year that separated the ice shelf into two sections (Wooton and Wooton East ice shelves; Table 5.4; Fig. 5.2e).

5.3.8 Other Ice Shelves

The two westernmost ice shelves in 1959 were found north of Henson Bay and adjacent to Cape Armstrong, respectively, and remained unaltered until 1980 (Serson 1983; Figs. 5.1 and 5.9; Table 5.4). By 1988 the western portion of the Henson Ice Shelf had calved and the Armstrong Ice Shelf broke into two separate ice shelves. There was some further attrition of the Armstrong Ice Shelf in 1992 and then no evidence of the existence of either ice shelf following that (Figs. 5.4h and 5.9; Table 5.4).

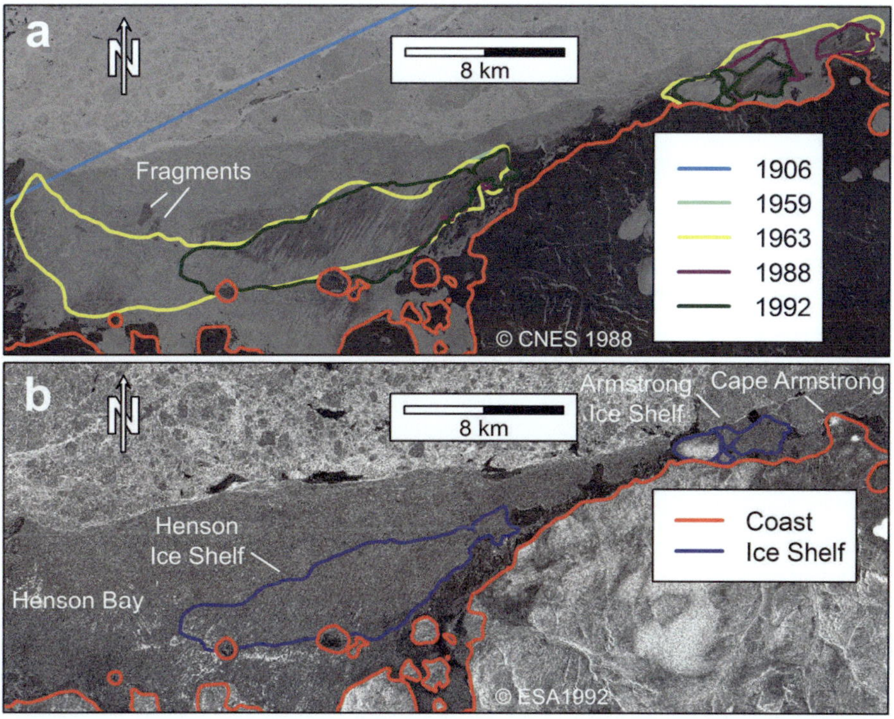

Fig. 5.9 The extent of the Henson and Armstrong ice shelves (**a**) from 1959 to 1992 and (**b**) in 1992. In (**a**) ice shelf extents are denoted by *coloured lines*; note that the extent in more recent years may cover the extent of previous years. Background images are (**a**) SPOT-1 true colour multispectral image acquired on August 8, 1988 and (**b**) an ERS-1 Standard Beam HH image acquired on January 24, 1992

Richards Ice Shelf, which occupied the area between Cape Richards and Bromley Island (Fig. 5.8), split in half between 1966 and 1988 (Hattersley-Smith 1967 and SPOT imagery). These two pieces were lost between 1992 and 1998, although two ice tongues which abutted against each other (but didn't appear to merge) remained and then began to disintegrate during the period 2011–2015.

There were two ice shelves, the Columbia Ice Shelf near Markham Ice Shelf and the Aldrich Ice Shelf (both named unofficially after adjacent capes), flanking the main headland of the northernmost coast of Ellesmere Island in 1959 (Fig. 5.10a). These two ice shelves were not imaged between 1963 and 1998 and, although it is difficult to be certain, due to low image contrast, it appears that they both melted or calved away completely during that interval.

To the east of Aldrich Ice Shelf there are four small ice shelves in the embayments to the west of Clements Markham Inlet (Fig. 5.10; Table 5.4). These are named (unofficially) after nearby features (W-E: Doidge Bay, Point Moss, Stuckberry Point and Cape Colan). They were not imaged between 1959 and 1998. Field measurements of ice thickness from 1989–90 indicated that MLSI, not ice

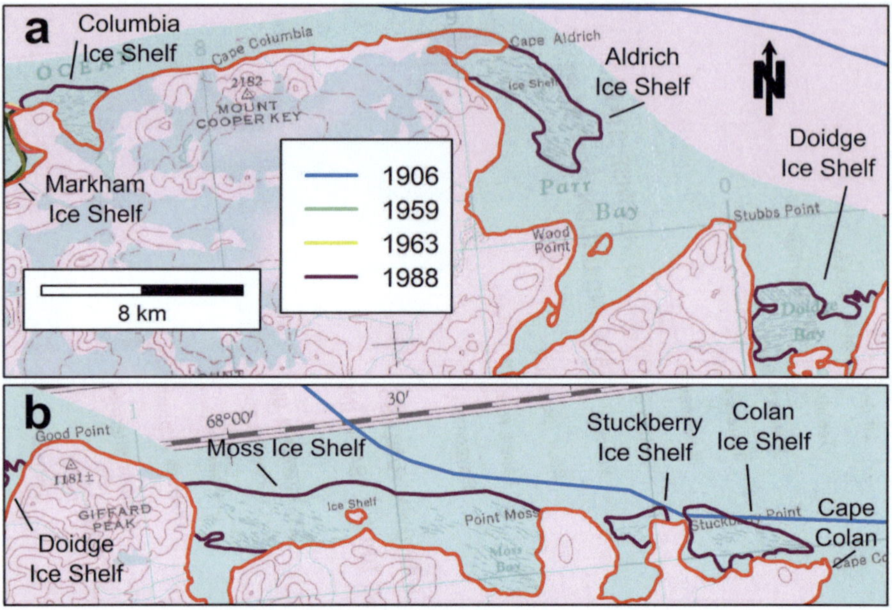

Fig. 5.10 The extent of the (**a**) Columbia, Aldrich and Doidge ice shelves, as well as the (**b**) Moss, Stuckberry and Colan ice shelves from 1959 to 1988. Ice shelf extents are denoted by *coloured lines*; note that the extent in more recent years may cover the extent of previous years. Background image is a scan of the Clements Markham (NTS 120F) 1:250,000 map sheet. The scale is the same in both (**a**) and (**b**)

shelf ice was present along this coast (Verrall and Todoeschuk unpublished, cited in Jeffries 2017). Furthermore, no ice shelves are apparent in 1998 imagery (Vincent et al. 2001). We therefore assume that these ice shelves thinned away by melting after 1988.

5.4 Discussion

5.4.1 Discoveries and Refinements

In this study, we documented ice shelves and ice tongues along the northernmost coast of Ellesmere Island, and used a systematic approach to digitize their entire but dwindling extent since the start of the twentieth century. As a result, we describe many small ice shelves and ice tongues, some of which were known previously but never properly digitized or described (e.g., Henson, Armstrong and Richards ice shelves along with the ice shelves east of Markham), and other ice shelves which were not recognized by previous researchers (e.g., East Ayles). In addition, large ice tongues such as Serson, Mitchell (south of Wooton Ice Shelf, not shown) Marine North (south of Petersen Ice Shelf, not shown), Milne, Ayles, Fanshawe and Richards glaciers have now been digitized throughout the entire data record in this project.

While the primary aim of this chapter was to inventory long-term changes in the ice shelves, we have also been able to constrain the timing of some key events with the addition of newly exploited imagery. The Cold War Corona spy satellite helped to refine the timing of calving of the M'Clintock (1963–66), Ayles (1959–63), and Milne (1959–63) ice shelves that have been described in Hattersley-Smith (1967) and Jeffries (1986a), but poorly constrained due to lack of observations. We also discovered new calving events that occurred between the last observation in the published records that end ca. 2009–2012 (Vincent et al. 2011; Pope et al. 2012; Derksen et al. 2012; Mortimer et al. 2012; White et al. 2015) and the winter of 2015. This was the case for Serson, Wooton, Milne (at both the northern and southern calving fronts), Ward Hunt Ice Shelf, East Ayles, and Petersen ice shelves. We also documented calving events that had not previously been recognized in the literature, for example the Markham Ice Shelf calving event that occurred around the turn of the twenty-first century. The recent calving and widespread disintegration of the Milne Glacier tongue has now also been documented, which builds on previous work by Jeffries (1984), Mortimer (2011) and Mortimer et al. (2012).

This work is more comprehensive in terms of spatial and temporal scope than previous studies and, with the exception of studies of single ice shelves (Jeffries 1986a, 1992b; Copland et al. 2007; Mortimer et al. 2012; White et al. 2015), it is also more detailed. For instance, we have delineated small gaps in the extent of ice shelves that had not been fully documented previously, notably on the Serson Ice Shelf, and also on the Milne Ice Shelf (after Mortimer et al. 2012). In conducting this research, we found several places where the disappearance of coastal ice has altered the NTDB 2009 coastline; this was dealt with by identifying and classifying 'inland' polygons to reconcile the difference between this data set and the imagery to properly account for the ice shelf and ice tongue extent within the study area. We did not evaluate more up to date products that are currently available and may have a more accurate coastline. The Government of Canada has not properly mapped the Ellesmere Island ice shelves since the first paper maps of the area were produced in the 1960s (the maps found in the Atlas of Canada were derived from previous versions of this work).

5.4.2 *Comparison of Ice Shelf Extents with Previous Literature*

In general, the extents that we have calculated are similar to those available in the published literature (Table 5.4). This underscores that the technique we employed along with our interpretation of ice types is repeatable. Percentage differences that are not in bold in Table 5.5 are normally distributed (Shapiro-Wilk $W = 0.96$; $p = 0.48$) and have a mean of 0.22%, suggesting very little systematic bias. The standard

Table 5.5 Comparison between ice shelf extents from this and other studies. Available extents within 5 years of our observation years were included. Values in bold are explained in the text. The Ellesmere Ice Shelf refers to the extent of all ice shelves in 1906 based on anecdotal evidence and the Marvin section is a surveyed subset of this (see Fig. 5.1)

	This study		Other study					
Ice shelf	Year	Extent (km^2)	Year	Extent (km^2)	Difference (km^2)	Difference (%)	Source	Comments
Ellesmere	1906	8529	1906	8900	−371	**−4.3**	Vincent et al. (2001)	
	1906	8529	1906	7450	1079	**12.7**	Spedding (1977)	A description of the estimated area is not available
Marvin section	1906	2370	1906	2327	43	1.8	Bushnell (1956)	Table 1 in Vincent et al. (2001)
	1959	1396	1954	1422	−26	−1.9	Crary (1956)	Table 1 in Vincent et al. (2001)
	1963	852	1962	906	−54	−6.3	Hattersley-Smith (1963)	Table 1 in Vincent et al. (2001)
	1988	632	1985	644	−12	−1.9	Jeffries and Serson (1983)	Table 1 in Vincent et al. (2001)
	1998	516	1998	490	26	5.0	Vincent et al. (2001)	Table 1 in Vincent et al. (2001)
Ward Hunt	1988	471	1984	440	31	6.5	Jeffries and Serson (1986)	
Petersen	1988	51.9	early 1980s	55.0	−3.1	−5.9	Jeffries and Serson (1986)	
Wooton	1988	21.1	early 1980s	20.0	1.1	5.4	Jeffries and Serson (1986)	
Serson	2003	200	2000	205	−5	−2.6	Mueller et al. (2006)	
Petersen	2003	53.8	2000	51.2	2.6	4.8	Mueller et al. (2006)	
Milne	2003	200	2001	206	−6	−2.7	Mueller et al. (2006)	
Ayles	2003	46.1	2003	84.1	−38.0	**−82.4**	Mueller et al. (2006)	Included multiyear landfast sea ice around Ayles Ice Shelf
Ward Hunt	2003	450	2002	448	2	0.5	Mueller et al. (2006)	

(continued)

Table 5.5 (continued)

	This study		Other study					
Ice shelf	Year	Extent (km^2)	Year	Extent (km^2)	Difference (km^2)	Difference (%)	Source	Comments
Markham	2003	49.2	2002	50	−0.8	−1.6	Mueller et al. (2006)	
Ayles	2003	44.8	2005	87.3	−42.5	**−94.9**	Copland et al. (2007)	Included multiyear landfast sea ice around Ayles Ice Shelf
Milne	1959	288	1959	281	7	2.3	Mortimer et al. (2012)	
	1988	200	1984	250	−50	**−25**	Mortimer et al. (2012)	Included the epishelf lake
	1992	199	1993	214	−15	−7.5	Mortimer et al. (2012)	
	1998	201	2001	206	−5	−2.5	Mortimer et al. (2012)	
	2009	197	2009	205	−8	−4.1	Mortimer et al. (2012)	
Petersen	1988	51.9	1984	48.4	3.5	6.7	White et al. (2015)	
	1998	53.8	1999	49.9	3.9	7.3	White et al. (2015)	
	2003	53.8	2003	50.3	3.5	6.5	White et al. (2015)	
	2006	40.9	2005	40.8	0.0	0.0	White et al. (2015)	
	2009	33.2	2009	32.1	1.1	3.3	White et al. (2015)	
	2011	27.6	2011	30.3	−2.7	−9.7	White et al. (2015)	
	2012	24.7	2012	24.8	−0.1	−0.4	White et al. (2015)	
	2013	19.6	2012	19.3	0.3	1.4	White et al. (2015)	

deviation is 4.8%, which indicates that the typical discrepancy is within ±5%. This aligns well with the worst case uncertainty in digitization (see Sect. 5.2.4). It is difficult to compare two estimates of an extent without any reference to a definitive measurement; however, some of the differences in estimates are discussed below and we suggest reasons for the discrepancies.

The small difference between the original estimate for the Ellesmere Ice Shelf in 1906 and our extent could be due to a difference in geographic projection, digital analysis capability, or a digitizing error. Whatever the source of the error, it is undoubtedly much smaller than the conversion of anecdotal descriptions to a geographical coverage. In particular, Vincent et al. (2001) acknowledge that the extension of the ice shelves across Nansen Sound to Axel Heiberg Island is controversial. The writings of Otto Sverdrup, who traversed from Ellesmere to Axel Heiberg islands in 1902, indicate the ice was "very much broken and covered with high pressure ridges" (Sverdrup et al. 1904), which is inconsistent with Peary's description. Koenig et al. (1952) suggest that a large ice island may have occupied Nansen Sound at that time. If this section of the Vincent et al. (2001) ice shelf polygon is excluded, the area of the Ellesmere Ice Shelf was 6373 km^2 in 1906, which fits better with Spedding's (1977) estimate of 7450 km^2, although he also assumed that the ice shelf reached out into Nansen Sound. Unfortunately there are no maps of this reconstruction available to evaluate the differences any further.

Mueller et al. (2006) published the extents of the six largest ice shelves using SAR imagery acquired between 2000 and 2003. These values are within ~5% of the extents obtained here in 2003, with the exception of Ayles Ice Shelf due to the inclusion by Mueller et al. (2006) of multiyear landfast sea ice that grew in around the ice shelf since the 1960s. Copland et al. (2007) had a similar area for the same reasons. Mortimer et al. (2012) included the epishelf lake as part of the Milne Ice Shelf in their 1984 digitization, which explains why our estimate in 1988 is much lower.

5.4.3 *Implications of Semantics*

The 2008 break-up of the Serson Ice Shelf caused this composite ice shelf to become separated into two constituent parts: the sea-ice ice shelf to the north and an ice tongue to the south. The event represented a 151 km^2 reduction in ice shelf extent, but 27% of this loss was due to the reclassification of parts of the former ice shelf to an ice tongue. For this reason, the scope of this chapter encompasses both thick landfast floating ice types and their break-up products (Table 5.3).

Another related issue is that the operational definition of an ice shelf is based on a minimum thickness. This presents difficulties when thinning reduces an expanse of ice shelf below this threshold. Since the remote sensing data used here don't quantify thickness, the ice shelf extent can only be gleaned from image interpretation and ancillary information. Interpretation of both optical and SAR data can include judgement of the thickness of ice from the wavelength of surface undulations (c.f. Jeffries and Serson 1986; Jeffries 2017) or by using available ice penetrating radar thickness measurements (Narod et al. 1988; Mortimer et al. 2012; White et al. 2015) or personal observations. It may be useful to consider ice that was formerly ice shelf, but has apparently thinned substantially as 'transition ice' (Mortimer 2011; Mortimer et al. 2012). The slow thinning of ice shelves to transition ice was likely the cause of the disappearance of several ice shelves along the northern coast of Ellesmere Island. Examples include Henson, Armstrong and Richards ice shelves along with all the ice shelves east of Markham Ice Shelf. Transition ice is definitely present today at the southeastern margin of Milne Ice Shelf (based on ice penetrating radar observations) and possibly throughout the eastern portion of the Petersen

Ice Shelf (also based on ice penetrating radar, but here reflections were attenuated due to saline ice) (Mortimer et al. 2012; White et al. 2015). Where ice thickness was well constrained, areas of transition ice were removed from the ice shelf extents. However, identifying when those ice areas became too thin to be considered ice shelf ice was a judgement call.

The converse may also be true, where multiyear landfast sea ice may thicken enough to qualify as ice shelf ice. The latter is difficult to explain over the twentieth century, given trends in air temperature and reductions in the average age and thickness of sea ice in the Arctic Ocean, but it may have played a role in some cases (for example, the western side of Markham Ice Shelf). In both cases, we were careful to leverage repeat observations so that the interpretation (thinning or thickening) is consistently applied across the dataset.

5.4.4 Other Known Sources of Data

The collection of ice shelf extents presented here is the most complete to date for anywhere in the Arctic, although it could be improved by incorporating other datasets. This includes using the trimetrogon oblique and vertical photographs taken in 1950, and the original, complete set of vertical air photographs from 1959 that were used to derive the first topographic maps of the region. Also, there is one documented instance where out of date imagery was used to make the map (Jeffries 1992b). There are more air photographs available at the National Air Photograph Library that were taken in 1974 and 1984, although these cover only certain ice shelves and concentrate on their northern edges.

Side-looking airborne real-aperture radar images were obtained in the early 1980s and some airborne X-band SAR images from the late 1980s (Canadian Coast Guard, Canarctic Shipping 1990). For the pre-SAR era, it may also be advantageous to scan and digitize maps from Crary (1954), Hattersley-Smith (1963) and Jeffries and Serson (1983) since they provide fairly detailed reproductions. There are optical satellite data, such as Landsat 5 to 8 imagery and products from the Advanced Spaceborne Thermal Emission and Reflection Radiometer (ASTER) on NASA's Terra satellite, for the more southerly ice shelves (those west of Ayles Fiord). Some high resolution Formosat-2 data exist from the summer of 2008 and there is also relatively good SAR coverage in 1993 (ERS-1), 1997, 2000 and 2008 (RADARSAT-1), and beyond (RADARSAT-2). We recommend that improvements to the current dataset be concentrated on the years before 2003, unless there is a need to better document the timing of individual calving events, in which case the temporal resolution since that date is unparalleled. MODIS optical imagery (250 m resolution) represents a viable option to detect major calving events provided that the imagery is cloud-free, that there is enough contrast between ice shelf and sea ice to detect change, and that fine spatial resolution is not required.

5.4.5 Implications of Ice Shelf Loss

From 1906 to 2015, there has been a monotonic decrease in the extent of the Ellesmere Island ice shelves (Fig. 5.3a; Table 5.3). However, in spite of irregular observations and uncertainty within this dataset, there is evidence of active periods of calving and more quiescent phases. For example, the first half of the twentieth century was a very active period, when 75% of the estimated ice shelf extent in 1906 was lost. In the period from 1959 to 1960 there was also a large loss and a further 30% of the shelf ice calved away. The period following this (1963–1988) saw the loss of ~20% by area of the ice shelves. After these initial years, the pace of loss (in terms of absolute extent lost, relative extent lost and attrition rate) slowed. A relatively quiescent period from the late 1980s to 2003 saw only the loss of 186 km^2, which was followed by a spate of large calving events in 2005, 2008 and 2010 (Table 5.3; Mueller et al. 2008; Vincent et al. 2011; Derksen et al. 2012). In recent years (since 2012), this rate has slowed again but it will likely increase sporadically in the future (see Copland et al. (2017) for a review on factors that contribute to calving). It is likely that if the rates of attrition in the past 10 years are representative of the future, the remaining Ellesmere Island ice shelves will not last more than 25 years (c.f. White et al. 2015).

The situation for ice tongues is somewhat different, given that they can replenish by ice flow across their grounding lines and their extent varies over time as a result (Fig. 5.3b; Table 5.3). However, ice tongues are also collapsing along the coastline of northern Ellesmere Island (e.g., Milne Glacier; Yelverton Glacier (Adrienne White, unpublished)) and the outlook for the remaining ice tongues is poor.

5.5 Conclusions

The goal of this chapter has been to document the changes of northern Ellesmere Island ice shelves and associated ice tongues since 1906. At present, only 6.3% of the original Ellesmere ice shelf remains, and it seems certain that there will be further attrition and potentially complete loss, in the coming decades. This study has provided the first quantitative assessment of Ellesmere ice shelf disintegration over a century time span with modern GIS and remote sensing techniques. The data are freely available via Nordicana D (Mueller et al. 2017) and can be improved or used for further analyses in the future.

Acknowledgements Corona imagery was acquired through the US Geological Survey's Earth Explorer (http://earthexplorer.usgs.gov) and SPOT imagery was obtained courtesy of SPOT Image Corporation. We are indebted to the Alaska Satellite Facility, University of Alaska Fairbanks for providing ERS-1 and RADARSAT-1 imagery from (1992 to 2006). Some supplemental RADARSAT-1 imagery (1998–2006) was accessed through the Polar Data Catalogue (http://www.polardata.org). RADARSAT-2 data from 2009 to 2015 were provided by MacDonald, Dettwiler and Associates (MDA) under the RADARSAT-2 Government Data Allocation administered by the Canadian Ice Service (CIS) and the Canadian Space Agency's Science and Operational Applications

Research – Education (SOAR-E) program (project #5054 and #5106). RADARSAT-2 Data and Products are copyright MacDonald, Dettwiler and Associates, Ltd., 2009–2015 – All Rights Reserved. Georeferencing of imagery was conducted in part by Laura Derksen and Mustafa Naziri. Spatial queries and image projections were performed by Sougal Bouh Ali. We are grateful for funding from the Natural Sciences and Engineering Research Council (NSERC), ArcticNet, the Canada Foundation for Innovation, Ontario Research Fund, Ontario Graduate Scholarships, Fonds de recherche du Québec – Nature et technologies (FRQNT) and the Northern Scientific Training Program, and for field support from the Polar Continental Shelf Program. A special thanks to our colleagues and students who have provided insights and valuable discussions over the years, including Adrienne White, Andrew Hamilton, Miriam Richer-McCallum, Sierra Pope and Colleen Mortimer. We also thank Alison Cook and Christian Haas for peer-reviewing this manuscript and providing useful suggestions for improvement.

References

Aldrich, P. (1877). Western sledge party 1876. In Journals and proceedings of the Arctic Expedition, 1875–6, under the command of Captain Sir George Nares, R.N., K.C.B. In continuation of Parliamentary Papers C 1153 of 1875 and C 1560 of 1875 (pp. 522). London: Harrison and Sons.

Antoniades, D. (2017). An overview of paleoenvironmental techniques for the reconstruction of past Arctic ice shelf dynamics. In L. Copland & D. Mueller (Eds.), *Arctic ice shelves and ice islands* (p. 207–226). Dordrecht: Springer. doi:10.1007/978-94-024-1101-0_8.

Belkin, I., & Kessel, S. (2017). Russian drifting stations on Arctic ice islands. In L. Copland & D. Mueller (Eds.), *Arctic ice shelves and ice islands* (p. 367–393). Dordrecht: Springer. doi:10.1007/978-94-024-1101-0_14.

Bushnell, V. C. (1956). Marvin's ice shelf journey, 1906. *Arctic, 9*, 166–177. doi:10.14430/arctic3791.

Canadian Coast Guard, Canarctic Shipping. (1990). *Canadian Arctic marine ice atlas – winter 1987/88 with perspectives from 1986/87 and 1988/89*. Ottawa: Canadian Coast Guard.

Copland, L., Mueller, D. R., & Weir, L. (2007). Rapid loss of the Ayles Ice Shelf, Ellesmere Island, Canada. *Geophysical Research Letters, 34*, L21501. doi:10.1029/2007GL031809.

Copland, L., Mortimer, C. A., White, A., et al. (2017). Factors contributing to recent Arctic ice shelf losses. In L. Copland & D. Mueller (Eds.), *Arctic ice shelves and ice islands* (p. 263–285). Dordrecht: Springer. doi:10.1007/978-94-024-1101-0_10.

Crary, A. P. (1954). Seismic studies on Fletcher's Ice Island, T-3. *EOS, Transactions of the American Geophysical Union, 35*, 293–300.

Crawford, A. J. (2013). *Ice island deterioration in the Canadian Arctic: Rates, patterns and model evaluation*. MSc thesis, Carleton University, Ottawa.

De Abreu, R., Arkett, M., Cheng, A., Zagon, T., Mueller, D. R., Vachon, P., & Wolfe, J. (2011). *RADARSAT-2 mode selection for maritime surveillance*. Ottawa: Defence Research and Development Canada.

Derksen, C., Smith, S. L., Sharp, M., et al. (2012). Variability and change in the Canadian cryosphere. *Climatic Change, 115*, 59–88. doi:10.1007/s10584-012-0470-0.

Dowdeswell, J. A., & Jeffries, M. O. (2017). Arctic ice shelves: An introduction. In L. Copland & D. Mueller (Eds.), *Arctic ice shelves and ice islands* (p. 3–21). Dordrecht: Springer. doi:10.1007/978-94-024-1101-0_1.

England, J. H., Evans, D. A., & Lakeman, T. R. (2017). Holocene history of Arctic ice shelves. In L. Copland & D. Mueller (Eds.), *Arctic ice shelves and ice islands* (p. 185–205). Dordrecht: Springer. doi:10.1007/978-94-024-1101-0_7.

Flett, D. (2003). Operational use of SAR at the Canadian ice service: Present operations and a look into the future. In *Proceedings of the 2nd Workshop on Coastal and Marine Applications of SAR*, Norway, Svalbard

Ghilani, C. D. (2000). Demystifying area uncertainty: More or less. *Surveying and Land Information Systems, 60*, 177–182.
Hattersley-Smith, G. (1957). The rolls on the Ellesmere Ice Shelf. *Arctic, 10*, 32–44. doi:10.14430/arctic3753.
Hattersley-Smith, G. (1963). The Ward Hunt Ice Shelf: Recent changes of the ice front. *Journal of Glaciology, 4*, 415–424.
Hattersley-Smith, G. (1967). Note on ice shelves off the north coast of Ellesmere Island. *Arctic Circular, 17*, 13–14.
Hobson, G. (1989). Ice island field station: New features of Canadian polar margin. *EOS, Transactions American Geophysical Union, 70*, 833, 835, 838–839.
Hoffman, M. J., Fountain, A. G., & Achuff, J. M. (2007). 20th-century variations in area of cirque glaciers and glacierets, Rocky Mountain National Park, Rocky Mountains, Colorado, USA. *Annals of Glaciology, 46*, 349–354. doi:10.3189/172756407782871233.
Jeffries, M. O. (1982). The Ward Hunt Ice Shelf, spring 1982. *Arctic, 35*, 542–544. doi:10.14430/arctic2363.
Jeffries, M. O. (1984). Milne Glacier, northern Ellesmere Island, N.W.T., Canada: A surging glacier? *Journal of Glaciology, 30*, 251–253. doi:10.3198/1984JoG30-105-251-253.
Jeffries, M. O. (1985). *Physical, chemical and isotopic investigations of Ward Hunt Ice Shelf and Milne Ice Shelf, Ellesmere Island, NWT*. PhD Dissertation, University of Calgary, Calgary.
Jeffries, M. O. (1986a). Ice island calvings and ice shelf changes, Milne Ice Shelf and Ayles Ice Shelf, Ellesmere Island, N.W.T. *Arctic, 39*, 15–19. doi:10.14430/arctic2039.
Jeffries, M. O. (1986b). Glaciers and the morphology and structure of the Milne Ice Shelf, Ellesmere Island, N.W.T., Canada. *Arctic and Alpine Research, 18*, 397–405. doi:10.2307/1551089.
Jeffries, M. O. (1987). The growth, structure and disintegration of Arctic ice shelves. *Polar Record, 23*, 631–649. doi:10.1017/S0032247400008342.
Jeffries, M. O. (1992a). Arctic ice shelves and ice islands: Origin, growth and disintegration, physical characteristics, structural-stratigraphic variability, and dynamics. *Reviews of Geophysics, 30*, 245–267. doi:10.1029/92RG00956.
Jeffries, M. O. (1992b). The source and calving of ice island ARLIS-II. *Polar Record, 28*, 137–144. doi:10.1017/S0032247400013437.
Jeffries, M. O. (2002). Ellesmere Island ice shelves and ice islands. In R. S. Williams & J. G. Ferrigno (Eds.), *Satellite image atlas of glaciers of the world: North America* (p. J147–J164). Washington, DC: United States Geological Survey.
Jeffries, M. O. (2017). The ice shelves of northernmost Ellesmere Island, Nunavut, Canada. In L. Copland & D. Mueller (Eds.), *Arctic ice shelves and ice islands* (p. 23–54). Dordrecht: Springer. doi:10.1007/978-94-024-1101-0_2.
Jeffries, M. O., & Sackinger, W. M. (1991). Detection of a calving event at the Milne Ice Shelf, N.W.T. and the contribution of offshore winds. In T. K. S. Murthy, J. G. Paren, W. M. Sackinger, & P. Wadhams (Eds.), *Ice technology for Polar operations* (p. 321–331). Southampton: Computation Mechanics Publications.
Jeffries, M. O., & Serson, H. V. (1983). Recent changes at the front of Ward Hunt Ice Shelf, Ellesmere Island, N.W.T. *Arctic, 36*, 289–290. doi:10.14430/arctic2278.
Jeffries, M. O., & Serson, H. (1986). Survey and mapping of recent ice shelf changes and landfast sea ice growth along the north coast of Ellesmere Island. *Annals of Glaciology, 8*, 96–99.
Jeffries, M. O., & Shaw, M. A. (1993). The drift of ice islands from the Arctic Ocean into the channels of the Canadian Arctic Archipelago: The history of Hobson's Choice Ice Island. *Polar Record, 29*, 305–312. doi:10.1017/S0032247400023950.
Jungblut, A. D., Mueller, D., & Vincent, W. F. (2017). Arctic ice shelf ecosystems. In L. Copland & D. Mueller (Eds.), *Arctic ice shelves and ice islands* (p. 227–260). Dordrecht: Springer. doi:10.1007/978-94-024-1101-0_9.
Koenig, L. S., Greenaway, K. R., Dunbar, M., & Hattersley-Smith, G. (1952). Arctic ice islands. *Arctic, 5*, 67–103. doi:10.14430/arctic3901.

Lee, J.-S. (2009). *Polarimetric radar imaging: From basics to applications*. Boca Raton: CRC Press.

Lemmen, D. S., Evans, D. J. A., & England, J. (1988). Ice shelves of northern Ellesmere Island, N.W.T., Canadian landform examples. *Canadian Geographic, 32*, 363–367.

Mortimer, C. (2011). *Quantification of volume changes for the Milne Ice Shelf, Nunavut, 1950–2009*. Ottawa: University of Ottawa, Department of Geography.

Mortimer, C. A., Copland, L., & Mueller, D. R. (2012). Volume and area changes of the Milne Ice Shelf, Ellesmere Island, Nunavut, Canada, since 1950. *Journal of Geophysical Research, 117*, F04011. doi:10.1029/2011JF002074.

Mueller, D. R., Vincent, W. F., & Jeffries, M. O. (2003). Break-up of the largest Arctic ice shelf and associated loss of an epishelf lake. *Geophysical Research Letters, 30*, 2031. doi:10.1029/2003GL017931.

Mueller, D. R., Vincent, W. F., & Jeffries, M. O. (2006). Environmental gradients, fragmented habitats and microbiota of a northern ice shelf cryoecosystem, Ellesmere Island, Canada. *Arctic, Antarctic, and Alpine Research, 38*, 593–607. doi:10.1657/1523-04300.

Mueller, D. R., Copland, L., Hamilton, A., & Stern, D. R. (2008). Examining Arctic ice shelves prior to 2008 breakup. *EOS. Transactions of the American Geophysical Union, 89*, 502–503.

Mueller, D., Copland, L., & Jeffries, M. O. (2017). Northern Ellesmere Island ice shelf and ice tongue extent, v. 1.0 (1906–2015). Nordicana D28. doi:10.5885/45455XD-24C73A8A736446CC.

Narod, B. B., Clarke, G. K. C., & Prager, B. T. (1988). Airborne UHF radar sounding of glaciers and ice shelves, northern Ellesmere Island, Arctic Canada. *Canadian Journal of Earth Sciences, 25*, 95–105. doi:10.1139/e88-010.

Onstott, R. G., & Shuchman, R. A. (2004). SAR measurements of sea ice. In C. R. Jackson & J. R. Apel (Eds.), *Synthetic aperture radar marine user's manual* (p. 81–115). Washington, DC: National Oceanic and Atmospheric Administration.

Paul, F., Barrand, N. E., Baumann, S., et al. (2013). On the accuracy of glacier outlines derived from remote-sensing data. *Annals of Glaciology, 54*, 171–182. doi:10.3189/2013AoG63A296.

Peary, R. E. (1907). *Nearest the pole: A narrative of the Polar expedition of the Peary Arctic club in the S.S. Roosevelt, 1905–1906*. London: Hutchinson.

Pope, S., Copland, L., & Mueller, D. (2012). Loss of multiyear landfast sea ice from Yelverton Bay, Ellesmere Island, Nunavut, Canada. *Arctic, Antarctic, and Alpine Research, 44*, 210–221. doi:10.1657/1938-4246-44.2.210.

Pope, S., Copland, L., & Alt, B. (2017). Recent changes in sea ice plugs along the northern Canadian Arctic Archipelago. In L. Copland & D. Mueller (Eds.), *Arctic ice shelves and ice islands* (p. 317–342). Dordrecht: Springer. doi:10.1007/978-94-024-1101-0_12.

Prager, B. T. (1983). *Digital signal processing of UHF radar system for airborne surveys of ice thicknesses*. Vancouver, BC, Canada: University of British Columbia, Department of Geophysics and Astronomy.

Ruffner, K. C. (Ed.). (1995). *Corona: America's first satellite program*. Washington, DC: Center for the Study of Intelligence, Central Intelligence Agency.

Sackinger, W. M., Serson, H. V., Jeffries, M. O., et al. (1985). Ice island generation and trajectories north of Ellesmere Island, Canada. In *Proceedings of the 8th International Conference on Port and Ocean Engineering under Arctic Conditions, Narssarssuaq* (p. 1009).

Serson, H. V. (1983). *Ice conditions off the north coast of Ellesmere Island, spring 1980*. Victoria: Defence Research Establishment Pacific.

Shokr, M., & Sinha, N. (2015). *Sea ice: Physics and remote sensing*. Washington, DC: American Geophysical Union.

Spedding, L. G. (1977). *Ice island count, southern Beaufort Sea 1976*. Calgary: Arctic Petroleum Operations Association.

Sverdrup, O. N., Bay, E., Schei, P., Simmons, H. G., & Hearn, E. H. (1904). *New land; four years in the Arctic regions*. London: Longmans Green and Co.

Veillette, J., Mueller, D. R., Antoniades, D., & Vincent, W. F. (2008). Arctic epishelf lakes as sentinel ecosystems: Past, present and future. *Journal of Geophysical Research – Biogeosciences, 113*, G04014. doi:10.1029/2008JG000730.

Vincent, W. F., Gibson, J. A. E., & Jeffries, M. O. (2001). Ice shelf collapse, climate change, and habitat loss in the Canadian high Arctic. *Polar Record, 37*, 133–142. doi:10.1017/S0032247400026954.

Vincent, W. F., Whyte, L. G., Lovejoy, C., Greer, C. W., Laurion, I., Suttle, C. A., Corbeil, J., & Mueller, D. R. (2009). Arctic microbial ecosystems and impacts of extreme warming during the International Polar Year. *Polar Science, 3*, 171–180. doi:10.1016/j.polar.2009.05.004.

Vincent, W. F., Fortier, D., Lévesque, E., Boulanger-Lapointe, N., Tremblay, B., Sarrazin, D., Antoniades, D., & Mueller, D. R. (2011). Extreme ecosystems and geosystems in the Canadian high Arctic: Ward Hunt Island and vicinity. *Ecoscience, 18*, 236–261. doi:10.2980/18-3-3448.

White, A., & Copland, L. (2015). Decadal-scale variations in glacier area changes across the southern Patagonian Icefield since the 1970s. *Arctic, Antarctic, and Alpine Research, 47*, 147–167. doi:10.1657/AAAR0013-102.

White, A., Copland, L., Mueller, D., & Van Wychen, W. (2015). Assessment of historical changes (1959–2012) and the causes of recent break-ups of the Petersen Ice Shelf, Nunavut, Canada. *Annals of Glaciology, 56*, 65–76. doi:10.3189/2015AoG69A687.

WMO. (1970). *Sea ice nomenclature: Terminology, codes, illustrated glossary and symbols.* Geneva: World Meteorological Organization.

Chapter 6
The Surface Mass Balance of the Ward Hunt Ice Shelf and Ward Hunt Ice Rise, Ellesmere Island, Nunavut, Canada

Carsten Braun

Abstract This chapter chronicles the mass balance and thickness of the Ward Hunt Ice Shelf and Ward Hunt Ice Rise, Ellesmere Island, Canada (1952–2008). The surface mass balance of the ice shelf and ice rise followed the mass balance changes of other monitored Canadian Arctic glaciers, but their overall mass losses have been comparatively low due to their proximity to the Arctic Ocean. Nevertheless, thinning of the Ward Hunt Ice Shelf via surface and basal melting has reduced its structural integrity as evidenced by the large-scale breakup of its eastern section between 2008 and 2011. In this context basal melting is considerably more significant in terms of overall ice shelf thickness and stability than the associated surface mass losses. The Ward Hunt Ice Shelf cannot readily reform again unless climatic conditions deteriorate for a prolonged period of time. The other remaining Canadian Arctic ice shelves appear to be in a similar situation as open water conditions on the Arctic Ocean become more prevalent and the dynamic stresses on the ice shelves related to wind, wave, and tidal action increase. This will eventually leave the Ward Hunt Ice Rise as one of the last remnants of Peary's 'Glacial Fringe' along the northern coast of Ellesmere Island, although its long-term survival is also threatened by current and predicted climatic change. A comprehensive measurement and modeling program is urgently needed to document and better understand the behaviour of these unique ice masses.

Keywords Arctic • Ice shelf • Ice rise • Mass balance • Climate change • Glacier

---

C. Braun (✉)
Geography and Regional Planning Department, Westfield State University, Westfield, MA, USA
e-mail: cbraun@westfield.ma.edu

L. Copland, D. Mueller (eds.), *Arctic Ice Shelves and Ice Islands*, Springer Polar Sciences, DOI 10.1007/978-94-024-1101-0_6

6.1 Introduction

The ice cover along the northern coast of Ellesmere Island today includes several small ice shelves and associated ice rises (Figs. 6.1 and 6.2) in topographically-suitable locations. These ice shelves were formed from in-situ accumulation of multi-year landfast sea ice, surface accumulations of snow/firn/superimposed ice, and basal freezing of seawater, with only minor mass input from associated upstream land glaciers (in contrast to a 'typical' Antarctic-type ice shelf). Lemmen et al. (1988), Jeffries (1987, 1992, 2002, 2017), and Vincent et al. (2001, 2011) provided comprehensive overviews of the Canadian Arctic ice shelves and Hattersley-Smith (1956, 1957a) included very readable summaries of the early scientific exploration history of northern Ellesmere Island.

Mass balance observations on the Ward Hunt Ice Shelf (WHIS) and Ward Hunt Ice Rise (WHIR) began indirectly more than 100 years ago with R.E. Peary's travels along the northern coast of Ellesmere Island during his quest for the North Pole

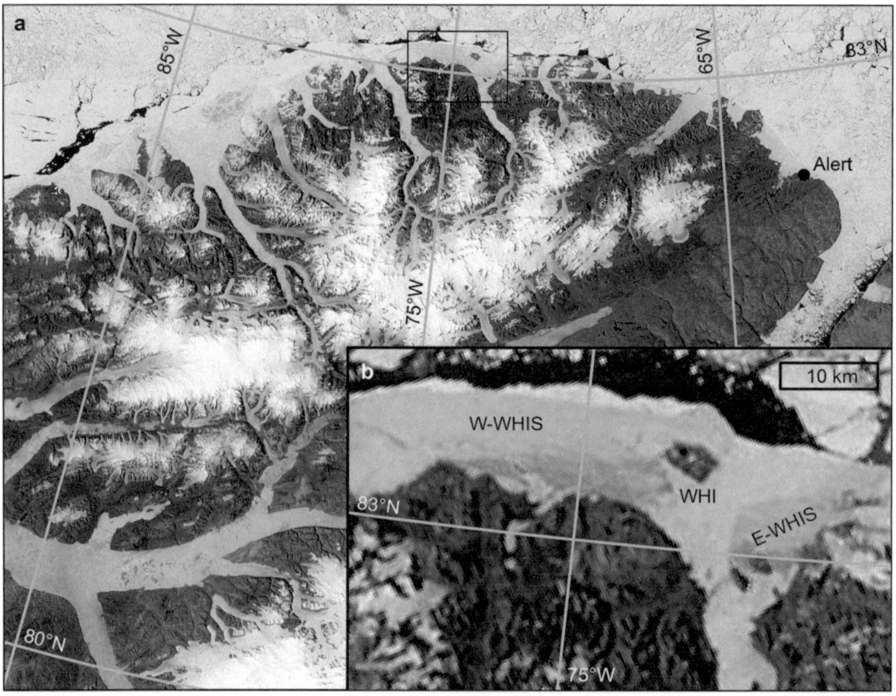

Fig. 6.1 (**a**) Northern Ellesmere Island seen as a Moderate Resolution Imaging Spectroradiometer (MODIS) image from 12 July 2015. The long-term climate station Alert is indicated by a *solid circle* and the *black rectangle* indicates the location of the inset map. (**b**) Close-up map using a MODIS image from 3 August 2015. The western Ward Hunt Ice Shelf (W-WHIS) extends westward from Ward Hunt Island (WHI) and the remnants of the eastern Ward Hunt Ice Shelf (E-WHIS) are visible east of WHI

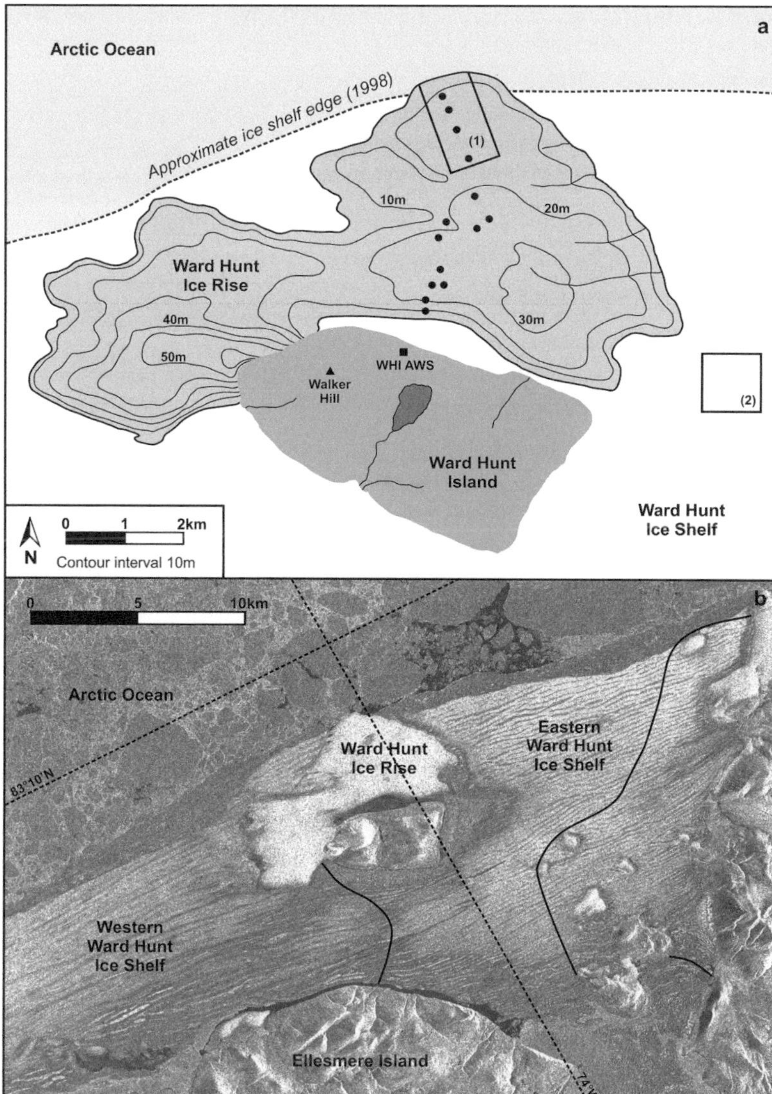

Fig. 6.2 (**a**) The Ward Hunt Ice Rise (WHIR) and surrounding Ward Hunt Ice Shelf (WHIS), Ellesmere Island, Canada, circa 1998. The *black circles* mark the ablation stake transect installed in 2002 and the approximate locations and extents of the original 1959 and 1966 stake grids on the ice rise (*1*) and the ice shelf (*2*) are indicated by *solid rectangles* (approximated from Serson 1979). The Ward Hunt Island Automated Weather Station (WHI AWS) (*black square*) is located at about 81°05'N and 74°09'W. The summit of Walker Hill (*black triangle*) is at 436 m asl. (**b**) RADARSAT-1 image of the WHIR and WHIS from 30 August 1998. The ice shelf surface shows the characteristic series of long, parallel ridges and troughs, which contain elongated meltwater lakes each summer. The eastern WHIS fractured and broke-apart between 2008 and 2011 (Mueller et al. 2017) and the *black lines* approximate the margins of the western WHIS and the eastern WHIS remnant in September 2015 based on 26 August 2012 Moderate Resolution Imaging Spectroradiometer (MODIS) image (Figure modified from Braun et al. (2004a) Copyright [2004] American Geophysical Union. Reproduced by permission of American Geophysical Union)

Table 6.1 The history of mass balance measurements on the Ward Hunt Ice Shelf (WHIS) and Ward Hunt Ice Rise (WHIR) since 1906

Time period	WHIS, WHIR
1906	R.E. Peary and companions travel along the northern coast of Ellesmere Island (Peary 1907; Hattersley-Smith et al. 1955).
1952 to 1958	Mass balance estimates available for the WHIS and WHIR based on limited measurements and observations on Ice Island T3, the WHIS (1954), and IGY Drifting Stations A and B.
1959 to 1986	WHIR: Parts of the data were presented by Sagar (1962), Hattersley-Smith and Serson (1970), Ommanney (1977), Serson (1979), and Koerner (1996).
1966 to 1986	WHIS: The complete records were compiled by Serson (1989, unpublished table) and discussed by Jeffries (1994).
1986 to 1989	15 (13) stakes measured on the WHIS (WHIR) on 9 March 1989 by R. Fiennes and members of the PUNS Expedition (Jeffries 1998). Cumulative net surface mass balance determined by H. Serson (Jeffries, pers. comm. 2002).
1989 to 2002	WHIS: Reconnaissance stake survey 27/28 July 2002; two (of six) remaining original stakes measured. WHIR: Comprehensive stake survey 23 to 27 July 2002; remaining original stakes measured; new stake transect installed (Braun et al. 2004a).
2003	WHIS: Comprehensive stake survey 10 August 2003 of all six remaining original stakes. WHIR: Measurement of original stakes and new stake transect 9 August 2003 (Braun et al. 2004a).
2004/2005	Complete stake measurements on the WHIS and WHIR each year in early August. Installation of five ablation stake clusters across the ice shelf (Mueller and Vincent 2006) in 2004.
2006 to 2008	Stake measurements on the WHIS and WHIR continue sporadically.
2008 to 2011	Eastern section of the WHIS disintegrates and stake network is lost. Condition of the WHIR stake network is unknown.

(Peary 1907). Scientific studies of the Canadian Arctic ice shelves and ice rises started over 50 years ago (e.g., Koenig et al. 1952; Miller 1954; Rodahl 1954; Hattersley-Smith et al. 1955), with surface mass balance measurements beginning on the ice rise in 1959 and in 1966 on the ice shelf (Table 6.1). Measurements continued on an annual basis until the mid-1970s and more intermittently until the spring of 1989. The University of Massachusetts Amherst (UMass Amherst) and Parks Canada re-initiated the mass balance measurements in July 2002. Remarkably, some of the original ablation stakes from the 1959/1966 observation grids had not melted out since last surveyed in 1989 and therefore provided a consistent record of relative ice surface height changes. The eastern section of the WHIS fractured and disintegrated between 2008 and 2011 resulting in the loss of the stake network and thereby ending the mass balance record for the time being.

This chapter chronicles the surface mass balance and thickness of the WHIS and WHIR, compares them to other mass balance data from the Canadian Arctic, and discusses implications in terms of ice shelf thickness and stability. Much of the collected glaciological data have been previously published as governmental or organizational reports not readily accessible to the international scientific community. This chapter represents an update and expansion of Braun et al. (2004a), supplemented with an extensive literature review, recent measurements, and additional data analysis included in Braun (2006). The record from the WHIR is currently the longest glacier mass balance series from the Canadian Arctic, and, together with the record from the WHIS, represents the world's two northernmost records of glacier mass balance. These records are of particular relevance today in light of public and scientific interest in the Arctic Ocean and its role in the ongoing environmental changes affecting the Arctic region as a whole and the glaciers and ice shelves of northern Ellesmere Island in particular (e.g., Vincent et al. 2001, 2009, 2011; Mueller and Vincent 2006; Mueller et al. 2003, 2006, 2008; Braun et al. 2004a; b; Koerner 1996, 2005; Copland et al. 2007; Veillette et al. 2008; Mortimer et al. 2012, 2016; Pope et al. 2012; Sharp et al. 2014; White et al. 2015a, b).

6.2 Background and Geographic Setting

6.2.1 Glaciation and Climate of Ellesmere Island's North Coast

The Canadian Arctic ice shelves formed 3000 to 5000 years ago (Lyons and Mielke 1973; Stewart and England 1983; Lemmen et al. 1988; Evans and England 1992; England et al. 2008; Antoniades et al. 2011) when climatic conditions deteriorated from the early/mid Holocene warm phase (Hattersley-Smith 1961; Bradley 1990; Jeffries 2002). The ice shelves were first described as 'paleocrystic ice' by the Nares expeditions, although the Eskimo word 'sikussak' (very old ice) appears to be roughly equivalent (Koch 1926; Koenig et al. 1952; Baird 1955; Bradley and England 2008). The entire northern coastline of Ellesmere Island was fringed by a continuous ice shelf some 500 km in length as late as the turn of the twentieth century (Hattersley-Smith 1960; Vincent et al. 2001) and the map by Koch (1926, his Fig. 1) shows paleocrystic ice continuing around the northern tip of Greenland at the time. There is also evidence suggesting that ice shelf-like features existed along the eastern coast of Ellesmere Island and western coast of Greenland in the early twentieth century (Koch 1926; Wright 1940). This large 'Ellesmere Ice Shelf' broke apart over the course of the twentieth century and today only about 10% remain (Vincent et al. 2001), the largest remnant being the WHIS (Fig. 6.1). Vincent et al. (2001) and Jeffries (1992, 2002) attributed the progressive disintegration of the 'Ellesmere Ice Shelf' to climate warming over the last 100 years. The WHIS fractured into two distinct pieces south of Ward Hunt Island between 2000 and 2002 after experiencing some 20 years of relative stability (Mueller et al. 2003). Ice shelf breakup accelerated with the collapse of the Ayles Ice Shelf in 2005, the loss of the Markham Ice Shelf in 2008, and the extensive fracturing and subsequent large-scale breakup of the eastern WHIS between 2008 and 2011 (Copland et al. 2007; Mueller et al. 2008, 2017; Vincent et al. 2011).

The WHIR probably formed within the last 1500 years when the ice shelf thickened and grounded on the isostatically-uplifted seafloor north of Ward Hunt Island (Crary 1960; Lyons and Ragle 1962; Lyons et al. 1971, 1972; Lyons and Mielke 1973). Gravity observations suggested a 'gently sloping sea floor north of Ward Hunt Island' suitable for ice shelf grounding, whereas water depths in the narrow channel between Ellesmere and Ward Hunt Island reach over 800 m (Lyons et al. 1972). Hattersley-Smith (1969) further speculated that the WHIR and the small Marvin Ice Rises near the mouth of Disraeli Fiord are grounded on glacial moraines accumulated 'in the lee of Ward Hunt Island and the Marvin Islands'. Ice rise thicknesses shown by Hattersley-Smith (1969) ranged between 45 and 100 m for an average of about 65 m and thickness of 55 m was confirmed in 1960 by ice core drilling into basal glacial till (Ragle et al. 1964). Ice rises and their associated shoals typically provide basal support and anchor points for otherwise floating ice shelves (e.g., Swithinbank 1955; Thomas 1979; Sanderson 1979; Doake and Wolff 1985; Cook and Vaughan 2010), but also create localized zones of weakness or 'nuclei' for ice shelf disintegration (e.g., Doake and Vaughan 1991; Vaughan 1993; Humbert and Braun 2008; Braun et al. 2009).

The lowest glaciation levels (i.e., the lowest elevation at which permanent ice can persist on a given landscape) and glacier equilibrium line altitudes (ELAs) in the Northern Hemisphere are found today along the northern coast of Ellesmere Island (Miller et al. 1975; Andrews and Barry, 1978), as manifested by low-elevation, coastal ice caps/ice rises, and marine ice shelves (such as the WHIR and WHIS). Frequent fog and low stratus clouds, associated with dominant airflow from the Arctic Ocean, lead to cooler conditions and reduced summer ablation in its immediate vicinity (e.g., Sagar 1962; Lister 1962a; Paterson 1969; Koerner 1979). At the same time, the Arctic Ocean also represents a local moisture source (Bradley and Eischeid 1985; Jeffries and Krouse 1984, 1987), leading to increased precipitation along the coast relative to the more interior parts of Ellesmere Island (e.g., Miller et al. 1975; Koerner 1979; Edlund and Alt 1989). The combination of reduced summer melt and enhanced snow accumulation causes the Arctic Ocean Effect (Braun et al. 2004a) that is responsible for the low long-term ELAs along the northern coast of Ellesmere Island (Miller et al. 1975; Koerner 1979) and the existence of the WHIS and WHIR (Crary 1960; Jeffries 2002; Braun et al. 2004a). Similarly, Koerner (2005) attributed melt suppression at the lower elevations of the Sverdrup Glacier (Devon Ice Cap) to increased open water conditions and resultant fog/cool air advection from nearby Jones Sound (Anda et al. 1985; Kinnard et al. 2008; Mair et al. 2009).

6.2.2 The Surface Topography of the WHIS

The characteristic corrugated ridge-trough surface of the Canadian Arctic ice shelves (i.e., 'The rolls on the Ellesmere Ice Shelf', Hattersley-Smith 1957b) was arguably responsible for their geographic and scientific re-discovery once their calved ice islands were found adrift in the Arctic Ocean (Koenig et al. 1952; Miller 1954, 1956, 1957; Rodahl 1954). Yet, similar ice surface features have been documented elsewhere, for example on lake ice (e.g., Montgomery 1952; Hattersley-Smith 1957b), on other ice shelves or ice-shelf like features around Greenland and Antarctica (e.g., Helk and Dunbar 1953; Swithinbank 1970; Higgins 1989; Cook and Vaughan 2010), on Arctic Ocean sea ice (e.g., Koenig et al. 1952; Baird 1955; Holt and Digby 1985; Fetterer and Untersteiner 1998; Luethje et al. 2006), and on multi-year landfast sea ice along the northern coast of Ellesmere Island (e.g., Hattersley-Smith 1957b; Jeffries and Serson 1986; Jeffries 1987). This implies that the ridge-trough surface topography is not (at least directly) related to the internal ice dynamics of a floating ice shelf (Hattersley-Smith 1957b), but rather that it can develop in a variety of horizontal locations, given suitable climatic conditions and the long-term presence of ice in that particular location. For example, the lack of systematic 'rolls' on the WHIR is presumably due to its vertical relief and associated sloped surface (Hattersley-Smith 1957b). At the same time, a view northwards from Walker Hill (Fig. 6.2) suggests that some of the ice shelf meltwater lakes are 'merging' with ice rise meltwater streams in an attempt to connect themselves across the ice rise (Lister 1962a).

The most plausible explanation (Jeffries 1992) for the ridge-trough surface topography was developed by Hattersley-Smith (1957b) and Crary (1960): Consistent winds, blowing predominately from west to east along the coastline, over time, elongate and coalesce existing random meltwater ponds into systematic, elongated, and parallel meltwater lakes (troughs), separated by dry ice ridges (Holt and Digby 1985). Jeffries et al. (1987, their Figs. 4 and 5) provided convincing visual documentation of melt pond development and elongation parallel to the prevailing wind direction and Miller (1954) included a beautiful photograph of the ice shelf rolls 'curling' into M'Clintock Inlet, following the shape of the coastline. Once established, melting within these lakes is enhanced relative to the surrounding ice shelf ridges by albedo differences and convection currents associated with radiative surface heating and wind (Crary 1960; Smith 1960, 1961; Untersteiner 1961; Hanson 1961). At the same time, winter snow drifts also tend to form about parallel to the coastline, aiding in the formation of already 'pre-oriented' meltwater ponds each spring (Hattersley-Smith 1957b).

6.2.3 Ice Shelf: Climate Relationships

Vincent et al. (2001, 2011), Mueller et al. (2003), Braun et al. (2004a), Copland et al. (2007), and White et al. (2015a) attributed the recent fracturing, calving, and thinning of the Canadian Arctic ice shelves to climate change, in particular to regional warming. The concept of an air temperature 'threshold' beyond which ice shelves are no longer viable is well-established for the Antarctic Peninsula (e.g., Mercer 1978; Vaughan and Doake 1996; van der Veen 2002; Vaughan 2006; Glasser and Scambos 2008; Cook and Vaughan 2010). However, the evidence also suggests that floating ice shelves have only a limited number of stable configurations (Doake et al. 1998) as dictated by the surrounding and/or underlying topography and are therefore susceptible to rapid retreat or break-up once the ice shelf front is forced to retreat behind such a stable configuration (e.g., Sanderson 1979; Weidick 1984; Warren 1991; Vaughan and Doake 1996; Cook et al. 2005). This implies that the rate of observable change for an ice shelf (e.g., retreat, fracturing, thinning, calving, breakup, etc.) is not a simple function of (concurrent) climate change, but also a dynamically-controlled process governed by the surrounding and/or underlying topography (Thomas 1973a, b; Warren 1991; Hindmarsh 1993, 1996; Doake et al. 1998; Fox and Vaughan 2005; Pfeffer 2007). On the other hand, Moon and Joughin (2008) stressed the importance of summer warming over other topographic or glacio-dynamic controls when evaluating the 1992 to 2007 retreat of tidewater glaciers around Greenland.

The surface mass balance record of the WHIS, furthermore, does not include possible mass gains or losses occurring at the ice shelf base (Crary 1956, 1960; Debenham 1954; Wexler 1960) and Hattersley-Smith et al. (1955) acknowledged that their 'prediction' for ice shelf disappearance by 2035 excluded any possible mass gains or additional mass losses at the underside of the floating ice shelf (Sect. 6.4.3). The recent scientific literature for the Antarctic ice shelves abounds with studies linking increased basal melting and resultant ice shelf thinning to warmer ocean water temperatures and ocean circulation changes (e.g., Rignot and Jacobs 2002; Shepherd et al. 2004; Smith et al. 2007; Jenkins and Jacobs 2008; Cook and Vaughan 2010; Wen et al. 2010; Cook et al. 2016) essentially broadening the scope of the ice shelf stability question to emphasize the ocean-ice shelf interface. A similar situation is apparent below the floating ice tongues around Greenland (e.g., Thomas et al. 2003; Rignot and Steffen 2008; Rignot et al. 2008; Holland et al. 2008; Falkner et al. 2011), where mass losses from basal melting far exceed those due to surface ablation and exert the dominant control on the overall floating ice thickness (to the point where ice-atmosphere interactions become essentially irrelevant, Thomas 1979; Pfeffer 2007; Nick et al. 2009; Rignot et al. 2010). A thinner ice shelf is also structurally weaker and thus more vulnerable to additional mechanical stresses such as crevasse-fracture, wind/wave/tidal action, internal oscillations, seismicity, and pack-ice pressure (e.g., Rott et al. 1996; Scambos et al. 2000; Glasser and Scambos 2008). All these factors have been proposed, in one way or another, as explanations for ice shelf breakup and ice island calving from the Canadian Arctic ice shelves (e.g., Koenig et al. 1952; Reeh 1968; Holdsworth 1971) as further discussed and synthesized by Copland et al. (2007, 2017) and White et al. (2015a).

6.3 Mass Balance Measurements on the WHIS and WHIR: History, Details, and Methods

6.3.1 Historical Overview

The surface mass balance measurements (Table 6.1) on the WHIS and WHIR began indirectly with R.E. Peary's quest for the North Pole at the turn of the twentieth century (Peary 1907) as G. Hattersley-Smith and companions discovered one of Peary's camp sites from 1906 during the first modern scientific exploration of Ellesmere Island's northern coastline in 1953 (Hattersley-Smith et al. 1955). Their finding implies that there was no net accumulation on the ice shelf (or adjacent ice rise) for the first half of the twentieth century (Hattersley-Smith and Serson 1970) consistent with the apparent warmth of the 1920s to 1940s (e.g., Bengtsson et al. 2004; Chylek et al. 2006). Crary et al. (1952) had previously interpreted a heavy dirt layer found on the surface of Ice Island T3 to indicate at least 100 years of net ablation. Tritium analysis of surface ice samples from the WHIR collected in August 1954 (Hattersley-Smith et al. 1955) confirmed that the ice rise surface was at least 25 years old in 1954 (Giletti and Kulp 1959), implying net surface mass losses between 1929 and 1954. Limited mass balance studies were conducted between 1952 and 1958 from Ice Island T3, the WHIS, and IGY Drifting Stations A and B (Crary 1960; Crary et al. 1952, 1955; Hattersley-Smith 1954, 1956; Hattersley-Smith et al. 1955; Hattersley-Smith and Serson 1970; Untersteiner 1961; Smith 1961). Crary (1960, his Fig. 4), for example, combined snow accumulation and ablation data measured on Ice Island T3 (1952, 1953, and 1955) with similar measurements from the WHIS in 1954.

Comprehensive mass balance measurements began in the spring of 1959 with the installation of 114 ablation stakes (Walker and Mattox 1961). 70 stakes were installed in a 5 by 14 grid pattern, covering about 820 by 2340 m, on the northern margin of the WHIR and extending onto the adjacent ice shelf (Fig. 6.2). Almost half of the stake grid (i.e., the part located on the floating ice shelf) was lost during the 1961/62 calving event (Hattersley-Smith 1963; Hattersley-Smith and Serson 1970) but 42 stakes remained useable on the ice rise after 1962 (Serson 1979). 44 additional ablation stakes were installed along a north-south survey line across the ice shelf, about 10 km west of Ward Hunt Island (Walker and Mattox 1961). The first measurements from 1959 and 1960 conducted at the 70 stake grid were analyzed by Sagar (1962), who noted greater surface ablation on the ice shelf compared to the adjacent ice rise (Lister 1962a, b). Ice shelf movement studies were initiated in 1964 just to the east of Ward Hunt Island (Konecny 1966; Konecny and Faig 1966) and repeated in 1968 (Dorrer 1971). Ice movement amounted to 30–50 cm year^{-1} in a generally northward direction, suggesting a more-or-less stagnant ice shelf (Konecny and Faig 1966). Ice movement on the WHIR was essentially zero (Dorrer 1971). 100 ablation stakes were installed in 1965/66 on the eastern WHIS in a 10 by 10 grid pattern, covering an area of about 0.9 km^2, just to the south of the original strain network used by Konecny, Faig, and Dorrer (Hattersley-Smith and

Serson 1970). This grid was thought to be representative of about 10 km^2 of the larger ice shelf surface (Ommanney 1977).

A number of different investigators and organizations have been involved in the field measurements and data analysis over the years (Table 6.1). The 1959 to 1968 ice rise and 1966 to 1968 ice shelf data were published by Hattersley-Smith and Serson (1970) and Serson (1979) summarized both records until 1976. In addition, Ommanney (1977) and Koerner (1996) compiled parts of both records. The late Harold Serson also recorded annual summaries of the mass balance measurements up until 25 May 1986 as a comprehensive, hand-written table (dated 1 May 1989) and those data were discussed by Jeffries (1994). The ablation stake networks were measured by R. Fiennes and members on the PUNS Expedition on 9 March 1989 (Jeffries 1998), allowing Serson to determine the cumulative surface mass balance of the WHIS and WHIR between 1986 and 1989 (Jeffries, pers. comm. 2002). In 2002, UMass Amherst and Parks Canada personnel installed a new north-south ablation stake transect across the ice rise consisting of 14 stakes (Fig. 6.2) and five clusters with six ablation stakes each were installed in 2004 to evaluate the spatial variability of ablation associated with different surface ice types and surficial materials (Mueller and Vincent 2006). One of those clusters (cluster WA) was deliberately placed within the area of the original 100 stake grid to ensure spatial and temporal overlap. The ice shelf stake network was lost between 2008 and 2011 during the breakup of the eastern WHIS. In addition, funding constraints and shifted research priorities curtail systematic glaciological research and monitoring on the WHIR and WHIS today.

6.3.2 Measurement Details

The original ablation stake grids (Fig. 6.2) were designed to measure surface mass balance changes using the direct glaciological method (Meier 1962; Østrem and Brugman 1991; Barry 2006) within a limited area (about 1 km^2) using a high-density of stakes (Serson 1979). The records do not, therefore, represent spatially-integrated values but rather a 'stake-farm' or 'index' mass balance approach (e.g., Koerner in Østrem and Brugman 1991). This is an important limitation, as Mueller and Vincent (2006) documented considerable spatial variability of surface ablation across the WHIS as a function of exposed ice type (e.g., marine vs. meteoric ice) and associated surficial materials (e.g., microbial mats and sediment). This also means that the interior and higher elevation parts of the WHIR were not included in the original measurement program (Ommanney 1977). It is also important to emphasize that any ice-surface lowering measured relative to the top of an ablation stake does not necessarily imply that the WHIS is thinning overall (Serson 1979) because of possible ice accretion processes occurring at the ice shelf base at the same time (Debenham 1954; Crary 1956; Lyons and Mielke 1973; Jeffries et al. 1988, 1991). This important caveat obviously does not apply to the grounded portion of the ice shelf, the WHIR.

The annual field measurements occurred as early as 9 March (1989) and as late as 24 June (1967). Measurements of winter snow accumulation thus represent an 8- to 10-month long window since the end of the previous summer's ablation season (Jeffries 1994). The winter balance (b_w) for each site was determined as the average of all available individual stake measurements (Serson 1979). The average change in ice-surface height from the previous year's measurement yielded the summer balance (b_s). Superimposed ice formation and summer snowfall were not explicitly measured or accounted for. The annual net surface balance was then calculated as $b_n = b_w - b_s$ (Serson 1979) and it is reasonable to consider ±20 to 50% as a conservative uncertainty estimate for the surface mass balance records (Hattersley-Smith and Serson 1970; Braun et al. 2004a).

On the WHIR, 19 of the original ablation stakes were recovered on 23 July 2002 in excellent condition and it was possible to match 16 of them with their last measurements on 9 March 1989. Seven stakes measured by Fiennes and members of the PUNS Expedition in 1989 were not found and presumably melted out at some point between 1989 and 2002 (Jeffries 1998). However, three additional stakes not measured in 1989 (presumably missed due to darkness and otherwise difficult circumstances) were recovered in 2002. In 2003, 2004, and 2005, the remaining stakes from the original grid were re-measured together with the new stake transect installed in 2002. The 2006 and 2008 stake measurements on the ice rise were inadequate to be included in the mass balance calculations, but allow for robust generalized statements and an estimate of the 2007/2008 winter snow accumulation after assuming a snow density of 0.31 g cm^{-3} (Hattersley-Smith and Serson 1970; Serson 1979).

On the WHIS, logistic constraints in 2002 prevented a comprehensive survey of the 1966 grid, but two of the original ablation stakes were recovered in useable condition. Four additional stakes were recovered during a more detailed survey on 10 August 2003. It was possible to match all six stakes recovered in 2003 with their respective 1989 measurements and it is reasonable to assume that the other nine stakes measured by Fiennes and other members of the PUNS Expedition melted out sometime between 1989 and 2003. Starting in 2005, ablation stake cluster WA installed by D. Mueller (Mueller and Vincent 2006) was used to continue the ice shelf surface mass balance record. Unfortunately, the 2003 and 2004 mass balance values shown in Fig. 6.3 are based on measurements at only two and one ablation stakes, respectively. The ice shelf stakes were not measured in 2006; therefore the spring 2008 measurements (representing the 2007 ice surface) yielded a combined 2-year balance. Ice surface height changes from all available ablation stakes for each balance year were averaged and multiplied by an ice density of 0.9 (Ragle et al. 1964; Hattersley-Smith and Serson 1970; Jeffries et al. 1988) to express the net surface mass balance changes in water equivalent (weq).

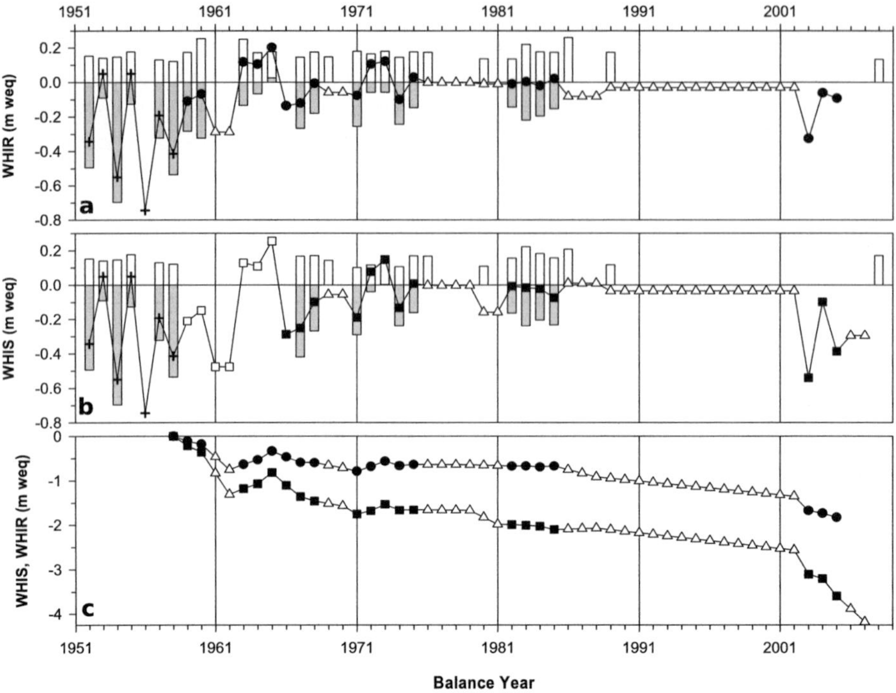

Fig. 6.3 The surface mass balance of the Ward Hunt Ice Shelf (WHIS) and Ice Rise (WHIR) (1952 to 2008). (**a**) WHIR: winter (*white bar*), summer (*grey bar*), and annual (*black circle*) surface mass balance. (**b**) WHIS: winter (*white bar*), summer (*grey bar*), and annual (*black square*) mass surface balance. (**c**) WHIR (*circle*, 1959 to 2005) and WHIS (*square*, 1959 to 2008) cumulative surface mass balance. Values shown prior to 1959 are based on limited measurements and estimates (Sect. 6.3.1, Table 6.1). Annual values calculated as averages of multi-year balances are indicated with *open triangles*. The 1959 to 1965 values for the ice shelf (*open squares*) are calculated from corresponding ice rise measurements using a linear regression (Sect. 6.4.1). Note: the records shown in Fig. 6.3 are not spatially-integrated values, but instead represent a 'stake-farm' or 'index' mass balance approach (Sect. 6.3.2) (Figure modified from Braun et al. (2004a) Copyright [2004] American Geophysical Union. Reproduced by permission of American Geophysical Union)

6.4 Surface Mass Balance of the WHIS and WHIR

6.4.1 Surface Mass Balance Records

Figure 6.3 presents the surface mass balance records compiled for the WHIS and WHIR (1952 to 2008). The records are continuous, but measurements after 1976 occurred much more intermittently (except for several years in the early 1980s), resulting in many multi-year balances. The striking absence of inter-annual variability between 1986 and 2002 is therefore largely an artifact and reflects the aforementioned gap in the observations. Separately measured winter and summer

Table 6.2 Surface mass balance measurements (1959 to 2003) on the WHIS and WHIR. Values in parentheses are based on years when measurements are available at both sites

Winter Balance	**WHIR**	**WHIS**
Mean snow depth (m)	0.52 (0.48)	0.50
Mean snow bulk density	0.35 (0.36)	0.31
Mean snow accumulation (m weq)	0.18 (0.17)	0.15
Number of years measured	21	16
Coefficient of determination (R^2; 16 years)	0.41 ($p = 0.008$)	
Summer Balance		
Mean ablation (m weq)	−0.17 (−0.18)	−0.20
Number of years measured	16	11
Coefficient of determination (R^2; 11 years)	0.84 ($p < 0.0001$)	
Annual Balance		
Mean annual balance (m weq)	−0.04	−0.07
Cumulative annual balance (m weq)	−1.68	−3.1
Number of years measured	45	45
Coefficient of determination (R^2; 12 years)[a]	0.89 ($p < 0.0001$)	

Measurements on the WHIS began in 1966; the 1959 to 1965 values were estimated from measurements on the WHIR

[a]Using only individually measured years (i.e., excluding those annual values determined based on averages of multi-year surface balances)

balances are available for about half of all years on record. The 1952 to 1958 values shown are based on sporadic measurements and observations found in the literature.

Winter snow accumulation has remained relatively constant from year-to-year (Fig. 6.3, Table 6.2) and compares well with values reported by Koerner (1973) during a traverse of the Arctic Ocean (Herbert and Koerner 1970) and by Jeffries and Krouse (1987) for snow surveys along the northern coast of Ellesmere Island between 1982 and 1985. The spatial variability of snow accumulation on the ice shelf was primarily related to its corrugated ridge-trough surface topography (Lister 1962a; Sagar 1962). Hattersley-Smith and Serson (1970) attributed the relatively large spatial variability of snow depth on the ice rise to snow drifting interacting with the surface topography as the relief on the WHIR, although subdued, is comparatively large in relation to overall snow depth. In contrast, summer ablation was considerably more variable from year-to-year and largely controlled the annual mass balance variations. The stake data reveal a high-degree of spatial variability, including sometimes net ablation and net accumulation at different stakes for the same mass balance year (Hattersley-Smith and Serson 1970). Both records show infrequent positive surface mass balance years (e.g., 1963 to 1965, 1972/73), but overall negative years dominate both records.

Summer and annual mass balances are consistently more negative on the ice shelf than on the ice rise (Table 6.2) because of the corrugated ridge-trough topography and associated formation of elongated meltwater lakes on the ice shelf surface (Lister 1962a; Sagar 1962; Hattersley-Smith and Serson 1970; Jeffries 1994).

Ablation within these lakes is enhanced relative to the surrounding ice shelf ridges (or the ice rise) by the continuous wind-driven circulation of meltwater (Hattersley-Smith 1957b; Crary 1960; Lister 1962a). In addition, albedo measurements on Ice Island T3 indicated a surface albedo of about 77% for the dry ice ridges and of about 38% for the meltwater lakes (Hanson 1961). Mueller and Vincent (2006) reported similar albedo values for the Ward Hunt and Markham ice shelves. Lotz (1961a, b) proposed a complementary meteorological explanation noting the higher incidence of fog suppressing melt on the ice rise compared to the adjacent ice shelf.

Annual and summer balances for individually-measured years (i.e., excluding those values determined as averages of multi-year balances) are highly correlated between the WHIS and WHIR, whereas the correlation for their respective winter balances is much lower (Table 6.2). This high degree of statistical association was used to extend the ice shelf record back to 1959 (using a simple linear regression) to allow a better comparison of the respective cumulative surface mass balances. Between 1959 and 2005, there has been an overall surface mass loss of 3.6 m weq (0.08 m weq year^{-1}) on the ice shelf and of 1.8 m weq (0.04 m weq year^{-1}) on the ice rise (Fig. 6.3c). Between 1989 and 2002, the ice rise lost 0.44 m weq ice (0.03 m weq year^{-1}), whereas the ice shelf experienced an overall surface mass loss of 1.03 m weq between 1989 and 2003 (0.07 m weq year^{-1}). Measurements at two stakes on the WHIS indicate that about 50% of this mass (0.54 m weq) was lost during the 2003 balance year. 2003 was also the most negative individually-measured year on record for the WHIR, with an annual surface mass loss of 0.33 m weq. If we consider the entire record (1952 to 2007), cumulative surface mass losses on the WHIS amounted to about 6.3 m weq (0.11 m weq year^{-1}), representing a relative ice surface lowering of about 7 m.

It is important to emphasize that the 1989 to 2003 cumulative balance (−1.03 m weq) calculated for the ice shelf is based on measurements at only six ablation stakes. The 1989 to 2002 cumulative balance of the ice rise (−0.44 m weq) is more reliable as it represents an average of 16 individual stake measurements. However, the latter parts of both records may progressively underestimate the actual mass losses, as the total number of ablation stakes contributing to each annual average decreased, with those at (locally) high melt locations (e.g., stakes located inside ice shelf meltwater lakes) likely to have been lost earlier. It is also reasonable to assume that ablation stakes were preferentially installed on localized topographic highs to minimize problems with meltwater ponds forming around the base of the stake each summer (Sagar 1962; Hanson 1980). This would systematically bias the stake data towards lower accummulation and higher ablation values. The records represent a relatively small horizontal area at both sites (Sect. 6.3.2) and in the past, any spatial mass balance variability across the WHIS was attributed solely to its ridge-trough surface topography. The data presented by Mueller and Vincent (2006), however, also indicated systematic differences across the ice shelf with, for example, about twice the ablation between 2004 and 2005 in marine vs. meteoric surface ice areas. Marine ice areas tend to contain large amounts of sediment and microbial mats (Mueller et al. 2006), thus decreasing their albedo. The original ice shelf stake grid (Fig. 6.2) was located in a relatively clean, meteoric surface ice area and the record is therefore biased towards an underestimation of overall surface ablation. These spatial ablation patterns are presumably also applicable to the other remaining Canadian Arctic ice shelves (Mueller et al. 2006).

Measurements along the new stake transect indicated that the WHIR remained entirely in the ablation zone between 2003 and 2008 and thus did not support an accumulation area. Mass losses were consistently greater at the lower elevations near the ice margin compared to the higher-elevation areas toward the center of the ice rise. This situation, in the absence of a meaningful elevation gradient across the ice rise, may be explained by subtle differences in ice surface albedo and localized climate (Lotz 1961a, b). Considerable accumulations of wind-blown dust, together with well-developed cryoconite holes, are characteristic at the lower elevations near the ice margin, whereas the ice surface towards the center and higher elevations of the ice rise is generally clean, white ice.

6.4.2 Comparison with Other Canadian Arctic Glaciers

The mass balance of other glaciers in the Canadian Arctic has been predominantly negative over the last five decades (e.g., Dowdeswell et al. 1997; Koerner 2005; Gardner and Sharp 2007; Sharp et al. 2011, 2014; Fisher et al. 2012; Derksen et al. 2012; Mortimer et al. 2016) with a consistent turn towards increasingly negative values since the 1990s (Fig. 6.4; Table 6.3). This generalized trend was interrupted from the mid-1960s to the mid-1970s when much of the Canadian Arctic experienced colder summers, more severe sea ice conditions, and positive glacier mass balances (e.g., Dunbar 1972, 1976; Bradley and Miller 1972; Bradley and England 1978; Alt 1979, 1987; Dowdeswell et al. 1997; Braun et al. 2004a; b). The mass balance of the WHIS and WHIR tracked this overall temporal pattern, but the magnitude of their surface mass losses has been comparatively low, especially between 1991 and 2000. The simple fact that ablation stakes installed in 1959 and 1966 had survived for four decades without melting out of the ice speaks to the limited amount of surface ablation on the WHIS and WHIR. This difference, and, more fundamentally, their existence and survival reflects the localized climatic influence of the Arctic Ocean (Sect. 6.2.1). In addition, superimposed ice formation plays an important role in the annual mass balance of the ice shelf and ice rise (Koenig et al. 1952; Hattersley-Smith 1954), allowing them to 'absorb' some amount of melting each year before any net surface mass losses occur (Schytt 1949; Ambach 1985).

A close glaciological analogue to the WHIS and WHIR is the Meighen Ice Cap, a low-elevation, coastal ice cap located about 500 km to the southwest on Meighen Island. The existence and continued survival of this ice cap have been linked to locally increased snow accumulation and reduced summer ablation due to its close proximity to the Arctic Ocean (Paterson 1969; Alt 1979, 1987; Koerner 1979, 2005) and superimposed ice formation (Arnold 1965). However, the 1990s and 2000s have been the most negative mass balance decades on record for Meighen Ice Cap (Table 6.3), and followed three decades with a weak trend towards less negative

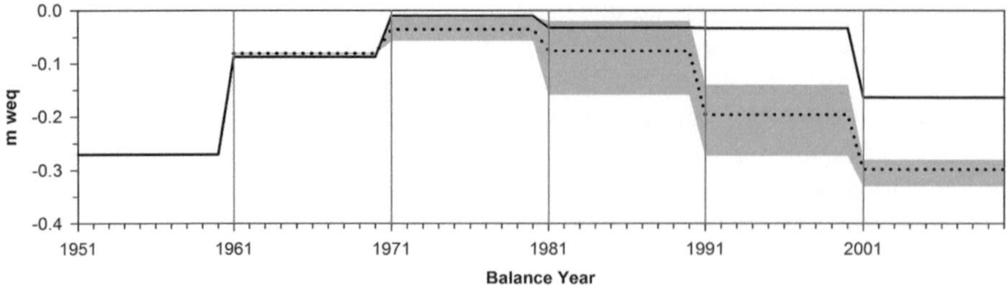

Fig. 6.4 Decadal mass balance means for selected Canadian Arctic Glaciers (Table 6.3). Shown in *grey* is the composite range in decadal values for the White Glacier, Devon Ice Cap, and Meighen Ice Cap, with the decadal mean for this group of glaciers shown as a *dotted line* (1961 to 2010). The decadal mean combined surface mass balance for the Ward Hunt Ice Shelf (WHIS) and Ice Rise (WHIR) is shown by a *solid black line* (1951 to 2010) (Figure modified from Braun et al. (2004a) Copyright [2004] American Geophysical Union. Reproduced by permission of American Geophysical Union)

Table 6.3 Selected glacier mass balance records (m weq per year) from the Canadian Arctic and decadal mean July air temperature at Ward Hunt Island

Glacier Name	1951–60 mean	1961–70 mean	1971–80 mean	1981–90 mean	1991–00 mean	2001–10 mean
WHIR	–0.27[a]	–0.05	+0.01	–0.03	–0.03	–0.11[b]
WHIS	–0.29[a]	–0.12[c]	–0.03	–0.03	–0.04	–0.22[b]
Meighen Ice Cap	NA	–0.08	–0.06	–0.02	–0.18	–0.28
Devon Ice Cap NW	NA	–0.08	–0.01	–0.05	–0.14	–0.28
White Glacier[d]	NA	–0.08	–0.03	–0.16	–0.27	–0.33
July air temperature[e]	1.44°C	1.19°C	0.88°C	1.05°C	1.09°C	1.46°C

Decadal mean annual mass balance values calculated based on annual data from the GMBAL Release 1501 database available at http://people.trentu.ca/~gcogley/glaciology/

[a]Sporadic observations began in 1952 (Sect. 6.3.1). Detailed measurements started on the ice rise in 1959. The 1959/1960 ice shelf values are estimated from ice rise measurements

[b]2001 to 2005 only

[c]Measurements started in 1966; the 1961 to 1965 values are estimated from ice rise measurements

[d]The White Glacier is considered a 'typical' High Arctic glacier (Müller 1962; Cogley et al. 1996; Adams et al. 1998)

[e]Mean July temperature at Ward Hunt Island, reconstructed from Alert climate data (Sect. 6.4.3)

mass balance values (Dowdeswell et al. 1997; Koerner, pers. comm. 2003). More recently, Mair et al. (2009) reported that the overall mass balance of the Prince of Wales Icefield on south-eastern Ellesmere Island has remained approximately balanced for the last several decades due to melt suppression and/or increased accumulation associated with the North Water polynya. This finding and its suggested explanation is similar to the situation documented here for the WHIS and WHIR (Braun et al. 2004a).

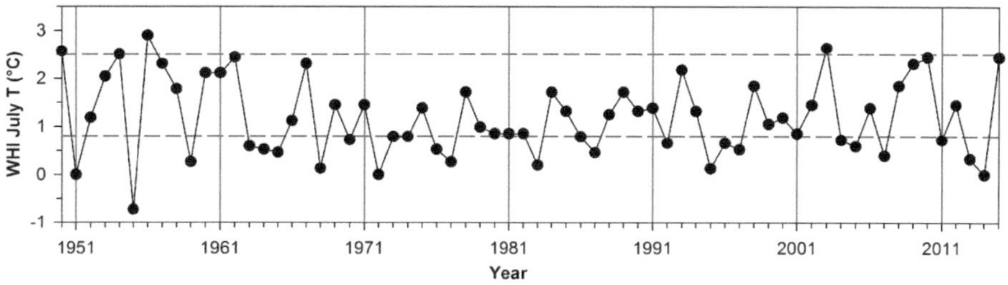

Fig. 6.5 Reconstructed mean July air temperature at Ward Hunt Island (WHI) (1950 to 2015) based on a transformation of the corresponding monthly temperatures at Alert. Data: http://climate.weather.gc.ca/prods_servs/cdn_climate_summary_e.html. The stability threshold for the Ward Hunt Ice Shelf (WHIS) (~2.5°C, Hattersley-Smith et al. 1955) and the positive mass balance threshold for the Ward Hunt Ice Rise (WHIR) (~0.8°C, Serson 1979) are indicated by dashed lines (Figure modified from Braun et al. (2004a) Copyright [2004] American Geophysical Union. Reproduced by permission of American Geophysical Union)

6.4.3 Climatic Considerations

Serson (1979) speculated, based on simple statistical comparisons, that a mean July air temperature of about 3°C or less at Alert (Fig. 6.1) is typically required for a positive mass balance year on the WHIR. Hattersley-Smith and Serson (1970), on the other hand, had remarked earlier that 'there is no close correlation' between summer ablation on the ice shelf or ice rise and mean monthly air temperatures at Alert, yet also noted the overall association of summer climate and surface mass balance for 'more than average warm summers and more than average cold summers in the two areas' (Serson 1979). Copland et al. (2007) extended this air temperature threshold concept (Sect. 6.2.3) to the more meaningful metric of cumulative positive degree-days (PDDs) (Reeh 1991; Vaughan 2006) and speculated that the Canadian Arctic ice shelves may no longer be viable beyond about 200 PDDs year^{-1}. Such simple temperature-based indexes can only serve as general estimates, as other factors (such as variations in winter snow accumulation or the amount/frequency of summer snowfall) also significantly influence their annual surface mass balance.

Parks Canada has operated a weather station on Ward Hunt Island (WHI AWS; Fig. 6.2) since June 1995. July is the only month of the year with a mean air temperature above freezing and a useful index for the surface mass balance of the ice shelf and ice rise (Mercer 1978; Serson 1979; Bradley and England 1978; Wang et al. 2005; Gardner and Sharp 2007). Mean monthly air temperatures at Alert and Ward Hunt Island are highly correlated, and a simple regression was used to reconstruct July air temperatures for Ward Hunt Island (using only June, July, and August; $n = 14$ months; $R^2 = 0.84$; $p < 0.0001$; SE = 0.69°C; Fig. 6.5). The 1950 to 2015 mean of 1.2°C is considerably higher than Serson's threshold temperature required for mass gains on the WHIR (~3°C at Alert; ~0.8°C on Ward Hunt Island). There is no statistically significant trend over the entire 66-year-long record, although there have been increases in local and Arctic-wide summer air temperatures if one evaluates trends since the end of the 'Little Ice Age' (100 to 150 years ago), trends over the last 20 to 30 years (e.g., Vincent et al. 2001; Mueller et al. 2003; Comiso 2003, 2006), or more recent trends between 1995 and 2005 (Przybylak 2007). It is also worth noting that fall and winter temperatures in the region have risen considerably over the last 60+ years (e.g., Zhang et al. 2000; Przybylak 2007; Lesins et al. 2010), a trend that enhances melting each subsequent summer (Copland et al. 2007).

The last few decades have also seen considerable reductions in Arctic Ocean sea ice cover (e.g., Comiso et al. 2008; Rothrock et al. 2008; Stroeve et al. 2008, 2014; Kwok and Rothrock 2009; Howell et al. 2013; Overland and Wang, 2013). Paradoxically, reduced sea ice actually favours, at least to some extent, more positive (or at least less negative) surface mass balances on the surrounding glaciers by increasing accumulation and reducing ablation via an enhanced Arctic Ocean Effect (e.g., Jonsson 1982; Krabill et al. 2000; Bamber et al. 2004; Koerner 2005; Mair et al. 2009; Sylvestre et al. 2013; Sect. 6.2.1). Increased summer snowfall may only contribute little additional mass, but indirectly reduces ablation by raising the albedo of the ice and surrounding land surface (Ahlmann 1929).

Hattersley-Smith et al. (1955) anticipated the disappearance of the WHIS by 2035 if summer conditions similar to those of 1954 (mean July air temperature at Alert 5.6°C; ~2.5°C on Ward Hunt Island) were to become common. 1954 was an exceptionally warm and long summer on northern Ellesmere Island, lasting from early July to early September (Crary 1956) and sea ice freeze-up around Alert was delayed until late September (Bilello 1961). Such warm summers have not recurred over the last ~40 years (Fig. 6.5) until July 2003 and consequently the WHIS and WHIR experienced one of their most negative surface mass balance years on record (Fig. 6.3). The WHIS experienced considerable physical changes since 2000 (Vincent et al. 2001, 2011; Mueller et al. 2003, 2008, 2017) after an extended phase of relative stability and it is interesting to note that the 2008 to 2011 breakup of the eastern WHIS coincided with a 3-year period of unusually warm summers (cf. Copland et al. 2007; White et al. 2015a). At the same time, it is important to emphasize that the (over)-simplified cause-and-effect sequence of climate change, glacier mass balance, and glacier advance or retreat does not necessarily apply to floating ice shelves—the influence of surrounding and underlying topography can 'decouple' ice shelf growth or retreat from direct climatic controls (Sect. 6.2.3).

6.5 The Thinning of the WHIS

The WHIS surface mass balance record represents only one side of the overall ice shelf thickness, but melting and freezing (i.e., mass losses and gains) at the base of the floating ice shelf also influence its overall thickness (Serson 1979). Jenkins and Jacobs (2008), for example, documented extensive surface melting on the northern parts of the George VI Ice Shelf (Antarctic Peninsula), yet that only represented an insignificant fraction of the overall mass budget, since basal melt rates ranged between 1 and 5 m year^{-1}. Mortimer et al. (2012) and White et al. (2015a) have documented similar circumstances for the Milne and Petersen ice shelves off northern Ellesmere Island. The processes at the ocean/ice shelf interface have fascinated scientists since their initial discovery (e.g., Wright 1925; Gould 1935; Debenham 1919, 1948, 1954), yet it remains essentially inaccessible. In the absence of direct measurements, processes at the base of the WHIS have been inferred through indirect means, such as ice coring, seismic stratigraphy, studies of surficial materials, and limited measurements of fiord oceanography (Jeffries 1987, 2002).

6.5.1 Ice Shelf Thickness Measurements

Table 6.4 compiles all thickness measurements and estimates for the WHIS shown graphically in Fig. 6.6. These measurements often represent discrete point values (using a variety of methods) from different parts of the ice shelf and therefore not a spatially-integrated or spatially-consistent data set (Debenham 1954). Hattersley-Smith (1969), for example, provided the much-quoted thickness range of 20 to 80 m for the WHIS, yet more recent surveys by Narod et al. (1988) revealed a much more consistent thickness between about 45 and 50 m for much of the eastern WHIS. Their airborne thickness measurements were subsequently confirmed by ice core drilling on 'Hobson's Choice' Ice Island (Jeffries et al. 1988, 1991).

Despite these limitations, Fig. 6.6 demonstrates that the thickness of the WHIS has remained more-or-less constant until the early-1990s and decreased considerably thereafter. Miller (1956, 1957) provided the first published ice shelf thickness estimate (50 to 60 m) based on his aerial observations of ice shelf freeboard on 1 August 1951 (Althoff 2007) and the first in situ seismic measurements by Crary et al. (1952) revealed a thickness of about 50 m for Ice Island T3. Crary (1958) presented ice thickness measurements and topographic survey data demonstrating a gradual thinning of the ice shelf west of Ward Hunt Island from up to 60 m near the coast of Ellesmere Island to only about 9 to 12 m near its northern seaboard margin for a mean ice shelf thickness of about 50 m. Measurements and estimates in the 1970s and 1980s ranged consistently between 40 and 50 m. Vincent et al. (2001) and Mueller et al. (2003, 2008) recently provided evidence for a substantial thinning of the eastern WHIS down to about 25 m or less.

6.5.2 Ice Shelf Thinning

In the past, 40 to 50 m was a reasonable estimate for the average thickness of the WHIS (e.g., Jeffries et al. 1988; Jeffries 1992). It is also reasonable to assume a more-or-less constant ice thickness for the unconfined, floating sections of an ice shelf (Wright 1925; Weertman 1957; Reeh 1968; Thomas 1973a, b), but ice thickness should increase away from the ice front, especially in sections where the ice

Table 6.4 Ice thickness measurements and estimates for the WHIS. 'Year' refers to the year measured (if specified); otherwise the publication year is used, recognizing that the actual measurement must have occurred earlier

Year	Thickness	Source
1951	50–60 m	Miller (1956, 1957), aerial freeboard observations on 1 August 1951.
1952	49 m	Crary et al. (1952). Single thickness measurement of Ice Island T3 using seismic methods. Ice Island T3 originated from an ice shelf disintegration in Yelverton Bay around 1935 (Jeffries and Serson 1986).
1954	50 m	Crary (1958). Mean of 7 measurements, using seismic methods, along a north-south survey line across the western WHIS (range 43 m to 54 m).
1958	60 m	Plouff et al. (1961). Electromagnetic survey of Ice Island T3.
1960	43 m	Lyons and Ragle (1962, and references therein). Mean thickness for the WHIS based on a variety of different methods.
1964	45 m	Konecny and Faig (1966). Presumably an assumed ice shelf thickness, based on unspecified previous studies.
1966	44.5 m	Hattersley-Smith (1969). Simple mean based on airborne radio echo sounding of the eastern WHIS (thickness range 20 m to 80 m).
1967	48 m	Vincent et al. (2001, and references therein). Salinity water column profile near the mouth of Disraeli Fiord.
1968	37 m	Thomas (1973b, after Dorrer 1971). Mean thickness of the eastern WHIS the strain network based on freeboard and density considerations.
1969	47 m	Jeffries and Sackinger (1989). Point thickness from ice core B recovered between Ward Hunt Island and Ellesmere Island (Lyons et al. 1971).
1971	44 m	Lyons et al. (1971). Average ice shelf thickness, no date or location specified.
1972	43 m	Lyons et al. (1972). Average ice shelf thickness, no date or location specified.
1973	43 m	Lyons and Mielke (1973). Average ice shelf thickness, no date or location specified.
1981	47.5 m	Narod et al. (1988). Airborne UHF radar sounding, thickness range 40 to 50 m consistent across sections of the eastern WHIS. Part of the ice shelf surveyed became 'Hobson's Choice' Ice Island upon calving in 1982/83 (Jeffries and Serson 1983; Jeffries and Shaw 1993).
1982	50.66 m	Jeffries (1994). Core 82–1, western WHIS.
1983	45 m	Vincent et al. (2001, and references therein). Salinity water column profile near the mouth of Disraeli Fiord. Jeffries (1991) estimated a thickness of 44.5 m for the western WHIS based on ice core 83–1 and an assumed basal ice temperature/ice temperature profile.
1985	42 m	Jeffries et al. (1988, 1991). Core 85–10 from 'Hobson's Choice' Ice Island. Thickness range for 'Hobson's Choice' Ice Island: 41 to 44 m.
1988	47.5 m	Jeffries et al. (1988). Average ice thickness, considered representative for the larger WHIS.
1992	47.5 m	Jeffries (1992). Average thickness of the WHIS, citing measurements between 1954 and 1985 using a variety of methods.
1999	35 m	Vincent et al. (2001). Salinity water column profile near the mouth of Disraeli Fiord.
2002	<25 m	Mueller et al. (2003). Localized thickness estimate from ice shelf freeboard and density considerations along cracks south of Ward Hunt Island.

Additional redundant measurements and estimates were reported in, for example: Koenig et al. (1952), Crary and Cotell (1952), Fletcher (1953), Baird and Sharp (1954), Miller (1954), Rodahl (1954), Baird (1955), Crary et al. (1955), Polunin (1955), Hattersley-Smith (1957a, b, 1985), Crary (1960), Ragle et al. (1964), Evans and Robin (1966), Jeffries (1987, 2002), Jeffries et al. 1990, Mueller and Vincent (2006), and Althoff (2007)

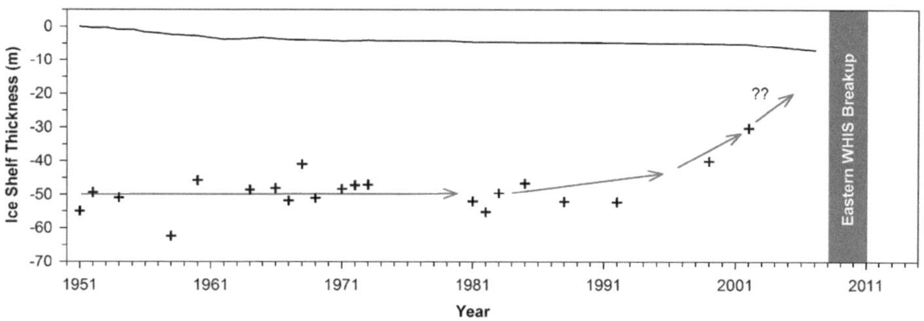

Fig. 6.6 Thickness changes of the Ward Hunt Ice Shelf (WHIS). The *top curve* shows the cumulative ice shelf lowering between 1952 and 2008 (about 7 m) due to surface mass losses (Sect. 6.3.2) and the ice thickness measurements compiled in Table 6.4 are shown as *black crosses* (Sect. 6.5)

shelf is confined by topography (Crary 1966; Thomas 1973a, b; Robin 1975, 1979; Sanderson 1979) and decrease where the ice shelf is locally-grounded in the vicinity of islands and ice rises (Crary 1960; Lyons and Leavitt 1961). Lyons and Ragle (1962), for example, reported a mean ice shelf thickness of about 43 m, decreasing from about 47 m between Ward Hunt Island and Ellesmere Island to about 20 m towards the ice front (Crowley 1961; Lyons and Leavitt 1961) consistent with the data published by Crary (1958). The majority of the overall thinning occurred over the last couple of kilometres towards the seaward margin (Crary 1958; Holdsworth 1971). It is interesting to note that this thinner section directly adjacent to the Arctic Ocean was also characterized by a strongly diminished ridge-trough surface topography (Crary 1958; Lister 1962a). Bushnell (1956), for example, noted that Marvin during his 1906 hydrographic survey along the northern coast of Ellesmere Island failed to mention the otherwise characteristic ‘rolls’ of the ice shelf surface, presumably because he traveled along the seaward margin of the ice shelf. Hattersley-Smith (1957b) also noted that the spacing between adjacent ice shelf ridges increases as a function of overall thickness, a relationship later discussed by Jeffries et al. (1990) and Jeffries (2017) as a simple tool to approximate ice shelf thickness. This thinner ice shelf ‘fringe’ broke free between August 1961 and April 1962 when the WHIS retreated back to its grounded portion (i.e., the WHIR), essentially losing its seaward section along almost its entire width (Hattersley-Smith 1963; Holdsworth 1971). The next major calving event did not occur until 1982/83, when some 35 to 40 km^2 broke free from the eastern WHIS to create ‘Hobson’s Choice’ Ice Island (Jeffries and Serson 1983, Jeffries and Shaw 1993), implying some 20 years of relative stability for the ice shelf. ‘Hobson’s Choice’ Ice Island, however, was about 41 to 44 m thick (Table 6.4), and thus presumably represented a section of the main body of the WHIS.

Crary (1960) discovered a large marine sponge on the ice shelf near the eastern end of Ward Hunt Island and attributed its presence to the ‘ice-elevator’ mechanism originally proposed by Debenham (1919, 1954): mass losses at the ice surface, when matched by concomitant mass gains through basal freezing, effectively create a vertical mass transport mechanism for ice, marine organisms, and other ocean floor material upwards through the ice shelf (Sverdrup 1931; Lyons and Ragle 1962; Swithinbank 1970; Hattersley-Smith 1970; Pedley et al. 1988; Jeffries 1992; Smith et al. 2007). The necessary 'downward thickening' (Jeffries 1985) occurs through the accretion of frazil ice crystals to the underside of the ice shelf. This process was first described for the WHIS by Crary (1956) and subsequently elaborated by upon Lyons et al. (1971) and Serson (1979). Jeffries (1987) and Jeffries et al. (1988, 1991) quantified this process and concluded that, at least between 1952 and 1982, mass additions to the base of the WHIS via basal freezing essentially compensated for the concomitant surface mass losses, leading to an overall stable ice shelf with respect to its thickness (Fig. 6.6). Furthermore, Jeffries and Serson (1986) discovered that calving losses from the ice front were often replaced through the accretion of multi-year landfast sea ice. This efficient 'self-healing' process was operational between about 1962 and 1985 (Jeffries 1987) but has apparently stopped in recent years, making calving losses today essentially permanent (Copland et al. 2007).

The recent evidence suggests that the aforementioned period of relative stability from the mid-1960s to the mid-1980s ended in the mid-1990s, setting the stage for the large-scale disintegration of the eastern WHIS between 2008 and 2011. Ice shelf thinning to 25 m or less by the early-2000s is difficult to reconcile with surface processes alone as only about 7 m cumulative thinning can be accounted for by surface mass losses since 1952 (Fig. 6.6) and implies that mass losses at the ice shelf base are considerably more significant than surface ablation in terms of overall ice shelf thickness. This interpretation remains somewhat speculative given the data limitations regarding ice shelf thickness, sub-ice processes, and ocean water temperatures but is consistent with findings from below ice shelves and floating ice tongues around Antarctica, Greenland, and Ellesmere Island (Sect. 6.2.3). There is also evidence for considerable basal melting of the WHIS during past warm periods (Jeffries et al. 1991).

6.6 Implications for Ice Shelf Stability

Hughes (1983) speculated that an ice shelf is metastable in the sense that it can withstand small temporary perturbations, but remains vulnerable to larger and prolonged ones (Goldberg et al. 2009). Vaughan and Doake (1996) argued that even small increases in air temperature can lead to the eventual collapse of an entire ice shelf as the ice margin is forced to calve backwards repeatedly from one stable configuration to the next as dictated by underlying and surrounding topography. The modeling experiments presented by Sanderson (1979) further suggest that a 'critical' ice thickness exists below which an ice shelf loses its structural integrity and abruptly disintegrates (hysteresis effect, Fig. 6.7). Once disintegrated, an ice shelf cannot readily reform again unless the climate cools considerably for an extended period of time. This, in essence, represents an extension of the simple air temperature-threshold concept where ice shelves are no longer viable once air temperatures (or

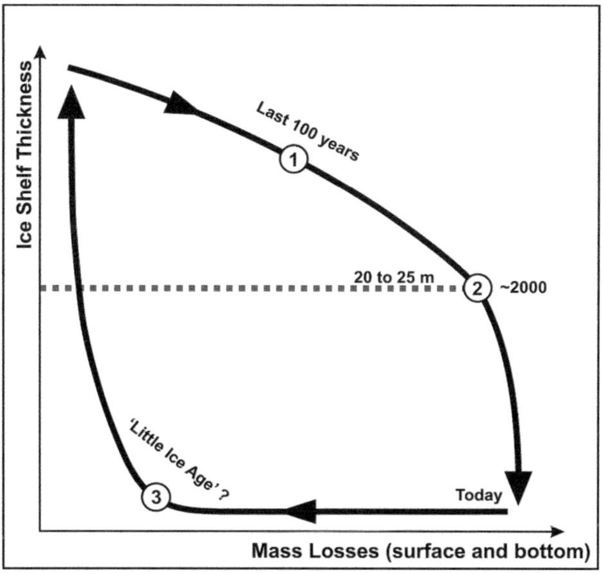

Fig. 6.7 Schematic hysteresis response of the Ward Hunt Ice Shelf (WHIS) to climate change (Sanderson 1979; Hughes 1983). Continuous thinning (*1*), even if slow and gradual, can lead to an abrupt breakup once a critical ice thickness is reached (*2*). The ice shelf cannot readily reform again (*3*) unless the climate cools for an extended length of time. A similar hysteresis argument can be made for the horizontal dimension of an ice shelf due to the influence of lateral and basal topographic buttressing (Vaughan and Doake 1996)

PDDs) exceed a critical value for a prolonged period of time. Air temperature, however, only directly influences surface ablation, while total ice shelf thickness is also influenced by thermal forcing from the underlying ocean water and associated basal melting (Doake 1976; Robin 1979; Sect. 6.5.2). Long-term ocean water temperature records and ocean circulation measurements, unfortunately, are not available from the northern coast of Ellesmere Island. Nevertheless, it is clear that the Arctic Ocean overall has been warming since the turn of the century, with a temporal pattern that broadly matches that of Arctic air temperatures (e.g., Polyakov et al. 2005; Steele et al. 2008) and sea ice extent (Lawrence et al. 2008).

The cumulative mass losses apparently thinned the eastern WHIS below its critical thickness of 20 to 25 m required for maintaining structural integrity by the beginning of the twenty-first century (= pre-weakened state; Copland et al. 2007; 2017; Mortimer et al., 2012; White et al. 2015a). Substantial fracturing within the main body of the eastern WHIS (i.e., excluding minor calving events from the thinner seaward margin) started between 2000 and 2002 (Mueller et al. 2003) and was followed by more extensive fracturing in late March/early April 2008 (Mueller et al. 2008) that lead to its large-scale breakup by September 2011 (Mueller 2011; Mueller et al. 2017). In contrast, the (larger) western section of the WHIS between Ward Hunt Island and Cape Discovery has remained stable and largely unchanged over the same time period. A small remnant of the eastern WHIS survives today 'anchored' between several small islands/ice rises and the coastline some distance east of Ward Hunt Island (Fig. 6.1). It is important to emphasize that there is no theoretical or modeling evidence in support of this particular critical ice thickness value, but it is interesting to note that White et al. (2015a) reported an average thickness of 23 to 29 m for the Petersen Ice Shelf during a series of major calving events between 2005 and 2012.

6.7 Summary and Conclusion

This chapter chronicled the surface mass balance and thickness of the WHIS and WHIR, Ellesmere Island, Canada since the beginning of the twentieth century. The surface mass balance of these two ice bodies tracked the mass balance of other monitored Canadian Arctic glaciers, but their surface mass losses over the last 60+ years have been comparatively low and this difference reflects the localized influence of the Arctic Ocean on the climatic conditions along the northern coast of Ellesmere Island. Nevertheless, the overall mass losses of the WHIS (including surface ablation, basal melting, and ice island calving) have reached critical levels as evidenced by the large-scale breakup of the WHIS between 2008 and 2011. In this context, basal mass losses appear to be considerably more significant in terms of ice shelf thickness and stability than the surface mass losses. Reduced Arctic Ocean sea ice cover serves as a critical amplification factor in ice shelf breakup, an idea first introduced by Moira Dunbar in 1953 (Helk and Dunbar 1953). Copland et al. (2007), for example, attributed the 2005 loss of the Ayles Ice Shelf to a juxtaposition of several factors (cf. Holdsworth 1971; Rott et al. 1996; Glasser and Scambos 2008), including record low Arctic sea ice cover, high wind speeds, loss of protective multi-year landfast sea ice 'armor', and an overall weaker ice shelf due to long-term thinning. The WHIS is particularly sensitive to these factors as its overall mass balance is not 'buffered' by mass input from upstream land glaciers. Once the WHIS disintegrates, it cannot readily reform again, unless climatic conditions deteriorate for a prolonged period of time. Eventually the WHIR may be one of the last remnants of Peary's 'Glacial Fringe', although its long-term survival is also threatened by current and predicted climatic change (Braun 2006).

Since their scientific rediscovery in the early 1950s, scientists have debated whether or not the Canadian Arctic ice shelves would be able to reform today under current climatic conditions (Hattersley-Smith and Serson 1970). The answer is quite clear: the WHIS is not in equilibrium with the present climate as there are no net mass gain processes operating today on, under, or around the ice shelf. This stands in contrast to much of the 1960s, 1970s, and 1980s when its size and thickness remained relatively stable due to two important mass addition processes (Jeffries and Serson 1983, 1986; Jeffries et al. 1988, 1991): horizontal ice shelf expansion by landfast sea ice accretion and vertical mass additions at the ice shelf base (i.e., thickening) by basal freezing. A similar conclusion can be reached for the other remaining Canadian Arctic ice shelves: they appear to be adjusting their geometry (i.e., their thickness and areal extent) to the changing environmental conditions, especially increases in air temperature, ocean water temperature, and reductions in Arctic Ocean sea ice cover. Unfortunately, this is not a gradual adjustment process that one might find for a typical land-based glacier, but rather abrupt as dictated by ice shelf thickness in combination with the underlying and surrounding topography.

The reality and impacts of climate change are ubiquitous and accelerating today throughout the entire Arctic and yet the demise of the only remaining ice shelves in the Northern Hemisphere is proceeding with only very limited glaciological observations. It is therefore important to follow the suggestions of Debenham (1954) and develop an accurate depiction of the three-dimensional geometry of the remaining ice shelves as baselines for assessing future ice shelf changes. In addition, a comprehensive glaciological monitoring program should be re-established on the Canadian Arctic ice shelves and adjacent ice rises to document and better understand their behaviour in the context of current pan-Arctic and global environmental change.

Acknowledgments Research was supported by a U.S. National Science Foundation Grant (OPP-9819362) to the University of Massachusetts, and by Parks Canada (Nunavut Field Unit). I thank Martin Jeffries (University of Alaska) for generously sharing H. Serson's original notes and tables. The thoughtful and detailed comments of two reviewers greatly improved an earlier version of this manuscript. Finally, my sincere thanks to the late Fritz Koerner for his support and encouragement over the years.

References

Adams, W. P., Cogley, J. G., Ecclestone, M. A., & Demuth, M. N. (1998). A small glacier as an index of regional mass balance: Baby Glacier, Axel Heiberg Island, 1959–1992. *Geografiska Annaler, 80A*(1), 37–50. doi:10.1111/j.0435-3676.1998.00025.x.

Ahlmann, H. W. (1929). On the formation of hoarfrost and its relation to glacial growth. *Journal of Geology, 37*(3), 275–280. doi:10.1086/623623.

Alt, B. T. (1979). Investigation of summer synoptic climate controls in the mass balance of Meighen Ice Cap. *Atmosphere-Ocean, 17*(3), 181–199. doi:10.1080/07055900.1979.9649060.

Alt, B. T. (1987). Developing synoptic analogs for extreme mass balance conditions on Queen Elizabeth Island Ice Caps. *Journal of Climate and Applied Meteorology, 26*(11), 1605–1623. doi:10.1175/1520-0450(1987)026<1605:DSAFEM>2.0.CO;2.

Althoff, W. F. (2007). *Drift station: Arctic outposts of superpower science*. Washington, DC: Potomac Books.

Ambach, W. (1985). Characteristics of the heat balance of the Greenland Ice Sheet for modeling. *Journal of Glaciology, 31*(107), 3–12. doi:10.3198/1985JoG31-107-3-12.

Anda, E., Orheim, O., & Mangerud, J. (1985). Late Holocene glacier variations and climate at Jan Mayen. *Polar Research, 3*(2), 129–140. doi:10.1111/j.1751-8369.1985.tb00501.x.

Andrews, J. T., & Barry, R. G. (1978). Glacial inception and disintegration during the last glaciation. *Annual Review of Earth and Planetary Sciences, 6*, 205–228. doi:10.1146/annurev.ea.06.050178.001225.

Antoniades, D., Francus, P., Pienitz, R., St-Onge, G., & Vincent, W. F. (2011). Holocene dynamics of the Arctic's largest ice shelf. *Proceedings of the National Academy of Sciences of the United States of America, 108*(47), 18899–18904. doi:10.1073/pnas.1106378108.

Arnold, K. C. (1965). Aspects of the glaciology of Meighen Island, Northwest Territories, Canada. *Journal of Glaciology, 5*(40), 399–410.

Baird, P. D. (1955). Glaciological research in the Canadian Arctic. *Arctic, 8*(2), 96–108. doi:10.14430/arctic3809.

Baird, P. D., & Sharp, R. P. (1954). Glaciology. *Arctic, 7*(3–4), 141–152.

Bamber, J., Krabill, W., Raper, V., & Dowdeswell, J. (2004). Anomalous recent growth of part of a large Arctic ice cap: Austfonna, Svalbard. *Geophysical Research Letters, 31*, L12402. doi:10.1029/2004GL019667.

Barry, R. G. (2006). The status of research on glaciers and global glacier recession: A review. *Progress in Physical Geography, 30*(3), 285–306. doi:10.1191/0309133306pp478ra.

Bengtsson, L., Semenov, V. A., & Johannessen, O. M. (2004). The early twentieth-century warming in the Arctic—A possible mechanism. *Journal of Climate, 17*, 4045–4057. doi:10.1175/1520-0442(2004)017<4045:TETWIT>2.0.CO;2.

Bilello, M. A. (1961). Formation, growth, and decay of sea-ice in the Canadian Arctic Archipelago. *Arctic, 14*(1), 2–24. doi:10.14430/arctic3658.

Bradley, R. S. (1990). Holocene paleoclimatology of the Queen Elizabeth Islands, Canadian High Arctic. *Quaternary Science Reviews, 9*(4), 365–384. doi:10.1016/0277-3791(90)90028-9.

Bradley, R. S., & Eischeid, J. K. (1985). Aspects of the precipitation climatology of the Canadian High Arctic. In R. S. Bradley (Ed.), *Glacial and glacio-climatic studies in the Canadian High Arctic, Contribution No. 49* (p. 240–271). Amherst: Department of Geology and Geography, University of Massachusetts.

Bradley, R. S., & England, J. H. (1978). Recent climatic fluctuations of the Canadian High Arctic and their significance for glaciology. *Arctic and Alpine Research, 10*, 715–731. doi:10.2307/1550739.

Bradley, R. S., & England, J. H. (2008). The Younger Dryas and the sea of ancient ice. *Quaternary Research, 70*, 1–10. doi:10.1016/j.yqres.2008.03.002.

Bradley, R. S., & Miller, G. H. (1972). Recent climatic change and increased glacierization in the eastern Canadian Arctic. *Nature, 237*, 385–387. doi:10.1038/237385a0.

Braun, C. (2006). Sensitivity of the Hazen Plateau and north coast, Ellesmere Island, Nunavut, Canada, to climate change. PhD dissertation, University of Massachusetts, Amherst.

Braun, C., Hardy, D. R., Bradley, R. S., & Sahanatien, V. (2004a). Surface mass balance of the Ward Hunt Ice Rise and Ice Shelf, Ellesmere Island, Nunavut, Canada. *Journal of Geophysical Research, 109*, D22110. doi:10.1029/2004JD004560.

Braun, C., Hardy, D. R., & Bradley, R. S. (2004b). Mass balance and area changes of four High Arctic plateau ice caps, 1959–2002. *Geografiska Annaler, 86A*(1), 43–52. doi:10.1111/j.0435-3676.2004.00212.x.

Braun, M., Humbert, A., & Moll, A. (2009). Changes of Wilkins Ice Shelf over the past 15 years and inferences on its stability. *The Cryosphere, 3*, 41–56. doi:10.5194/tc-3-41-2009.

Bushnell, V. C. (1956). Marvin's ice shelf journey, 1906. *Arctic, 9*(3), 166–177. doi:10.14430/arctic3791.

Chylek, P., Dubey, M. K., & Lesins, G. (2006). Greenland warming of 1920–1930 and 1995–2005. *Geophysical Research Letters, 33*, L11707. doi:10.1029/2006GL026510.

Cogley, J. G., Adams, W. P., Ecclestone, M. A., Jung-Rothenhäusler, F., & Ommanney, C. S. L. (1996). Mass balance of White Glacier, Axel Heiberg Island, N.W.T., Canada, 1960–91. *Journal of Glaciology, 42*(142), 548–563. doi:10.3198/1996JoG42-142-548-563.

Comiso, J. C. (2003). Warming trends in the Arctic from clear sky satellite observations. *Journal of Climate, 16*(21), 3498–3510. doi:10.1175/1520-0442(2003)016<3498:WTITAF>2.0.CO;2.

Comiso, J. C. (2006). Arctic warming signals from satellite observations. *Weather, 61*(3), 70–76. doi:10.1256/wea.222.05.

Comiso, J. C., Parkinson, C. L., Gersten, R., & Stock, L. (2008). Accelerated decline in the Arctic sea ice cover. *Geophysical Research Letters, 35*, L01703. doi:10.1029/2007GL031972.

Cook, A. J., & Vaughan, D. G. (2010). Overview of areal changes of the ice shelves on the Antarctic Peninsula over the past 50 years. *The Cryosphere, 4*, 77–98. doi:10.5194/tc-4-77-2010.

Cook, A. J., Fox, A. J., Vaughan, D. G., & Ferrigno, J. G. (2005). Retreating glacier fronts on the Antarctic Peninsula over the past half-century. *Science, 308*, 541–544. doi:10.1126/science.1104235.

Cook, A. J., Holland, P. R., Meredith, M. P., Murray, T., Luckman, A., & Vaughan, D. G. (2016). Ocean forcing of glacier retreat in the western Antarctic Peninsula. *Science, 353*(6296), 283–286. doi:10.1126/science.aae0017.

Copland, L., Mueller, D. R., & Weir, L. (2007). Rapid loss of the Ayles Ice Shelf, Ellesmere Island, Canada. *Geophysical Research Letters, 34*, L21501. doi:10.1029/2007GL031809.

Copland, L., Mortimer, C. A., White, A., Richer McCallum, M., & Mueller, D. R. (2017). Factors contributing to recent Arctic ice shelf losses. In L. Copland & D. Mueller (Eds.), *Arctic ice shelves and ice islands* (p. 263–285). Dordrecht: Springer. doi:10.1007/978-94-024-1101-0_10.

Crary, A. P. (1956). Geophysical studies along Northern Ellesmere Island. *Arctic, 9*(3), 154–165. doi:10.14430/arctic3790.

Crary, A. P. (1958). Arctic ice island and ice shelf studies—Part I. *Arctic, 11*(1), 2–42. doi:10.14430/arctic3731.

Crary, A. P. (1960). Arctic ice island and ice shelf studies—Part II. *Arctic, 13*(1), 32–50. doi:10.14430/arctic3687.

Crary, A. P. (1966). Mechanism for fjord formation indicated by studies of an ice-covered inlet. *Bulletin Geological Society of America, 77*, 911–930. doi:10.1130/00167606(1966)77[911:MFFFIB]2.0.CO;2.

Crary, A. P., & Cotell, R. D. (1952). Ice islands in Arctic research. *The Scientific Monthly, 75*(5), 298–302. doi:10.1111/j.1541-0064.1957.tb01812.x.

Crary, A. P., Cotell, R. D., & Sexton, T. F. (1952). Preliminary report on scientific work on 'Fletcher's' Ice Island', T3. *Arctic, 5*(4), 211–233. doi:10.14430/arctic3913.

Crary, A. P., Kulp, J. L., & Marshall, E. W. (1955). Evidences of climatic change from ice island studies. *Science, 122*, 1171–1173. doi:10.1126/science.122.3181.1171.

Crowley, F. A. (1961). Density distribution for a two-layered shelf. *Proceedings of the third annual Arctic planning session*. Air Force Cambridge Research Laboratories, Geophysics Research Directorate, Cambridge (p. 31–33).

Debenham, F. (1919). A new mode of transportation by ice: The raised marine muds of South Victoria Land. *Quarterly Journal of the Geological Society, 75*(2), 51–76. doi:10.1144/GSL.JGS.1919.075.01-04.08.

Debenham, F. (1948). The problem of the Great Ross Barrier. *Journal of Geographical, 112*(4/6), 196–212. doi:10.2307/1789698.

Debenham, F. (1954). The ice islands of the Arctic: A hypothesis. *Geographical Review, 44*(4), 495–507. doi:10.2307/212156.

Derksen, C., Smith, S. L., Sharp, M., et al. (2012). Variability and change in the Canadian cryosphere. *Climatic Change, 115*, 59–88. doi:10.1007/s10584-012-0470-0.

Doake, C. S. M. (1976). Thermodynamics of the interaction between ice shelves and the sea. *Polar Record, 18*(112), 37–41. doi:10.1017/S0032247400028692.

Doake, C. S. M., & Vaughan, D. G. (1991). Rapid disintegration of the Wordie Ice Shelf in response to atmospheric warming. *Nature, 350*, 328–330. doi:10.1038/350328a0.

Doake, C. S. M., & Wolff, E. W. (1985). Flow law for ice in polar ice sheets. *Nature, 314*, 255–257. doi:10.1038/318082b0.

Doake, C. S. M., Corr, H. F. J., Rott, H., Skvarca, P., & Young, N. W. (1998). Breakup and conditions for stability of the northern Larsen Ice Shelf, Antarctica. *Nature, 391*, 778–780. doi:10.1038/35832.

Dorrer, E. (1971). Movement of the Ward Hunt Ice Shelf, Ellesmere Island, N.W.T., Canada. *Journal of Glaciology, 10*(59), 211–224. doi:10.3198/1971JoG10-59-211-225.

Dowdeswell, J. A., Hagen, J. O., Bjoernsson, H., Glazovsky, A. F., Harrison, W. D., Holmlund, P., Jania, J., Koerner, R. M., Lefauconnier, B., Ommanney, C. S. L., & Thomas, R. H. (1997). The mass balance of circum-Arctic glaciers and recent climate change. *Quaternary Research, 48*(1), 1–14. doi:10.1006/qres.1997.1900.

Dunbar MJ (1972) Increasing severity of ice conditions in Baffin Bay and Davis Strait. In T. Karlsson (Ed), *Sea Ice (Proceedings of an International Conference)* (p. 87–93). Reykjavik: National Research Council.

Dunbar, M. J. (1976). Climatic change and northern development. *Arctic, 29*(4), 182–193. doi:10.14430/arctic2803.

Edlund, S. A., & Alt, B. T. (1989). Regional congruence of vegetation and summer climate patterns in the Queen Elizabeth Islands, Northwest Territories, Canada. *Arctic, 42*(1), 3–23. doi:10.14430/arctic1635.

England, J. H., Lakeman, T. R., Lemmen, D. S., Bednarski, J. M., Stewart, T. G., & Evans, D. J. A. (2008). A millennial-scale record of Arctic Ocean sea ice variability and the demise of the Ellesmere Island ice shelves. *Geophysical Research Letters, 35*, L19502. doi:10.1029/2008GL034470.

Evans, D. J. A., & England, J. (1992). Geomorphological evidence of Holocene climatic change from northwest Ellesmere Island, Canadian High Arctic. *Holocene, 2*, 148–158. doi:10.1177/095968369200200206.

Evans, S., & Robin, G. Q. (1966). Glacier depth-sounding from the air. *Nature, 210*(5039), 883–885. doi:10.1038/210883a0.

Falkner, K. K., Melling, H., Münchow, A. M., Box, J. E., Wohlleben, T., Johnson, H. L., Gudmandsen, P., Pamelson, R., Copland, L., Steffen, K., Rignot, E., & Higgins, A. K. (2011). Context for the recent massive Petermann Glacier calving event. *EOS. Transactions of the American Geophysical Union, 92*(14), 117–118. doi:10.1029/2011EO140001.

Fetterer, F., & Untersteiner, N. (1998). Observations of melt ponds on Arctic sea ice. *Journal of Geophysical Research, 103*(C11), 24821–24835. doi:10.1029/98JC02034.

Fisher, D., Zheng, J., Burgess, D., Zdanowicz, C., Kinnard, C., Sharp, M., & Bourgeois, J. (2012). Recent melt rates of Canadian Arctic ice caps are the highest in many millennia. *Global and Planetary Change, 84-85*, 3–7. doi:10.1016/j.gloplacha.2011.06.005.

Fletcher, J. O. (1953). Three months on an Arctic ice island. *National Geographic, 103*, 489–504.

Fox, A. J., & Vaughan, D. G. (2005). The retreat of Jones Ice Shelf, Antarctic Peninsula. *Journal of Glaciology, 51*(175), 555–560. doi:10.3189/172756505781829043.

Gardner, A. S., & Sharp, M. (2007). Influence of the arctic circumpolar vortex on the mass balance of Canadian High Arctic glaciers. *Journal of Climate, 20*, 4586–4598. doi:10.1175/JCLI4268.1.

Giletti, B. J., & Kulp, J. L. (1959). Tritium tracer in Arctic problems. *Science, 129*, 901–903. doi:10.1126/science.129.3353.901.

Glasser, N. F., & Scambos, T. A. (2008). A structural glaciological analysis of the 2002 Larsen B Ice Shelf collapse. *Journal of Glaciology, 54*(184), 3–16. doi:10.3189/002214308784409017.

Goldberg, D., Holland, D. M., & Schoof, C. (2009). Grounding line movement and ice shelf buttressing in marine ice sheets. *Journal of Geophysical Research, 114*, F04026. doi:10.1029/2008JF001227.

Gould, L. M. (1935). The Ross Shelf ice. *Bulletin Geological Society of America, 46*, 1367–1394. doi:10.1130/GSAB-46-1367.

Hanson, K. J. (1961). The albedo of sea-ice and ice islands in the Arctic Ocean basin. *Arctic, 14*(3), 188–196. doi:10.14430/arctic3673.

Hanson, A. M. (1980). The snow cover of sea ice during the Arctic Ice Dynamics Joint Experiment, 1975 to 1976. *Arctic and Alpine Research, 12*(2), 215–226. doi:10.2307/1550518.

Hattersley-Smith, G. (1954). Glaciological reconnaissance in Northern Ellesmere Island. In *Comptes rendus et rapports de la commission des neiges et des glaces* (p. 229–235). Rome: International Association of Hydrological Sciences.

Hattersley-Smith, G. (1956). Northern Ellesmere Island. *Geograph Journal, 122*(1), 13–23.

Hattersley-Smith, G. (1957a). The Ellesmere Ice Shelf and the ice islands. *Canadian Geographic, 9*, 65–70. doi:10.1111/j.1541-0064.1957.tb01812.x.

Hattersley-Smith, G. (1957b). The rolls on the Ellesmere Ice Shelf. *Arctic, 10*(1), 32–44. doi:10.14430/arctic3753.

Hattersley-Smith, G. (1960). Some remarks on glaciers and climate in northern Ellesmere Island. *Geografiska Annaler, 43*(1), 45–48. doi:10.2307/520176.

Hattersley-Smith, G. (1961). The ice cover of northern Ellesmere Island. *Annals of the New York Academy of Sciences, 95*, 282–289. doi:10.1111/j.1749-6632.1961.tb50039.x.

Hattersley-Smith, G. (1963). The Ward Hunt Ice Shelf: Recent changes of the ice front. *Journal of Glaciology, 4*(34), 415–424. doi:10.3198/1963JoG4-34-415-424.

Hattersley-Smith, G. (1969). Results of radio echo sounding in Northern Ellesmere Island, 1966. *The Geographical Journal, 135*(4), 553–557. doi:10.2307/1795101.

Hattersley-Smith, G. (1970). Ice exploration in Antarctica: Review. *The Geographical Journal, 136*(4), 597–600. doi:10.2307/1796190.

Hattersley-Smith, G. (1985). Spreading rate of an Arctic ice shelf. *Nature, 315*, 462.

Hattersley-Smith, G., & Serson, H. (1970). Mass balance of the Ward Hunt Ice Rise and Ice Shelf: A 10-year record. *Journal of Glaciology, 9*(56), 247–252. doi:10.3198/1970JoG9-56-247-252.

Hattersley-Smith, G., Crary, A. P., & Christie, R. L. (1955). Northern Ellesmere Island, 1953 and 1954. *Arctic, 8*(1), 3–36. doi:10.14430/arctic3802.

Helk, J. V., & Dunbar, M. (1953). Ice Islands: Evidence from North Greenland. *Arctic, 6*(4), 263–271. doi:10.14430/arctic3881.

Herbert, W., & Koerner, R. M. (1970). The first surface crossing of the Arctic Ocean. *The Geographical Journal, 126*(4), 511–533.

Higgins, A. K. (1989). North Greenland ice islands. *Polar Record, 25*(14), 207–212. doi:10.1017/S0032247400010809.

Hindmarsh, R. (1993). Modelling the dynamics of ice sheets. *Progress in Physical Geography, 17*(4), 291–312. doi:10.1177/030913339301700401.

Hindmarsh, R. (1996). Stability of ice-rises and uncoupled marine ice-sheets. *Annals of Glaciology, 23*, 105–115. doi:10.3198/1996AoG23-105-115.

Holdsworth, G. (1971). Calving from Ward Hunt Ice Shelf, 1961–1962. *Canadian Journal of Earth Sciences, 8*, 299–305. doi:10.1139/e71-028.

Holland, D. M., Thomas, R. H., De Young, B., & Ribergaard, M. H. (2008). Acceleration of Jakobshavn Isbrae triggered by warm subsurface ocean waters. *Nature Geoscience, 1*, 659–664. doi:10.1038/ngeo316.

Holt, B., & Digby, S. A. (1985). Processes and imagery from of first-year fast sea ice during the melt season. *Journal of Geophysical Research, 90*, 5045–5062. doi:10.1029/JC090iC03p05045.

Howell, S. E. L., Wohlleben, T., Komarov, A., Pizzolato, L., & Derksen, C. (2013). Recent extreme light sea ice years in the Canadian Arctic Archipelago: 2011 and 2012 eclipse 1998 and 2007. *The Cryosphere, 7*, 1753–1768. doi:10.5194/tc-7-1753-2013.

Hughes, T. (1983). On the disintegration of ice shelves: The role of fracture. *Journal of Glaciology, 29*(101), 98–117. doi:10.3198/1983JoG29-101-98-117.

Humbert, A., & Braun, M. (2008). The Wilkins Ice Shelf, Antarctica: Break-up along failure zones. *Journal of Glaciology, 54*(188), 943–944. doi:10.3189/002214308787780012.

Jeffries, M. O. (1985). Ice shelf studies off Northern Ellesmere Island, spring 1983. *Arctic, 38*(3), 174–177. doi:10.14430/arctic2130.

Jeffries, M. O. (1987). The growth, structure, and disintegration of Arctic ice shelves. *Polar Record, 23*(147), 631–649. doi:10.1017/S0032247400008342.

Jeffries, M. O. (1991). Massive, ancient sea-ice strata and preserved physical-structural characteristics in the Ward Hunt Ice Shelf. *Annals of Glaciology, 15*, 125–131. doi:10.3198/1991AoG15-1-125-131.

Jeffries, M. O. (1992). Arctic ice shelves and ice islands: Origin, growth and disintegration, physical characteristics, structural-stratigraphic variability and dynamics. *Reviews of Geophysics, 30*(3), 245–267. doi:10.1029/92RG00956.

Jeffries, M. O. (1994). Marine ice. In: *Resource description and analysis—Ellesmere Island National Park Reserve*, Chap. 6, Natural Resource Conservation Section, Parks Canada, Winnipeg.

Jeffries, M. O. (1998). Surface mass balance measurements at the stake networks on the Ward Hunt Ice Shelf and Ice Rise. In R. Fiennes (Ed.), *The 1998 report on scientific work of the North and South Polar Expeditions of Fiennes, Shepard, Howell and Stroud, 1985–1993* (p. 90–92). Exford: Ranulph Fiennes.

Jeffries, M. O. (2002). Ellesmere Island ice shelves and ice islands. In R. S. Williams & J. G. Ferrigno (Eds.), *Satellite image atlas of glaciers of the world, US Geological Survey Professional Paper 1386-J-1* (p. J147–J164). Reston: US Geological Survey.

Jeffries, M. O. (2017). The ice shelves of northernmost Ellesmere Island, Nunavut, Canada. In L. Copland & D. Mueller (Eds.), *Arctic Ice Shelves and Ice Islands* (p. 23–54). Dordrecht: Springer. doi:10.1007/978-94-024-1101-0_2.

Jeffries, M. O., & Krouse, H. R. (1984). Arctic ice shelf growth, fjord oceanography and climate. *Zeitschrift für Gletscherkunde und Glazialgeologie, 20*, 147–153.

Jeffries, M. O., & Krouse, H. R. (1987). Snowfall and oxygen-isotope variations off the north coast of Ellesmere Island, N.W.T., Canada. *Journal of Glaciology, 33*(114), 195–199. doi:10.3198/1987JoG33-114-195-199.

Jeffries, M. O., & Sackinger, W. M. (1989). Some measurements and observations of very old sea ice and brackish ice, Ward Hunt Ice Shelf, N.W.T. *Atmosphere-Ocean, 27*(3), 553–564. doi:10.1080/07055900.1989.9649352.

Jeffries, M. O., & Serson, H. (1983). Recent changes at the front of the Ward Hunt Ice Shelf, Ellesmere Island, N.W.T. *Arctic, 36*(3), 289–290. doi:10.14430/arctic2278.

Jeffries, M. O., & Serson, H. (1986). Survey and mapping of recent ice shelf changes and landfast sea ice growth along the north coast of Ellesmere Island. *Annals of Glaciology, 8*, 96–99. doi: 10.3198/1986AoG8-1-96-99.

Jeffries, M. O., & Shaw, M. A. (1993). The drift of ice islands from the Arctic Ocean into the channels of the Canadian Arctic Archipelago: The history of Hobson's Choice Ice Island. *Polar Record, 29*(171), 305–312. doi:10.1017/S0032247400023950.

Jeffries, M. O., Sackinger, W. M., & Serson, H. V. (1987). Remote sensing of sea-ice growth and melt-pool evolution, Milne Ice Shelf, Ellesmere Island, Canada. *Annals of Glaciology, 9*, 145–150.

Jeffries, M. O., Sackinger, W. M., Krouse, H. R., & Serson, H. V. (1988). Water circulation and ice accretion beneath Ward Hunt Ice Shelf (northern Ellesmere Island, Canada), deduced from salinity and isotope analysis of ice cores. *Annals of Glaciology, 10*, 68–72. doi:10.3198/1988AoG10-68-72.

Jeffries, M. O., Krouse, H. R., Sackinger, W. M., & Serson, H. V. (1990). Surface topography, thickness and ice core studies of multiyear landfast sea ice and Ward Hunt Ice Shelf, northern Ellesmere Island, N.W.T. In C. R. Harington (Ed.), *Canada's missing dimension—Science and history in the Canadian Arctic islands* (Vol. 1, p. 229–254). Ottawa: Canadian Museum of Nature.

Jeffries, M. O., Serson, H. V., Krouse, H. R., & Sackinger, W. A. (1991). Ice physical properties, structural characteristics and stratigraphy in Hobson's Choice Ice Island and implications for the growth history of the East Ward Hunt Ice Shelf, Canadian High Arctic. *Journal of Glaciology, 37*(126), 247–260. doi:10.3198/1991JoG37-126-247-260.

Jenkins, A., & Jacobs, S. (2008). Circulation and melting beneath George VI Ice Shelf, Antarctica. *Journal of Geophysical Research, 113*, C04013. doi:10.1029/2007JC004449.

Jonsson, S. (1982). On the present glaciation of Storöya, Svalbard. *Geografiska Annaler, 64A*(1–2), 53–79.

Kinnard, C., Koerner, R. M., Zdanowicz, C. M., Fisher, D. A., Zheng, J., Sharp, M. J., Nicholson, L., & Lauriol, B. (2008). Stratigraphic analysis of an ice core from the Prince of Wales Icefield, Ellesmere Island, Arctic Canada, using digital image analysis: high-resolution density, past summer warmth reconstruction, and melt effect on ice core solid conductivity. *Journal of Geophysical Research, 113*, D24120. doi:10.1029/2008JD011083.

Koch, L. (1926). Ice cap and sea ice in North Greenland. *Geographical Review, 16*(1), 98–107.

Koenig, L. S., Greenaway, K. R., Dunbar, M., & Hattersley-Smith, G. (1952). Arctic ice islands. *Arctic, 5*(2), 67–103. doi:10.14430/arctic3901.

Koerner, R. M. (1973). The mass balance of the sea ice of the Arctic Ocean. *Journal of Glaciology, 12*(65), 173–185. doi:10.3198/1947JoG12-65-173-185.

Koerner, R. M. (1979). Accumulation, ablation and oxygen isotope variations on the Queen Elizabeth Islands ice caps, Canada. *Journal of Glaciology, 22*(86), 25–41. doi:10.3198/1979JoG22-86-25-41.

Koerner, R. M. (1996). Canadian Arctic. In J. Jania & J. O. Hagen (Eds.), *Report on mass balance of Arctic glaciers* (p. 5–8). Sosnowiec/Oslo: International Arctic Science Committee, Working Group on Arctic Glaciology.

Koerner, R. M. (2005). Mass balance of glaciers in the Queen Elizabeth Islands, Nunavut, Canada. *Annals of Glaciology, 42*, 417–423. doi:10.3189/172756405781813122.

Konecny, G. (1966). Applications of photogrammetry to surveys of glaciers in Canada and Alaska. *Canadian Journal of Earth Sciences, 3*(6), 783–798. doi:10.1139/e66-061.

Konecny, G., & Faig, W. (1966). Studies of ice movements on the Ward Hunt Ice Shelf by means of triangulation-trilateration. *Arctic, 19*(4), 337–342. doi:10.14430/arctic3439.

Krabill, W., Abdalati, W., Frederick, E., Manizade, S., Martin, C., Sonntag, J., Swift, R., Thomas, R., Wright, W., & Yungel, J. (2000). Greenland ice sheet: High-elevation balance and peripheral thinning. *Science, 289*, 428–430. doi:10.1126/science.289.5478.428.

Kwok, R., & Rothrock, D. A. (2009). Decline in Arctic sea ice thickness from submarine and ICESat records: 1958–2008. *Geophysical Research Letters, 36*, L15501. doi:10.1029/2009GL039035.

Lawrence, D. M., Slater, A. G., Tomas, R. A., Holland, M. M., & Deser, C. (2008). Accelerated Arctic land warming and permafrost degradation during rapid sea ice loss. *Geophysical Research Letters, 35*, L11506. doi:10.1029/2008GL033985.

Lemmen, D. S., Evans, D. J. A., & England, J. (1988). Canadian landform examples—10: Ice shelves of northern Ellesmere Island, NWT. *The Canadian Geographer, 32*(4), 363–367.

Lesins, G., Duck, T. J., & Drummond, J. R. (2010). Climate trends at Eureka in the Canadian High Arctic. *Atmosphere-Ocean, 48*(2), 59–80. doi:10.3137/AO1103.2010.

Lister, H. (1962a). *Heat and mass balance at the surface of the Ward Hunt Ice Shelf, 1960, Research Paper No. 19*. Washington, DC: Arctic Institute of North America.

Lister, H. (1962b). Mass balance studies on the Ellesmere Ice Shelf. *Journal of Glaciology, 4*(23), 298–299. doi:10.3198/1970JoG9-56-247-252.

Lotz, J. R. (1961a). *Meteorological observations in Northern Ellesmere Island—1959, Scientific Report No. 11*. Washington, DC: Arctic Institute of North America.

Lotz, J. R. (1961b). *Analysis of meteorological and micrometeorological observations, Northern Ellesmere Island, 1959, Scientific Report No. 12*. Washington: Arctic Institute of North America.

Luethje, M., Feltham, D. L., Taylor, P. D., & Worster, M. G. (2006). Modeling the summertime evolution of sea-ice melt ponds. *Journal of Geophysical Research, 102*, C02001. doi:10.1029/2004JC002818.

Lyons, J. B., & Leavitt, F. G. (1961). *Structural-Stratigraphic studies on the Ward Hunt Ice Shelf. Final Report on Contract AF 19 (604)-6188*. Cambridge: Geophysics Research Directorate, Air Force Cambridge Research Laboratories.

Lyons, J. B., & Mielke, J. E. (1973). Holocene history of a portion of northernmost Ellesmere Island. *Arctic, 26*(4), 314–323. doi:10.14430/arctic2930.

Lyons, J. B., & Ragle, R. H. (1962). Thermal history and growth of the Ward Hunt Ice Shelf. In *Variations of the regime of existing glaciers* (pp. 88–97). Obergurgl: International Association Hydrology Science, Commission des neiges et glaces.

Lyons, J. B., Mavin, S. M., & Tamburi, A. J. (1971). Basement ice, Ward Hunt Ice Shelf, Ellesmere Island, Canada. *Journal of Glaciology, 10*(58), 93–100. doi:10.3198/1971JoG10-58-85-92.

Lyons, J. B., Ragle, R. H., & Tamburi, A. J. (1972). Growth and grounding of the Ellesmere Island ice rises. *Journal of Glaciology, 11*(61), 43–52. doi:10.3198/1972JoG11-61-43-52.

Mair, D., Burgess, D., Sharp, M., Dowdeswell, J. A., Benhan, T., Marshall, S., & Cawkwell, F. (2009). Mass balance of the Prince of Wales Icefield, Ellesmere Island, Nunavut, Canada. *Journal of Geophysical Research, 114*, F02011. doi:10.1028/2008JF001082.

Meier, M. F. (1962). Proposed definitions for glacier mass budget terms. *Journal of Glaciology, 4*(33), 252–263. doi:10.3198/1962JoG4-33-252-263.

Mercer, J. H. (1978). West Antarctic ice sheet and CO_2 greenhouse effect: A threat of disaster. *Nature, 271*, 321–325. doi:10.1038/271321a0.

Miller, M. (1954). Beyond the pole. *Appalachia, 20*(7), 10–15.

Miller, M. M. (1956). Floating islands. *Natural History Magazine, 65*, 233–239, 274, 276.

Miller, M. (1957). Vanishing lands of the polar sea. *Science World, 2*, 4–7.

Miller, G. H., Bradley, R. S., & Andrews, J. T. (1975). The glaciation level and lowest equilibrium line altitude in the High Canadian Arctic: Maps and climatic interpretation. *Arctic and Alpine Research, 7*(2), 155–168. doi:10.2307/1550318.

Montgomery, M. R. (1952). Further notes on ice islands in the Canadian Arctic. *Arctic, 5*(3), 183–187. doi:10.14430/arctic3910.

Moon, T., & Joughin, I. (2008). Changes in ice front position on Greenland's outlet glaciers from 1992 to 2007. *Journal of Geophysical Research, 113*, F02022. doi:10.1029/2007JF000927.

Mortimer, C. A., Copland, L., & Mueller, D. R. (2012). Volume and area changes of the Milne Ice Shelf, Ellesmere Island, Nunavut, Canada, since 1950. *Journal of Geophysical Research, 117*, F04011. doi:10.1029/2011JF002074.

Mortimer, C. A., Sharp, M., & Wouters, B. (2016). Glacier surface temperatures in the Canadian High Arctic, 2000–15. *Journal of Glaciology, 62*(235), 963–975. doi:10.1017/jog.2016.80.

Mueller, D. R. (2011). Summer 2011 loss of Arctic ice shelves. https://wirl.carleton.ca/research/calving-2011/. Accessed 26 July 2016.

Mueller, D. R., & Vincent, W. F. (2006). Microbial habitat dynamics and ablation control on the Ward Hunt Ice Shelf. *Hydrological Processes, 20*, 857–876. doi:10.1002/hyp.6113.

Mueller, D. R., Vincent, W. F., & Jeffries, M. O. (2003). Break-up of the largest Arctic ice shelf and associated loss of an epishelf lake. *Geophysical Research Letters, 30*(20), 2031. doi:10.1029/2003GL017931.

Mueller, D. R., Vincent, W. F., & Jeffries, M. O. (2006). Environmental gradients, fragmented habitats, and microbiota of a northern ice shelf cryoecosystem, Ellesmere Island, Canada. *Arctic, Antarctic, and Alpine Research, 38*(4), 593–607. doi:10.1657/1523-0430(2006)38[593:EGFHAM]2.0.CO;2.

Mueller, D. R., Copland, L., Hamilton, A., & Stern, D. (2008). Examining Arctic ice shelves prior to the 2008 breakup. *EOS. Transactions of the American Geophysical Union, 89*(49), 502–503. doi:10.1029/2008EO490002.

Mueller, D., Copland, L., & Jeffries, M. O. (2017). Changes in Canadian Arctic ice shelf extent since 1906. In L. Copland & D. Mueller (Eds.), *Arctic ice shelves and ice islands* (p. 109–148). Dordrecht: Springer. doi:10.1007/978-94-024-1101-0_5.

Müller, F. (1962). Glacier mass-budget studies on Axel Heiberg Island, Canadian Arctic Archipelago. In *Variations of the regime of existing glaciers* (p. 131–142). Obergurgl: International Association of Hydrology Science, Commission des neiges et glaces.

Narod, B. B., Clarke, G. K., & Prager, B. T. (1988). Airborne UHF radar sounding of glaciers and ice shelves, northern Ellesmere Island, Arctic Canada. *Canadian Journal of Earth Sciences, 25*, 95–105. doi:10.1139/e88-010.

Nick, F. M., Vieli, A., Howat, I. M., & Joughin, I. (2009). Large-scale changed in Greenland outlet glacier dynamics triggered at the terminus. *Nature Geosciences, 2*, 110–114. doi:10.1038/ngeo394.

Ommanney, C. S. L. (1977). *Quadrennial report to the Permanent Service on the Fluctuations of Glaciers on Canadian glacier variations and mass balance changes*. Ottawa: Glaciology Division, Inland Waters Directorate, Fisheries and Environment Canada.

Østrem, G., & Brugman, M. (1991). *Glacier mass-balance measurements. A manual for field and office work, Science Report 4*. Saskatoon: National Hydrology Research Institute, Environment Canada.

Overland, J. E., & Wang, M. (2013). When will the summer Arctic be nearly sea ice free? *Geophysical Research Letters, 40*, 2097–2101. doi:10.1029/grl.50316.

Paterson, W. S. B. (1969). The Meighen Ice Cap, Arctic Canada: Accumulation, ablation and flow. *Journal of Glaciology, 8*(54), 341–352. doi:10.3198/1969JoG8-53-341-352.

Peary, R. E. (1907). *Nearest the Pole: A narrative of the polar expedition of the Peary Arctic Club in the S.S. Roosevelt, 1905–1906*. New York: Doubleday, Page & Company.

Pedley, M., Paren, J. G., & Potter, J. R. (1988). Localized basal freezing within the George VI Ice Shelf, Antarctica. *Journal of Glaciology, 34*(116), 71–77. doi:10.3198/1988JoG34-116-71-77.

Pfeffer, W. T. (2007). A simple mechanism for irreversible tidewater glacier retreat. *Journal of Geophysical Research, 112*, F03S25. doi:10.1029/2006JF000590.

Plouff, D., Keller, G. V., Frischknecht, F. C., & Wahl, R. R. (1961). Geophysical studies on IGY drifting station Bravo, T-3, 1958 to 1959. In G. O. Raasch (Ed.), *Geology of the Arctic* (Vol. I, p. 709–716). Toronto: University of Toronto Press.

Polunin, N. (1955). Attempted dendrochronological dating of ice island T-3. *Science, 122*, 1184–1186. doi:10.1126/science.122.3181.1184.

Polyakov, I. V., Beszczynska, A., Carmack, E. C., et al. (2005). One more step toward a warmer Arctic. *Geophysical Research Letters, 32*, L17605. doi:10.1029/2005GL023740.

Pope, S., Copland, L., & Mueller, D. (2012). Loss of multiyear landfast sea ice from Yelverton Bay, Ellesmere Island, Nunavut, Canada. *Arctic, Antarctic, and Alpine Research, 44*(2), 210–221. doi:10.1657/1938-4246-44.2.210.

Przybylak, R. (2007). Recent air-temperature changes in the Arctic. *Annals of Glaciology, 46*, 316–324. doi:10.3189/172756407782871666.

Ragle, R. H., Blair, R. G., & Persson, L. E. (1964). Ice core studies of Ward Hunt Ice Shelf, 1960. *Journal of Glaciology, 5*(37), 39–59.

Reeh, N. (1968). On the calving of ice from floating glaciers and ice shelves. *Journal of Glaciology, 7*(50), 215–232. doi:10.3198/1968JoG7-50-215-232.

Reeh, N. (1991). Parameterization of melt rate and surface temperature on the Greenland Ice Sheet. *Polarforschung, 59*(3), 113–128. doi:10.2312/polarforschung.59.3.113.

Rignot, E., & Jacobs, S. S. (2002). Rapid bottom melting widespread near Antarctic ice sheet grounding lines. *Science, 296*, 2020–2023. doi:10.1126/science.1070942.

Rignot, E., & Steffen, K. (2008). Channelized bottom melting and stability of floating ice shelves. *Geophysical Research Letters, 35*, L02503. doi:10.1029/2007GL031765.

Rignot, E., Box, J. E., Burgess, E., & Hanna, E. (2008). Mass balance of the Greenland ice sheet from 1958 to 2007. *Geophysical Research Letters, 35*, L20502. doi:10.1029/2008GL035417.

Rignot, E., Koppes, M., & Velicogna, I. (2010). Rapid submarine melting of the calving faces of West Greenland glaciers. *Nature Geoscience, 3*(3), 187–191. doi:10.1038/ngeo765.

Robin, G. Q. (1975). Ice shelves and ice flow. *Nature, 253*, 168–172. doi:10.1038/253168a0.

Robin, G. Q. (1979). Formation, flow, and disintegration of ice shelves. *Journal of Glaciology, 24*(90), 259–271. doi:10.3198/1979JoG24-90-259-271.

Rodahl, K. (1954). *North—The nature and drama of the polar world*. London: William Heinemann Ltd.

Rothrock, D. A., Percival, D. B., & Wensnahan, M. (2008). The decline in Arctic sea-ice thickness: Separating the spatial, annual, and interannual variability in a quarter century of submarine data. *Journal of Geophysical Research, 113*, C05003. doi:10.1029/2007JC004252.

Rott, H., Skvarca, P., & Nagler, T. (1996). Rapid collapse of northern Larsen Ice Shelf, Antarctica. *Science, 271*, 788–792. doi:10.1126/science.271.5250.788.

Sagar, R. B. (1962). *Meteorological and glaciological studies ice rise station, Ward Hunt Island, May to September 1960, Research Paper No. 24*. Washington: Arctic Institute of North America.

Sanderson, T. J. O. (1979). Equilibrium profile of ice shelves. *Journal of Glaciology, 22*(88), 435–460. doi:10.3198/1979JoG22-88-435-460.

Scambos, T. A., Hulbe, C., Fahnestock, M., & Bohlander, J. (2000). The link between climate warming and break-up of ice shelves in the Antarctic Peninsula. *Journal of Glaciology, 46*(154), 516–530. doi:10.3189/172756500781833043.

Schytt, V. (1949). Refreezing of melt water on the surface of glacier ice. *Geografiska Annaler, 31*, 222–227.

Serson, H. V. (1979). *Mass balance of the Ward Hunt Ice Rise and Ice Shelf: An 18-year record.* Victoria: Defence Establishment Pacific.

Sharp, M., Burgess, D. O., Cogley, J. G., Ecclestone, M., Labine, C., & Wolken, G. (2011). Extreme melt on Canada's Arctic ice caps in the 21st century. *Geophysical Research Letters, 38*, L11501. doi:10.1029/2011GL047381.

Sharp, M., Burgess, D. O., Cawkwell, F., Copland, L., Davis, J. A., Dowdeswell, E. K., Dowdeswell, J. A., Gardner, A. S., Mair, D., Wang, L., Williamson, S. N., Wolken, G. J., & Wyatt, F. (2014). Remote sensing of recent glacier changes in the Canadian Arctic. In J. S. Kargel, G. J. Leonard, M. P. Bishop, A. Kääb, & B. Raup (Eds.), *Global Land Ice Measurements from Space (GLIMS)* (p. 205–228). Berlin: Springer. doi:10.1007/978-3-540-798187_9.

Shepherd, A., Wingham, D., & Rignot, E. (2004). Warm ocean is eroding the West Antarctic Ice Sheet. *Geophysical Research Letters, 31*, L23402. doi:10.1029/2004GL021106.

Smith, D. D. (1960). *Origin of parallel pattern of meltwater lakes on Fletcher's Ice Island, T3.* International Geological Congress, Report of the Twenty-First Session, Part XXI, 51–58.

Smith, D. D. (1961). Sequential development of surface morphology on Fletcher's Ice Island T3. In G. O. Raasch (Ed.), *Geology of the Arctic* (Vol. II, p. 896–914). Toronto: University of Toronto Press.

Smith, J. A., Bentley, M. J., Hodgson, D. A., & Cook, A. J. (2007). George VI Ice Shelf: Past history, present behaviour, and potential mechanisms for future collapse. *Antarctic Science, 19*(1), 131–142. doi:10.1017/S0954102007000193.

Steele, M., Ermold, W., & Zhang, J. (2008). Arctic Ocean surface warming trends over the past 100 years. *Geophysical Research Letters, 35*, L002614. doi:10.1029/2007GL031651.

Stewart, T. G., & England, J. (1983). Holocene sea-ice variations and paleoenvironmental change, Northernmost Ellesmere Island, N.W.T, Canada. *Arctic and Alpine Research, 15*(1), 1–17. doi:10.2307/1550979.

Stroeve, J., Serreze, M., Drobot, S., Gearheard, S., Holland, M., Maslanik, J., Meier, W., & Scambos, T. (2008). Arctic sea ice extent plummets in 2007. *EOS. Transactions of the American Geophysical Union, 89*(2), 13–14. doi:10.1029/2008EO020001.

Stroeve, J. C., Markus, T., Boisvert, L., Miller, J., & Barrett, A. (2014). Changes in Arctic melt season and implications for sea ice loss. *Geophysical Research Letters, 41*(4), 1216–1225. doi:10.1002/2013GL058951.

Sverdrup, H. U. (1931). The transport of material by pack-ice. *Geographical Journal, 77*(4), 399–400.

Swithinbank, C. (1955). Ice shelves. *Geographical Journal, 121*(1), 64–76. doi:10.2307/1791807.

Swithinbank, C. (1970). Ice movement in the McMurdo area of Antarctica. *International Association of Hydrological Sciences, 86*, 472–487.

Sylvestre, T., Copland, L., Gray, L., Demuth, M. N., & Sharp, M. (2013). Spatial patterns of snow accumulation across Belcher Glacier, Devon Ice Cap. *Journal of Glaciology, 59*, 874–882. doi:10.3189/2013JoG12J227.

Thomas, R. B. (1973a). The creep of ice shelves: Theory. *Journal of Glaciology, 12*(64), 45–53. doi:10.3198/1973JoG12-64-45-53.

Thomas, R. B. (1973b). The creep of ice shelves: Interpretation of observed behaviour. *Journal of Glaciology, 12*(64), 55–70. doi:10.3198/1973JoG12-64-55-70.

Thomas, R. B. (1979). Ice shelves: A review. *Journal of Glaciology, 24*(90), 273–286. doi:10.3198/1979JoG24-90-273-286.

Thomas, R. B., Abdalati, W., Frederick, E., Krabill, W. B., Manizade, S., & Steffen, K. (2003). Investigation of surface melting and dynamic thinning on Jakobshavn Isbrae, Greenland. *Journal of Glaciology, 49*(165), 231–239. doi:10.3189/172756503781830764.

Untersteiner, N. (1961). On the mass and heat budget of Arctic sea ice. *Meteorology and Atmospheric Physics, 12*(2), 151–182. doi:10.1007/BF02247491.

Vaughan, D. G. (1993). Implications of the break-up of Wordie Ice Shelf, Antarctica, for sea level. *Antarctic Science, 5*(4), 403–408. doi:10.1017/S0954102093000537.

Vaughan, D. G. (2006). Recent trends in melting conditions on the Antarctic Peninsula and their implications for ice-sheet mass balance and sea level. *Arctic, Antarctic, and Alpine Research, 38*(1), 147–152. doi:10.1657/1523-0430(2006)038[0147:RTIMCO]2.0.CO;2.

Vaughan, D. G., & Doake, C. S. M. (1996). Recent atmospheric warming and retreat of ice shelves on the Antarctic Peninsula. *Nature, 379*, 328–331. doi:10.1038/379328a0.

van der Veen, C. J. (2002). Calving glaciers. *Progress in Physical Geography, 26*(1), 96–122. doi: 10.1191/0309133302pp327ra.

Veillette, J., Mueller, D. R., Antoniades, D., & Vincent, W. F. (2008). Arctic epishelf lakes as sentinel ecosystems: Past, present and future. *Journal of Geophysical Research, 113*, G04014. doi :10.1029/2008JG000730.

Vincent, W. F., Gibson, J. A. E., & Jeffries, M. O. (2001). Ice-shelf collapse, climate change, and habitat loss in the Canadian high Arctic. *Polar Record, 37*(201), 133–142. doi:10.1017/S0032247400026954.

Vincent, W. F., Whyte, L. G., Lovejoy, C., et al. (2009). Arctic microbial ecosystems and impacts of extreme warming during the International Polar Year. *Polar Science, 3*, 171–180. doi:10.1016/j.polar.2009.05.004.

Vincent, W. F., Fortier, D., Lévesque, E., Boulanger-Lapointe, N., Tremblay, B., Sarrazin, D., Antoniades, D., & Mueller, D. R. (2011). Extreme ecosystems and geosystems in the Canadian High Arctic: Ward Hunt Island and vicinity. *Écoscience, 18*, 236–261. doi:10.2980/18-3-3448.

Walker, P. W., & Mattox Jr., W. G. (1961). *Glaciological observations in northern Ellesmere Island—1959, Scientific Report No. 139.* Washington, DC: Arctic Institute of North America.

Wang, L., Sharp, M. J., Rivard, B., Marshall, S., & Burgess, D. (2005). Melt season duration on Canadian Arctic ice caps, 2000–2004. *Geophysical Research Letters, 32*, L19502. doi:10.1029/2005GL023962.

Warren, C. R. (1991). Terminal environment, topographic control and fluctuations of West Greenland glaciers. *Boreas, 20*, 1–15. doi:10.1111/j.1502-3885.1991.tb00453.x.

Weertman, J. (1957). Deformation of floating ice shelves. *Journal of Glaciology, 3*(21), 38–42. doi:10.3198/1957JoG3-21-38-42.

Weidick, A. (1984). Studies of glacier behavior and glacier mass balance in Greenland—A review. *Geografiska Annaler, 66A*(3), 183–195. doi:10.2307/520693.

Wen, J., Wang, Y., Wang, W., Jezek, K. C., Liu, H., & Allison, I. (2010). Basal melting and freezing under the Amery Ice Shelf, East Antarctica. *Journal of Glaciology, 56*(195), 81–90. doi:10.3189/002214310791190820.

Wexler, H. (1960). Heating and melting of floating ice shelves. *Journal of Glaciology, 3*(27), 626–645.

White, A., Copland, L., Mueller, D., & Van Wychen, W. (2015a). Assessment of historical changes (1959-2012) and the causes of recent break-ups of the Petersen Ice Shelf, Nunavut, Canada. *Annals of Glaciology, 56*(69), 65–76. doi:10.3189/2015AoG69A687.

White, A., Mueller, D., & Copland, L. (2015b). Reconstructing hydrographic change in Petersen Bay, Ellesmere Island, Canada, inferred from SAR imagery. *Remote Sensing of Environment, 165*, 1–13. doi:10.1016/j.rse.2015.04.017.

Wright, C. S. (1925). The Ross Barrier and the mechanisms of ice movement. *Geographical Journal, 65*(3), 198–213. doi:10.2307/1782887.

Wright, J. (1940). South-East Ellesmere Island. *Geographical Journal, 95*(4), 278–291. doi:10.2307/1788464.

Zhang, X., Vincent, L. A., Hogg, W. D., & Niitsoo, A. (2000). Temperature and precipitation trends in Canada during the 20th century. *Atmosphere-Ocean, 38*(3), 395–429. doi:10.1080/07055900.2000.9649654.

Chapter 7
Holocene History of Arctic Ice Shelves

John H. England, David J.A. Evans, and Thomas R. Lakeman

Abstract The modern Ellesmere ice shelves constitute the oldest landfast sea ice in the Northern Hemisphere, originally described by late nineteenth century sledging expeditions. At that time, the 'Ellesmere Ice Shelf' formed a contiguous, coastal apron that extended hundreds of kilometres in length and covered up to 9000 km^2. By the mid-twentieth century the Ellesmere Ice Shelf was reduced to several large remnants and half its original area. Early efforts to document the age of ice shelf inception used conventional radiocarbon dating of internal, aeolian layers and marine organisms incorporated by basal accretion, all providing problematic results. The most widely cited age (3000 ^{14}C yr before present (BP)) was provided by a radiocarbon date on driftwood stranded behind the Ward Hunt Ice Shelf.

Here we constrain the age of ice shelf inception based on 69 accelerator mass spectrometry radiocarbon dates on driftwood collected from raised marine shorelines inland of the Ward Hunt Ice Shelf, including adjacent fiords deglaciated ~9500 cal yr BP. Driftwood entered Disraeli Fiord, behind the Ward Hunt Ice Shelf, as well as Phillips Inlet to the east, continuously from 9500 to 5500 cal yr BP, after which it abruptly terminated until present. This termination is interpreted to record blockage of the coast by establishment of multiyear landfast sea ice, rather than the lack of driftwood availability due to changing ocean currents, because driftwood delivery continued farther east at several locations through the same interval. This driftwood delivery was favoured by the seasonal expansion of the Lincoln Sea polynya that extends from northern Nares Strait to Clements Markham Inlet.

The radiocarbon chronology of entrapped driftwood behind the former Ellesmere Ice Shelf provides a proxy for the severity of Arctic Ocean pack ice with which it co-varies. This record extends back at least five millennia, far beyond the four decades of satellite surveillance, demonstrating that modern sea ice reduction extends back to the early twentieth century. The dramatic acceleration of seasonal

The original version of this chapter was revised. An erratum to this chapter can be found at DOI 10.1007/978-94-024-1101-0_16

J.H. England (✉) • T.R. Lakeman
Department of Earth and Atmospheric Sciences, University of Alberta, Edmonton, AB, Canada
e-mail: john.england@ualberta.ca; thomas.lakeman@ngu.no

D.J.A. Evans
Department of Geography, Durham University, Durham, UK
e-mail: d.j.a.evans@durham.ac.uk

L. Copland, D. Mueller (eds.), *Arctic Ice Shelves and Ice Islands*, Springer Polar Sciences, DOI 10.1007/978-94-024-1101-0_7

pack ice reduction in the Arctic Ocean during the early twenty-first century heralds the imminent demise of the remaining ice shelf remnants across northern Ellesmere Island, considered an unprecedented event on the scale of millennia. Fiords adjacent to the former Ellesmere Ice Shelf—historically occupied by mulit-year landfast sea ice—have also become seasonally ice-free within the last decade.

Keywords Sea ice • Ice shelves • Ellesmere Island • Driftwood • Holocene • Arctic Ocean

7.1 Introduction

We describe the Holocene history of the oldest sea ice in the Northern Hemisphere, the Ellesmere Island ice shelves (Fig. 7.1), significantly extending the record of sea ice reduction in the Arctic Ocean based solely on the last four decades of satellite surveillance (Nghiem et al. 2007; Serreze et al. 2007; Comiso et al. 2008). In addition to the historical documentation of ice shelf reduction (Hattersley-Smith 1963; Hattersley-Smith et al. 1969; Holdsworth 1971; Mueller et al. 2003, 2017; Copland et al. 2007), a longer chronology of ice shelf growth and decay is now provided by the radiocarbon dating of driftwood, marine mammal remains and marine shells

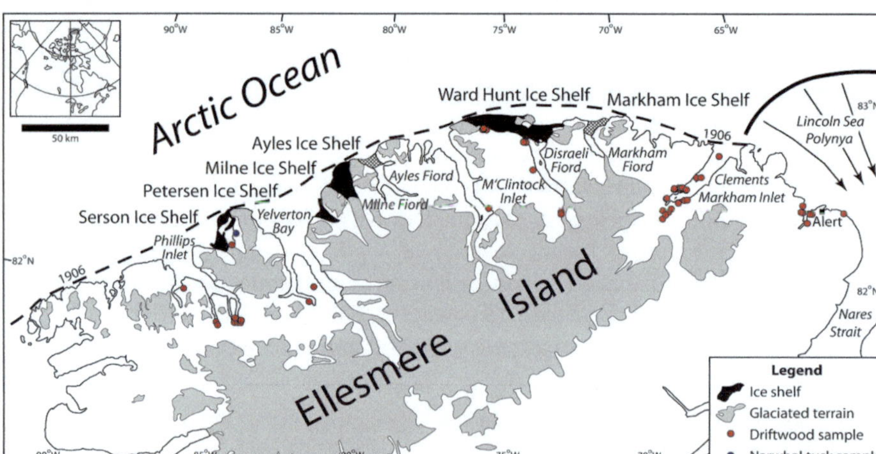

Fig. 7.1 The ice shelves of northern Ellesmere Island, Nunavut, Canada. The *dashed black line* approximates the probable extent of the "Ellesmere Ice Shelf" around 1906, after the end of the LIA (Vincent et al. 2001). The remaining Ellesmere ice shelves (as of 2009) are shown in *black*, whereas the recently detached Ayles and Markham ice shelves are cross-hatched (Copland et al. 2007; Mueller et al. 2008). Note also approximate limit of modern day Lincoln Sea polynya (*top right*) (Reproduced from England et al. (2008), with permission of the American Geophysical Union)

associated with raised shorelines that are widespread on the coast of Ellesmere Island, which has experienced continuous and ongoing postglacial (Holocene) emergence. Here we present a substantially expanded radiocarbon database on driftwood preserved landward of the ice shelves, as well as within neighbouring fiords occupied historically by multiyear landfast sea ice. Collectively, these dates provide important insights concerning the evolution and variability of sea ice along the northern margin of the Canadian Arctic Archipelago (CAA; Crary 1960; Lyons and Mielke 1973; Lemmen et al. 1988; Lemmen 1989; England et al. 2008). The substantial, post twentieth century reduction of the northern Ellesmere Island ice shelves is additionally relevant because such landfast sea ice is now recognized to co-vary with, and hence also reflect, regional reductions in the buttressing by Arctic Ocean pack ice, thereby extending our understanding of sea ice variability prior to satellite surveillance (England et al. 2008).

7.2 The Ellesmere Island Ice Shelves

The modern ice shelves of Ellesmere Island (Fig. 7.1) are remnants of a much more extensive 'Ellesmere Ice Shelf' first described by late nineteenth century sledging parties (Nares 1878; Peary 1907; Marvin 1906, *in* Bushnell 1956). These explorers referred to a 600 km-long platform of sea ice (~9000 km^2) attached to the coast and possessing a surface of "long prairie-like swells", a description consistent with the surface topography of modern ice shelves with their distinctive, parallel ridges and troughs (Hattersley-Smith 1957; Jeffries 1987, 2017; Ommanney 1982a, b; Lemmen et al. 1988; Fig. 7.2). Significant with respect to the longer-term evolution of the ice shelves, the nineteenth century sledging trips across northern Ellesmere Island coincided with the end of the Little Ice Age (LIA; 1600–1900 A.D.) when climate deterioration resulted in widespread expansion of snow and ice across the Canadian High Arctic (Evans and England 1992; Levesque and Svoboda 1999; Wolken et al. 2005). The LIA essentially constituted the culmination of a preceding mid- to late-Holocene climate deterioration during which glaciers throughout the region readvanced to positions beyond their early Holocene minima when a warmer climate prevailed (Blake 1981; Bradley 1990; Evans and England 1992).

Evidence for ice shelf breakup since the late nineteenth century is fragmentary, but Crary (1960) states that the perceived landmasses of Crocker Land (Peary 1910) and Bradley Land (Cook 1911) may have actually represented sizeable parts of the ice shelf that had broken off towards the end of the LIA. Subsequently, Crary (1960) noted that several parts of the outer Ward Hunt Ice Shelf had been removed between the time of Marvin's original survey (1906) and the time of his own observations (1954); this breakup possibly producing a variety of ice islands noted during the 1950s (T-1, T-2, T-3, NP 6). With respect to Ice Island T-3 (utilized by American researchers in the 1950s), Crary (1960) states that it could not have been dislodged from Yelverton Inlet "earlier than 1935" based on growth rings of Arctic willow detached from land along with the ice island (Polunin 1955).

Fig. 7.2 Oblique aerial view across the Ward Hunt Ice Shelf (July 1985). View is eastward across the mouth of Disraeli Fiord towards Cape Columbia. Prominent ridges and troughs ("rolls") that distinguish the Ellesmere ice shelves have a wavelength of ~200–300 m (Reproduced from England et al. (2008), with permission of the American Geophysical Union)

Much improved information concerning the ice shelves followed the mid twentieth century when aircraft surveillance (1946) over the Arctic Ocean discovered gigantic blocks of sea ice, so-called "ice islands", whose surface topography was later matched in jigsaw fashion with the parallel ridges and troughs on the northern Ellesmere Island ice shelves (Koenig et al. 1952). Aerial photographs obtained in 1959/60 indicated that the ice shelves then covered approximately 2100 km^2 and extended up to 16 km into the Arctic Ocean. At this time, the combined area of ice shelves, and ~80 ice islands (2500 km^2; Crary 1958) then entrapped within the CAA, was calculated to be approximately 4600 km^2, half of the estimated area of the "Ellesmere Ice Shelf" at the end of the LIA (Vincent et al. 2001). Since 1960 significant, further reductions have been widely reported (Hattersley-Smith 1963; Holdsworth 1971; Vincent et al. 2001; Copland et al. 2007; England et al. 2008; Mueller et al. 2003, 2008, 2017).

7.3 Driftwood Delivery and Sea Ice/Ice Shelf Chronology

The age of the ice shelves has been constrained by radiocarbon dating of samples of Holocene driftwood on northern Ellesmere Island after long-distance transport across the Arctic Ocean, primarily from northern Russia (Häggblom 1982; Stewart and England 1983; Evans and England 1992; Dyke et al. 1997; England et al. 2008). During the course of fieldwork on the Late Quaternary history of northern Ellesmere Island, widespread driftwood samples were collected from raised marine shorelines located inland of the Ward Hunt Ice Shelf and from five neighbouring fiords previously occupied by multiyear landfast sea ice, including Clements Markham Inlet, M'Clintock Inlet, Yelverton Inlet, Phillips Inlet and Disraeli Fiord (Fig. 7.1). This research concerning the chronology of Holocene glacier retreat and related environmental change inseparably involves the documentation of postglacial emergence occasioned by glacial unloading across northern Ellesmere Island (Bednarski 1986; Lemmen 1989; Evans 1989, 1990). During the last deglaciation of northern Ellesmere Island, ice retreated inland from the fiord mouths starting ~8500–9000 ^{14}C yr BP (~9500 calendar years before present), when relative sea level was 60–90 m above (modern) sea level (asl) (Bednarski 1986; Lemmen 1989; Evans 1990; England et al. 2006). Due to the geologically recent and sizeable glacial unloading, the coastline of northern Ellesmere Island has experienced continuous emergence since deglaciation, resulting in widespread preservation of the post-9 ka BP shorelines (and their marine records) above modern sea level. Such continuous emergence optimizes the potential for driftwood stranding throughout the Holocene, especially along the Ellesmere Island coastline during intervals when ice shelves were absent (see below).

As long as a piece of driftwood does not become reworked (upslope or downslope after stranding), its elevation and radiocarbon age provide a reliable control point for constructing a relative sea level curve (i.e., for northern Ellesmere Island, these dated shorelines record the successive fall of relative sea level from deglacial marine limit to modern). Furthermore, radiocarbon dates on driftwood samples, regardless of their elevation above present sea level, still serve to identify intervals when both Arctic Ocean surface currents and seasonally ice-free coastal waters permitted driftwood delivery and deposition along the northern Ellesmere Island coast. This delivery and deposition of driftwood are affected by four variables. Firstly, fallen trees must enter northward flowing rivers in the circum-Arctic region in order to provide the influx of driftwood into the Arctic Ocean (Dyke et al. 1997). Because the circumpolar boreal forest has occupied the Arctic Ocean terrestrial drainage basin throughout the Holocene, it is reasonable to assume that the rate of driftwood incursion into the Arctic Ocean has remained similar throughout that period. Identification of radiocarbon-dated driftwood collected throughout the CAA has been done to the generic level, indicating that most of the trees originated from northern Russia as opposed to North America (Dyke et al. 1997). Secondly, based on average surface current velocities, the successful transport of driftwood across the Arctic Ocean involves a 3-year voyage from Russia to northern Ellesmere Island. This duration requires that the driftwood must be rafted by, or entrained within, drifting sea ice, otherwise it becomes water-logged and sinks (Häggblom 1982). Thirdly, the journey from the rivers of northern Russia to the coastline of northern Ellesmere Island further depends upon a favourable trajectory of ocean currents to convey the sea ice and its driftwood across the Arctic Ocean (Dyke et al. 1997; see below). Finally, in order for driftwood to successfully strand on the coast of northern Ellesmere Island, there must be seasonally open water to allow free passage of the drifting pack ice versus its exclusion by ice shelves and multiyear landfast sea ice that has characterized this region during the late Holocene (Crary 1960; England et al. 2008).

In addition to providing a chronology for ice shelf development, driftwood ages also facilitate a better understanding of the long-term variability in Arctic Ocean currents. Most driftwood is transported across the Arctic Ocean by sea ice entrained in the Transpolar Drift (TPD), which originates on the continental shelves of northern Russia (Fig. 7.3; Dyke et al. 1997). During the Holocene, the TPD has oscillated between three different configurations, ultimately controlled by atmospheric circulation (Dyke et al. 1997; Tremblay et al. 1997). The first is an eastward routing of the TPD through Fram Strait (between Greenland and Svalbard; Fig. 7.3) when driftwood is delivered to the European Arctic (excluding the CAA). The second is a westward routing of the TPD when driftwood is delivered to the CAA, including northern Ellesmere Island. The third is a split routing when driftwood is delivered less abundantly to both Fram Strait and the CAA. Dyke et al. (1997) reported that driftwood was deposited on northern Ellesmere Island since the onset of deglaciation (approximately 9500 cal yr BP), but more abundantly since 4500 cal yr BP when a westward shift in the TPD is proposed. Thus, from at least 4500 cal yr BP to present, the northern Ellesmere Island coastline was geographically disposed to driftwood delivery, rendering its history of deposition a useful proxy for the presence or absence of ice shelves. Periods characterized by a lack of driftwood deposition across northern Ellesmere Island imply either: (a) the non-availability of driftwood-bearing sea ice by ocean currents reaching this area; and/or (b) blockage of the northern Ellesmere Island coast by multiyear landfast sea ice. The youngest driftwood samples stranded inland of the modern sea-ice ice shelves, and within nearby fiords occupied (until recently) by multiyear landfast sea ice on northern Ellesmere Island, therefore provide maximum-limiting ages for driftwood blockage from the Arctic Ocean.

7.4 Ice Shelf Chronology

Above we have made the case for the utility of a driftwood radiocarbon chronology for determining ice shelf inception and its maintenance prior to their original description by late nineteenth century expeditions to northern Ellesmere Island. We now turn to the details of that chronology, adding to the driftwood dates a small number of ages on the remains of marine mammals stranded on raised beaches during the Holocene that help to constrain the sea level histories from several localities on northern Ellesmere Island. All ages are reported in calendar years before present (cal yr BP) at the 95% confidence level (Table 7.1). The ages are calibrated with the OxCal v3.7 programme, which uses the IntCal04 calibration curve (Reimer et al. 2004).

The radiocarbon dates first used to establish a chronology for the inception of the northern Ellesmere Island ice shelves have been questioned by England et al. (2008)

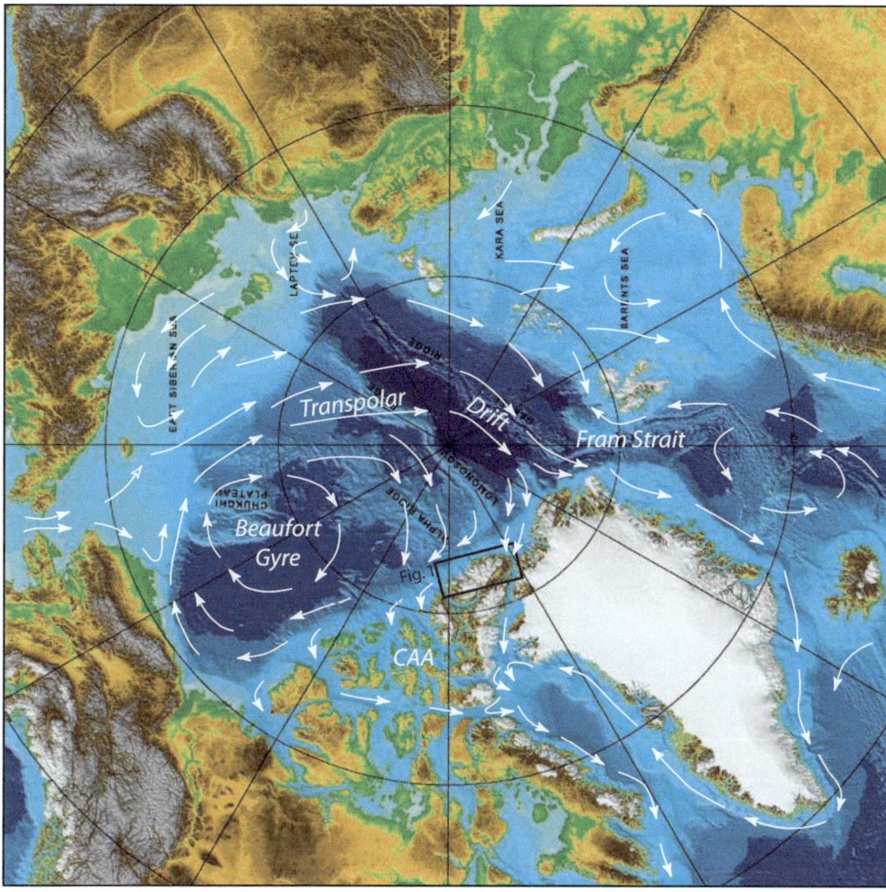

Fig. 7.3 Map of the Arctic Ocean and prominent modern surface currents (Beaufort Gyre and Transpolar Drift) that are responsible for the transport of sea ice and embedded driftwood from Russia to northern Ellesmere Island. Note that the Transpolar Drift (*TPD*) can oscillate between three different configurations (Dyke et al. 1997; Tremblay et al. 1997) that effect the trajectory of sea ice transport and therefore the nature of driftwood delivery to the CAA and to the European Arctic via Fram Strait (see text). *Box* denotes the location of Fig. 7.1 (Modified from International Bathymetric Chart of the Arctic Ocean with permission of the National Oceanic and Atmospheric Administration)

for a variety of reasons. Firstly, radiocarbon dates obtained on "dirt layers" enclosed within Ice Island T-3, presumed to have originated from Yelverton Inlet (Crary et al. 1955; Crary 1958, 1960) have been dismissed because of their allochthonous (aeolian) origin. Furthermore, the dates were further compromised by likely contamination during their original sampling. For example, several dates on the dirt layers are stratigraphically reversed (older over younger) whereas another dirt layer was reportedly contaminated by soot from diesel fuel used to melt the ice (requiring 38 meltings of 3–4 h each to produce 1 g of carbon, a minimal amount for conventional

Table 7.1 Radiocarbon ages cited in this paper

Location	Laboratory number[a]	^{14}C age[b] (yr BP)	Calibrated age range[c] (yr BP)	Dated material	Coordinates (Lat., Long.)	Elevation (m asl)	Relative sea level (m asl)	Reference
Ward Hunt Ice Shelf and Disraeli Fiord	L254D	3000 ± 200	3700–2700	Driftwood	83°00′N, 74°13′W	?	?	Crary (1960)
	L254A	3400 ± 150	4100–3300	"	"	4	≥4	Crary (1960)
	TO-12887	4850 ± 80	5750–5300	"	82°36′N, 72°43′W	14	≥14	England et al. (2008)
	TO-12881	5070 ± 60	5930–5660	"	83°03′N, 75°52′W	0.5	≥0.5	England et al. (2008)
	TO-12876[d]	5330 ± 60	6280–5940	"	83°03′N, 75°56′W	0.5	≥0.5	England et al. (2008)
	UCIAMS-34329[d]	5400 ± 25	6290–6120	"	"	0.5	≥0.5	England et al. (2008)
	TO-12880	5340 ± 60	6280–5990	"	"	0.5	≥0.5	England et al. (2008)
	TO-12884	5460 ± 60	6410–6020	"	"	0.5	≥0.5	England et al. (2008)
	TO-12882	5690 ± 60	6640–6320	"	"	0.5	≥0.5	England et al. (2008)
	L254B	5740 ± 200	7250–5950	"	83°00′N, 74°13′W	?	?	Crary (1960)
	TO-12883	5830 ± 60	6790–6490	"	83°03′N, 75°56′W	0.5	≥0.5	England et al. (2008)
	TO-12875	5900 ± 60	6890–6560	"	"	0.5	≥0.5	England et al. (2008)
	TO-12886	6110 ± 60	7170–6790	"	82°49′N, 73°44′W	20	≥20	England et al. (2008)
	L254C	6120 ± 150	7450–6600	"	83°00′N, 74°13′W	?	?	Crary (1960)
	SI-568	6280 ± 140	7500–6800	"	"	?	?	Blake (1972)
	TO-12878	6880 ± 60	7850–7590	"	83°03′N, 75°56′W	0.5	≥0.5	England et al. (2008)
	TO-12879	7850 ± 60	8980–8510	"	"	0.5	≥0.5	England et al. (2008)
	TO-12877	8230 ± 90	9440–9000	"	"	0.5	≥0.5	England et al. (2008)
	TO-12888	32,310 ± 350	~	"	82°37′N, 72°43′W	1	≥1	England et al. (2008)

Clements Markham Inlet	TO-12858	modern	~	”	82°43′N, 67°44′W	0.8	≥0.8	England et al. (2008)
	TO-12859[d]	200 ± 60	430–(−11)	”	”	0.8	≥0.8	England et al. (2008)
	UCIAMS-34326[d]	185 ± 25	300–(−11)	”	”	0.8	≥0.8	England et al. (2008)
	GSC-3031	2180 ± 60	2340–2000	”	82°39′N, 67°46′W	5.5	7	Stewart (1981)
	L261B	2190 ± 150	2700–1800	”	82°39′N, 67°30′W	7	7	Crary (1960)
	L251B	2190 ± 180	2750–1700	”	82°39′N, 67°46′W	7	7	Bednarski (1986)
	TO-12864[d]	3230 ± 60	3590–3340	”	82°42′N, 67°40′W	12	≥12	England et al. (2008)
	UCIAMS-34327[d]	3545 ± 30	3920–3710	”	”	12	≥12	England et al. (2008)
	TO-12865	3380 ± 60	3830–3460	”	”	12	≥12	England et al. (2008)
	TO-12861	3850 ± 60	4430–4090	”	82°35′N, 67°35′W	0.6	≥0.6	England et al. (2008)
	SI-4763	4215 ± 75	4960–4520	”	82°46′N, 67°10′W	21	21	Bednarski (1986)
	SI-4766	4340 ± 65	5300–4800	”	82°35′N, 67°35′W	1.2	>1.2	Bednarski (1986)
	TO-12874	4580 ± 60	5470–5040	”	”	1.2	≥1.2	England et al. (2008)
	GSC-2973	4660 ± 60	5590–5280	”	82°34′N, 68°43′W	21	21	Stewart (1981)
	TO-12862	5350 ± 70	6290–5940	”	82°40′N, 68°32′W	31	≥31	England et al. (2008)
	SI-4764	5515 ± 65	6440–6190	”	82°45′N, 67°56′W	8	8	Bednarski (1986)
	TO-12860	5710 ± 70	6670–6320	”	82°43′N, 67°53′W	32	≥32	England et al. (2008)
	TO-12863	5750 ± 70	6730–6400	”	82°40′N, 68°32′W	33	≥33	England et al. (2008)
	S-2083	5780 ± 100	6850–6300	”	82°34′N, 68°43′W	22.5	≥22.5	Bednarski (1986)
	TO-12873	6370 ± 60	7430–7170	”	82°33′N, 68°44′W	59	≥59	England et al. (2008)
	SI-4315	6445 ± 65	7480–7250	”	82°35′N, 68°37′W	43	45	Stewart (1981)
	UQ-282	6730 ± 80	7710–7430	”	82°36′N, 68°25′W	59	59	Bednarski (1986)
	TO-12870	6860 ± 70	7850–7580	”	82°43′N, 68°01′W	53	≥53	England et al. (2008)
	TO-12866	6880 ± 70	7920–7580	”	”	53	≥53	England et al. (2008)

(continued)

Table 7.1 (continued)

Location	Laboratory number[a]	^{14}C age[b] (yr BP)	Calibrated age range[c] (yr BP)	Dated material	Coordinates (Lat., Long.)	Elevation (m asl)	Relative sea level (m asl)	Reference
	TO-12869	7170 ± 70	8170–7850	”	”	53	≥53	England et al. (2008)
	TO-12867	7390 ± 80	8370–8020	”	”	53	≥53	England et al. (2008)
	TO-12868	7470 ± 80	8420–8050	”	”	53	≥53	England et al. (2008)
	GSC-2975	7830 ± 80	9000–8400	”	82°36′N, 68°25′W	72	≥82	Stewart (1981)
	TO-12872	7870 ± 70	8990–8540	”	82°35′N, 68°45′W	57	≥57	England et al. (2008)
	TO-12871	8270 ± 70	9440–9030	”	82°43′N, 67°47′W	79	≥79	England et al. (2008)
	S-2210	8545 ± 110	9900–9250	”	82°38′N, 68°00′W	93	>93	Bednarski (1986)
	S-2211	8915 ± 115	10,250–9600	”	”	84	>84	Bednarski (1986)
Phillips Inlet	TO-12893	4700 ± 60	5590–5310	”	82°20′N, 85°25′W	1	≥1	England et al. (2008)
	TO-12890	5230 ± 60	6190–5890	”	81°53′N, 84°20′W	10	≥10	England et al. (2008)
	TO-12892	5510 ± 60	6420–6190	”	82°00′N, 86°35′W	2	≥2	England et al. (2008)
	TO-12900	5760 ± 70	6730–6400	”	81°54′N, 84°12′W	31	≥31	England et al. (2008)
	TO-12889	6180 ± 70	7260–6900	”	81°53′N, 84°35′W	1	≥1	England et al. (2008)
	TO-12891	6410 ± 80	7470–7170	”	81°52′N, 84°40′W	4	≥4	England et al. (2008)
	TO-12896	6540 ± 80	7580–7290	”	81°55′N, 84°25′W	3	≥3	England et al. (2008)
	TO-12899[d]	7210 ± 80	8190–7860	”	81°55′N, 84°20′W	28	≥28	England et al. (2008)
	UCIAMS-34328[d]	7415 ± 25	8330–8180	”	”	28	≥28	England et al. (2008)
	TO-260	7450 ± 80	8410–8050	”	81°54′N, 84°42′W	55	82.5	Evans (1988)
	TO-12897	7500 ± 70	8420–8180	”	81°48′N, 85°05′W	40	≥40	England et al. (2008)
	TO-12898	8170 ± 80	9450–8950	”	81°51′N, 85°18′W	21	≥21	England et al. (2008)
	TO-12894	39,270 ± 500	~	”	82°20′N, 85°25′W	1	≥1	England et al. (2008)
	TO-12895	43,240 ± 1350	~	”	”	1	≥1	England et al. (2008)
M’Clintock Inlet	TO-12885	7180 ± 70	8170–7860	”	82°38′N, 75°37′W	1	≥1	England et al. (2008)

Yelverton Bay	GSC-1534	8160 ± 140	9500–8650	”	82°02′N, 81°57′W	76?	≥76?	Blake (1972)
	GSC-1603	6410 ± 250	7750–6650	”	82°08′N, 81°57′W	53?	≥53?	Blake (1972)
Serson Ice Shelf	TO-477	4310 ± 70	5300–4600	”	82°16′N, 85°20′W	9	9	Evans (1988)
	TO-476	6830 ± 50	7790–7580	Narwhal tusk	82°16′N, 85°17′W	35	35	Evans (1988)
Alert	L261A	980 ± 100	1125–685	Driftwood	82° 33′N, 63° 01′W	6	≥6	Crary (1960)
	St 4341	985 ± 180	1285–650	”	82°27′N, 63°00′W	2	≥2	England (1976)
	GSC-1770	1070 ± 270	1525–550	”	82°30′N, 63°07′W	11.5	≥11.5	Hattersley-Smith (1973)
	St 4339	1830 ± 200	2305–1340	”	82°27′N, 63°00′W	5	≥5	England (1976)
	S-1993	2130 ± 115	2355–1835	”	82°28′N, 62°49′W	6	≥6	Stewart and England (1983)
	DIC-552	3650 ± 80/−70	4230–3720	”	82°30′N, 62°55′W	10	≥10	Bradley and England (1977)
	S-1986	4770 ± 125	5885–5065	”	82°31′N, 63°06′W	23	≥23	Stewart and England (1983)
	S-1991	5500 ± 125	6565–5950	”	82°31′N, 63°06′W	35	≥35	Stewart and England (1983)
	L261C	6050 ± 200	7415–6465	”	82°28′N, 61°35′W	30	≥30	Crary (1960)

Modified from England et al. (2008), reproduced by permission of American Geophysical Union

[a]*DIC* Dicarb Radioisotope Company, *GSC* Geological Survey of Canada, *L* Lamont-Doherty, *S* Saskatchewan, *SI* Smithsonian Institution, *St* Stockholm, *TO* IsoTrace Laboratory (University of Toronto), *UCIAMS* University of California, Irvine, *UQ* University of Quebec at Montreal

[b]Laboratory-reported errors are 1-sigma

[c]Calibrated ages determined using the program OxCal v3.7, which uses the IntCal04 calibration curve (Reimer et al. 2004). The range represents the 95% confidence interval (±2 sigma)

[d]Denotes re-dated sample

beta counting used at the time; see Crary 1960, p. 41). Other dates have been obtained from marine shells and sponges entrained by basal accretion to the bottom of the Ward Hunt Ice Shelf, and range from 3400 to 13,200 ^{14}C yr BP (Lowdon and Blake 1970; Lyons and Mielke 1973). The most commonly cited estimate for the establishment of the Ward Hunt Ice Shelf was previously based on radiocarbon-dated driftwood samples collected landward of the ice shelf that yielded ages of 3400 ± 150 (L254A) and 3000 ± 200 (L254D) ^{14}C yr BP (Crary 1960; equivalent to 4100–3300, and 3700–2700 cal yr BP, Table 7.1). However, these dates are now called into question because they have not been replicated within the larger database reported here (Table 7.1), many determined by modern accelerator mass spectrometry, to which we now turn.

The most recent compilation of radiocarbon dated driftwood by England et al. (2008) was obtained on samples collected landward of the Ward Hunt Ice Shelf and from two fiords, Clements Markham Inlet and Phillips Inlet, then occupied by multiyear landfast sea ice during Quaternary fieldwork of the 1980s (Table 7.1). Nineteen samples collected inland of the Ward Hunt Ice Shelf range in age from 9200 to 5500 cal yr BP, after which the entry of driftwood appears to have terminated (Fig. 7.4). The majority of these samples (12) date between 7000 and 5500 cal yr BP, recording an interval of widespread driftwood deposition. Fourteen samples from Phillips Inlet, west of the Ward Hunt Ice Shelf, show a similar age distribution, with the last entry of driftwood at around 5400 cal yr BP, while seven of the ages are younger than 7000 cal yr BP. In contrast, Clements Markham Inlet, to the east of the Ward Hunt Ice Shelf, has 31 dated samples recording nearly continuous driftwood deposition from 9900 to 3500 cal yr BP (Fig. 7.4). After 3500 cal yr BP driftwood is rare in Clements Markham Inlet but intervals of sea ice removal are recorded nonetheless by three driftwood ages ranging from 2800 to 1700 cal yr BP, and by two modern samples (Fig. 7.4 and Table 7.1).

Prior to ice shelf formation, driftwood was stranded along northern Ellesmere Island sporadically from the onset of deglaciation until about 7000 cal yr BP, when its abundance increased (Fig. 7.4). This trend is consistent with a westward shift of the TPD towards the CAA after 7000 cal yr BP (Dyke et al. 1997). From 5500 cal yr BP to present, the currently available dates indicate that the previous succession of driftwood entry terminated abruptly both landward of the Ward Hunt Ice Shelf and within the multiple fiords comprising Phillips Inlet (Fig. 7.4). This cessation contrasts with driftwood deposition in Clements Markham Inlet that extended until at least 3500 cal yr BP, after which it was intermittent and sparse. Farther east, in the vicinity of Alert, northeast Ellesmere Island, the driftwood record indicates that its delivery continued to the present (England 1983; Stewart and England 1983). This regional driftwood chronology around northern Ellesmere Island indicates that the absence of driftwood younger than 5500 cal yr BP behind Ward Hunt Ice Shelf and within Phillips Inlet is due to its exclusion by the establishment of multiyear landfast sea ice and ice shelves rather than a widespread termination of its availability controlled by atmospherically-driven ocean currents (Tremblay et al. 1997).

In contrast, the continued delivery of driftwood to Clements Markham Inlet from 5500 to 3500 cal yr BP requires ongoing summer sea-ice clearance following the establishment of the Ward Hunt Ice Shelf to the west. This may reflect the greater width of Clements Markham Inlet, favouring sea ice evacuation and driftwood entry, augmented by its proximity to an adjoining local polynya. At present, this polynya expands northwestward from the north end of Nares Strait, extending across the Lincoln Sea to the mouth of Clements Markham Inlet (Fig. 7.1; Kwok 2006), facilitating evacuation of Arctic Ocean sea ice southward into Nares Strait, and including the seasonal clearance of Clements Markham Inlet. Radiocarbon dates indicate that after the establishment of the Ward Hunt Ice Shelf (5500 cal yr BP), Clements Markham Inlet still received frequent driftwood until 3500 cal yr BP, suggesting the persistence of the Lincoln Sea polynya, whose western limit may have been shortened thereafter when driftwood entered the inlet infrequently until present (Table 7.1, Fig. 7.4).

That the coastline of northern Ellesmere Island has remained blocked by the Ward Hunt Ice Shelf for the past 5500 cal yr is also reinforced by the geomorphology along its landward margin, where many sites are occupied by steep embankments of wind-drifted snow that essentially form a sea-level glacier or 'ice-rise' extending to >100 m asl. Crary (1958, 1960) noted that the junction of the modern (floating) ice shelf with its adjoining (grounded) ice rise is often marked by tidal/strand cracks, and that some of the older tidal cracks now rise up to 22 m above their modern counterparts, embedded in what is now grounded ice. Crary (1960) recognized that the preservation of such older and elevated (abandoned) tidal cracks would require the maintenance of the ice shelf throughout the intervening interval of postglacial emergence, likely for several thousands of years after their initial formation. Crary (1960) attempted to calculate the ages of the uppermost tidal cracks, estimating that their emergence along the south side of Ward Hunt Island occurred at an average rate of 0.5 m/century, which would provide an age of 4400 yr for tidal cracks at 22 m asl. This chronology supporting the antiquity of the elevated tidal cracks is reinforced by the profile of the ~5500 cal yr BP shoreline (equivalent to the proposed age of the Ward Hunt Ice Shelf) surveyed throughout Clements Markham Inlet; this descends northeastwards to ~18 m asl at the fiord mouth (e.g., the 4700 ^{14}C yr BP shoreline; Fig. 7 *in* Bednarski 1986). Assuming that the coastline behind the Ward Hunt Ice Shelf has experienced a similar amount of postglacial emergence as surveyed in outer Clements Markham Inlet (i.e., if both sites occupy the same isobase) then the elevated tidal cracks (22 m asl) would be of similar mid-Holocene age. Alternatively, if the Ward Hunt Ice Shelf had been removed intermittently since 5500 cal yr BP, then uplifted tidal cracks at 22 m asl would be absent and periods of open water would have favoured development of raised beaches (<22 m asl) along this coast that have not been reported inland of the ice shelf (Lemmen 1989). Elsewhere, the landward margin of the ice shelves exhibits small fresh water moats (Lyons et al. 1971) but no description of their geomorphic characteristics has been reported. Based on the current evidence of raised tidal cracks and the limiting ages of the available radiocarbon-dated driftwood, we propose a mid-Holocene age for the initiation of the Ward Hunt Ice Shelf, followed by its subsequent persistence.

Recently collected sediment cores from Disraeli Fiord have been used to reconstruct the former history of establishment and maintenance of the Ward Hunt Ice Shelf (Antoniades et al. 2011). The cores were analyzed for both biological and geochemical proxies that were used to differentiate between two oscillating conditions: (1) intervals when Disraeli Fiord was occupied entirely by marine waters (precluding blockage by an ice shelf across the fiord mouth); and (2) intervals when the marine water was capped by a relatively fresh epishelf lake requiring ice shelf blockage of the fiord mouth. Evidence for the most recent drainage of the epishelf lake in Disraeli Fiord occurred in 2001 when the Ward Hunt Ice Shelf fractured, but remained otherwise intact. The sediment core record spans most of the Holocene and indicates that the Ward Hunt Ice Shelf first formed 4000 cal yr BP and persisted until 1400 cal yr BP when the epishelf lake again drained until reforming 800 cal yr BP (Antoniades et al. 2011). Currently, it is difficult to reconcile the different ice shelf histories proposed by the available driftwood and sediment core records from Disraeli Fiord. However, both records emphasize the great antiquity of the Ward Hunt Ice Shelf that spans all, or close to, the last half of the Holocene, underscoring the significance of its recent demise. Furthermore, the proposed drainage of the epishelf lake between 1400 and 800 cal yr BP, based on the core interpretation, may not be incompatible with the driftwood record if the proposed fracturing of the ice shelf was not accompanied by ice shelf clearance, hence maintaining driftwood blockage from Disraeli Fiord initiated after 5500 cal yr BP (Fig. 7.4). In contrast, the continuous and abundant entry of driftwood into nearby Clements Markham Inlet persisted for 2000 cal yr after the termination of driftwood entry into Disraeli Fiord; this indicates that these different histories cannot be ascribed to the lack of driftwood availability, as might be caused by a redirection or cessation of delivery by the TPD within the Arctic Ocean.

Periods of open water in the Canadian High Arctic are nonetheless recorded in some locations by the remains of marine mammals at or below marine limit. For example, Evans (1989) made a case for open water along the NW Ellesmere Island coast until at least 6830 ^{14}C yr BP based upon the occurrence of a narwhal tusk *(Monodon monoceros* L.) stranded at 32 m asl on Wootton Peninsula, landward of the previous Serson Ice Shelf ('Cape Alfred Ernest Ice Shelf' in Lemmen et al. 1988; Evans 1990). The Serson Ice Shelf was mostly removed by calving in 2008 (Mueller et al. 2008, 2017). Corroborative geomorphological evidence for this earlier period of open water (6830 ^{14}C yr BP) is marked by the occurrence of sea-ice pushed ridges in raised marine gravels along the western shoreline of Wootton Peninsula (Evans and England 1992). Nonetheless, similar surveys of raised marine shorelines along thousands of kilometres of coastline in the Canadian High Arctic demonstrate that the remains of postglacial bowhead whales are rare. Summarizing radiocarbon dates on these collections, Dyke and England (2003) conclude that their ages record opportunistic 'strays' rather than intervals of widespread summer sea ice clearance. Examples include a rare bowhead whale (*Balaena mysticetus*) skeleton dating to 7475 ^{14}C yrs. BP collected from Nansen Sound (between Axel Heiberg and Ellesmere islands; Bednarski 1990) and another bowhead whale skull

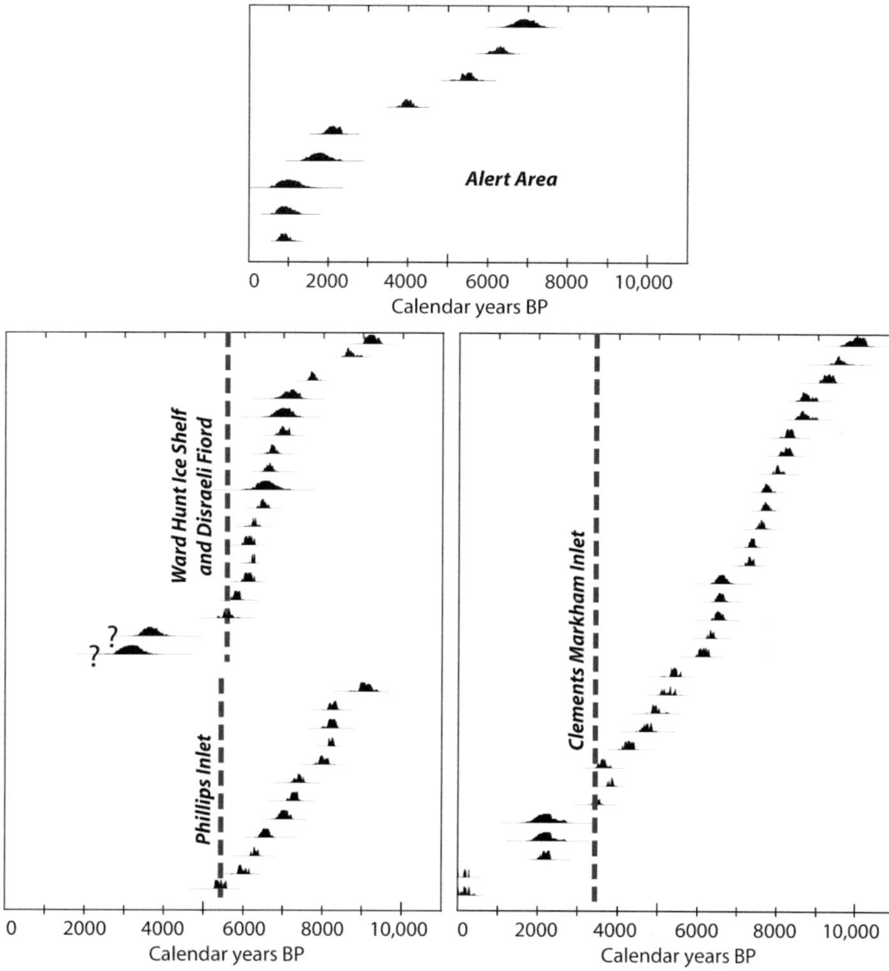

Fig. 7.4 Plot of individual samples of radiocarbon-dated driftwood. Each date is plotted showing its calibrated probability distribution, which was determined using OxCal v3.7 (Reimer et al. 2004). See Table 7.1 for sample details and Fig. 7.1 for site locations. The available dates discussed in text were collected from four sites on northern Ellesmere Island: (1) behind the Ward Hunt Ice Shelf and within adjoining Disraeli Fiord (*left frame*); (2) within Phillips Inlet historically occupied by multiyear landfast sea ice (*left frame*); (3) within Clements Markham Inlet, currently bordered by the Lincoln Sea polynya that increases the likelihood of its seasonal clearance (*right frame*); and (4) around the Alert area within the influence of the Lincoln Sea polynya. Vertical dashed lines (*blue*) mark widespread termination of driftwood stranding for each locality, except the Alert area. On *left frame*, both sites show widespread driftwood entry from the time of deglaciation to ~5500 cal yr BP, after which driftwood is precluded, coinciding with ice shelf inception and maintenance to the present. However, younger dates reported by Crary (1960) are also shown, designated by question marks. Note that cessation of driftwood entry into Clements Markham Inlet (*right frame*) is delayed until 3500 cal yr BP after which it continues intermittently. For the Alert area, driftwood entry and deposition was continuous throughout the Holocene (Reproduced from England et al. (2008), with permission of the American Geophysical Union)

dating to 10,400 ^{14}C yrs. BP on Ellef Ringnes Island (Atkinson 2009), both sites characterized by severe (multiyear) summer sea ice in the northern CAA at the time of their collection.

Although open water during the Early Holocene in the Canadian High Arctic was insufficient for its occupation by whales, driftwood entry to the north coast of Ellesmere Island required the mobilization of pack ice rather than landfast sea ice. Such conditions are consistent with the widely recognized, regional signal of climatic amelioration (Koerner 1977; Koerner and Fisher 1990) that accompanied glacier recession inside modern ice mergins that readvanced in the mid to late Holocene (Blake 1972, 1981; Bradley 1990; Evans and England 1992; La Farge et al. 2013). Furthermore, based on ice core analyses, Koerner and Paterson (1974) suggest that the entire Meighen Ice Cap (~85 km^2, 80°N) regrew sometime after the mid Holocene (~3000–4500 cal yr BP) and recognized that this history resembled the inception of the Ward Hunt Ice Shelf, ~500 km to the east. The climatic deterioration responsible for widespread glacier advances throughout the QEI after the mid Holocene is clearly recorded in palaeotemperature records in cores retrieved from the Agassiz Ice Cap on northern Ellesmere Island (Fisher et al. 1995). This deterioration culminated in the worst sea ice conditions in the Canadian Arctic Archipelago, which coincided with many ill-fated, late nineteenth century expeditions towards the end of the LIA (Alt et al. 1985). The driftwood ages from northern Ellesmere Island are consistent with this paleoenvironmental record, indicating that the Ellesmere Ice Shelf and the last of its modern remnants began to form shortly after 5500 cal yr BP. Interestingly, Crary (1960) estimated that Ice Island T-3 was 5000 years old, based on an extrapolated age for its lowermost dirt layers. He attributed the origin of Ice Island T-3 to an ancestral ice shelf in Yelverton Inlet that was older than his estimated age for the Ward Hunt Ice Shelf (~3000 ^{14}C yr BP).

7.5 Conclusion

The age distribution of radiocarbon dates on driftwood deposited landward of the modern Ward Hunt Ice Shelf and inside the multiyear landfast sea ice in Phillips Inlet, record its exclusion from these coastlines for more than the past five millennia. Stranded driftwood of the circumpolar region represents a chronological record of a sea-ice buoy system (Häggblom 1982) whose trajectory varies through time due to changes in atmospherically driven surface currents (Dyke et al. 1997; Tremblay et al. 1997). Driftwood stranding is widely available around the Arctic Ocean at sites that have undergone postglacial emergence and remains largely underutilized *vis a vis* reconstructing past ocean currents and sea ice history.

The lack of driftwood deposition at multiple sites across northern Ellesmere Island during the past 5500 years is the outcome of coastal blockage by sea ice rather than the lack of driftwood availability due to unfavourable surface current patterns. This conclusion is supported by the continued arrival of younger driftwood in nearby fiords like Clements Markham Inlet (Bednarski 1986) and along northeast

Ellesmere Island (Stewart and England 1983). Radiocarbon dates on driftwood behind these ice shelves and within neighbouring fiords occupied historically by multiyear sea ice allow us to reconstruct a chronology of regional sea ice variability. This chronology resonates with other proxy indicators of climate deterioration widely recognized across the Canadian Arctic since the mid-Holocene, including the re-growth of the Meighen Ice Cap (Paterson and Koerner 1974). The magnitude of recent and ongoing ice shelf breakup across northern Ellesmere Island (especially post twenty-first century, Mueller et al. 2017) is clearly in lockstep with the dramatic seasonal reduction of Arctic Ocean sea ice, documented by satellite surveillance during the last four decades. The driftwood record serves to place this loss in a much longer perspective (England et al. 2008). For example, it has been demonstrated that the stability of seasonal landfast sea ice along the coast of northern Alaska is strongly tied to the maintenance of buttressing pack ice (Mahoney et al. 2007). Therefore, assuming that similar sea ice processes prevail along the north coast of Ellesmere Island (i.e., that pack ice and landfast sea ice co-vary), it is apparent that the substantial reduction in the size of Ellesmere ice shelves since the onset of the twentieth century must also record a corresponding and previously unrecognized reduction in Arctic Ocean pack ice that began prior to the onset of now widely published satellite surveillance (post-1966; Vinnikov et al. 1999; Johannessen et al. 1999; Serreze et al. 2007; Comiso et al. 2008). This conclusion is strengthened by the observation by Copland et al. (2007) that the detachment of the Ayles Ice Shelf in August 2005 coincided with an unusually large area of adjacent open water (15 km wide) as opposed to buttressing pack ice. Bushnell (1956, p.175) reports that severe pack ice was "of a permanent nature and badly hummocked" along the outer margin of the Ward Hunt Ice Shelf during both Marvin's surveys of 1906, and those of Hattersley-Smith and Crary during 1954. Assuming that the ongoing reduction of Arctic Ocean sea ice will continue apace with global warming during the next several decades, the final demise of the remaining Ellesmere Island ice shelves is imminent. Indeed, based on the rate of breakup of the Ellesmere ice shelves, England et al. (2008) concluded that the seasonal clearance of the fiords of northern Ellesmere Island, historically occupied by multiyear landfast sea ice, also beckoned. This prediction was documented only 4 years later with the complete clearance of multiyear landfast sea ice from Yelverton Inlet and Fiord (Pope et al. 2012). Crary (1960, p. 49) had already anticipated this possibility when he stated: "it would appear that we are indeed approaching a period when the ice shelves might very well become extinct".

The removal of the remaining Ellesmere Island ice shelves would herald the disappearance of the oldest sea ice in the Northern Hemisphere and would be unprecedented in more than 5000 years, underscoring the importance of ongoing, post-twentieth century warming. We recognize, however, that large stretches of the northern Ellesmere Island and Greenland coasts remain to be investigated for their driftwood chronologies, which are widespread and underutilized in the areas we have investigated. Important collections could be made landward of the recently detached Ayles Ice Shelf (Copland et al. 2007) and along the recently reduced and remnant margins of all the others outlined by Mueller et al. (2017). Profitable driftwood dates could also be obtained from samples still inland of the most resistant core of the remnant Ward Hunt Ice Shelf separating Disraeli and Markham fiords (Fig. 7.1). When widespread driftwood samples are increasingly collected and dated, throughout the Arctic Ocean basin, these records will provide new insights concerning past trajectories of the TPD and will further refine our understanding of circumpolar sea ice variability throughout the Holocene. Based on the currently available samples of radiocarbon-dated driftwood (69) from northern Ellesmere Island, the hypothesis that ice shelf initiation was widespread by 5500 cal yr BP can be readily tested and improved. This driftwood record will provide an important complement to other glaciological and biological environmental proxies (Antonaides et al. 2011) as well as coupled ocean-atmospheric modelling designed to clarify our understanding of the sea ice system (Tremblay et al. 1997) that is undergoing rapid reduction.

Acknowledgements Fieldwork responsible for the collection of the driftwood reported here was logistically supported by the Polar Continental Shelf Project, NRCan, Ottawa, throughout the 1970s and 1980s. This northern Ellesmere Island fieldwork was conducted by former graduate students at the University of Alberta: Jan Bednarski, Tom Stewart, Don Lemmen and David Evans whose collections now reside within Earth and Atmospheric Sciences (U of A). Additional former students from the University of Alberta are also acknowledged for assistance in the field: Doug Calvert, Tim Fisher, Philip Friend, Ulrika Hawkins, Brian Szuster, Maria Matishak, Val Sloan, and Richard England. Throughout this field research, financial support was provided by NSERC Discovery Grants to J. England and by grants awarded to graduate students (above) by the Canadian Circumpolar Institute (University of Alberta). The opportunity to finally analyze many undated samples stored for several decades was facilitated by the award of an NSERC Northern Research Chair to J. England (2002–2012). We thank Jan Bednarski, Pacific Geoscience Centre, Geological Survey of Canada (Sydney, B.C.), and Art Dyke, Terrain Sciences Division, Geological Survey of Canada (Ottawa), for helpful formal reviews on an earlier version of this manuscript.

References

Alt, B. T., Koerner, R. M., Fisher, D. A., & Bourgeois, J. C. (1985). Arctic climate during the Franklin era as deduced from ice cores. In P. D. Sutherland (Ed.), *The Franklin era in Canadian Arctic history, 1845–1859, National Museum of Man Symposium, Ottawa, June 7–8, 1984*.

Antoniades, D., Francus, P., Pienitz, R., St. Onge, G., & Vincent, W. F. (2011). Holocene dynamics of the Arctic's largest ice shelf. *Proceedings of the National Academy of Science, 108*, 18899–18904.

Atkinson, N. (2009). A 10400-year-old bowhead whale (*Balaena mysticetus*) skull from Ellef Ringnes Island, Nunavut: Implications for sea-ice conditions in High Arctic Canada at the end of the last glaciation. *Arctic, 62*(1), 38–44.

Bednarski, J. (1986). Late Quaternary glacial and sea-level events, Clements Markham Inlet, northern Ellesmere Island, Arctic Canada. *Canadian Journal of Earth Sciences, 23*, 1343–1355.

Bednarski, J. (1990). An early Holocene bowhead whale *(Balaena mysticetus)* in Nansen Sound, Canadian Arctic Archipelago. *Arctic, 43*, 50–54.

Blake Jr., W. (1972). Climatic implications of radiocarbon-dated driftwood in the Queen Elizabeth Islands, Arctic Canada. In Y. Vasari, H. Hyvarinen, S. Hicks (Eds.), *Climatic changes in Arctic areas during the last ten thousand years* (Acta Universitatis Ouluensis Series A, No. 3, Geologica No. 1, p. 77–104). Proceedings of a symposium held in Oulanka-Kevo, Finland, 1971.

Blake Jr., W. (1981). Neoglacial fluctuations of glaciers, southeastern Ellesmere Island, Canadian Arctic Archipelago. *Geografiska Annaler, 63A*, 201–218.

Bradley, R. S. (1990). Holocene paleoclimatology of the Queen Elizabeth Islands, Canadian High Arctic. *Quaternary Science Reviews, 9*, 365–384.

Bradley, R. S., & England, J. H. (1977). *Past glacial activity in High Arctic*. Amherst: University of Massachusetts, Department of Geology and Geography, Contribution, No. 31, 184pp.

Bushnell, V. C. (1956). Marvin's ice shelf journey, 1906. *Arctic, 9*, 166–177.

Comiso, J. C., Parkinson, C. L., Gersten, R., & Stock, L. (2008). Accelerated decline in the Arctic sea ice cover. *Geophysical Research Letters, 35*, L01703. doi:10.1029/2007GL031972.

Cook, F. A. (1911). *My attainment of the Pole*. New York: Polar Publishing Company, 604pp.

Copland, L., Mueller, D. R., & Weir, L. (2007). Rapid loss of the Ayles Ice Shelf, Ellesmere Island, Canada. *Geophysical Research Letters, 34*, L21501. doi:10.1029/2007GL031809.

Crary, A. P. (1958). Arctic ice islands and ice shelf studies, part I. *Arctic, 11*, 2–42.

Crary, A. P. (1960). Arctic ice islands and ice shelf studies, part II. *Arctic, 13*, 32–50.

Crary, A. P., Kulp, J. L., & Marshall, E. W. (1955). Evidence of climatic change from ice island studies. *Science, 122*, 1171–1173.

Dyke, A. S., & England, J. H. (2003). Canada's most northerly postglacial bowhead whales (*Balaena mysticetus*): Holocene sea ice conditions and polynya development. *Arctic, 56*, 14–20.

Dyke, A. S., England, J. H., Reimnitz, E., & Jetté, H. (1997). Changes in driftwood delivery to the Canadian Arctic Archipelago: The hypothesis of postglacial oscillations of the transpolar drift. *Arctic, 50*, 1–16.

England, J. H. (1976). Postglacial isobases and uplift curves from the Canadian and Greenland High Arctic. *Arctic and Alpine Research, 8*, 61–78.

England, J. H. (1983). Isostatic adjustments in a full glacial sea. *Canadian Journal of Earth Sciences, 20*, 895–917.

England, J. H., Atkinson, N., Bednarski, J., Dyke, A. S., Hodgson, D. A., & Ó Cofaigh, C. (2006). The Innuitian Ice Sheet: Configuration, dynamics and chronology. *Quaternary Science Reviews, 25*, 689–703.

England, J. H., Lakeman, T. R., Lemmen, D. S., Bednarski, J. B., Stewart, T. G., & Evans, D. J. A. (2008). A millennial-scale record of Arctic Ocean sea ice variability and the demise of the Ellesmere Island ice shelves. *Geophysical Research Letters, 35*, L19502. doi:10.1029/2008GL034470.

Evans, D. J. A. (1988). *Glacial geomorphology and late Quaternary history of Phillips Inlet and the Wootton Peninsula, northwest Ellesmere Island, Canada* (PhD thesis). Edmonton: University of Alberta.

Evans, D. J. A. (1989). An early Holocene narwhal tusk from the Canadian High Arctic. *Boreas, 18*, 43–50.

Evans, D. J. A. (1990). The last glaciation and relative sea level history of northwest Ellesmere Island, Canadian High Arctic. *Journal of Quaternary Science, 5*, 67–82.

Evans, D. J. A., & England, J. H. (1992). Geomorphological evidence of Holocene climate change from northwest Ellesmere Island, Canadian High Arctic. *The Holocene, 2*, 148–158.

Fisher, D. A., Koerner, R. M., & Reeh, N. (1995). Holocene climatic records from Agassiz Ice Cap, Ellesmere Island, NWT, Canada. *The Holocene, 5*, 19–24.

Häggblom, A. (1982). Driftwood in Svalbard as an indicator of sea ice conditions. *Geografiska Annaler, 64A*, 81–94.

Hattersley-Smith, G. (1957). The rolls on the Ellesmere Ice Shelf. *Arctic, 10*, 32–44.

Hattersley-Smith, G. (1963). The Ward Hunt Ice Shelf: Recent changes of the ice front. *Journal of Glaciology, 4*, 415–424.

Hattersley-Smith, G. (1973). An archeological site on the north coast of Ellesmere Island. *Arctic, 26*, 255–256.

Hattersley-Smith, G., Fuzesy, A., & Evans, S. (1969). *Glacier depths in northern Ellesmere Island: airborne radio-echo sounding in 1966, Technical Note 69-6*. Ottawa: Defence Research Board, 23pp.

Holdsworth, G. (1971). Calving from the Ward Hunt Ice Shelf 1961–1962. *Canadian Journal of Earth Sciences, 8*, 299–305.

Jeffries, M. O. (1987). The growth, structure and disintegration of Arctic ice shelves. *Polar Record, 23*, 631–640.

Jeffries, M. O. (2017). The Ellesmere ice shelves, Nunavut, Canada. In L. Copland & D. Mueller (Eds.), *Arctic ice shelves and ice islands* (p. 23–54). Dordrecht: Springer. doi:10.1007/978-94-024-1101-0_2.

Johannessen, O. M., Shalina, E. V., & Miles, M. W. (1999). Satellite evidence for an Arctic ice cover in transition. *Science, 286*, 1937–1939.

Koenig, L. S., Greenaway, K. R., Dunbar, M., & Hattersley-Smith, G. (1952). Arctic ice islands. *Arctic, 5*, 67–103.

Koerner, R. M. (1977). Devon Island Ice Cap: Core stratigraphy and paleoclimate. *Science, 196*, 15–18.

Koerner, R. M., & Fisher, D. A. (1990). A record of Holocene summer climate from a Canadian High-Arctic ice core. *Nature, 343*, 630–631.

Koerner, R. M., & Paterson, W. S. B. (1974). Analysis of a core through the Meighen Ice Cap, Arctic Canada, and its paleoclimatic implications. *Quaternary Research, 4*, 253–263.

Kwok, R. (2006). Near zero replenishment of the Arctic multiyear sea ice cover at the end of 2005 summer. *Geophysical Research Letters, 32*, L24502. doi:10.1029/2005GL024768.

La Farge, C., Williams, K. H., & England, J. H. (2003). Regeneration of Little Ice Age bryophytes emerging from a polar glacier with implications of totipotency in extreme environments. *Proceedings of the National Academy of Science, 110*, 9839–9844.

Lemmen, D. S. (1989). The last glaciation of Marvin Peninsula, northern Ellesmere Island, High Arctic, Canada. *Canadian Journal of Earth Sciences, 26*, 2578–2590.

Lemmen, D. S., Evans, D. J. A., & England, J. H. (1988). Ice shelves of northern Ellesmere Island, N.W.T., Canadian landform examples. *Canadian Geographer, 32*, 363–367.

Levesque, E., & Svoboda, J. (1999). Vegetation re-establishment in polar "lichen-kill" landscapes: A case study of the Little Ice Age impact. *Polar Research, 18*, 221–228.

Lowdon, J. A., & Blake Jr., W. (1970). Geological Survey of Canada radiocarbon dates IX. *Radiocarbon, 12*, 46–86.

Lyons, J. B., & Mielke, J. E. (1973). Holocene history of a portion of northernmost Ellesmere Island. *Arctic, 26*, 314–323.

Lyons, J. B., Savin, S. M., & Tamburi, A. J. (1971). Basement ice, Ward Hunt Ice Shelf, Ellesmere Island, Canada. *Journal of Glaciology, 10*, 93–100.

Mahoney, A., Eicken, H., Gaylord, A. G., & Shapiro, L. (2007). Alaska landfast sea ice: Links with bathymetry and atmospheric circulation. *Journal of Geophysical Research, 112*, C02001. doi:10.1029/2006JC003559.

Mueller, D. R., Vincent, W. F., & Jeffries, M. O. (2003). Break-up of the largest Arctic ice shelf and associated loss of an epishelf lake. *Geophysical Research Letters, 30*(20), 2031. doi:10.1029/2003GL017931.

Mueller, D. R., Copland, L., Hamilton, A., & Stern, D. (2008). International Polar Year scientists and Canadian Rangers visit Arctic ice shelves just before massive loss in 2008. *EOS, Transactions, American Geophysical Union, 89*(49), 502–503.

Mueller, D., Copland, L., & Jeffries, M. (2017). Changes in Canadian Arctic ice shelf extent since 1906. In L. Copland & D. Mueller (Eds.), *Arctic ice shelves and ice islands* (p. 109–148). Dordrecht: Springer. doi:10.1007/978-94-024-1101-0_5.

Nares, G. S. (1878). *Narrative of a voyage to the Polar Sea during 1875–6 in HM Ships 'Alert' and 'Discovery'*. London: Sampson Low, Marston, Searle, and Rivington.
Nghiem, S. V., Rigor, I. G., Perovich, D. K., Clemente-Colon, P., Weatherly, J. W., & Neumann, G. (2007). Rapid reduction of Arctic perennial sea ice. *Geophysical Research Letters, 34*, L19504. doi:10.1029/2007GL031138.
Ommanney, C. S. L. (1982a). *Bibliography of Canadian glaciology, 1982 – Bibliography No. 2, Ellesmere Island glaciers and ice shelves. Paper No. 20*. Saskatoon: Environment Canada, National Hydrology Research Centre.
Ommanney, C. S. L. (1982b). *Bibliography of Canadian glaciology, 1982 – Bibliography No. 3, Ice Islands of the Arctic Ocean. Environment Canada, Paper No. 21*. Saskatoon: National Hydrology Research Centre.
Paterson, W. S. B., & Koerner, R. M. (1974). Radio echo sounding on four ice caps in Arctic Canada. *Arctic, 27*, 225–233.
Peary, R. E. (1907). *Nearest the Pole*. London: Hutchinson.
Peary, R. E. (1910). *The North Pole*. London: Hodder and Stoughton, 326pp.
Polunin, N. (1955). Attempted dendrochronological dating of ice island T-3. *Science, 122*, 1184–1186.
Pope, S., Copland, L., & Mueller, D. (2012). Loss of multiyear landfast sea ice from Yelverton Bay, Ellesmere Island, Nunavut, Canada. *Arctic, Antarctic, and Alpine Research, 44*(2), 210–221.
Reimer, P. J., Baillie, M. G. L., Bard, E., Bayliss, A., Beck, J. W., Betrand, C. J. A. & 23 others. (2004). IntCal04 terrestrial radiocarbon age calibration, 0–26 cal kyr BP. *Radiocarbon, 46*, 1029–1058.
Serreze, M. C., Holland, M. M., & Stroeve, J. (2007). Perspectives on the Arctic's shrinking sea-ice cover. *Science, 315*, 1533–1536.
Stewart, T. G. (1981). *The Holocene paleoenvironment of Clements Markham Inlet, northern Ellesmere Island, NWT, Canada* (MSc thesis). Edmonton: University of Alberta.
Stewart, T. G., & England, J. H. (1983). Holocene sea-ice variations and paleoenvironmental change, northernmost Ellesmere Island, NWT, Canada. *Arctic and Alpine Research, 15*, 1–17.
Tremblay, L.-B., Mysak, L. A., & Dyke, A. S. (1997). Evidence from driftwood records for century- to millennial-scale variations of the high latitude atmospheric circulation during the Holocene. *Geophysical Research Letters, 24*(16), 2027–2030.
Vincent, W. A., Gibson, J. A. E., & Jeffries, M. O. (2001). Ice-shelf collapse, climate change, and habitat loss in the Canadian High Arctic. *Polar Record, 37*, 133–142.
Vinnikov, K. Y., Robock, A., Stouffer, R. J., Walsh, J. E., Parkinson, C. L., Cavalieri, D. J., Mitchell, J. F. B., Garrett, D., & Zakharov, V. F. (1999). Global warming and Northern Hemisphere sea ice extent. *Science, 286*, 1934–1937.
Wolken, G. J., England, J. H., & Dyke, A. S. (2005). Re-evaluating the relevance of vegetation trimlines in the Canadian Arctic as an indicator of Little Ice Age paleoenvironments. *Arctic, 58*, 341–353.

Chapter 8
An Overview of Paleoenvironmental Techniques for the Reconstruction of Past Arctic Ice Shelf Dynamics

Dermot Antoniades

Abstract Information about the long-term responses of Arctic ice shelves to changes in climate is critical for understanding the significance of their recent declines. However, the history of these systems is poorly understood, and few methods exist to reconstruct these missing datasets. Paleoenvironmental studies of sediment cores can provide valuable information about past ice shelf fluctuations and provide the context in which to evaluate recent changes. This chapter discusses the methods available for the reconstruction of ice shelf dynamics from sediment cores. It reviews the proxy indicators that provide insights into different ice shelf characteristics and their effects on sedimentation and water column properties, and gives examples where these have been applied in studies of ice shelves from both polar regions.

Keywords Paleoenvironmental techniques • Epishelf lake • Proxy indicators • Arctic fiords • Sediment cores

8.1 Introduction

Recent fracturing and reduction in extent of Arctic ice shelves (Mueller et al. 2017) and sea ice (Pope et al. 2017) has led to renewed interest in the history of these systems. Although the recent disintegration and disappearance of ice shelves in the Canadian Arctic is a source of concern, the lack of records of their dynamics prior to the twentieth century means that it is impossible to place the significance of these changes in a long-term perspective. The earliest efforts to constrain the ages of these systems were associated with the studies of ice islands (e.g., Crary 1956, 1960). Several indirect approaches were used, including the dating of driftwood found landward of ice shelves (Crary 1960; Mielke and Long 1969; Lemmen 1989;

D. Antoniades (✉)
Department of Geography and Centre for Northern Studies (Centre d'études nordiques)
Université Laval, Québec, QC, Canada
e-mail: dermot.antoniades@cen.ulaval.ca

L. Copland, D. Mueller (eds.), *Arctic Ice Shelves and Ice Islands*, Springer Polar Sciences, DOI 10.1007/978-94-024-1101-0_8

England et al. 2008; reviewed in England et al. 2017), dating dirt layers and shells contained in the ice (Crary 1960; Lyons and Mielke 1973), and thermal modeling (Lyons and Ragle 1962). These studies provided largely indirect evidence that contributed to the understanding of the history of Arctic ice shelf complexes. The ages that they provide are considered to be maximum limiting ages (Bradley 1990), and so much information about ice shelf dynamics since formation remains to be elucidated. One potential approach for better understanding past changes in ice shelves is the use of paleoenvironmental methods in studies of Arctic fiord ecosystems.

A potential application of paleoenvironmental techniques to studies of ice shelves is reconstruction based on epishelf lake sediments. Arctic ice shelves are typically situated at the mouths of fiords, in many cases blocking them completely. As a result, they form dams that trap inflowing meltwater, creating epishelf lakes (see Veillette et al. 2008). These lakes consist of a stratified water column where lighter freshwater sits atop denser marine water, and contain an unusual combination of marine and freshwater biota (e.g., Van Hove et al. 2001; Jungblut et al. 2017). The thickness of the freshwater layer is generally interpreted to reflect the draft of the ice dam, and therefore the depth of the freshwater/marine transition has been used as a proxy of ice shelf thickness (Mueller et al. 2003; Veillette et al. 2008). In the absence of ice shelves, these systems behave as typical fiords, with a freshwater cap occurring only seasonally and occupying a small fraction of the total volume of water in the system. Clearly, temporal shifts from freshwater to marine conditions result in profound differences in the fiords' surface waters, including chemical characteristics and biotic communities, among others. Given the utility of bottom sediments as an archive of past environmental conditions (see Smol 2008), Arctic fiords and their sedimentary records have significant, largely untapped potential for the reconstruction of the past dynamics of the ice shelves at their mouths.

Epishelf and ice-dammed lakes are analogous in that their existence and characteristics are controlled by glacial processes. The value of sediments for reconstructing past proglacial lake environments has been recognized since studies of Lake Agassiz beginning in the nineteenth century (e.g., Upham 1895; Teller and Clayton 1983). More recently, paleoenvironmental studies have illustrated the utility of Arctic proglacial lake sediments in the reconstruction of several aspects of past environmental change. Numerous sedimentary properties and constituents have been used as proxy indicators, including particle size, varve thickness, organic carbon, carbonate, iron and manganese content, fossil diatom analysis, and x-ray analysis of sedimentary structure, in studies that reconstructed characteristics such as glacial activity (Snyder et al. 2000), deglaciation history (Mangerud and Svendsen 1990), and climate and biological productivity (Lemmen et al. 1988; Lewis et al. 2002; Lamoureux and Gilbert 2004).

Epishelf lakes, however, are distinct from other proglacial lake environments in that sedimentation occurs in a marine or mixed marine/freshwater setting. Arctic epishelf lakes maintain a hydraulic connection with the ocean, with the proportion of the water column occupied by fresh and marine water varying between ecosystems as well as over time (Mueller et al. 2003; Veillette et al. 2008). There are important fundamental differences in sedimentary processes between freshwater and marine environments that control sedimentation in Arctic fiords (Gilbert 1983; Syvitski 1989). Moreover, if ice shelves have fractured and re-established during their histories, then the sediments of epishelf lakes will contain phases of combined marine/freshwater influence and marine-only sedimentation. A complete understanding of epishelf lake sediments must therefore draw analogies to linkages between the environment and the sedimentary record that occur in both freshwater and marine settings.

There are important differences between Arctic and Antarctic ice shelves that have clear implications for paleoenvironmental reconstructions (Veillette et al. 2008). In the Antarctic, ice shelves form as the floating extensions of glacier tongues that project into the ocean. They therefore act as a vector for the transport of terrestrial material to fiord sediments. In the Arctic, however, ice shelves form primarily as progressive thickenings of multiyear landfast sea ice. While they contain aeolian sediment both on their surfaces and within, they do not act as direct agents of terrestrial sediment transport. Sediments are delivered from land to Arctic fiords by fluvial transport, ensuring that the sedimentary record of both polar regions contains terrestrial components.

In studies where paleoenvironmental techniques have been used to reconstruct ice shelf histories, two approaches have been used: those considering the sediments below ice shelves, and those examining the sediments of epishelf lakes. Studies that examine sediments below existing ice shelves are constrained by the difficulty and expense of sediment coring below extant ice shelves. As a result, very few such studies exist, and to date these have only been carried out in the Antarctic (Webb et al. 1979; Kolobov and Savatyugin 1983; Raiswell and Tan 1985; Kellogg and Kellogg 1986; Hemer and Harris 2003; Naish et al. 2007; Smith et al. 2016). The main source of information about below-ice sedimentation is from studies performed shortly after the collapse of ice shelves (Pudsey and Evans 2001; Brachfield et al. 2003; Gilbert and Domack 2003; Domack et al. 2005; Pudsey et al. 2006). Not only have these studies reconstructed past ice shelf dynamics, they have also greatly advanced our understanding of the sedimentary regimes under conditions of stable, retreating, and absent ice shelves, allowing the application of this methodology elsewhere.

8.2 Sediment Classifications in Ice Shelf Glaciomarine Environments

While authors have attempted to classify the depositional environments near individual ice shelves, more general classifications are challenging because the sedimentation rates beneath ice shelves are highly variable and depend on several factors including the velocity of inflowing glaciers, catchment geology and the degree to which rocks are exposed in the catchment, and the mechanism of ice shelf formation. In studying sediments collected below the former positions of the Larsen and

Prince Gustav ice shelves on the Antarctic Peninsula, Gilbert and Domack (2003) recognized three stages in below-ice shelf sedimentation related to growth and stability. Under a stable or advancing ice shelf, small amounts of sediments were periodically released by melting at the base of the ice shelf. This sediment consisted of a mixture of large clasts and fine-grained sediment, and was also characterized by relatively low accumulation rates. At the same time, stable ice shelves accumulated aeolian sediments on their surface. As ice shelves ablated, higher sediment accumulation rates resulted from increases in the rate of basal melting and the release of ice shelf-borne sediment. Throughout this stage, sediments on the ice shelf became concentrated in ponds and depressions due to increased surface melt and transport. In ice shelves where they occur, surface ponds provide water to crevasses, and this meltwater is instrumental in further weakening ice shelves by expanding and propagating fissures in the ice shelf surface (Scambos et al. 2000). When these surface ponds melted through the ice, they drained catastrophically and the aeolian sediment accumulated within the ponds was released to the ocean floor in deposits laid down intermittently over several years prior to ice shelf break up. However, this mode of sedimentation was only local in extent, and therefore the different patterns of sediment accumulation that existed were related to nearby melt conditions on the ice shelf above. During the final disintegration of the ice shelf, the sediment flux from the ice shelf increased greatly. In addition to sediments from the ice shelf surface and those released by basal melting, englacial sediment may contribute significant amounts of material at this time. Overall sedimentation rates were observed to increase by two to four times during ice shelf disintegration in Antarctica (Gilbert and Domack 2003); these sediments can also be recognized by the higher gravel content that results from the higher contribution of glacial sources due to ice rafting.

In a study of Antarctic Peninsula sediments, Evans and Pudsey (2002) determined that three depositional environments related to proximity to ice shelves can be recognized based on a series of facies characteristics, including grain size, sorting, shear strength, and sediment structure. Subglacial sediments (basal tills deposited below grounded ice sheets) were massive, comprised largely of diamicton, and contained abrupt upper bed contacts. Proximal ice shelf sediments were deposited at or near the grounding line below existing ice shelves. These sediments were generally coarse-grained, dominated by diamicton and gravel, with deposition that was at times size-sorted and/or laminated. Dropstones were present in this facies as a result of rafting of larger material by icebergs. Ice shelf distal facies sediments were deposited at or near the ice margin, and contained massive diatom-rich muds and silty clays that showed evidence of bioturbation. These sediments were characterized by a transition to finer grain sizes relative to the coarse gravels and diamicton of the ice-proximal deposits.

8.3 Proxy Indicators

A wide variety of proxy indicators in sediment can be used to reconstruct ice shelf histories. These include physical, biological and geochemical proxies that can provide a remarkable amount of information about past glaciomarine environments. Each proxy has strengths and limitations, and the applicability of individual indicators may depend on the characteristics of individual ecosystems. As inferences based on single indicator groups can, at times, leave room for ambiguity, it is important that multi-proxy approaches be used in ice shelf paleoenvironmental studies to reinforce and corroborate inferences of past change.

What follows is a summary of several proxy indicators from the sedimentary record that have been employed to reconstruct ice shelf dynamics. Undoubtedly, new approaches and proxies will be developed in the future. Although the majority of studies cited below were carried out in Antarctica, the similarities between ice shelf-dominated ecosystems in both polar regions imply that the underlying principles hold equally true for Northern Hemisphere sites. While caution must be exercised when drawing parallels between processes and individual ecosystems in different polar regions, similar applications of paleoenvironmental techniques in the Arctic have the potential to augment our understanding of the history of northern ice shelves.

8.3.1 Sedimentation and Grain Size

The transition from ice shelf presence to absence is marked by changes in physical sediment properties, including mean grain size and gravel content and changes in sediment accumulation rates. The mean particle size of sediments may differ depending on the sediment source as well as the depositional environment. As a general rule, particle sorting is higher under ice shelves and stable ice covers because these environments are sediment-starved and extremely low energy, which results in the predominance of suspension settling as a sedimentation mechanism (Syvitski 1989; Lemmen 1990; Curry and Pudsey 2007). Sedimentation rates are low since ice shelves often contain little debris and sedimentation in some cases may be impeded by basal freeze-on of underflowing meltwater, particularly in Arctic ice shelves (Pudsey et al. 2006; Jeffries 1992). During ice retreat, sedimentation rates typically increase markedly due to the release of sediments from the ice and to the greatly augmented supply of particle-laden meltwater (Powell 1981; Lemmen 1990; Gilbert and Domack 2003). At the onset of open water conditions, the sand and silt fractions of sediment typically become poorly sorted, while gravel clasts appear and increase in abundance (Curry and Pudsey 2007).

Grain size becomes finer with distance from the ice shelf margin due to the longer time in suspension of smaller particles (Domack et al. 2005). Very fine-grained, massively bedded deposits from both polar regions have been interpreted to indicate the deposition by suspension settling of water-borne sediment below ice shelves; coarsening sequences are therefore representative of ice retreat (Lemmen 1990; Curry and Pudsey 2007). Increases in gravel content occur during the retreat of ice shelves as melting releases supraglacial and englacial sediments to the sea floor (Gilbert and Domack 2003; Domack et al. 2005). Changes in the position of the ice margin have been inferred from changes in the amount of well sorted aeolian sand deposited in the sedimentary record (Domack et al. 1995). Fining upwards in sediment cores has also been used to infer increasingly distal sediment sources and therefore a retreat of floating ice fronts away from the core position (Lemmen 1990; Doran et al. 2000; Domack et al. 2005, Curry and Pudsey 2007).

8.3.2 Ice-Rafted Debris and Clast Analysis

The terrigenous fraction of the sedimentary record can be used to infer a wealth of information about past environmental conditions. With respect to ice shelves, two of the most important characteristics that have been used are the analysis of ice-rafted debris (IRD) content and mineral analysis of clasts.

The principle behind IRD analysis of sediments is based on the predominance of a solid, stable ice layer (i.e., ice shelf conditions). In this scenario there is limited melting, no open water, and no mobile icebergs; reduced quantities of IRD will therefore be released to the sediments, with an increase in the proportion of locally derived IRD. With the removal of ice shelves, open water is more common; icebergs calve more frequently and are more mobile. IRD is then released from melting ice in regions previously covered by ice shelves, and IRD grains may be drawn from a much larger source region (Roberts et al. 2008). Changes in the frequency of IRD in sediments should therefore record transitions between ice shelf and open marine conditions. Numerous studies have identified such patterns, including lack of IRD in below-ice shelf sediments (Hemer and Harris 2003; Curry and Pudsey 2007) and increased frequency of IRD during periods of ice shelf absence (Pudsey and Evans 2001; Bentley et al. 2005; Smith et al. 2006, 2007). It has also been suggested that a wider range of IRD clast sizes from a variety of sources is deposited in the absence of an ice shelf (Smith et al. 2007; Roberts et al. 2008).

In addition to analysis of the frequency of IRD, an understanding of mineral composition permits the deduction of further information about provenance, depositional processes and environmental conditions. Studies such as Roberts et al. (2008) have used geochemical and isotopic composition, supported by visual identification, to characterize shifting IRD provenance. By analyzing the mineral makeup of individual IRD grains and comparing them to the geological characteristics of both local and regional deposits, it is possible to ascertain the source regions of IRD (Darby and Bischof 1996). The presence of IRD from specific source regions, as well as changes in diversity of IRD provenance, implies different depositional pathways that result from changing ice conditions. In environments that are without ice shelves but still strongly frozen, limited melting and ice mobility imply that, in addition to low IRD concentrations, strata will contain only IRD grains derived from local rocks (Pudsey and Evans 2001; Brachfield et al. 2003; Roberts et al. 2008). With continued ice reduction, icebergs are able to arrive from increasingly diverse and distant source areas, transporting exotic IRD grains of greater diversity. The presence of exotic IRD is therefore indicative of open marine conditions and the absence of ice shelves, and the compositional diversity of grains can be indicative of wider patterns of iceberg mobility (Pudsey and Evans 2001; Brachfield et al. 2003; Bentley et al. 2005; Domack et al. 2005; Roberts et al. 2008). These principles have also been applied to reconstructions of Arctic glacial dynamics (Darby and Zimmerman 2008), illustrating their potential in studies of northern ice shelves.

8.3.3 Foraminifera

Foraminifera are a group of single-celled marine organisms that are typically present in virtually all habitats of the world's oceans (Sen Gupta 1999). Below extant ice shelves they have, however, been observed to be either absent or present in greatly reduced abundances (Bentley et al. 2005; Domack et al. 2005, Antoniades et al. 2011). Consequently, changes in foraminiferal abundances in the sedimentary record can provide information about past ice shelf characteristics. An absence of foraminifera in the sedimentary record often indicates the existence of ice shelves, while their presence implies a seasonally open marine environment and thus ice shelf absence (Bentley et al. 2005; Domack et al. 2005; Hillenbrand et al. 2005; Smith et al. 2006), although these scenarios may be complicated somewhat by the advection of material beneath ice shelves (e.g., Hemer and Harris 2003). Changes in foraminiferal concentrations may also be indicative of the nearby advance or retreat of ice shelves (Domack et al. 1995).

Foraminiferal community composition can also be a useful tool for inferring environmental differences, as foraminiferal diversity and morphology are related to ocean conditions (Ishman and Szymcek 2003; Pudsey et al. 2006). Increases in calcareous benthic and planktonic forms have been taken to indicate open marine conditions (Domack et al. 1995), as well as productivity increases caused by ice shelf retreat (Brachfield et al. 2003). Arenaceous forms (those with tests (shells) formed of clay particles cemented with calcium carbonate) were suggested to be indicative of cold, saline bottom waters (Brachfield et al. 2003), while those with tests formed from agglutinated sediments implied ice shelf conditions (Domack et al. 1995). In certain regions calcite dissolution may occur; it has been suggested that in such circumstances paleoenvironmental inferences are best drawn from agglutinated forms (Murray and Pudsey 2004), although the taxonomy and ecology of agglutinated foraminifera are less well known relative to other groups.

8.3.4 *Diatoms and Biogenic Silica*

Diatoms are unicellular siliceous algae that are commonly used as environmental proxy indicators in both freshwater and marine ecosystems (Stoermer and Smol 1999). Diatom communities display rapid and pronounced responses to environmental changes such as those implicit in the advance, retreat or collapse of ice shelves. Several characteristics of diatom communities have therefore been used to infer changes in ice shelf history. Numerous studies have concluded that the presence of diatoms in the sedimentary record can be used to infer the absence of ice shelves in the area, given that diatom productivity implies seasonally open water conditions (Brachfield et al. 2003; Bentley et al. 2005; Domack et al. 2005; Smith et al. 2007). Near the margin of ice shelves sedimentary deposits can be composed of siliceous ooze when little clastic sedimentation occurs due to limited meltwater supply, where they are indicative of proximity to open water productivity (Brachfield et al. 2003). It should be noted, however, that sedimentation of diatoms below existing ice shelves has been recorded 80 km from the open ocean, where it was attributed to advection due to ocean currents (Hemer and Harris 2003). Additionally, diatoms have been noted to be nearly absent in modern environments with heavy sea ice cover, but that did not have an ice shelf (Curry and Pudsey 2007). Accordingly, interpretations of ice shelf presence based on a lack of diatoms in the sedimentary record should be made with caution, and while considering the specific characteristics of each sedimentary environment.

Absolute abundances of diatoms in sediments are used as an indicator of past productivity. Given that ice shelves generally create sedimentary environments with no diatom deposition (with the exception of advected diatoms; Hemer and Harris 2003), increases in the concentration of diatom microfossils in the sedimentary record can be used to infer the retreat of ice shelves. As a rule, diatom abundances increase moderately in perennial sea ice regimes (from absence or near zero concentrations under ice shelves), and increase greatly under seasonal open water conditions (Bentley et al. 2005; Domack et al. 2005). By examining core transects across an area of interest, it is possible to develop detailed records of ice shelf retreat from the timing of diatom abundance changes over space and time (Brachfield et al. 2003; Hemer and Harris 2003; Domack et al. 2005; Heroy et al. 2008). The sedimentary concentration of biogenic silica is another indicator of the past overall abundance of siliceous microfossils (largely diatoms); these analyses have also been used to reconstruct ice shelf dynamics using the same principles (Pudsey and Evans 2001; Domack et al. 2005).

Differences in diatom community composition can also be indicative of varying characteristics of the glaciomarine environment. In epishelf lake scenarios, changes in the balance of freshwater and marine diatoms can record changes in water column conditions. Near the George VI Ice Shelf, the diatom community was composed of strictly marine taxa under open marine conditions, while during epishelf phases freshwater to brackish taxa were dominant (Smith et al. 2006; Smith et al. 2007). Decreases in the relative proportion of planktonic taxa were also taken to indicate lower planktonic production due to increased perennial sea ice (Bentley et al. 2005). In addition to overall community composition, individual taxa may be key indicators of specific conditions. Certain diatom species are characteristic of sea ice productivity, including the taxa *Fragilariopsis cylindrus* and *Melosira arctica*, among others (Zielinski and Gersonde 1997; Melnikov 2009). The abundance of such taxa in sediments therefore indicates that sea ice, and not ice shelves, was present at the time of deposition (Brachfield et al. 2003; Hemer and Harris 2003; Domack et al. 2005; Heroy et al. 2008).

8.3.5 Sedimentary Organic Matter

Analyses of sedimentary organic matter have applications relevant to paleoecological questions in both freshwater and marine environments (Meyers 1997; Meyers and Teranes 2001). Several different measurements of the sediment organic matter can provide complementary information that can be used to infer paleoenvironmental change, including bulk carbon content, carbon to nitrogen ratios (C/N), and stable isotopes. These data may be used to infer the existence of glaciers and ice shelves, changes in the terrestrial/aquatic nature and depositional pathways of organic material, and the presence of open water, among other ecosystem characteristics.

8.3.5.1 Total Organic Carbon

Total organic carbon (TOC) content in sediments is a measure of the initial biomass as well as of decomposition during deposition, although it may also be influenced by changes in inorganic sedimentation rates (Meyers and Teranes 2001). In general, increased sedimentary TOC reflects higher ecosystem biomass. Higher TOC values in glaciomarine ecosystems, therefore, should be indicative of greater productivity when ecosystems are less dominated by ice. Consequently, sedimentary TOC has been observed to decrease under ice shelf advance (Domack et al. 1995), and to rise during open marine periods (Heroy et al. 2008; Smith et al. 2007). However, some studies have also measured TOC concentrations with little change across horizons where ice shelf retreat and reformation were inferred (Pudsey and Evans 2001, Brachfield et al. 2003), or even higher TOC values during periods of inferred increases in sea ice cover (McMinn et al. 2001).

8.3.5.2 Stable Isotope Ratios

Carbon isotopic ratios can provide valuable information about rates of past algal dynamics and organic matter sources, particularly when viewed together with C/N ratios (Meyers 1997; Meyers and Teranes 2001; Maslin and Swann 2006). Among the major controls of $\delta^{13}C$ in the marine sedimentary record is surface productivity, although ocean circulation, allochthonous inputs, nutrient limitation, and other factors can play important roles (Meyers 1997; Maslin and Swann 2006). Measurements of carbon isotopes from marine environments are often made on the organic fraction of bulk sediment ($\delta^{13}C_{Org}$) or, in some cases, on foraminiferal tests, which use source carbon from pools with different $\delta^{13}C$ signatures.

Isotopic ratios of organic matter are controlled by the $\delta^{13}C$ of the carbon source and fractionation during assimilation. An awareness of these processes is essential when interpreting stratigraphic changes in $\delta^{13}C$, particularly if the material being examined contains a mixture of marine and terrestrial or planktonic and benthic contributions. Terrestrial plants use atmospheric CO_2 as their carbon source; this CO_2 has a $\delta^{13}C$ signature of −7‰. After fractionation effects, the typical $\delta^{13}C$ signature of land plants is between −22‰ and −31‰ for C3 plants (average −27‰), and between −8‰ and −15‰ for C4 plants (average −14‰) (Meyers 1997). Marine phytoplankton use dissolved bicarbonate as an inorganic source, which has a $\delta^{13}C$ signature of 1‰, and organic matter derived from marine phytoplankton typically has $\delta^{13}C$ values between −17‰ and −22‰, although it may range between −10‰ and −31‰ (Maslin and Swann 2006). Temperature-dependent variation in the solubility of CO_2 may also cause significant differences in carbon fractionation, greatly affecting $\delta^{13}C_{Org}$ values based on phytoplankton productivity (Maslin and Swann 2006). It is noteworthy that there is significant overlap between the $\delta^{13}C$ ranges of C3 plants and phytoplankton, and distinguishing between organic carbon produced by these two groups from $\delta^{13}C$ alone is often impossible. In this case, C/N ratios provide invaluable data, as phytoplankton have values below 10, while terrestrial plants have ratios in excess of 20 (Meyers 1997).

Ice shelf paleoenvironmental studies from Antarctic epishelf lakes have indicated that, with the retention of meltwater behind ice shelves, sediment $\delta^{13}C_{Org}$ values reflect catchment influences and freshwater algae. Due to low in situ productivity under perennial lake ice, the overall $\delta^{13}C_{Org}$ signature may be dominated by inflow streams and local catchment geology (Smith et al. 2007). In the absence of ice shelves, however, the signature will be similar to that of marine algae (Smith et al. 2007). $\delta^{13}C_{Org}$ values could therefore be used to delineate shifts between epishelf lake and open marine conditions. In a study of Moutonnée Lake, an epishelf lake behind the George VI Ice Shelf, $\delta^{13}C_{Org}$ signatures in organic material more depleted than −25‰ were taken to indicate marine conditions, while those less depleted than −25‰ indicated that an epishelf lake was present (Bentley et al. 2005; Smith et al. 2007).

Foraminiferal $\delta^{13}C$ can also be used to infer environmental change given that it reflects changes in the isotopic composition of seawater dissolved inorganic carbon. The $\delta^{13}C$ of seawater can vary locally, and is influenced by factors such as local photosynthetic rates and the mixing of water masses with different $\delta^{13}C$ signatures (Ravelo and Hillaire-Marcel 2007). Numerous species-specific fractionation factors also affect the $\delta^{13}C$ of foraminiferal tests, and $\delta^{13}C$ values within species can also increase with size (Ravelo and Hillaire-Marcel 2007). Limiting isotopic analyses to tests from a particular size fraction of a single species is a common strategy to restrict $\delta^{13}C$ variation to that caused by environmental change (Ravelo and Hillaire-Marcel 2007). Clearly the applicability of foraminiferal $\delta^{13}C$ may be limited in studies of ice shelves, given that ice shelves typically preclude the presence of foraminifera (see discussion above). However, in an Antarctic epishelf lake, changes in foraminiferal $\delta^{13}C$ in response to the removal of the ice dam were more equivocal than those in $\delta^{13}C_{Org}$ and other proxies (Smith et al. 2007). Further work is therefore necessary to determine the utility of foraminiferal $\delta^{13}C$ in ice shelf multi-proxy studies. Although less frequently applied, shifts in $\delta^{18}O$ have also been used to infer ice shelf variability. For example, Domack et al. (2005) examined $\delta^{18}O$ in foraminifera and used gradual shifts towards lower values that indicated long-term thinning of the Larsen B Ice Shelf.

8.3.6 *Photosynthetic Pigments*

Fossil pigments are used in paleoenvironmental studies as a tool to detect changes in phototrophic communities. Sedimentary pigments may be used as proxy indicators of past algal biomass and changes in photosynthetic community composition, among others (Leavitt and Hodgson 2001; Guilizzoni and Lami 2002). Studies from the Arctic Ocean and elsewhere have shown that sedimentary pigment concentrations are reflective of productivity from phytoplankton and sea ice communities above (Sun et al. 1991; Ambrose et al. 2005; Cooper et al. 2008; Pirtle-Levy et al. 2009). Polar freshwater and marine phytoplankton communities have divergent ratios of chlorophylls *b*:*a*, with marine signatures generally below ~0.3 and freshwater ratios of 0.4 and greater (Trees et al. 2000; Morgan-Kiss et al. 2006; Antoniades et al. 2009). These differences likely reflect the greater importance of Chlamydomonads and other Chlorophyta in freshwater communities, similar to those observed in Antarctic lakes with thick perennial ice cover (Morgan-Kiss et al. 2006), and may be used to detect the presence or absence of ice shelves in Arctic fiord mouths, given that ice shelves may dam freshwater and lead to changes in fiord salinity (Antoniades et al. 2011).

8.4 Chronologies

Developing reliable chronologies is often the most challenging aspect of high latitude paleoenvironmental studies (Wolfe et al. 2004). The relevancy of individual methods is dependent on the time scale of the study. On short time scales, persistent ice cover regimes and low precipitation imply that there is very limited influx of atmospherically-deposited radioisotopes such as ^{210}Pb and ^{137}Cs. These radioisotopes have been applied with mixed results in perennially ice-covered high Arctic lakes, and their potential in such systems appears to be dependent on site-specific characteristics such as catchment size and rates of inorganic and organic sedimentation (Zolitschka 1996; Antoniades et al. 2007; Tomkins et al. 2009).

On longer time scales, caution needs to be exercised with ^{14}C dating, as terrestrial organic matter is typically in short supply due to climatic limitations on vegetation, and also because many Arctic landscapes contain carbonate rocks and glacial materials that may contribute old, radiocarbon-depleted carbon to the source pool (Wolfe et al. 2004). The difficulty in obtaining reliable radiocarbon dates is a well-established problem in Antarctic paleoceanography. Large and variable reservoir corrections are often applied due to the presence of ancient carbon in the carbon pool from glacial and deep water sources (Domack et al. 1999; Doran et al. 1999; Hillenbrand et al. 2010). For the Arctic, the geographic variability of reservoir corrections has recently been addressed by Coulthard et al. (2010). The severity of these issues is typically dependent on the type of material analyzed. The picking of foraminifera from the sediment matrix is one method to circumvent such problems and has been shown to produce reliable radiocarbon ages, although some caveats exist (Antoniades et al. 2011; Broecker et al. 1989). Other methods, including compound-specific radiocarbon dating (Ohkouchi and Eglinton 2008; Mollenhauer and Rethemeyer 2009), and step-combustion dating (Rosenheim et al. 2008; Subt et al. 2016) have been developed recently to address these issues in polar sediment cores.

Another technique with the potential for developing chronologies is paleomagnetic analysis (King and Peck 2001). This technique involves the measurement of sediment magnetic properties, followed by correlating their variations with geomagnetic field models and/or correlations to cores with independently established chronologies. With recent advances in the understanding of past Arctic geomagnetic variation (St-Onge and Stoner 2011), the application of paleomagnetism to chronologies in Arctic sediments has become more common in both lakes and marine environments (e.g., Cook et al. 2009; Barletta et al. 2010; Antoniades et al. 2011; Simon et al. 2012). Paleomagnetism has also been directly applied to the study of past ice shelf change in both the Northern and Southern hemispheres (Brachfield et al. 2003; Antoniades et al. 2011). Generally, due to the limitations of most geochronological techniques in these environments it is preferable to apply at least two independent methods to develop chronologies in order to build confidence in the accuracy of sediment ages.

8.5 Applications of Paleoenvironmental Techniques to the Reconstruction of Arctic Ice Shelves

Although the methodology is relatively well-established in Antarctic research, only one study to date has used a sediment core approach to address the issue of past ice shelf dynamics in the Arctic. Antoniades et al. (2011; Fig. 8.1) examined past changes of the state of the Ward Hunt Ice Shelf in the Canadian Arctic in a

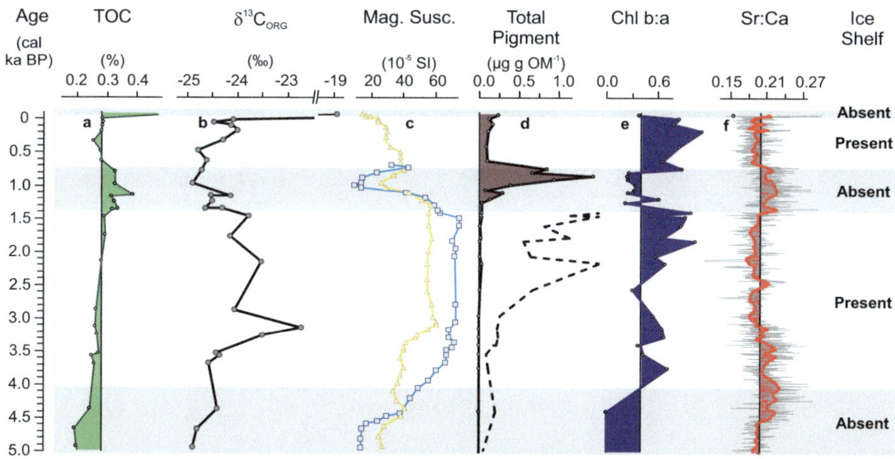

Fig. 8.1 Stratigraphy of Disraeli Fiord showing the concurrent shifts in indicators of ice shelf presence and absence: (**a**) Total organic carbon (TOC); (**b**) $\delta^{13}C_{Org}$; (**c**) magnetic susceptibility; (**d**) Total pigment concentration; (**e**) Chl *b*:*a*; and (**f**) Ratio of Strontium:Calcium activity (Sr:Ca) (Data from Antoniades et al. 2011)

multi-proxy study that included several of the indicators described above. The Ward Hunt Ice Shelf sits in the mouth of Disraeli Fiord and, in colder periods, dams the fiord completely leading to the formation of an epishelf lake. Antoniades et al. (2011) used the differences in the freshwater/marine balance, geochemistry, and sedimentary environments that result from the presence or absence of this epishelf lake to infer the past conditions of Disraeli Fiord, and thus of changes in the Ward Hunt Ice Shelf.

Indicators were grouped in several classes that responded to changes in salinity, changes in ice cover, or changes in the retention of sediment behind the ice dam. Changes in the ratio of chlorophyll *b*:*a* over time were used to infer periods of presence and breakup of the Ward Hunt Ice Shelf, illustrating the utility of pigment analyses for reconstructions of ice shelf history. Total pigment concentrations, Sr:Ca ratios, and TOC also varied concomitantly with the presence/absence of the ice shelf, as they were reflective of primary productivity and thus responded to changes in the severity of ice conditions under warming and cooling climates (Fig. 8.2). Foraminifera concentrations in sediments varied together with these variables, indicating that their abundance was controlled by food supply and that the bottom waters of Disraeli Fiord remained marine during epishelf lake stages. Magnetic susceptibility and titanium (Ti) and total inorganic carbon concentrations (not shown; Antoniades et al. 2011) increased when the epishelf lake was present due to the retention of allochthonous minerogenic material that was released to the Arctic Ocean in the absence of an ice shelf. $\delta^{13}C_{Org}$, meanwhile, increased in the presence of the ice shelf and decreased in its absence due to shifts between the influence of ^{13}C-rich meltwater (presence) and marine-derived organic matter (absence). This study thus illustrated the potential for reconstructing past ice shelf dynamics in

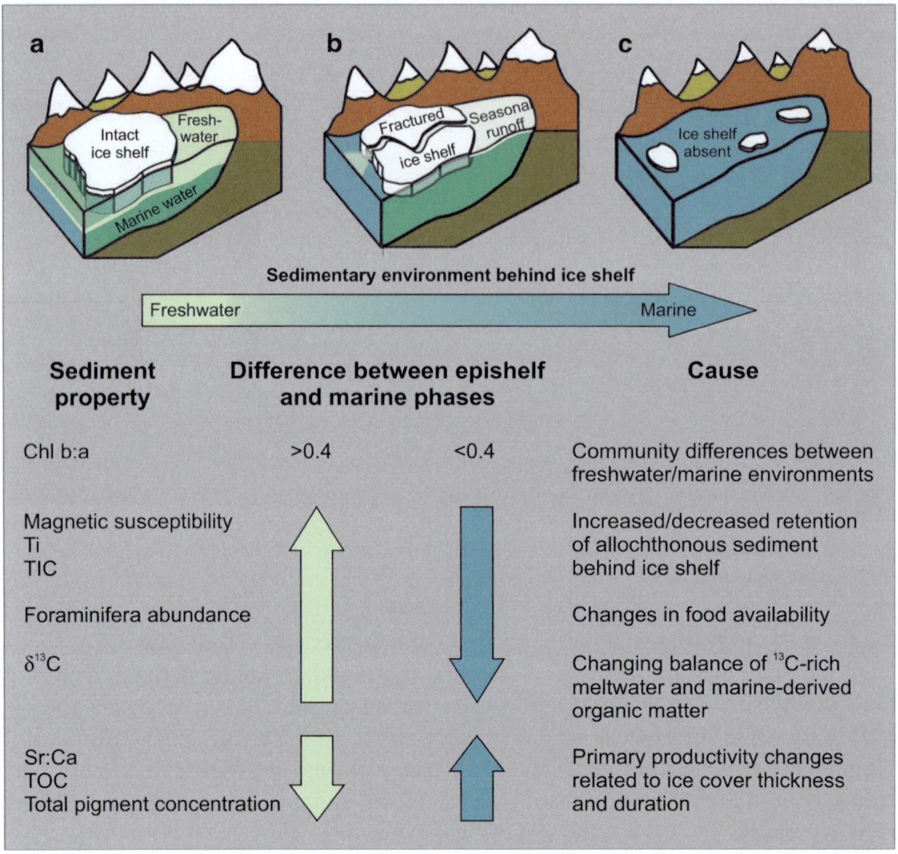

Fig. 8.2 Summary diagram of Arctic fiords and their ice/water properties with (**a**) intact, (**b**) fractured and (**c**) absent ice shelves, along with changes in sediment variables observed under these regimes that have been used to reconstruct ice shelf histories using paleoenvironmental techniques. *Chl* chlorophyll, *Ti* titanium, *TIC* total inorganic carbon, *Sr* strontium, *Ca* calcium, *TOC* total organic carbon

the Arctic using paleoecological techniques, however the logistical challenges involved in accessing and sampling the fiords of northern Ellesmere Island and Greenland have implied that, until the present, further such studies remain to be completed.

Acknowledgments This work was supported by le Fonds de recherche du Québec – Nature et technologies (FRQNT) and the Network of Centres of Excellence program ArcticNet. Thanks also to Warwick Vincent for valuable comments on a draft of the manuscript.

References

Ambrose, W. G., von Quillfeldt, C., Clough, L. M., Tilney, P. V. R., & Tucker, T. (2005). The sub-ice algal community in the Chukchi Sea: Large-and small-scale patterns of abundance based on images from a remotely operated vehicle. *Polar Biology, 28*, 784–795. doi:10.1007/s00300-005-0002-8.

Antoniades, D., Crawley, C., Douglas, M. S., Pienitz, R., Andersen, D., Doran, P. T., Hawes, I., Pollard, W., & Vincent, W. F. (2007). Abrupt environmental change in Canada's northernmost lake inferred from fossil diatom and pigment stratigraphy. *Geophysical Research Letters, 34*, L18708. doi:10.1029/2007GL030947.

Antoniades, D., Veillette, J., Martineau, M.-J., Belzile, C., Tomkins, J., Pienitz, R., Lamoureux, S., & Vincent, W. F. (2009). Bacterial dominance of phototrophic communities in a High Arctic lake and its implications for paleoclimate analysis. *Polar Science, 3*, 147–161. doi:10.1016/j.polar.2009.05.002.

Antoniades, D., Francus, P., Pienitz, R., St-Onge, G., & Vincent, W. F. (2011). Holocene dynamics of the Arctic's largest ice shelf. *Proceedings of the National Academy of Sciences of the United States of America, 108*, 18899–18904. doi:10.1073/pnas.1106378108.

Barletta, F., St-Onge, G., Channell, J. E. T., & Rochon, A. (2010). Dating of Holocene western Canadian Arctic sediments by matching paleomagnetic secular variation to a geomagnetic field model. *Quaternary Science Reviews, 29*, 2315–2324. doi:10.1016/j.quascirev.2010.05.035.

Bentley, M. J., Hodgson, D. A., Sugden, D. E., Roberts, S. J., Smith, J. A., Leng, M. J., & Bryant, C. (2005). Early Holocene retreat of the George VI Ice Shelf. Antarctic Peninsula. *Geology, 33*, 173–176. doi:10.1130/G21203.1.

Brachfield, S., Domack, E., Kissel, C., Laj, C., Leventer, A., Ishman, S., Gilbert, R., Camerlenghi, A., & Eglinton, L. B. (2003). Holocene history of the Larsen-A Ice Shelf constrained by geomagnetic paleointensity dating. *Geology, 31*, 749–752. doi:10.1130/G19643.1.

Bradley, R. S. (1990). Holocene paleoclimatology of the Queen Elizabeth Islands, Canadian High Arctic. *Quaternary Science Reviews, 9*, 365–384. doi:10.1016/0277-3791(90)90028-9.

Broecker, W. S., Trumbore, S. E., Bonani, G., Wolfli, W., & Klas, M. (1989). Anomalous AMS radiocarbon ages for foraminifera from high-deposition-rate ocean sediments. *Radiocarbon, 31*, 157–162. doi:10.2458/azu_js_rc.31.1131.

Cook, T. L., Bradley, R. S., Stoner, J. S., & Francus, P. (2009). Five thousand years of sediment transfer in a high Arctic watershed recorded in annually laminated sediments from Lower Murray Lake, Ellesmere Island Nunavut Canada. *Journal of Paleolimnology, 41*, 77–94. doi:10.1007/s10933-008-9252-0.

Cooper, L. W., Lalande, C., Pirtle-Levy, R., Larsen, I. L., & Grebmeier, J. M. (2008). Seasonal and decadal shifts in particulate organic matter processing and sedimentation in the Bering Strait Shelf region. *Deep Sea Research, Part II, 56*(17), 1316–1325. doi:10.1016/j.dsr2.2008.10.025.

Coulthard, R. D., Furze, M. F. A., Pieńkowski, A. J., Nixon, F. C., & England, J. H. (2010). New marine ΔR values for Arctic Canada. *Quaternary Geochronology, 5*, 419–434. doi:10.1016/j.quageo.2010.03.002.

Crary, A. P. (1956). Geophysical studies along northern Ellesmere Island. *Arctic, 9*, 154–165. doi:10.14430/arctic3790.

Crary, A. P. (1960). Arctic ice island and ice shelf studies, Part II. *Arctic, 13*, 32–50. doi:10.14430/arctic3687.

Curry, P., & Pudsey, C. J. (2007). New Quaternary sedimentary records from near the Larsen C and former Larsen B ice shelves; evidence for Holocene stability. *Antarctic Science, 19*, 355–364. doi:10.1017/S0954102007000442.

Darby, D. A., & Bischof, J. F. (1996). A statistical approach to source determination of lithic and Fe oxide grains: An example from the Alpha Ridge, Arctic Ocean. *Journal of Sedimentary Research, 66*(3), 599–607. doi:10.1306/D42683BA-2B26-11D7-8648000102C1865D.

Darby, D. A., & Zimmerman, P. (2008). Ice-rafted detritus events in the Arctic during the last glacial interval, and the timing of the Innuitian and Laurentide ice sheet calving events. *Polar Research, 27*, 114–127. doi:10.1111/j.1751-8369.2008.00057.x.

Domack, E. W., Ishman, S. E., Stein, A. B., McClennen, C. E., & Jull, A. T. (1995). Late Holocene advance of the Müller Ice Shelf, Antarctic Peninsula: Sedimentological, geochemical and palaeontological evidence. *Antarctic Science, 7*(02), 159–170.

Domack, E. W., Jacobson, E. A., Shipp, S., & Anderson, J. B. (1999). Late Pleistocene–Holocene retreat of the West Antarctic Ice-Sheet system in the Ross Sea: Part 2 – sedimentologic and stratigraphic signature. *Geological Society of America Bulletin, 111*, 1517–1536. doi:10.1130/0016-7606(1999)111<1517:LPHROT>2.3.CO;2.

Domack, E., Duran, D., Leventer, A., Ishman, S., Doane, S., McCallum, S., Amblas, D., Ring, J., Gilbert, R., & Prentice, M. (2005). Stability of the Larsen B Ice Shelf on the Antarctic Peninsula during the Holocene epoch. *Nature, 436*, 681–685. doi:10.1038/nature03908.

Doran, P. T., Berger, G. W., Lyons, W. B., Wharton, R. A., Davisson, M. L., Southon, J., & Dibb, J. E. (1999). Dating quaternary lacustrine sediments in the McMurdo Dry Valleys, Antarctica. *Palaeogeography Palaeoclimatology, 147*, 223–239. doi:10.1016/S0031-0182(98)00159-X.

Doran, P. T., Wharton Jr., R. A., Lyons, W. B., Des Marais, D. J., & Andersen, D. T. (2000). Sedimentology and geochemistry of a perennially ice-covered epishelf lake in Bunger Hills Oasis, East Antarctica. *Antarctic Science, 12*, 131–140. doi:10.1017/S0954102000000171.

England, J. H., Lakeman, T. R., Lemmen, D. S., Bednarski, J. M., Stewart, T. G., & Evans, D. J. A. (2008). A millennial-scale record of Arctic Ocean sea ice variability and the demise of the Ellesmere Island ice shelves. *Geophysical Research Letters, 35*, L19502. doi:10.1029/2008GL034470.

England, J. H., Evans, D. A., & Lakeman, T. R. (2017). Holocene history of Arctic ice shelves. In L. Copland & D. Mueller (Eds.), *Arctic ice shelves and ice islands* (p. 185–205). Dordrecht: Springer. doi:10.1007/978-94-024-1101-0_7.

Evans, J., & Pudsey, C. J. (2002). Sedimentation associated with Antarctic Peninsula ice shelves: Implications for palaeoenvironmental reconstructions of glaciomarine sediments. *Journal of the Geological Society of London, 159*, 233–237. doi:10.1144/0016-764901-125.

Gilbert, R. (1983). Sedimentary processes of Canadian Arctic fjords. *Sedimentary Geology, 36*, 147–175. doi:10.1016/0037-0738(83)90007-6.

Gilbert, R., & Domack, E. W. (2003). Sedimentary record of disintegrating ice shelves in a warming climate, Antarctic Peninsula. *Geochemistry, Geophysics, Geosystems, 4*, 1038. doi:10.1029/2002GC000441.

Guilizzoni, P., & Lami, A. (2002). Paleolimnology: Use of algal pigments as indicators. In G. Bitton (Ed.), *The encyclopedia of environmental microbiology* (p. 2306–2317). New York: Wiley and Sons. doi:10.1002/0471263397.env313.

Hemer, M. A., & Harris, P. T. (2003). Sediment core from beneath the Amery Ice Shelf, East Antarctica, suggests mid-Holocene ice-shelf retreat. *Geology, 31*, 127–130. doi:10.1130/0091-7613.

Heroy, D. C., Sjunneskog, C., & Anderson, J. B. (2008). Holocene climate change in the Bransfield Basin, Antarctic Peninsula: Evidence from sediment and diatom analysis. *Antarctic Science, 20*, 69–87. doi:10.1017/S0954102007000788.

Hillenbrand, C. D., Baesler, A., & Grobe, H. (2005). The sedimentary record of the last glaciation in the western Bellingshausen Sea (West Antarctica): Implications for the interpretation of diamictons in a polar-marine setting. *Marine Geology, 216*, 191–204. doi:10.1016/j.margeo.2005.01.007.

Hillenbrand, C. D., Smith, J. A., Kuhn, G., Esper, O., Gersonde, R., Larter, R. D., Maher, B., Moreton, S. G., Shimmield, T. M., & Korte, M. (2010). Age assignment of a diatomaceous ooze deposited in the western Amundsen Sea Embayment after the Last Glacial Maximum. *Journal of Quaternary Science, 25*, 280–295. doi:10.1002/jqs.1308.

Ishman, S. E., & Szymcek, P. (2003). Foraminiferal distributions in the former Larsen – A Ice Shelf and Prince Gustav Channel region, eastern Antarctic Peninsula margin: A baseline for

Holocene paleoenvironmental change. In E. Domack, A. Levente, A. Burnet, R. Bindschadler, P. Convey, & M. Kirby (Eds.), *Antarctic Peninsula climate variability: Historical and paleoenvironmental perspectives* (p. 239–260). Washington: American Geophysical Union. doi:10.1029/AR079p0239.

Jeffries, M. O. (1992). Arctic ice shelves and ice islands: Origin, growth and disintegration, physical characteristics, structural-stratigraphic variability, and dynamics. *Reviews of Geophysics, 30*, 245–267. doi:10.1029/92RG00956.

Jungblut, A. D., Mueller, D., & Vincent, W. F. (2017). Arctic ice shelf ecosystems. In L. Copland & D. Mueller (Eds.), *Arctic ice shelves and ice islands* (p. 227–260). Dordrecht: Springer. doi:10.1007/978-94-024-1101-0_9.

Kellogg, D. E., & Kellogg, T. B. (1986). Diatom biostratigraphy of sediment cores from beneath the Ross Ice Shelf. *Micropaleontology, 32*, 74–94. doi:10.2307/1485703.

King, J., & Peck, J. (2001). Use of paleomagnetism in studies of lake sediments. In W. M. Last & J. P. Smol (Eds.), *Tracking environmental change using lake sediments, Basin analysis, coring, and chronological techniques* (Vol. 1, p. 371–389). Dordrecht: Kluwer Academic Publishers. doi:10.1007/0-306-47669-X_14.

Kolobov, D. D., & Savatyugin, L. M. (1983). Bottom sediments under the Novolazarevskiy Ice Shelf. *Polar Geography, 6*, 267–271. doi:10.1080/10889378209377176.

Lamoureux, S. F., & Gilbert, R. (2004). Physical and chemical properties and proxies of high latitude lake sediments. In R. Pienitz, M. S. V. Douglas, & J. P. Smol (Eds.), *Long-term environmental change in Arctic and Antarctic lakes* (p. 53–87). Dordrecht: Springer Academic Publishers. doi:10.1007/978-1-4020-2126-8_3.

Leavitt, P. R., & Hodgson, D. A. (2001). Sedimentary pigments. In J. P. Smol, H. J. B. Birks, & W. M. Last (Eds.), *Tracking environmental change using lake sediments, Terrestrial, algal, and siliceous indicators* (Vol. 3, p. 1–31). Dordrecht: Kluwer Academic Publishers. doi:10.1007/0-306-47668-1_15.

Lemmen, D. S. (1989). The last glaciation of Marvin Peninsula, northern Ellesmere Island, High Arctic, Canada. *Canadian Journal of Earth Sciences, 26*, 2578–2590. doi:10.1139/e89-220.

Lemmen, D. S. (1990). Glaciomarine sedimentation in Disraeli Fiord, High Arctic Canada. *Marine Geology, 94*, 9–22. doi:10.1016/0025-3227(90)90100-X.

Lemmen, D. S., Gilbert, R., Smol, J. P., & Hall, R. (1988). Holocene sedimentation in glacial Tasikutaaq Lake, Baffin Island. *Canadian Journal of Earth Sciences, 25*, 810–823. doi:10.1139/e88-080.

Lewis, T., Gilbert, R., & Lamoureux, S. F. (2002). Spatial and temporal changes in sedimentary processes at proglacial Bear Lake, Devon Island, Nunavut, Canada. *Arctic, Antarctic, and Alpine Research, 34*, 119–129. doi:10.2307/1552463.

Lyons, J. B., & Mielke, J. E. (1973). Holocene history of a portion of northernmost Ellesmere Island. *Arctic, 26*, 314–323. doi:10.14430/arctic2930.

Lyons, J. B., & Ragle, R. H. (1962). Thermal history and growth of the Ward Hunt Ice Shelf. *The International Association of Hydrological Sciences (IAHS) Publications, 58*, 88–97.

Mangerud, J., & Svendsen, J. I. (1990). Deglaciation chronology inferred from marine sediments in a proglacial lake basin, western Spitsbergen, Svalbard. *Boreas, 19*, 249–272. doi:10.1111/j.1502-3885.1990.tb00450.x.

Maslin, M. A., & Swann, G. E. A. (2006). Isotopes in marine sediments. In M. J. Leng (Ed.), *Developments in paleoenvironmental research, Isotopes in palaeoenvironmental research* (Vol. 10, p. 227–290). Dordrecht: Springer. doi:10.1007/1-4020-2504-1_06.

McMinn, A., Heijnisj, H., Harle, K., & McOrist, G. (2001). Late-Holocene climatic change recorded in sediment cores from Ellis Fjord eastern Antarctica. *Holocene, 11*(3), 291–300. doi:10.1191/095968301671577682.

Melnikov, I. A. (2009). Recent sea ice ecosystem in the Arctic Ocean: A review. In J. C. J. Nihoul & A. G. Kostianoy (Eds.), *Influence of climate change on the changing Arctic and sub-Arctic conditions* (p. 57–71). Dordrecht: Springer. doi:10.1007/978-1-4020-9460-6_6.

Meyers, P. A. (1997). Organic geochemical proxies of paleoceanographic, paleolimnologic, and paleoclimatic processes. *Organic Geochemistry, 27*(5), 213–250.

Meyers, P. A., & Teranes, J. L. (2001). Sediment organic matter. In W. M. Last & J. P. Smol (Eds.), *Tracking environmental change using lake sediments, Physical and geochemical methods* (Vol. 2, p. 239–269). Dordrecht: Kluwer Academic Publishers. doi:10.1007/0-306-47670-3_9.

Mielke, J. E., & Long, A. (1969). Smithsonian Institution radiocarbon measurements V. *Radiocarbon, 11*(1), 163–182.

Mollenhauer, G., & Rethemeyer, J. (2009). Compound-specific radiocarbon analysis–analytical challenges and applications. *IOP Conference Series: Earth and Environmental Science (EES), 5*, 012006. doi:10.1088/1755-1307/5/1/012006.

Morgan-Kiss, R. M., Priscu, J. C., Pocock, T., Gudynaite-Savitch, L., & Huner, N. P. A. (2006). Adaptation and acclimation of photosynthetic microorganisms to permanently cold environments. *Microbiology and Molecular Biology Reviews, 70*, 222–252. doi:10.1128/MMBR.70.1.222-252.2006.

Mueller, D. R., Vincent, W. F., & Jeffries, M. O. (2003). Break-up of the largest Arctic ice shelf and associated loss of an epishelf lake. *Geophysical Research Letters, 30*(20), 2031. doi:10.1029/2003GL017931.

Mueller, D., Copland, L., & Jeffries, M. O. (2017). Changes in Canadian Arctic ice shelf extent since 1906. In L. Copland & D. Mueller (Eds.), *Arctic ice shelves and ice islands* (p. 109–148). Dordrecht: Springer. doi:10.1007/978-94-024-1101-0_5.

Murray, J. W., & Pudsey, C. J. (2004). Living (stained) and dead foraminifera from the newly ice-free Larsen Ice Shelf, Weddell Sea, Antarctica: Ecology and taphonomy. *Marine Micropaleontology, 53*, 67–81. doi:10.1016/j.marmicro.2004.04.001.

Naish, T., Powell, R., Levy, R., Florindo, F., Harwood, D., Kuhn, G., Niessen, F., Talarico, F., & Wilson, G. (2007). A record of Antarctic climate and ice sheet history recovered. *EOS. Transactions of the American Geophysical Union, 88*, 557–558. doi:10.1029/2007EO500001.

Ohkouchi, N., & Eglinton, T. I. (2008). Compound-specific radiocarbon dating of Ross Sea sediments: A prospect for constructing chronologies in high-latitude oceanic sediments. *Quaternary Geochronology, 3*, 235–243. doi:10.1016/j.quageo.2007.11.001.

Pirtle-Levy, R., Grebmeier, J. M., Cooper, L. W., & Larsenetal, I. L. (2009). Chlorophyll *a* in Arctic sediments implies long persistence of algal pigments. *Deep Sea Research, Part II, 56*(17), 1326–1338. doi:10.1016/j.dsr2.2008.10.022.

Pope, S., Copland, L., & Alt, B. (2017). Recent changes in sea ice plugs along the northern Canadian Arctic Archipelago. In L. Copland & D. Mueller (Eds.), *Arctic ice shelves and ice islands* (p. 317–342). Dordrecht: Springer. doi:10.1007/978-94-024-1101-0_12.

Powell, R. D. (1981). A model for sedimentation by tidewater glaciers. *Annals of Glaciology, 2*, 129–134.

Pudsey, C. J., & Evans, J. (2001). First survey of Antarctic sub-ice shelf sediments reveals mid-Holocene ice shelf retreat. *Geology, 29*, 787–790. doi:10.1130/0091-7613(2001)029<0787:FSOASI>2.0.CO;2.

Pudsey, C. J., Murray, J. W., Appleby, P., & Evans, J. (2006). Ice shelf history from petrographic and foraminiferal evidence, Northeast Antarctic Peninsula. *Quaternary Science Reviews, 25*, 2357–2379. doi:10.1016/j.quascirev.2006.01.029.

Raiswell, R., & Tan, M. M. (1985). Diagenesis of sediments beneath the Ross Ice Shelf and their sedimentary history. *Nature, 315*, 483–485. doi:10.1038/315483a0.

Ravelo, A. C., & Hillaire-Marcel, C. (2007). The use of oxygen and carbon isotopes of foraminifera in paleoceanography. In C. Hillaire-Marcel & A. de Vernal (Eds.), *Proxies in late Cenozoic paleoceanography* (p. 735–763). Amsterdam: Elsevier. doi:10.1016/S1572-5480(07)01023-8.

Roberts, S. J., Hodgson, D. A., Bentley, M. J., Smith, J. A., Millar, I. L., Olive, V., & Sugden, D. E. (2008). The Holocene history of George VI Ice Shelf, Antarctic Peninsula from clast-provenance analysis of epishelf lake sediments. *Palaeogeography Palaeoclimatology Palaeoecology, 259*, 258–283. doi:10.1016/j.palaeo.2007.10.010.

Rosenheim, B. E., Day, M. B., Domack, E., Schrum, H., Benthien, A., & Hayes, J. M. (2008). Antarctic sediment chronology by programmed-temperature pyrolysis: Methodology and data treatment. *Geochemistry, Geophysics, Geosystems, 9*(4), Q04005. doi:10.1029/2007GC001816.

Scambos, T. A., Hulbe, C. L., Fahnestock, M. A., & Bohlander, J. (2000). The link between climate warming and break-up of ice shelves in the Antarctic Peninsula. *Journal of Glaciology, 46*, 516–530. doi:10.3189/172756500781833043.

Sen Gupta, B. K. (Ed.). (1999). *Modern foraminifera*. Dordrecht: Kluwer Academic Publishers.

Simon, Q., St-Onge, G., & Hillaire, M. C. (2012). Late Quaternary chronostratigraphic framework of deep Baffin Bay glaciomarine sediments from high-resolution paleomagnetic data. *Geochemistry, Geophysics, Geosystems, 13*, Q0AO03. doi:10.1029/2012GC004272.

Smith, J. A., Hodgson, D. A., Bentley, M. J., Verleyen, E., Leng, M. J., & Roberts, S. J. (2006). Limnology of two Antarctic epishelf lakes and their potential to record periods of ice shelf loss. *Journal of Paleolimnology, 35*, 373–394. doi:10.1007/s10933-005-1333-8.

Smith, J. A., Bentley, M. J., Hodgson, D. A., Roberts, S. J., Leng, M. J., Lloyd, J. M., Barrett, M. S., Bryant, C., & Sugden, D. E. (2007). Oceanic and atmospheric forcing of early Holocene ice shelf retreat, George VI Ice Shelf, Antarctica Peninsula. *Quaternary Science Reviews, 26*, 500–516. doi:10.1016/j.quascirev.2006.05.006.

Smith, J. A., Andersen, T. J., Shortt, M., Gaffney, A. M., Truffer, M., Stanton, T. P., Bindschadler, R., Dutrieux, P., Jenkins, A., Hillenbrand, C.-D., Ehrmann, W., Corr, H. F. J., Farley, N., Crowhurst, S., & Vaughan, D. G. (2016). Sub-ice-shelf sediments record history of twentieth-century retreat of Pine Island Glacier. *Nature*. doi:10.1038/nature20136.

Smol, J. P. (2008). *Pollution of lakes and rivers: A paleoenvironmental perspective* (2nd ed.). Oxford: Blackwell Publishing.

Snyder, J. A., Werner, A., & Miller, G. H. (2000). Holocene cirque glacier activity in western Spitsbergen, Svalbard: Sediment records from proglacial Linnévatnet. *Holocene, 10*, 555–563. doi:10.1191/095968300667351697.

St-Onge, G., & Stoner, J. S. (2011). Paleomagnetism near the North Magnetic Pole: A unique vantage point for understanding the dynamics of the geomagnetic field and its secular variations. *Oceanography, 24*, 42–50. doi:10.5670/oceanog.2011.53.

Stoermer, E. F., & Smol, J. P. (1999). *The diatoms: Applications for the earth and environmental sciences*. Cambridge: Cambridge University Press.

Subt, C., Fangman, K. A., Wellner, J. S., & Rosenheim, B. E. (2016). Sediment chronology in Antarctic deglacial sediments: Reconciling organic carbon ^{14}C ages to carbonate ^{14}C ages using Ramped PyrOx. *Holocene, 26*, 265–273. doi:10.1177/0959683615608688.

Sun, M., Aller, R. C., & Lee, C. (1991). Early diagenesis of chlorophyll-*a* in Long Island Sound sediments: A measure of carbon flux and particle reworking. *Journal of Marine Research, 49*, 379–401. doi:10.1357/002224091784995927.

Syvitski, J. P. M. (1989). On the deposition of sediment within glacier-influenced fjords: Oceanographic controls. *Marine Geology, 85*, 301–329. doi:10.1016/0025-3227(89)90158-8.

Teller, J. T., & Clayton, L. (Eds). (1983). *Glacial Lake Agassiz*. Geological Association of Canada Special Paper 26, pp. 451.

Tomkins, J. D., Lamoureux, S. F., Antoniades, D., & Vincent, W. F. (2009). Sedimentology of perennial ice-covered, meromictic Lake A, Ellesmere Island, at the northern extreme of Canada. *Canadian Journal of Earth Sciences, 46*, 83–100. doi:10.1139/E09-008.

Trees, C. C., Clarke, D. K., Bidigare, R. R., Ondrusek, M. E., & Mueller, J. L. (2000). Accessory pigments versus chlorophyll *a* concentration within the euphotic zone: A ubiquitous relationship. *Limnology and Oceanography, 45*, 1130–1143. doi:10.4319/lo.2000.45.5.1130.

Upham, W. (1895). The glacial Lake Agassiz. *US Geological Survey Monograph, 25*, 1–658.

Van Hove, P., Swadling, K., Gibson, J. A. E., Belzile, C., & Vincent, W. F. (2001). Farthest north lake and fjord populations of calanoid copepods in the Canadian High Arctic. *Polar Biology, 24*, 303–307. doi:10.1007/s003000000207.

Veillette, J., Mueller, D. R., Antoniades, D., & Vincent, W. F. (2008). Arctic epishelf lakes as sentinel ecosystems: Past, present and future. *Journal of Geophysical Research, 113*, G04014. doi:10.1029/2008JG000730.

Webb, P. N., Ronan Jr., T. E., Lipps, J. H., & DeLaca, T. E. (1979). Miocene glaciomarine sediments from beneath the southern Ross Ice Shelf, Antarctica. *Science, 203*, 435–437. doi:10.1126/science.203.4379.435.

Wolfe, A. P., Miller, G. H., Olsen, C. A., Forman, S. L., Doran, P. T., & Holmgren, S. U. (2004). Geochronology of high latitude lake sediments. In R. Pienitz, M. S. V. Douglas, & J. P. Smol (Eds.), *Developments in paleoenvironmental research, Long-term environmental change in Arctic and Antarctic Lakes* (Vol. 8, p. 19–52). Dordrecht: Springer. doi:10.1007/978-1-4020-2126-8_2.

Zielinski, U., & Gersonde, R. (1997). Diatom distribution in Southern Ocean surface sediments (Atlantic sector): Implications for paleoenvironmental reconstructions. *Palaeogeography Palaeoclimatology Palaeoecology, 129*, 213–250. doi:10.1016/S0031-0182(96)00130-7.

Zolitschka, B. (1996). Recent sedimentation in a high Arctic lake, northern Ellesmere Island, Canada. *Journal of Paleolimnology, 16*, 169–186. doi:10.1007/BF00176934.

Chapter 9
Arctic Ice Shelf Ecosystems

Anne D. Jungblut, Derek Mueller, and Warwick F. Vincent

Abstract Arctic ice shelves are microbial ecosystems with a rich biodiversity. Until recently, polar ice shelves were seen as mostly abiotic glaciological features, however they are oases for life, with snow, meltwater pools and sediments providing cryohabitats for microbiota. The biological communities are composed of diverse forms of microscopic life, including cyanobacteria, heterotrophic bacteria, viruses, algae, other protists and microfauna, and occupy a variety of habitats: supraglacial meltwater lakes, englacial microhabitats within the ice and snow and planktonic environments in ice-dammed, epishelf lakes. These habitats are defined by seasonal light availability, cold temperatures and nutrient poor conditions. In the supraglacial pools, production is dominated by benthic microbial mat assemblages that have diverse stress adaptation systems and that use internal nutrient recycling and scavenging strategies. Despite short growth periods and perennial low temperatures, biomass accumulations are considerable, with a striking diversity of light-harvesting, UV-protection and other accessory pigments. The chemical characteristics such as conductivity and origin of salts are defined by the underlying ice types, and microbial mat studies from adjacent habitats show a high resilience to solute concentration during freeze-up. The structural integrity of these cryoecosystems is dependent on ice, and they are therefore vulnerable to climate change. Many of these unique Arctic ecosystems have been lost by ice shelf collapse over the last two decades, and they are now on the brink of complete extinction.

A.D. Jungblut (✉)
Centre d'Études Nordiques (CEN), Institut de biologie intégrative et des systèmes (IBIS) and Département de Biologie, Laval University, Quebec City, QC, Canada

Life Sciences Department, The Natural History Museum, London, UK
e-mail: a.jungblut@nhm.ac.uk

D. Mueller
Department of Geography and Environmental Studies, Carleton University, Ottawa, ON, Canada
e-mail: derek.mueller@carleton.ca

W.F. Vincent
Centre d'Études Nordiques (CEN), Takuvik Joint International Laboratory and Département de Biologie, Laval University, Quebec City, QC, Canada
e-mail: warwick.vincent@bio.ulaval.ca

L. Copland, D. Mueller (eds.), *Arctic Ice Shelves and Ice Islands*, Springer Polar Sciences, DOI 10.1007/978-94-024-1101-0_9

Keywords Arctic ecosystems • Benthic • Biological production • Cryosphere • Ice shelf • Microbial biodiversity

9.1 Introduction

Polar ice including snow, glaciers, lake-ice and sea-ice is no longer considered an abiotic feature of the environment, but instead it is now realized to provide a broad suite of habitats for biological communities that survive and sometimes thrive in these extreme conditions (Priscu and Christner 2004; Boetius et al. 2015). Ice shelves have long been recognized as a compelling example of the biological richness of polar cryohabitats, as dynamic ecosystems that contain a remarkable diversity of life, most of it microscopic (Vincent 1988). Biological communities occur within and at the base of meltwater ponds (supraglacial ecosystems) and in meltwater microhabitats within the ice and snow (englacial ecosystems). Epishelf lakes are an additional ecosystem type that occur where ice shelves block the head of fiords and embayments, retaining a layer of freshwater (that has flowed in from terrestrial sources) over the marine water below. Given the complete dependence on ice for their structural integrity, each of these ecosystem types (supraglacial, englacial and epishelf) is highly vulnerable to climatic change.

The study of Arctic ice shelf ecology has been ongoing since the discovery of novel microbial communities on the Ward Hunt Ice Shelf in 1998 (Vincent and Howard-Williams 2000). In this chapter we review what has been learned to date about these ecosystems, and where scientific investigation has not yet occurred, we draw on literature from analogous environments (e.g., Arctic freshwater and marine ecosystems, Antarctic ice shelves and other cryoecosystems) to fill in these knowledge gaps. First, we introduce the general physical and chemical characteristics of the Canadian Arctic ice shelves as habitats for life. We review the biology of the supraglacial environments of ice shelves, from microscopic to macroscopic biota, and then describe the environmental and biological characteristics of epishelf lakes. We conclude this review by examining the significance of Arctic ice shelf ecosystems for insights into biological adaptation to environmental extremes, for an improved understanding of biotic-physical interactions in the cryosphere, and as ecosystems on the brink of extinction as a result of human induced climate warming.

9.2 Supraglacial Meltwater Lakes

9.2.1 Physical Characteristics

The surface of Arctic ice shelves is marked by an undulating topography or ‘rolls’ (Holdsworth 1987; Jeffries 2017), which created difficulty for travel by the early explorers to the region who referred to them as a ‘long fringe of large and troublesome hummocks' (Nares 1878, cited in Vincent et al. 2001). In summer, the troughs between these hummocks fill with meltwater, and form long (1–15 km), narrow (10–100 m) and shallow (<3 m) lakes (Fig. 9.1). These Arctic supraglacial meltwater lakes are roughly parallel to each other and to the coastline, but become more chaotic in form and orientation behind any substantial obstruction (i.e., in fiords or behind islands). This parallel orientation has led some to conclude that these features are maintained by prevailing westerly winds (Holdsworth 1987). At times, lakes occupying bifurcating troughs have been observed to drain (or spill over) into each other. These long lakes can also drain completely if deep cracks form in the ice shelf and provide a conduit to the ocean below (e.g., during the 2002 Ward Hunt Ice Shelf break-up event; Mueller et al. 2003a).

Depressions in the ice surface, such as these long lakes, meltponds of variable size and small melt holes, termed cryoconite holes, collect sediment over time. This sediment is colonized by microbiota forming a matrix known as a microbial mat. Microbial mats on these ice shelves are dominated by cyanobacteria, vary greatly in their thickness and extent and may be emergent and exposed to the air, or covered by snow, ice or water (Mueller et al. 2006). Research on the physical and chemical aspects of supraglacial meltwater lakes, ponds and cryoconite holes has focused on variables likely to be of importance to the biology of the benthic microbial mats and, to a lesser extent, water column plankton.

Photosynthesis is the primary means of capturing energy within microbial mats, and *in situ* irradiance levels are therefore of critical importance for these consortia. The irradiance experienced by the mats may not increase until long after polar

Fig. 9.1 Elongated meltwater ponds on the Ward Hunt Ice Shelf in the Canadian High Arctic on July 18, 2003

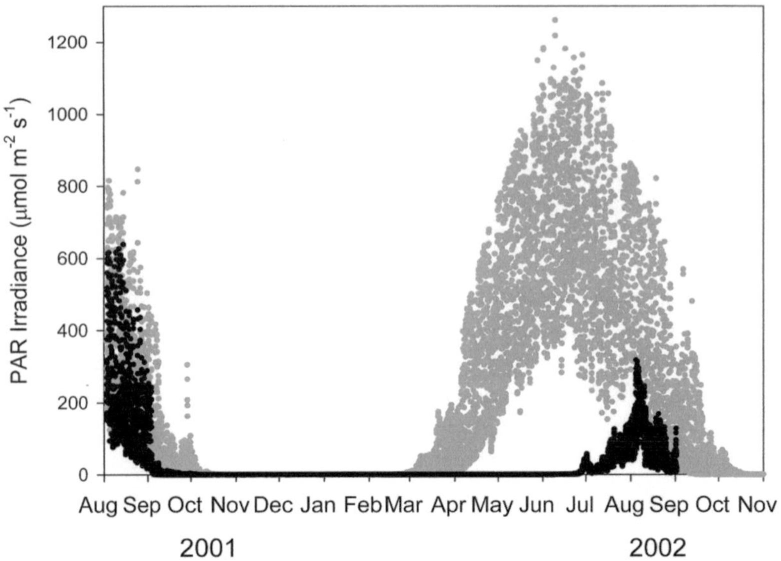

Fig. 9.2 Downwelling PAR irradiance (*grey*) from August 2002 to August 2003 on Ward Hunt Ice Shelf compared with irradiance 14 cm above the surface of a microbial mat (*black*) (From Mueller and Vincent (2006), © John Wiley & Sons. Ltd., used with permission)

dawn, once the snow and ice covering the lakes begins to melt (e.g., June or July, Fig. 9.2; Mueller and Vincent 2006). Solar energy is transmitted through the ice and absorbed by the microbial consortia and associated sediment, and the resultant heat begins to melt the ice above. The ice cover may persist throughout the entire melt season, but with a change in form from solid ice to vertically-oriented ice prisms ('candles') via preferential melting at the crystal boundaries. This ice candling contributes to light scattering and prevents much of the light from reaching the microbial mats (Fig. 9.2). At the height of the melt season, shading by the overlying ice and snow can reduce Photosynthetically Active Radiation (PAR) levels to 15% of ambient. However, this reduction in irradiance does not appear to affect the productivity of microbial mats, owing to their ability to acclimate to reduced irradiance (see Sect. 9.3.6; Mueller et al. 2005; Mueller and Vincent 2006; Hawes et al. 2008).

If the lake ice cover melts completely, then the downwelling irradiance is usually a little below ambient in-air values due to the albedo of the water surface and absorption of light in the water column. However, it is possible for the irradiance experienced by the benthic microbial mats to exceed atmospheric levels due to reflection from ice and snow that surround the lake. During a typical summer, a thin ice cover can form at any time on these lakes, and by late August or early-September the seasonal ice cover is re-established and thickens. In the presence of clear congelation ice, irradiance values at the lake bottom will drop somewhat, but if snow covers the ice, then light levels below will plummet further, long before the polar night begins.

Research on Arctic ice shelf biota has also characterized the temperatures that organisms are subjected to in their cryo-habitats. During the winter on the Ward Hunt Ice Shelf, the mean temperature in benthic mats in frozen ponds (−17°C; Mueller and Vincent 2006) is typically higher than the mean temperature of the air (−26°C; 01 September 2001 to 31 May 2002, Mueller unpublished). This relative warmth comes from the release of latent heat as the lake freezes and the insulation by snow from the colder air above. Some deep meltwater lakes on Antarctic ice shelves are even known retain a liquid core as a result of these effects (Hawes et al. 1999). Temperatures in the frozen microbial mats at the bottom of several ponds on the Ward Hunt Ice Shelf descended to an average minimum of −17°C in April after a long period of cumulative heat loss to the ice above and below (Mueller and Vincent 2006). Even at these temperatures, it is possible that metabolic activity can continue within the mats (Rivkina et al. 2000), albeit at a minimal rate (Price and Sowers 2004; see Sect. 9.3).

The snow and ice that insulates and protects benthic microbial mats from the Arctic winter air also retards their melt-out in the spring. For example, in one study, surface air temperatures on the Ward Hunt Ice Shelf exceeded 0°C in the first week of June 2002, yet several sub-nival microbial mat environments remained below 0°C until late June and early July (Mueller and Vincent 2006).

Due to their relatively low albedo, microbial mats begin to warm when solar radiation penetrates their overlying ice and snow cover. This can occur rapidly if the snow layer is removed, leaving only transparent ice between the microbial mat and the atmosphere. Once melted, the temperature of the microbial mats in the meltwater lakes and ponds does not rise appreciably above 0°C, since the absorbed solar energy and sensible heat flux are dissipated via melting of the surrounding ice. In contrast, microbial mats that are exposed on thick beds of sediment can warm to several degrees above freezing due to solar heating. Using temperature observations, an average growth season of 65 days was inferred for microbial mat communities on the Ward Hunt Ice Shelf in 2002 (Mueller and Vincent 2006, Fig. 9.3). This was slightly longer than the 61 days in 2002 when air temperature was above 0°C, likely due to enhanced absorption of solar radiation and meltwater production in the microbial mats. Temperatures in the microbial mats decrease below 0°C once the water column above freezes completely.

9.2.2 Conductivity and Origin of Salts

Arctic ice shelves are composed of two fundamental ice types: ice of marine origin and ice of meteoric origin (Mueller et al. 2006). Consequently, meltwater chemistry depends on which ice type is exposed at the ice shelf surface. For example, ice that is formed from melted and refrozen precipitation has a very low conductivity (on average 41 μS cm^{-1}) in contrast to areas where ancient sea ice is exposed at the surface (on average 1260 μS cm^{-1}; Mueller et al. 2006). The ionic composition of

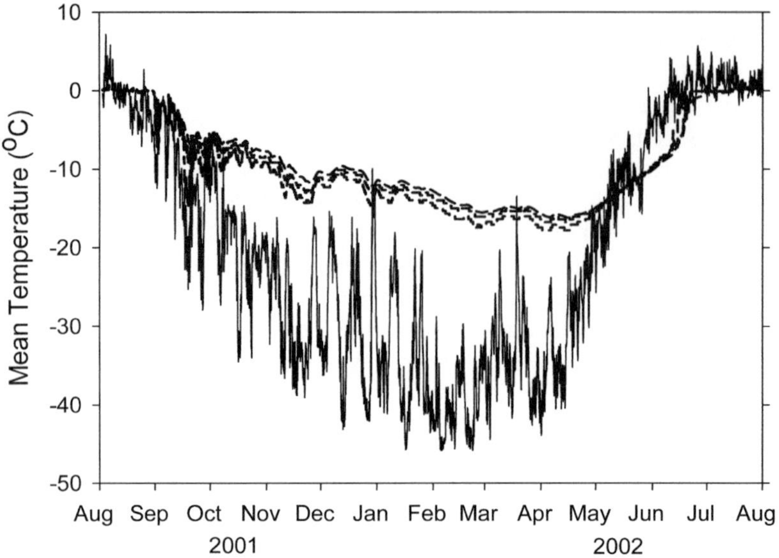

Fig. 9.3 Mean surface air and microbial mat temperatures on Ward Hunt Ice Shelf. Air temperature (*solid line*) versus the temperature within three microbial mats (*dashed line*) between August 2001 and August 2002 (From Mueller and Vincent (2006), © John Wiley & Sons. Ltd., used with permission)

surface meltwater can also be influenced by wind-blown sediments, which alters the relative concentration of ions (Mueller et al. 2006).

Meltwater ponds on the Arctic ice shelves have conductivities that typically range from 50–5700 μS cm^{-1}, with bottom water conductivities that can reach up to 10,200 μS cm^{-1}, based on measurements on the Ward Hunt and Markham ice shelves (Mueller et al. 2005; Mueller and Vincent 2006). However, these values are much lower than conductivities found on the McMurdo Ice Shelf in Antarctica which range from 100–50,000 μS cm^{-1} (Howard-Williams et al. 1990). Similar to the McMurdo Ice Shelf, the ratio of potassium to chloride in meltwater from certain regions of the Ward Hunt Ice Shelf indicates a marine origin, while the sulphate to chloride ratio is much higher; this is consistent with the redissolution of mirabilite precipitated during the freezing of seawater (de Mora et al. 1994; Vincent et al. 2000). A marine origin of parts of the Ward Hunt and Markham ice shelves is further supported by the presence of marine sediments and remains of sponges, molluscs and benthic crustaceans on the surface of the ice shelves, due to the freeze-on of these materials to the bottom of the ice shelf followed by eventual release at the surface of the ice shelf following many years of surface ablation (Vincent et al. 2000, 2004).

The presence of an ice cover on these lakes and ponds determines to a great extent whether their water columns will be mixed. Even when a lake itself mixes, water in cryoconite holes on the lake bottom may remain stratified, with conditions at the sediment/water interface being vastly different than the lake in general. This

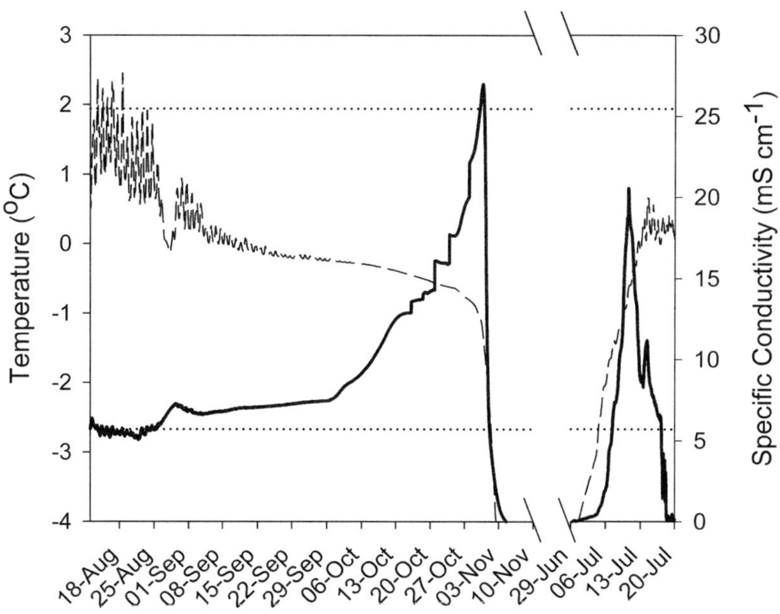

Fig. 9.4 Freeze-up and thaw of a meltwater pond on the Ward Hunt Ice Shelf. Specific conductivity (*solid line*) and temperature (*dashed line*) measured at the bottom of a 9.4 m^3 meltwater lake. Horizontal *dotted lines* indicate the mean of half-hourly observations for temperature (*upper line*) and specific conductivity (*lower line*) during open-water conditions (23 July to 10 August 2002) (From Mueller and Vincent (2006), © John Wiley & Sons. Ltd., used with permission)

discrepancy is more pronounced in areas where marine ice underlies the lakes and provides a source of salts, which promotes density stratification (Mueller and Vincent 2006).

As meltwater lakes freeze, solutes in the water are rejected from the freezing front and concentrate in the water column below. In one study lake on the Ward Hunt Ice Shelf, this phenomenon increased the conductivity of the remaining water over time, which caused a slight freezing point depression (−1.3°C) before the phase change occurred (Mueller and Vincent 2006; Fig. 9.4). This increase in salinity is a function of the subjacent ice type (marine or meteoric), freezing rate and elevation relative to sea level of the lake in question. In the case of one 9.4 m^3 pond, the bottom froze at the end of October 2002, 52 days after the air temperature went below freezing (Mueller and Vincent 2006). The conductivity near the microbial mats increased fourfold (from 5700 to 27,000 $\mu S\ cm^{-1}$) during freeze-up which suggests that salinity is not only spatially heterogeneous in the ice shelf cryo-ecosystem but also varies substantially with time (Mueller and Vincent 2006). Studies on microbial mats in the nearby Ward Hunt Lake have shown that the photosynthetic communities are highly resistant to the major increases in salinity that may be associated with solute concentration during freeze-up (Lionard et al. 2012).

9.2.3 Nutrients and Organic Matter

Polar freshwater ecosystems are usually considered to be ultra-oligotrophic, with the phytoplankton community biomass severely constrained by limited nutrients (Vézina and Vincent 1997). However, the microbial mat communities may escape this constraint because nutrients may become trapped over time within the diffusion-limited mat matrix and made available through decomposition of dead biomass (Vincent et al. 1993; Bonilla et al. 2005). In addition to recycling nutrients, there is evidence based on $\delta^{15}N$ signatures that nitrogen may be fixed by the benthic heterocyst-forming cyanobacteria that exist within these mats (Mueller and Vincent 2006).

Analysis of nutrient concentrations within the microbial mats illustrates how the mat microenvironment is dissimilar from the rest of the ice shelf habitats. Studies of Ward Hunt and Markham ice shelf mats on northern Ellesmere Island show they have two to five orders of magnitude more dissolved inorganic carbon (DIC) and ammonium-N than found in the overlying water column (Mueller and Vincent 2006). Other nutrients, such as dissolved organic carbon (DOC), total dissolved nitrogen, nitrate, nitrite, total dissolved phosphorus and soluble reactive phosphorus are two orders of magnitude more concentrated in the microbial mat than in the overlying water. A metagenomic analysis of microbial mats from the Ward Hunt and Markham ice shelves (Varin et al. 2010) showed that they contained diverse nutrient scavenging systems including genes for transport proteins and enzymes converting larger molecules into more readily assimilated inorganic forms (allantoin degradation, cyanate hydrolysis, exophosphatases, phosphonatases). These molecular results underscored the capability of ice shelf mats to retain and recycle nutrients in the benthic microenvironment.

Analysis of nutrient concentrations in the water column of ponds of the Markham Ice Shelf showed that dissolved reactive phosphorus levels were high enough to exclude P-limitation, however inorganic nitrogen concentrations were relatively low and the supply of nitrogenous nutrients may have limited phytoplankton biomass (Mueller and Vincent 2006; Mueller et al. 2006). Thus ice shelf supraglacial ponds contain two disparate communities that differ in nutrient status: nutrient-limited plankton and the nutrient-replete benthic mats, consistent with observations on land-based water bodies in the High Arctic (Bonilla et al. 2005).

In one study, coloured dissolved organic matter (CDOM) in the benthic mat pore water was similar in composition to that in the overlying water, based on the McKnight ratio (fluorescence index, McKnight et al. 2001). The CDOM was composed of complex molecules likely derived from terrestrial vegetation debris, such as *Salix arctica* leaves, that have occasionally been observed within the ice shelf microbial mats. The presence of other water-soluble compounds, such as exopolymeric substances (EPS) and oligosaccharide mycosporine-like amino acids (MAAs), may have also been responsible for such a low McKnight ratio (Mueller et al. 2005).

CDOM strongly absorbs in the ultraviolet (UV) region of the solar spectrum and in the process undergoes photochemical degradation to products that can have both positive (e.g., increased C, N, P and Fe availability to microbiota), and negative effects (e.g., production of reactive oxygen species such as peroxides) on aquatic communities (Vincent and Neale 2000). Photochemical reactions may also promote the loss of carbon from these ecosystems via photochemical degradation of DOC to CO_2 and CO during continuous summer daylight and subsequent evasion (transfer from the water to the atmosphere) of these gases. The implications of stratospheric ozone depletion for surface water photochemistry are of particular interest for northern lake environments given the potential for increased levels of UV-B in this region (ACIA 2005).

9.3 Biological Properties of Supraglacial Lakes

9.3.1 Habitat Distribution

Until summer 2005, the six largest remnant ice shelves in the Canadian High Arctic totaled 1043 km^2 and 8% of their surface area provided sediment cryohabitats for microbiota (Fig. 9.5; Mueller et al. 2006; Copland et al. 2007). Ecological studies to date have focused primarily on the Markham and Ward Hunt ice shelves with Markham Ice Shelf being the richest in biomass per unit area of all Arctic ice

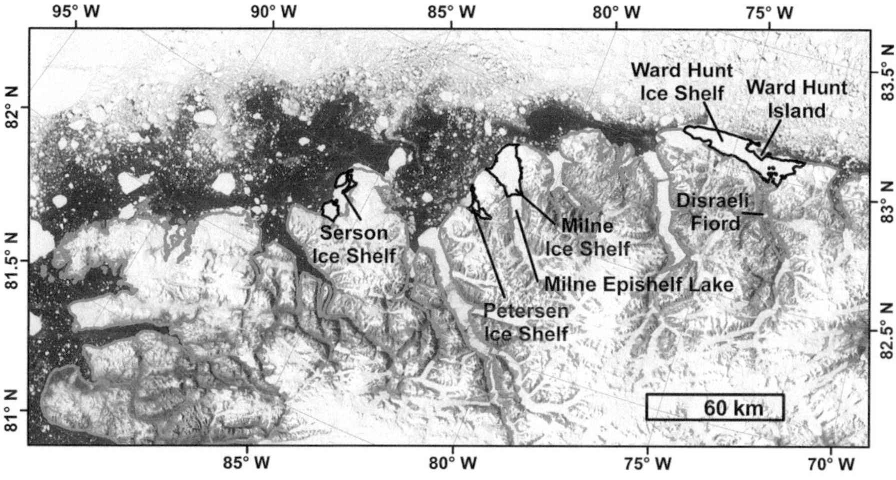

Fig. 9.5 Open water stretching along the northwest coast of Ellesmere Island for 250 km at the end of August, 2008. This unusually wide (up to 55 km) region of ice-free water began to form in the third week of July 2008, at the same time as the ice shelves began to calve. The ice shelves are outlined in *black* and the coast is outlined in *grey* (MODIS image acquired on 2008-08-29 at 19:55 UTC, from the Rapid Response Project at NASA/GSFC)

shelves, until it broke up completely in the summer of 2008 (Mueller et al. 2008; Vincent et al. 2009).

On the Ward Hunt and Markham ice shelves 10% and 44%, respectively, of their surface areas were covered with sediments and microbial mat communities (Mueller et al. 2006). Meltwater ponds on the ice shelves are dominated by benthic communities that represent most of the biomass. A total of 34 Gg of organic matter was estimated for the entire Canadian Ellesmere ice shelves' ecosystem with average per unit area productivities (129 mg C m^{-2} d^{-1}) that are well above values in the Central Arctic pack ice (Mueller et al. 2006).

Arctic ice shelf cryohabitats are heterogeneous at the sub-ice shelf scale with sediments and associated microbial mats distributed in patches. However, species richness and diversity among ice shelves and habitat patches does not differ significantly. Therefore, it is plausible that there is considerable connectivity and dispersal between the individual ice shelves as well as between habitat patches, likely in the direction of the prevailing wind (Mueller et al. 2006). Most of the organisms found in ice shelf mats are present in nearby microbial mats from terrestrial ponds (Villeneuve et al. 2001) and glacial cryohabitats (Mueller et al. 2001), which serve as sources of organisms for ice shelf cryo-ecosystems.

9.3.2 Benthic Biology

The abundance of microbial mats is highly variable across the surface of Arctic ice shelves, ranging from lakes with a continuous layer of thick accumulated benthic microbial biomass and sediment (e.g., Markham Ice Shelf; Fig. 9.6), to lakes with little or no microbial mats, containing only small patches of mats or sparse cryoconite holes. In the latter case, the cryoconite holes seem to be preferentially distributed along the strand line or former strand lines (high water mark), parallel to the lake edge (Mueller, personal observations). Microbial mats may also be found in the numerous water-filled caverns and ice interstices that occur within the ice shelves (Mueller, personal observations).

Microbial mats on the Canadian Arctic ice shelves have been macroscopically classified into three categories: "orange", "matlet" and "sediment" mat (Mueller et al. 2006). Orange mats are luxuriant, thick (>0.5 cm) loosely cohesive mats with a thin surface layer containing orange pigmentation (Fig. 9.7a). The "matlet" type mats are 1- to 2- mm wide flakes that lie loosely on the ice but can accumulate to a thickness of 3 cm, and the "sediment" mats have no visible biological aggregate and occur in a fine to coarse sediment (Mueller et al. 2006). These mats differ greatly from benthic communities in nearby lakes such as Ward Hunt Lake (on Ward Hunt Island) and Antoniades Pond (northern Ellesmere Island), where the mats are more cohesive with a macroscopic stratification of different pigment groupings (Bonilla et al. 2005, Jungblut et al. 2010). Arctic ice shelf mats are less diverse in their macrostructure than similar assemblages on Antarctic ice shelves, such as the McMurdo Ice Shelf. This could be due the greater range of physical and chemical properties

Fig. 9.6 Sampling a meltwater lake on the Markham Ice Shelf, Canadian High Arctic (Date: 12 July 2007; Photograph: W.F. Vincent)

present in the meltwater ponds of the McMurdo Ice Shelf (Howard-Williams et al. 1990) where higher variability of pond size, depth and conductivity likely supports a higher phenotypic and genotypic diversity and hence, mat macrostructure.

The mat-containing sediments of the Markham Ice Shelf occurred on raised mounds of ice and also at the base of meltwater ponds and lakes that occupied the parallel troughs that are characteristic of all of the Canadian High Arctic ice shelves. The microbial mats on Markham Ice Shelf could be separated into two layers with the lower layer being directly in contact with the ice. Algal counts (protists and cyanobacteria) identified slightly different abundances in these layers with 4.58×10^5 and 2.51×10^5 cells cm^{-2} in the upper and lower layer respectively (Vincent et al. 2004). Microbial mats on the Ward Hunt Ice Shelf do not have macroscopic strata, with the exception of an orange surface layer up to 220 μm thick. Total standing stock of microbial mats on the Markham Ice Shelf in 2001 was estimated to be 16.5 Gg of organic matter, about 50% of the total estimated organic matter on the Ellesmere Ice Shelves at that time (34 Gg; Mueller et al. 2006).

The biomass of the aquatic ecosystems on Arctic ice shelves is dominated by benthic assemblages, whereas the planktonic assemblages in the overlying oligotrophic water occur in dilute concentrations. Chlorophyll *a* in the water ranges from 0.02–0.68 μg L^{-1} indicating a sparse phytoplankton communities and limited resuspension of benthic communities (Vincent et al. 2004). In comparison, chlorophyll *a*

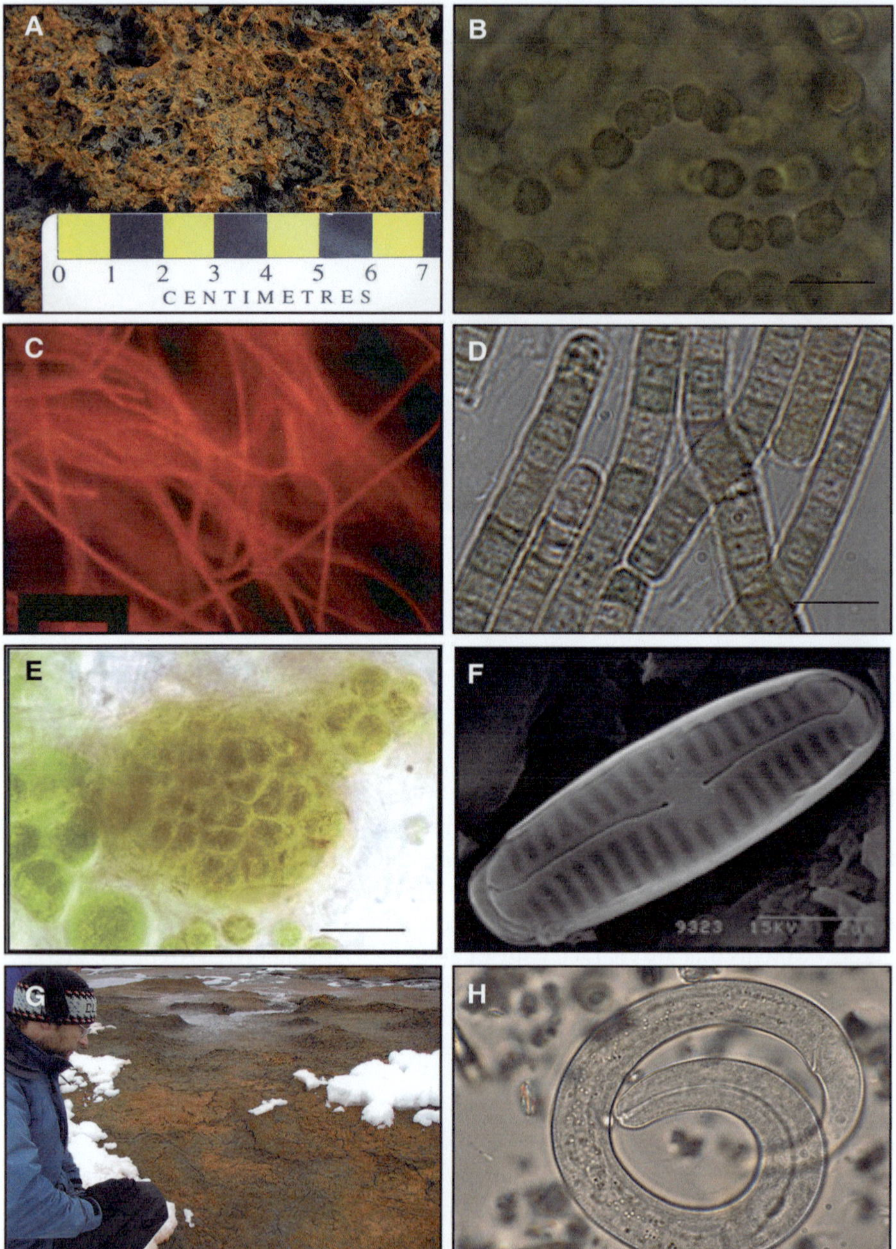

Fig. 9.7 Microbial mat diversity on Ward Hunt and Markham ice shelves. (**a**) "*Orange*" mat (>0.5 cm thick) with characteristic *thin layer* of *orange* pigmentation (Photograph: D. Sarrazin). (**b**) N_2-fixing filamentous cyanobacterium *Nostoc* sp.; the scale bar represents 10 μm (Photograph: A.D. Jungblut). (**c**) Epifluorescence micrograph of filamentous cyanobacteria (Oscillatoriales); the scale bar represents 10 μm (Photograph: V. Villeneuve). (**d**) Filamentous cyanobacterium

concentrations in the benthic communities ranges from 5.4 to 448 chl *a* mg m^{-2} with an average of 147 chl *a* mg m^{-2}, while carotenoid levels range from 18–6460 mg m^{-2} with an average of 1307 mg m^{-2} (Vincent et al. 2004; Mueller et al. 2005). However, benthic chlorophyll *a* concentrations can vary greatly among meltwater ponds due to the patchy distribution of mats (Vincent et al. 2004; Mueller et al. 2005).

Benthic microbial mat consortia are often characterized by vertical physical and biochemical stratification including gradients in variables such as irradiance, redox potential, pH, and concentrations of dissolved oxygen, carbon dioxide, methane and nutrients (Stal 2000). These gradients are generated in part by external physical factors such as light, sediment composition and water characteristics, but also through the zonation of metabolic activities such as sulphate-reduction and oxidation, photosynthesis, respiration, nitrification, denitrification, nitrogen fixation, fermentation and methanogenesis (Stal 2000), which arise from the functionally diverse microbial consortia. These gradients have received little attention in the ice shelf mats and more research should be undertaken to understand them.

9.3.3 *Prokaryotic Diversity*

9.3.3.1 Cyanobacterial Diversity

Cyanobacteria dominate the autotrophic biomass of the Arctic ice shelf mats, comprising at least 88% of total algal cell counts (Vincent et al. 2004). Cyanobacteria are important for mat cohesion and structure due to their production of exopolymeric substances (EPS) and the filamentous morphology of the dominant taxa, which intertwine to form the mats. Commonly found genera are *Nostoc* (Fig. 9.7b), *Phormidium, Oscillatoria* (Fig. 9.7c, d), *Leptolyngbya* and *Gloeocapsa* as revealed by microscopy on the Markham and Ward Hunt ice shelves (Mueller et al. 2006). However, further molecular investigations such as 16S rRNA gene analysis are needed to fully understand their diversity, distribution between sites and their biogeographic relationship to cyanobacterial communities elsewhere in the Arctic, or Antarctic shelves such as the McMurdo Ice Shelf (Jungblut et al. 2005) The first molecular analyses of this type on the Ellesmere Island ice shelves indicated the presence of cyanobacterial ecotypes that are distributed throughout the cold biosphere, including Antarctica (Jungblut et al. 2010). A subsequent detailed meta-analysis has corroborated these findings, but also draws attention to the urgent need for full genomic analysis of cold-tolerant cyanobacteria (Chrismas et al. 2015).

Fig 9.7 (continued) (Oscillatoriales) isolated from Markham Ice Shelf mats; the scale bar represents 10 μm (Photograph: A.D. Jungblut). (**e**) Mat algae from Markham Ice Shelf; the scale bar represents 10 μm (Photograph: S. Bonilla). (**f**) Electron micrograph of a mat diatom; the scale bar represents 2 μm (Photograph: V. Villeneuve). (**g**) Highly pigmented microbial mats on the Markham Ice Shelf (Photograph: W.F. Vincent). (**h**) Nematode, common in Arctic ice shelf mats; the scale bar represents 20 μm (Photograph: A.D. Jungblut)

Cyanobacteria dominate the benthic communities of many polar freshwater ecosystems due to multiple factors, including low grazing-pressure and their ability to tolerate the extreme conditions found in these environments (Vincent 2007, 2009). Tang et al. (1997) showed that many high latitude mat-forming cyanobacteria tend to be psychrotrophs (cold-tolerant microorganisms) not psychrophiles (cold-loving microorganisms). This strategy of general tolerance to a broader range of temperatures allows a faster acclimatization in the presence of frequent freeze-thawing cycles, rapid changes in temperatures and high UV radiation despite suboptimal growth under low temperatures. For example, freeze-dried cyanobacterial mats from Antarctic ice shelves can resume photosynthesis within minutes to hours after thawing and rehydration, which is important in an ecosystem with a limited growing season (Vincent 1988; Hawes et al. 1999).

Cyanobacteria have also developed an array of mechanisms to ensure cell integrity during freezing such as the synthesis of extracellular compounds to reduce ice nucleation around the cells (Vincent 1988). Furthermore, polyunsaturated fatty acids are incorporated into membranes to retain fluidity at low temperatures (Laybourn-Parry 2002) and compatible solutes such as trehalose are produced to reduce the freezing point of intracellular fluids (Oren 2000). Some of these organic osmolytes including glycine betaine can also be used as a long-term strategy to balance extracellular ions at higher salinities, as found in some Arctic ice shelf waters and particularly during freeze-up.

High UV radiation is another major stress factor for microorganisms in Arctic aquatic ecosystems (Roos and Vincent 1998). In cyanobacteria, it can lead to photoinhibition, phycobiliprotein degradation and chlorophyll-bleaching (Castenholz 1992; Ehling-Schulz and Scherer 1999). Furthermore, exposure of DNA to UVB radiation can lead to DNA lesions and mutagenesis, including dimerisation of adjacent pyrimidine bases (Vincent and Neale 2000, and references therein). However, cyanobacteria have evolved a variety of DNA repair mechanisms, such as excision repair and photo-reactivation, to cope with UV induced DNA damage (Garcia-Pichel and Castenholz 1991). Another strategy observed in motile cyanobacteria is migration to deeper layers within the microbial mats to avoid radiation (Vincent et al. 1993; Quesada and Vincent 1997). Cyanobacteria are also able to synthesize a variety of pigments such as carotenoids, scytonemin, and MAAs for protection against UV radiation (Hodgson et al. 2004b; Squier et al. 2004).

9.3.3.2 Heterotrophic Bacterial, Archaean and Viral Diversity

Other bacterial phyla are also well represented in Arctic ice shelf communities, and studies using epifluorescence microscopy showed the presence of viruses and heterotrophic bacteria, with abundances of 2.3–16.5 × 10^7 viruses and 0.7–7.5 × 10^7 bacteria per square centimetre in mat systems on the Ward Hunt Ice Shelf (Vincent 2000a). Bottos et al. (2008) identified heterotrophic bacterial communities that were dominated by the Bacteroidetes, Proteobacteria (α-, β-, δ- and γ-) and Actinobacteria on the Markham and Ward Hunt ice shelves using 16S rRNA gene analysis. They also found Firmicutes, Verrucomicrobia and Gemmatimonadetes on the Markham Ice Shelf and Fibrobacteres on Ward Hunt Ice Shelf, albeit in lower abundances. Interestingly, no community differences were noted between upper and lower layers in ice shelf microbial mats using Denaturating Gradient Gel Electrophoresis (DGGE) analyses. The presence of both α- and β-Proteobacteria implies a community with some marine affinities (β-Proteobacteria usually occur more frequently in marine plankton) as well as a strong freshwater influence (β-Proteobacteria are found in freshwater). Most of the archaeal taxa found within the ice shelf mats grouped within the phyla Euryarcheaota and Crenarcheota (Bottos et al. 2008; Varin et al. 2010). The microbial abundances of these bacterial communities are similar to those reported in microbial mats from Antarctic ice shelves communities (Van Trappen et al. 2002; Sjöling and Cowan 2003; Archer et al. 2015). In addition, Archer et al. (2014) identified that bacterioplankton is influenced by geochemical characteristics of Antarctic meltwater ponds, and that distinct populations can be present in the upper and lower part of highly stratified water columns of meltwater ponds on the McMurdo Ice Shelf based on ARISA finger print (Automated Ribosomal Intergenic Spacer Analysis). Further analyses are required of Arctic bacterioplankton to determine if similar heterotrophic bacterial communities are dominating meltwater ponds, however they will likely lack stratification as Arctic meltwater ponds are usually shallow and more edaphic then McMurdo Ice shelf aquatic ecosystems.

It is difficult to infer metabolic functionality based solely on current 16S rRNA gene analyses, therefore further metagenomic analysis is required for a better understanding of functional ecology. The first application of this approach to microbial mats from the Ward Hunt and Markham ice shelves revealed a large functional as well as taxonomic diversity, with proteobacteria contributing the largest number of gene sequences (Varin et al. 2010). This proteobacterial diversity in Arctic ice shelf mats would facilitate a broad suite of aerobic and anaerobic processes, including sulphur-reduction activities as observed in Antarctic microbial mats communities from the McMurdo Ice Shelf (Mountfort et al. 1999; Mountfort et al. 2003). A variety of stress tolerance strategies has been identified from subsequent metagenomic analysis of Arctic ice shelf mats, with similarities but also certain differences relative to Antarctic ice shelf mats (Varin et al. 2012).

9.3.4 Eukaryotic Diversity

To date, the study of eukaryotes in Arctic ice shelf mats is mostly limited to morphological identification and enumeration. Eukaryotic algae are commonly found within the matrix of cyanobacteria. In Ward Hunt Ice Shelf mats, these include the Chlorophytes *Palmellopsis, Chlorosarcinopsis, Pleurastrum, Chlamydomonas, Chlamydocapsa, Chlorella, Chlorococcum, Klebsormidium, Palmellopsis, cf. Chlorokybus* and solitary cells of *Bracteacoccus* (Vincent 2000a; Mueller et al. 2006). On Markham Ice Shelf similar groups of mainly terrestrial and subaerial palmelloid chlorophyte genera occur (Fig. 9.7e; Vincent et al. 2004). Phylogenetic analysis of 18S rRNA genes grouping within Chlorophyceae suggested that sequences from Arctic ice shelf microbial mat cluster together with ripotypes that had been previously been reported to be endemic to Antarctica (Jungblut et al. 2012).

Benthic and aerophilic diatom communities on the Ward Hunt Ice Shelf are dominated by *Chamaepinnularia begeri* (Krasske) Lange-Bertalot (syn. *Navicula begeri* Krasske) accounting for 90% of frustules (Fig. 9.7f; Vincent et al. 2000). Mats from this ice shelf also contain *Nitzschia palea* (Kützing) W. Smith (2.3%), *Navicula phylleptosoma* Lange-Bertalot (1%), *C. krookii* (Grunow) Lange-Bertalot, *C. gandrupii* (Petersen) Lange-Bertalot and Krammer, *Pinnularia borealis* Ehrenberg, *Luticola palaearctica* (Hustedt) Mann, *Achnanthes petersenii* Hustedt and *Nitzschia cf. pusilla* Grunow (all <1%). In comparison, Antarctic mats from the McMurdo Ice Shelf similarly contain coccoid chlorophytes and a diatom assemblage dominated by the genera *Navicula, Nitzschia, Pinnularia* and *Achnanthes* (Howard-Williams et al. 1990). Chrysophyte cysts are also present in Ward Hunt Ice Shelf mats (Vincent et al. 2000, 2004), but are not found on the McMurdo Ice Shelf.

The first 18S rRNA gene assessment of the microbial eukaryote diversity in microbial mats of High Arctic ice shelves such as the Ward Hunt and Markham ice shelves highlighted that their assemblages do not only contain various microalgae and ciliates but also but also fungi including Ascomycota, Basidiomycetes and Chytridiomycota, as well as rhizaria such as Cercozoa with many having the highest similarity to uncultured environmental sequences (Jungblut et al. 2012). Although the study was only preliminary due to incomplete sampling, it is still noteworthy a limit overlap was found between microbial eukaryote composition in microbial mats from ice- and land-based aquatic ecosystems.

9.3.5 Primary and Heterotrophic Productivity

On the Ward Hunt Ice Shelf, maximum photosynthesis rates range from 0.059 to 0.17 g C g chl a^{-1} h^{-1} (27.3–105 mg C m^{-2} h^{-1}), whereas bacterial productivity is three orders of magnitude lower, ranging from 0.085 to 0.38 µg C g biomass $^{-1}$ h^{-1} (0.037–0.21 mg C m^{-2} h^{-1}). The measured chlorophyll-specific primary productivity in Ward Hunt Ice Shelf mats, as well as the heterotrophic bacterial production lie within the range of polar microbial ice mat literature such as the McMurdo Ice Shelf (Vincent and Howard-Williams 1989). However primary productivity on a per unit area basis is much higher in the Arctic mats, due to the large standing stocks of photosynthetic pigments (Mueller et al. 2005).

Using the above productivity rates and based on a growing season of ~70 days, the annual primary production of the communities on the Ward Hunt Ice Shelf is approximately 108 g C m^{-2}. Of this, bacterial heterotrophy would recycle an estimated 4.3 g m^{-2} of biomass (or approximately 2.15 g C m^{-2}) within the microbial mats given an average efficiency of 30% (Mueller et al. 2005). This further reflects the overwhelming dominance of autotrophic biomass relative to heterotrophic bacteria in the mats (Mueller et al. 2005), and indicates the tendency of the mats to accumulate rather than fully degrade.

Primary productivity assays of Ward Hunt Ice Shelf mats identified broad tolerances optima outside the ambient range of key stressors such as salinity, irradiance and temperatures in the autotrophic community. This tolerance was related to their diverse suite of pigments (see Sect. 9.3.6; Mueller et al. 2005), which can assist the organisms to acclimatize to changing conditions. Experimental manipulation of salinity had only limited effects on primary productivity, which also suggests a broad growth optimum between 3000 and 29,000 μS cm^{-1} for these Arctic ice-mat autotrophs (Mueller et al. 2005). Similarly, in their studies on Antarctic microbial mats, Hawes et al. (1999) found no change in photosynthetic rates (relative to control) up to 20,000 μS cm^{-1}.

Heterotrophic bacterial productivity on Ward Hunt Ice Shelf does not appear to respond favourably to increased salinity and temperature (Mueller et al. 2005), and physiological studies on community isolates from Ward Hunt and Markham Ice Shelves showed that most of them were cold adapted, with growth at temperatures as low as −10°C (Bottos et al. 2008). Therefore, in contrast to the broad tolerance of the autotrophic community, Arctic ice shelf heterotrophic bacterial communities appeared to be extremophilic (specifically adapted to the 'extreme' ambient conditions in their habitat; Mueller et al. 2005). The large concentrations of viruses found in these mats (Vincent et al. 2000) and the diverse viral sequences in the metagenomic analysis of ice shelf mats (Varin et al. 2010) suggest that phage attacks may limit the standing stock of bacterial biomass, and thereby also limit the extent of bacterial decomposition processes.

9.3.6 Pigment Content and Photoprotection

A striking feature of Arctic ice shelf communities is their richness and diversity of pigments (Fig. 9.7g). These pigments can be classed into screening compounds (e.g., MAAs and scytonemins that absorb UV-A and UV-B radiation), light harvesting and accessory pigments (chlorophylls, phycobiliproteins and certain carotenoids), and anti-oxidants (other carotenoids and perhaps MAAs).

High Arctic planktonic and benthic communities exist in shallow waters with low concentrations of UV-filtering coloured dissolved organic matter (CDOM), and must therefore contend with a high UV exposure (Bonilla et al. 2009). The pigment assemblage suggests that these microbial communities can spectrally modify their environment for both photo-protection, and photosynthetic efficiency (Vincent et al. 2004; Mueller et al. 2005; Bonilla et al. 2009). On ice shelves, planktonic communities apparently use a different strategy to counter the effects of high UV radiation than benthic microbial mat communities. A study of pigments in Arctic lakes and ponds at several locations showed that planktonic communities favour quenching reactive oxygen species with carotenoids, particularly under high UV radiation and at low temperatures while benthic consortia employ a screening approach using scytonemin and reduced scytonemin, which accumulate in high concentrations (Bonilla et al. 2009).

Chlorophyll *a* levels in mats in high biomass areas of the Ward Hunt and Markham ice shelves are higher relative to microbial mats from many ponds and lakes throughout the polar regions, including meltwater ponds on the McMurdo Ice Shelf, Antarctica. For example, on Markham Ice Shelf mean chl *a* biomass value for microbial mats was 147 mg chl *a* m^{-2} while concentrations in the Ward Hunt Ice Shelf mats range from 5.6–31 mg chl *a* m^{-2}. A possible explanation for extremely high values might be the enhanced preservation of pigments due to the inhibitory effects of low temperatures on bacterial processes. McMurdo Ice Shelf microbial mats often lie atop relatively thick sediment, which likely warm to temperatures not experienced on the northern ice shelves (Mueller et al. 2005; Mueller and Vincent 2006).

Similar to land-based Arctic freshwater systems, the cyanobacterial-dominated mats on ice shelves are structured with respect to their pigment characteristics. The upper layer of mats may be optimized to filter out UV wavelengths and/or quench reactive oxygen species in response to the photo-oxidative stress imposed by the large cumulative UV dose in summer (Quesada et al. 1999). In Markham Ice Shelf mats, the carotenoid/chlorophyll *a* ratio was found to be twice as high in the surface layer than in the bottom, even with chl *a* and chl *b* concentrations being higher in the upper layer than at the bottom of the mat (Vincent et al. 2004). These results underscore the use of carotenoids as anti-oxidants in the microbial mats, whereas the high concentrations of scytomenin and presence of MAAs indicate the importance of screening deleterious radiation from the microbial consortia throughout the mat (Mueller et al. 2005). However, the high concentration of screening pigments and the low concentration of light-harvesting pigments deeper within the mat may also represent pigment accumulation and the relatively slow degradation of dead cells in this cold environment (Vincent et al. 2004).

9.3.7 Invertebrates and Other Animals

Observations of the faunal composition of meltwater lakes on Arctic ice shelves are very limited, and most records are based on anecdotal reports rather than detailed studies. Fish, such as the slimy sculpin (*Cottus cognatus*), and amphipods, such as gammaridae, have been observed in sections of the ice shelves which have some connection to the sea (Jungblut, personal observations). However, chironomid larvae and fairy shrimp that are regularly found in nearby lakes on Ward Hunt and Ellesmere islands have not been observed in meltwater ponds of the ice shelves. Metazoan diversity is commonly limited to nematodes (Fig. 9.7h), rotifers and tardigrades, which compares with findings from the McMurdo Ice Shelf, Antarctica. For example, one of the only molecular studies of eukaryotic diversity on ice shelves showed that Arctic ice shelf tardigrades had the highest 18S rRNA gene similarity to *Isohypsibius granulifer* (99%), a species with a cosmopolitan distribution (Jungblut et al. 2012). However, platyhelminth worms have also been found in microbial mats on the Ward Hunt Ice Shelf, but have not been found to date on Antarctic ice shelves (Vincent et al. 2000).

9.4 Other Biota Associated with Ice Shelves

9.4.1 *Macrobiology of the Ice Shelves*

Knowledge of the macrobiota inhabiting Arctic ice shelves is very limited, and records are based on anecdotal reports rather than detailed studies. However, it can be assumed that Arctic ice shelves are infrequently visited by animals and birds due to the limited food sources there. Arctic foxes (*Vulpes lagopus*) and polar bear (*Ursus maritimus*) tracks have been reported on the Ward Hunt Ice Shelf and polar bear, Arctic fox, ermine (*Mustela erminea*) and lemming (*Dicrostonyx greenlandicus*) tracks have been observed on the Milne Ice Shelf. Seals have been observed at the seaward edge and within fractures on the Ward Hunt and Milne ice shelves. Arctic hares (*Lepus arcticus*) and Peary caribou (*Rangifer tarandus pearyi*) inhabit the northern part of Ellesmere Island and could potentially cross the ice shelves. Muskox (*Ovibos moschatus*) excrement has been found on the Ward Hunt Ice Shelf (Mueller and Stern, personal observations) near Ward Hunt Island and caribou antlers were recovered from the Ice Island T-3 (Jeffries 1992). On Ellesmere Island approximately 34 different species of birds can be seen, with 21 confirmed to nest on the island (Parks Canada, unpublished data). Birds such as Brant geese, short tail jaegers, snow buntings, red-throated loons, red knots, gulls (species not identified) and shorebirds such as ruddy turnstone have been observed (Sarrazin, personal communication) flying over Ward Hunt Island and Ward Hunt Ice Shelf during summer. Ivory gulls have been spotted over Milne Ice Shelf on several occasions. Red-throated loons (*Gavia stellata*) have been found nesting and feeding on fish in the lakes with open water in the summer at the northern tip of Ellesmere Island (Jungblut, personal observation), however it is not known if they use the ice shelves as habitats as well. Recent changes in climate and ecosystem structure such as the opening of polynyas near the edge of the ice shelves due to sea ice melt and reduction of ice shelf size may redefine the role of the remaining ice shelves as habitats for Arctic macrobiota if they can gain access to the ocean from the ice edge.

9.4.2 *Microbial Biota of Snow*

Microbial diversity of snow on Arctic ice shelves is likely enhanced by propagules from the rich microbial mat communities and sediment on the ice shelf, as well as from meltwater lakes and terrestrial soil. Long-range aeolian dispersal of microbes could connect ice shelf ecosystems with not only other Arctic ecosystems but other cold environments such as temperate alpine regions and Antarctica (Christner et al. 2008). Snow microbial communities are often dominated by highly adapted and relatively fast growing biota such as heterotrophs and protists, whereas cyanobacteria are typically found in low abundances. For example, the snowpack in Spitsbergen, Svalbard was found to contain α-Proteobacteria, β-Proteobacteria and δ-Proteobacteria, Firmicutes and Actinobacteria using 16S rRNA gene analysis (Amato et al. 2007). Snow algae have been studied in detail in the Antarctic and genera such as *Chlamydomonas*, *Ochromonas, Raphidonema* and *Chlorosphaera* are found at different locations there (Vincent 1988). Taxa such as *Chlamydomonas* spp. and *Chloromonas* spp. are also reported from Spitsbergen at such high concentrations that they visibly coloured the snow owing to their pigmentation (Müller et al. 1998). This 'watermelon snow' has not been found on Arctic ice shelves and the abundance of these red pigmented snow algae is likely to be low. However, blooms of *Ancyclonema nordenskioldii* have been noted in the weathering crust (a slushy mixture of candled ice and snow) of the Ward Hunt Ice Rise during August 2002 (Mueller, unpublished data).

A molecular microbiological analysis of snow on the Ward Hunt Ice Shelf along with other sites in the northern Ellesmere Island region was undertaken by Harding et al. (2011). This revealed a diverse microbiota of bacteria and eukaryotes, including taxa found in the Arctic Ocean and in freshwater microbial mat environments, implying the importance of local aerial processes for microbial dispersion. Some sequences were most similar to those found outside the Canadian Arctic but previously reported in snow, lake ice, sea ice, glaciers and permafrost, alpine regions, Antarctica, and elsewhere in the Arctic, supporting the view that certain microbial ecotypes are globally distributed throughout the cold biosphere. As noted in Pointing et al. (2015), further evaluation of this view and of the contrary likelihood of a certain degree of microbial endemism in the polar regions will require more detailed genomic analysis of isolates in the future.

9.4.3 Microbial Life Within the Ice

Diverse microbial consortia are found in meltwater lakes on Arctic ice shelves and they also likely inhabit the ice itself. Arctic ice shelf ice has not yet been analyzed for microbial life, however the presence of microbes is plausible based on the availability of space, nutrients and liquid water. In systems such as the ice-covers of perennially frozen lakes in the Dry Valleys of Antarctica, pockets of liquid water that support microbial ecosystems form during the warmer months in ice where dark-coloured material has been incorporated into the ice (Paerl and Pinckney 1996; Fritsen and Priscu 1998; Hawes et al. 2008). These pockets of sediment are known to occur within the ice shelf, both near the surface ('buried' cryoconite holes) and at great depths within the ice (Crary et al. 1955). However, microbes that are not associated with sediments could also be buried within the ice shelf. Molecular analysis of ice shelf cryoconite holes are lacking by several sequencing surveys of glacial cryoconite communities in the Arctic and Antarctica have now shown that they contain diverse communities of cyanobacteria, bacteria, archaea and microbial eukaryotes and that both environmental variables and geography play role in shaping these cryospheric communities (Cameron et al. 2012; Edwards et al. 2013; Webster-Brown et al. 2015).

Microbes can survive unfavourable conditions in a dormant state for extended periods of time or, if they are surrounded by even minute quantities of water (such as in brine channels, veins or pore spaces in the ice), they can theoretically maintain metabolic activity (Price and Sowers 2004). Microorganisms could potentially survive at crystal boundaries in pure polycrystalline ice by using water and nutrients provided by aqueous ionic solutions (Price 2007). These organisms would have to be small enough to fit into a network of veins with diameters in the micron range, which form during freezing processes through solute and salt exclusion from the freezing front (Mader et al. 2006). Other microhabitats within ice include the surface of ice-entrained mineral grains covered with thin films of unfrozen water. Others studies were able to isolate aerobic bacteria grouping belonging to high-G + C gram-positives, low-G + C gram-positives, *Proteobacteria*, and *Cytophaga-Flavobacterium-Bacteroides* groups (Miteva et al. 2004), from a silty ice core obtained from Greenland that included metabolically active cells. Similarly, Brinkmeyer et al. (2003) identified a diverse assemblage of bacteria dominated by *α-, β-Proteobacteria* and *Cytophaga-Flavobacterium* group from Arctic and Antarctic pack ice. Based on the physical properties of the ice shelf ice and the high microbial diversity and activities present on the surface of the ice, bacterial assemblages are also expected to be present within the ice.

9.5 Epishelf Lakes

9.5.1 Physical Properties

Epishelf lakes are ice-dependent stratified ecosystems that form between an ice shelf and land with an upper freshwater layer, derived from ice and snow melt that is dammed behind an ice shelf (Gibson and Andersen 2002). The depth of the freshwater layer is constrained by the minimum draft of the ice shelf (Keys 1978) and the mixing of the freshwater and marine layer is prevented through strong density stratification and the lack of wind-induced turbulence due to a perennial ice-cover. The saline bottom waters are hydrologically connected to the ocean and therefore the entire water column is influenced by tidal forces (Veillette et al. 2008, and references therein).

The last remaining deep epishelf lake in the Arctic (as of July 2016) is located behind Milne Ice Shelf in Milne Fiord. This is a large reduction from the 17 deep epishelf lakes that may have formerly existed behind the "Ellesmere Ice Shelf", which up until the early twentieth century fringed the northern coast of Ellesmere Island from Cape Colan to Nansen Sound. These deep freshwater ecosystems were most likely lost following calving or draining through cracks in the former ice shelf. This latter case was documented for the Disraeli Fiord epishelf lake, where the freshwater layer drained through a fracture in the adjacent Ward Hunt Ice Shelf between 1999 and 2002 (Vincent et al. 2001; Mueller et al. 2003a, b), although the ice shelf remained in place. In contrast, numerous deep epishelf lakes exist in Antarctica such as at the edge of the Amery Ice Shelf (Beaver Lake; Laybourn-Parry et al. 2006), adjacent to the Shackleton Ice Shelf in the Bunger Hills Oasis, East Antarctica (including White Smoke Lake; Doran et al. 2000; Gibson and Andersen 2002) and between George VI Ice Shelf and Alexander Island on the western side of the Antarctic Peninsula (Ablation Lake and Moutonnée Lake; Heywood 1977; Hodgson et al. 2004a; Smith et al. 2006).

Shallow 'ice-dammed' lakes, with a freshwater layer of less than 5 m depth, also exist along the northern coast of Ellesmere Island and are intermediate systems between deep epishelf lakes with a complete ice shelf dam and open fiords without any ice dam. Some of the shallow ice-dammed lakes including Taconite Inlet, Ayles Fiord and the Serson ice-dammed lake, cannot accumulate a thick freshwater layer because their ice shelf dam is incomplete and/or they are impounded by relatively thin, multiyear landfast sea ice. A total of eight of these lakes were still present along the northern coast of Ellesmere Island as detected by RADARSAT-1 synthetic aperture radar imagery in the winter of 2007 (Veillette et al. 2008).

The epishelf lake in Milne Fiord is characterized by a 16 m thick freshwater layer overlying the marine layer beneath (Fig. 9.8). It is dammed by the Milne Ice Shelf, the second largest of the Ellesmere ice shelves with a length of 20 km. The central unit of this ice shelf is influenced by glacial ice and has been estimated in the past to be up to 100 m thick (Narod et al. 1988). However, measurements of the depth of the epishelf lake behind this ice shelf, suggest the Milne Ice Shelf has a minimum draft of 16 m. The persistence of the Milne Ice Shelf and its epishelf lake is likely aided by the long, protected fiord in which the ice shelf is located. Furthermore, the ice shelf is glacially thickened by inflowing ice from both sides, which keeps the ice shelf pinned in place (Veillette et al. 2008).

Profiling of Milne Fiord epishelf lake and several shallow ice-dammed systems in the Arctic shows that there is inter-annual variation in their freshwater layers, with slight differences in upper layer thickness and salinity. Salinities in the freshwater layers of these ice dammed lakes ranges from 0.6–3 ppt, and seasonal variations, as seen in Markham and Ayles Fiord, likely reflect ice melt and runoff during the summers, whereas salinities below the halocline are constant and match Arctic Ocean values (Ludlam 1996; Veillette et al. 2008).

Temperatures in the seawater layer of the water column range from −1.5°C to 0.3°C whereas the freshwater layer temperature varies from 0°C to 3.6°C. Summer temperature profiles are often complex with two temperature maxima. The first maximum is located below the ice cover as result of heating by solar radiation. The second, less pronounced temperature peak is found just above the thermocline and is caused by heat released by the formation of frazil ice in supercooled water at the bottom of the freshwater layer. This frazil ice then floats up and contributes to the ice cover of

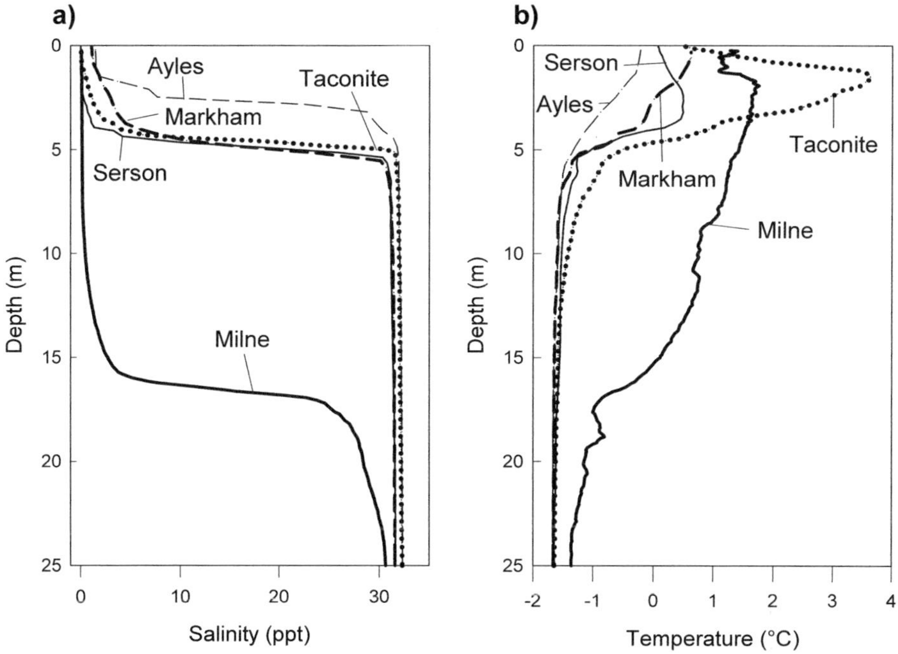

Fig. 9.8 Stratification of the Milne Fiord epishelf lake and ice-dammed lakes in Serson Bay, Taconite Inlet, Ayles Fiord and Markam Fiord. (**a**) Salinity profiles; (**b**) temperature profiles. Profiles were taken between 2004 and 2007 (From Veillette et al. (2008), Copyright 2008 American Geophysical Union. Reproduced with permission of American Geophysical Union)

the epishelf lake (Keys 1977). The correspondence between halocline and thermocline depths in Milne Fiord, as well as Serson Bay, Taconite Inlet and Ayles and Markham fiords is linked to the presence of cooler seawater (Keys 1977) that acts as a heat sink and prevents the freshwater layer from reaching higher temperatures.

In the long-term, ice-dammed lake ecosystems of Ellesmere Island could regain their original size and depths via a thickening of multiyear landfast sea ice; however, this is not likely to occur as Arctic ice shelf and multiyear landfast sea ice loss appear to be irreversible for the foreseeable future (Copland et al. 2007; Mueller et al. 2008; White et al. 2015). Therefore, epishelf lakes are highly vulnerable to climate change and can serve as sentinel ecosystems for the monitoring of regional and global climate change because their presence is entirely dependent on ice shelf integrity and their depths provide a proxy measure of the minimum draft of their ice dam.

9.5.2 Biology of Epishelf Lakes

Scientific knowledge of Arctic epishelf and ice-dammed lake biology is limited to only a few published studies. Nutrient concentrations in the former freshwater layer of Disraeli Fiord and Taconite Inlet were oligotrophic to ultra-oligotrophic based on total phosphorus, and values in the deeper layer seawater layer were also in the oligotrophic range (Van Hove et al. 2006). DIC increased from 10 mg L^{-1} in the freshwater layer of Disraeli Fiord to concentrations of 25 mg L^{-1} in the marine layer, whereas DOC concentrations remained low (0.3–0.8 mg L^{-1}) throughout the entire water column (Van Hove et al. 2001). A chlorophyll *a* maximum of 0.3 μg L^{-1} was found in the upper section of the freshwater layer of Disraeli Fiord, prior to its loss in 2002. This peak corresponded with the sub-ice temperature maximum and is comparable with the lower limit of chlorophyll *a* concentration measured in the central Arctic Ocean (Wheeler et al. 1996). The highest abundance of picocyanobacteria corresponded with the chlorophyll *a* maximum in Disraeli Fiord and Taconite Inlet with concentrations of up to 18×10^3 cells ml^{-1} and 6×10^3 cells ml^{-1}, respectively (Van Hove et al. 2008).

Initial 16S rRNA gene analyses of picocyanobacterial communities in Taconite Inlet and Disraeli Fiord identified one phylogenetic group (referred to in the molecular ecology literature as an Operational Taxonomic Unit, OTU), related to *Synechococcus*. This OTU was also identified in freshwater ecosystems in nearby lakes such as lakes A, C1, C2 and C3 and is characteristic of water layers around or under the oxycline of the different High Arctic lakes, which could mean that it is a microbial generalist with microaerophlic and haltotolerant characteristics (Van Hove et al. 2008). In a phylogenetic analysis, this OTU clustered with strains of temperate origin and OTUs from the Beaufort Sea that had been attributed to riverine inputs.

A more detailed analysis of the microbiota in an epishelf lake was undertaken in the Milne Fiord system (Veillette et al. 2011). Pigment profiles showed that there were pronounced floristic differences in the phytoplankton through its highly stratified water column, with chlorophytes dominating in the freshwater layer, prasinophytes in the halocline, and mainly fucoxanthin groups in the bottom marine layer. Detailed microscopy analysis of their freshwater layer at 5 m identified Chlorophyceae (e.g. *Chlamydomonas*), Cryptophyceae (cf. *Rhodomonas*), Dinophyceae (*Gymnodinium*, *Peridinium*) Choanoflagellates (cf. *Monosiga*) as well as various other unidentified autotrophs and heterotrophs. In the work by Veillette et al. (2011) it was also possible to retrieve 18S rRNA gene sequences from the freshwater layer of Milne Fiord, which had highest similarities to ciliates, stramenophiles, euglenozoa, and fungi. Eukaryotes in the upper layer were freshwater taxa while only marine Archaea were retrieved from below the halocline. Bacteria belonging to Alpha-, Beta, and Gammaproteobacteria as well as Actinobacteria, Bacteroidetes, and Planctomycetes were identified in the epishelf lake. Many of these bacterial taxa were characteristic of cold, freshwater environments, while bacterial genotypes known from the Pacific and Arctic oceans were found in the marine layer. Similarly, molecular analyses of T-4 like bacteriophages showed completely different viral assemblages in the upper and lower layers, further indicating the biologically diverse, highly stratified nature of epishelf lake ecosystems.

The zooplankton assemblages in the freshwater layer of Disraeli Fiord contained different growth stages for the species *Limnocalanus macrurus*, *Drepanopus bungei*, *Oithano similis* and *Oncaea borealis* (Van Hove et al. 2001). However, the genus

Calanus, which generally dominates zooplankton assemblages in the Arctic Ocean and has been found in nearby Nansen Sound (Cairns 1967; Thibault et al. 1999), was absent from the Disraeli Fiord sample. Due to the presence of *L. macrurus* and *D. bungei*, which are common in High Arctic lakes such as Lake A on Ellesmere Island, and the absence of *Calanus,* the Disraeli Fiord epishelf lake ecosystem provides an intermediate state between freshwater and marine zooplankton assemblages. Additional research efforts are needed to fully understand the significance of this unique ecosystem type, before it is lost due to the continued breakup of Arctic ice shelves.

9.6 Wider Applications of the Study of Ice Shelf Ecosystems

9.6.1 Astrobiology

The widespread distribution of microbial communities in cold habitats such as ice shelf ecosystems makes them of great interest for the reconstruction of microbial life and diversification on early Earth (Vincent and Howard-Williams 2000; Vincent et al. 2004). Polar microbes, including cyanobacteria, are also of interest to astrobiologists studying the prospects for life beyond our planet as ice-based ecosystems have been proposed as an analogue to an early stage of the development of Mars and Jupiter's frozen moon Europa. On Mars, where liquid water occurred long ago, it is proposed that life could have evolved at a similar time to the development of cyanobacteria on early Earth (Friedmann 1986). An understanding of conditions required for life to originate and evolve in the cryosphere assists in interpreting biogeochemical "fingerprints of life" in the study of life beyond Earth.

There is also evidence that the Precambrian biosphere experienced extreme low temperature conditions several times during the Paelo- and Neoproterozoic. Ice-based habitats with their sustainable microbial communities are thought to be potential analogues for biotopes present during these major glaciation events, such as the hypothesized "Snowball Earth" (Vincent and Howard-Williams 2000; Vincent et al. 2004). Fossil records suggest that cyanobacteria, in particular oscillatorian taxa, were present throughout these Proterozoic events, and perhaps during earlier periods of global cooling (Schopf and Walter 1982). The cold tolerance combined with growth optima at higher temperatures found in polar oscillatorian cyanobacteria would present ideal characteristics to survive global "ice house/hothouse" cycles during the Proterozoic (Tang et al. 1997). Furthermore, cyanobacteria and other microorganisms are able to maintain prolonged dormancy under freeze-up conditions as well as have a variety of adaptive mechanisms to withstand high UV radiation, desiccation and hypersaline conditions (Vincent 2009; Zakhia et al. 2008). Therefore, cyanobacteria-dominated mats in supraglacial cryoconite holes and other ice-based ecosystems would have presented a potential refuge for survival, growth and diversification for a variety of organisms, including multicellular eukaryotes with protection from the extreme low temperature conditions provided by the vertical stratification (Vincent and Howard-Williams 1989; Vincent 2000b; Vincent et al. 2004; Hoffman 2016). Furthermore, due to the large micro-and nanoscale variations in microhabitat properties, these ecosystems could have been conducive to evolutionary diversification due to inter- and intra-species interaction such as competition for resources and communication as well as homologous and heterologous gene transfer.

9.6.2 Biotic-Physical Interactions in the Cryosphere

Arctic ice shelves, with their surface marked by microbial mats covering vast areas or in more patchy distributions (Mueller et al. 2006), provide an intriguing opportunity to evaluate the influence of microorganisms on their physical environment as well as the reverse. Diverse life tolerates and even thrives in the environmental conditions found on Arctic ice shelves but the presence of these microbiota also impacts the ice shelf surface geomorphology. One example of this biotic-physical coupling is the effect microbial mats have on the inversion of topography in certain areas of the ice shelf (Fig. 9.8). Where microbial mat cover is extensive and meltwater ponds are less regularly shaped, there is a progression of the relative surface elevation due to differences in albedo between mats, water, ice and snow (Schraeder 1968). Frozen meltwater ponds with benthic microbial mats accumulate snow preferentially due to their lower relative elevation via wind re-distribution. In spring and summer, the lower albedo of the surrounding ice causes melting of the more exposed and higher 'icescape' features, leading to a plateau of snow and candled ice (the former meltwater pond) and new meltwater ponds in place of the former shoreline (Mueller and Vincent 2006). Microbial mats contain exopolymeric substances (EPS), a type of mucilage that enhances their cohesion (Stal 2003). They then tend to slide down-slope *en masse* from the plateau areas towards the meltwater ponds. This cycle, known as 'inversion of relief' (Smith 1961), appears to repeat itself although not necessarily on an annual basis. This geomorphological process affects the surface ablation of the shelf in areas where it occurs and it is doubtful that it would continue without the cohesive qualities provided by the ice shelf microbial mats (Mueller and Vincent 2006). Similarly, Yallop et al. (2012) identified an ice algae community that grew directly on the bare ice of the Greenland Ice Sheet that appeared to affect the melting of the ice due to their highly pigmented biomass.

9.6.3 Ecosystems on the Brink of Extinction

The ice shelf surface where microbial mats are found and the epishelf lakes that harbour freshwater microbial and zooplankton communities are ice-dependent ecosystems. These habitats are being extirpated due to the loss of Arctic ice shelves (Mueller et al. 2008). The scientific exploration of these ecosystems is in its infancy

Fig. 9.9 Microbial mat-mediated 'inversion of relief' on the Ward Hunt Ice Shelf. The photographs taken in: (**a**) August 2001 and (**b**) July 2002 show a change in surface cover (from microbial mat to snow and vice versa) and relative elevation over time (Modified from Mueller and Vincent (2006), © John Wiley & Sons. Ltd., used with permission)

and therefore it is not possible to fully gauge the implications of their imminent loss. To date, no endemic organisms have been described in these habitats, but these environments are unique and so this possibility cannot be ruled out. Each taxon found in the microbial mats and epishelf lakes has also been found in analogous habitats in the region (Van Hove et al. 2001; Mueller et al. 2006; Bottos et al. 2008); however, the consortia that are found on the ice shelf surface appear to be unique to Arctic ice shelves (Vincent et al. 2000). Habitat fragmentation does not appear to influence the biodiversity of algal taxa in ice shelf microbial mats (Mueller et al. 2006), but the ongoing loss of ice shelf habitat is leading to the extinction these cryo-ecosystems along with their unique microbial consortia. Given the high tolerance to cold and light stresses in these microbiota and the suite of pigments and other compounds that they can synthesize (Mueller et al. 2005), including bioactive compounds, the loss of their cryo-habitat could also have implications for bioprospecting for the biotechnology and biomedical industries (Priscu and Christner 2004). As analogues for life on Snowball Earth (Vincent et al. 2000; Vincent and Howard-Williams 2000) and model systems for investigating physical-biotic interactions, the extinction of these ecosystems also represents loss of potential scientific knowledge.

In a changing environment, some ecosystems may be eliminated, but others are created. Over recent summers (2005, 2007 and 2008) there have been extensive open water areas for periods of up to five weeks along the northern coast of Ellesmere Island (Fig. 9.9; Copland et al. 2007; Mueller et al. unpublished). Little is known about the characteristics of this novel habitat or the implications of its recent expansion. The North Water Polynya (in northern Baffin Bay) is the nearest permanent open water area to the northern coast of Ellesmere. This is an exceedingly productive area and may be, in a limited sense, an analogue for the nascent flaw lead polynya north of Ellesmere Island. These open water areas also facilitate the calving of ice shelves by removing the ice pressure that kept potentially broken pieces of ice shelf in place (Mueller et al. 2008; Copland et al. 2017). The development of new open water ecosystems along the northern coast of Ellesmere Island will likely impact coastal environment such as ice push from pack ice, disruption and thawing of permafrost, and wave-induced coastal erosion.

9.7 Conclusions

Until recently, Arctic ice shelves were seen as mostly abiotic glaciological features. However, in recent years it has become clear that their aquatic ecosystems are oases for life. Arctic ice shelves are now seen as dynamic structures that harbour a range of ecosystems with diverse biology. They contain ecosystems with different physical and chemical properties located on the surface and within the ice shelves. Based on current knowledge, much of the biomass is dominated by microbial assemblages, with the microbiota attaining biomass stocks comparable to lakes in temperate climatic zones. They represent analogues for the survival, growth and diversification of life on early Earth, and models for understanding properties of life beyond Earth. All the ice shelf ecosystems reviewed here are completely dependent on the continuous presence of ice, and are therefore highly vulnerable to ongoing climate change. With the accelerated break-up and loss of Arctic ice shelves (Mueller et al. 2008; Vincent et al. 2009), there may be little time left to explore their remarkable biology.

Acknowledgements This work has been supported by the Natural Sciences and Engineering Research Council of Canada, the Canada Research Chair program, le Fonds québécois de la recherche sur la nature et les technologies, the Northern Scientific Training Program, and the Network of Centre of Excellence program ArcticNet, with logistics support from the Polar Continental Shelf Program and Parks Canada. We thank Dominic Hodgson and an anonymous reviewer for their insightful comments on the manuscript.

References

ACIA. (2005). *Arctic climate impact assessment* (pp. 1042). Cambridge: Cambridge University Press.

Amato, P., Hennebelle, R. I., Maganda, O., Sancelme, M., Delort, A.-M., Barbante, C., Boutron, C., & Ferrari, C. (2007). Bacterial characterization of the snow cover at Spitzberg, Svalbard. *FEMS Microbiology Ecology, 59*, 255–264.

Archer, S., McDonald, I., Herbold, C., & Cary, S. (2014). Characterization of bacterioplankton communities in the meltwater ponds of Bratina Island, Victoria Land, Antarctica. *FEMS Microbiology Ecology, 89*, 451–464.

Archer, D. J., McDonald, I. R., Herbold, C. W., Lee, C. K., & Cary, C. S. (2015). Benthic microbial communities of coastal terrestrial and ice shelf Antarctic meltwater ponds. *Frontiers in Microbiology, 6*, 485. doi:10.3389/fmicb.2015.00485.

Boetius, A., Anesio, A. M., Deming, J. W., Mikucki, J. A., & Rapp, J. Z. (2015). Microbial ecology of the cryosphere: Sea ice and glacial habitats. *Nature Reviews Microbiology, 13*, 677–690.

Bonilla, S., Villeneuve, V., & Vincent, W. F. (2005). Benthic and planktonic algal communities in a High Arctic lake: Pigment structure and contrasting responses to nutrient enrichment. *Journal of Phycology, 41*, 1120–1130.

Bonilla, S., Rautio, M., & Vincent, W. F. (2009). Phytoplankton and phytobenthos pigment strategies: Implications for algal survival in the changing Arctic. *Polar Biology, 28*, 846–861.

Bottos, E. M., Vincent, W. F., Greer, C. W., & Whyte, L. G. (2008). Prokaryotic diversity of arctic ice shelf microbial mats. *Environmental Microbiology, 10*, 950–966.

Brinkmeyer, R., Knittel, K., Jürgens, J., Weyland, H., Amann, R., & Helmke, E. (2003). Diversity and structure of bacterial communities in Arctic versus Antarctic pack ice. *Applied and Environmental Microbiology, 69*, 6610–6619.

Cairns, A. A. (1967). The zooplankton of Tanquary Fjord, Ellesmere Island, with special reference to calanoid copepods. *Journal of Fisheries Research Board of Canada, 24*, 555–568.

Cameron, K. A., Hodson, A. J., & Osborn, A. M. (2012). Structure and diversity of bacterial, eukaryotic and archaean communities in glacial cryoconite holes from the Arctic and the Antarctic. *FEMS Microbiology Ecology, 82*, 254–267.

Castenholz, R. W. (1992). Species usage, concept, and evolution in the cyanobacteria (blue-green algae). *Journal of Phycology, 28*, 737–745.

Chrismas, N. A. M., Anesio, A. M., & Sánchez-Baracaldo, P. (2015). Multiple adaptations to polar and alpine environments within cyanobacteria: A phylogenomic and Bayesian approach. *Frontiers in Microbiology, 6*, 1070.

Christner, B. C., Cai, R., Morris, C., McCarter, K. S., Foreman, C. M., Skidmore, M. L., Montross, S. N., & Sands, D. C. (2008). Geographic, seasonal, and precipitation chemistry influence on the abundance and activity of biological ice nucleators in rain and snow. *Proceedings of the National Academy of Sciences of the United States of America, 105*, 18854–18859.

Copland, L., Mueller, D. R., & Weir, L. (2007). Rapid loss of the Ayles Ice Shelf, Ellesmere Island, Canada. *Geophysical Research Letters, 34*, L21501.

Copland, L., Mortimer, C., White, A., Richer McCallum, M., & Mueller, D. (2017). Factors contributing to recent Arctic ice shelf losses. In L. Copland & D. Mueller (Eds.), *Arctic ice shelves and ice islands* (p. 263–285). Dordrecht: Springer. doi:10.1007/978-94-024-1101-0_10.

Crary, A. P., Kulp, J. L., & Marshall, E. W. (1955). Evidences of climatic change from ice island studies. *Science, 122*, 1171–1173.

de Mora, S. J., Whitehead, R. F., & Gregory, M. (1994). The chemical composition of glacial melt water ponds on the McMurdo Ice Shelf, Antarctica. *Antarctic Science, 6*, 17–27.

Doran, P. T., Wharton Jr., R. A., Lyons, J. B., Des Marais, D. J., & Andersen, D. T. (2000). Sedimentology and geochemistry of a perennially ice-covered epishelf lake in Bunger Hills Oasis, East Antarctica. *Antarctic Science, 11*, 131–140.

Edwards, A., Douglas, B., Anesio, A. M., Rassner, S. M., Irvine-Flynn, T. D. L., Sattler, B., & Griffith, F. W. (2013). A distinctive fungal community inhabiting cryoconite holes on glaciers in Svalbard. *Fungal Ecology, 6*(2), 168–176.

Ehling-Schulz, M., & Scherer, S. (1999). UV protection in cyanobacteria. *European Journal of Phycology, 34*, 329–338.

Friedmann, E. I. (1986). The Antarctic cold desert and the search for traces of life on Mars. *Advances in Space Research, 6*, 265–268.

Fritsen, C. H., & Priscu, J. C. (1998). Cyanobacterial assemblages in permanent ice covers on Antarctic lakes: Distribution, growth rate, and temperature response of photosynthesis. *Journal of Phycology, 34*, 587–597.
Garcia-Pichel, F., & Castenholz, R. W. (1991). Characterization and biological implications of scytonemin, a cyanobacterial sheath pigment. *Journal of Phycology, 27*, 395–409.
Gibson, J. A. E., & Andersen, D. T. (2002). Physical structure of epishelf lakes of the southern Bunger Hills, East Antarctica. *Antarctic Science, 14*, 253–261.
Harding, T., Jungblut, A. D., Lovejoy, C., & Vincent, W. F. (2011). Microbes in high Arctic snow and implications for the cold biosphere. *Applied and Environmental Microbiology, 77*, 3234–3243.
Hawes, I., Smith, R., Howard-Williams, C., & Schwarz, A. M. (1999). Environmental conditions during freezing, and response of microbial mats in ponds of the McMurdo Ice Shelf, Antarctica. *Antarctic Science, 11*, 198–208.
Hawes, I., Howard-Williams, C., & Fountain, A. G. (2008). Ice-based freshwater ecosystems. In W. F. Vincent & J. Laybourn-Parry (Eds.), *Polar lakes and rivers – limnology of Arctic and Antarctic aquatic ecosystems* (p. 103–118). Oxford: Oxford University Press.
Heywood, R. B. (1977). A limnological survey of the ablation point area, Alexander Island, Antarctica. *Philosophical Transactions of the Royal Society B: Biological Sciences, 279*, 39–54.
Hodgson, D., Gibson, J., & Doran, P. T. (2004a). Antarctic paleolimnology. In R. Pienitz, M. S. V. Douglas, & J. P. Smol (Eds.), *Long-term environmental change in Arctic and Antarctic lakes* (p. 419–474). Dordrecht: Springer.
Hodgson, D., Vyverman, W., Verleyen, E., Sabbe, K., Leavitt, P., Taton, A., Squier, A., & Keely, B. (2004b). Environmental factors influencing the pigment composition of in situ benthic microbial communities in East Antarctic lakes. *Aquatic Microbial Ecology, 37*, 247–263.
Hoffman, P. F. (2016). Cryoconite pans on snowball earth: Supraglacial oases for Cryogenian eukaryotes? *Geobiology, 14*, 531–542.
Holdsworth, G. (1987). The surface waveforms on the Ellesmere Island ice shelves and ice islands. In *Workshop on Extreme Ice Features, Banff, Alberta, November 3–5, 1986, National Research Council of Canada* (p. 385–403).
Howard-Williams, C., Pridmore, R. D., Broady, P. A., & Vincent, W. F. (1990). Environmental and biological variability in the McMurdo Ice Shelf ecosystem. In K. R. Kerry & G. Hempel (Eds.), *Antarctic ecosystems: Ecological change and conservation* (p. 23–31). Berlin: Springer.
Jeffries, M. O. (1992). Arctic ice shelves and ice islands: Origin, growth and disintegration, physical characteristics, structural-stratigraphic variability, and dynamics. *Reviews of Geophysics, 30*, 245–267.
Jeffries, M. O. (2017). The Ellesmere ice shelves, Nunavut, Canada. In L. Copland & D. Mueller (Eds.), *Arctic ice shelves and ice islands* (p. 23–54). Dordrecht: Springer. doi:10.1007/978-94-024-1101-0_2.
Jungblut, A., Hawes, I., Mountfort, D., Hitzfeld, B., Dietrich, D., Burns, B., & Neilan, B. (2005). Diversity within cyanobacterial mat communities in variable salinity meltwater ponds of McMurdo Ice Shelf, Antarctica. *Environmental Microbiology, 7*, 519–529.
Jungblut, A., Lovejoy, C., & Vincent, W. F. (2010). Global distribution of cyanobacterial ecotypes in the cold biosphere. *The ISME Journal, 4*, 191–202.
Jungblut, A. D., Vincent, W. F., & Lovejoy, C. (2012). Eukaryotes in Arctic and Antarctic cyanobacterial mats. *FEMS Microbiology Ecology, 82*, 416–428.
Keys J. E. (1977). *Water regime of ice-covered fiords and lakes*. Ph.D. Thesis, Marine Sciences Centre, McGill University, Montreal, pp. 75
Keys J. E. (1978). *Water regime of Disraeli Fiord, Ellesmere Island Report Number 792*. Ottawa: Defence Research Establishment Ottawa, pp. 58
Laybourn-Parry, J. (2002). Survival mechanisms in Antarctica lakes. *Philosophical Transactions of the Royal Society B: Biological Sciences, 357*, 863–869.

Laybourn-Parry, J., Madan, N. J., Marshall, W. A., Marchant, H. J., & Wright, S. W. (2006). Carbon dynamics in an ultra-oligotrophic epishelf lake (Beaver Lake, Antarctica) in summer. *Freshwater Biology, 51*, 1116–1130.

Lionard, M., Péquin, B., Lovejoy, C., & Vincent, W. F. (2012). Benthic cyanobacterial mats in the high Arctic: Multi-layer structure and fluorescence responses to osmotic stress. *Frontiers in Aquatic Microbiology, 3*, 140. doi:10.3389/fmicb.2012.00140.

Ludlam, S. D. (1996). Stratification patterns in Taconite Inlet, Ellesmere Island, N.W.T. *Journal of Paleolimnology, 16*, 205–215.

Mader, H. M., Pettitt, M., Wadham, J. L., Wolff, E., & Parkes, J. (2006). Subsurface ice as a microbial habitat. *Geology, 34*, 169–172.

McKnight, D. M., Boyer, E. W., Westerhoff, P. K., Doran, P. T., Kulbe, T., & Andersen, D. T. (2001). Spectrofluorometric characterization of dissolved organic matter for indication of precursor organic material and aromaticity. *Limnology and Oceanography, 46*, 38–48.

Miteva, V. I., Sheridan, P. P., & Brenchley, J. E. (2004). Phylogenetic and physiological diversity of microorganisms isolated from a deep Greenland glacier ice core. *Applied and Environmental Microbiology, 70*, 202–213.

Mountfort, D., Kaspar, H. F., Downes, M. T., & Asher, R. (1999). Partitioning effects during terminal carbon and electron flow in sediments of a low-salinity meltwater pond near Bratina Island, McMurdo Ice Shelf, Antarctica. *Applied and Environmental Microbiology, 65*, 5493–5499.

Mountfort, D., Kaspar, H., Asher, R., & Sutherland, D. (2003). Influences of pond geochemistry, temperature, and freeze-thaw on terminal anaerobic processes occurring in sediments of six ponds of the McMurdo Ice Shelf, near Bratina Island, Antarctica. *Applied and Environmental Microbiology, 69*, 583–592.

Mueller, D. R., & Vincent, W. F. (2006). Microbial habitat dynamics and ablation control on the Ward Hunt Ice Shelf. *Hydrological Processes, 20*, 857–876.

Mueller, D. R., Vincent, W. F., Pollard, W. H., & Fritsen, C. H. (2001). Glacial cryoconite ecosystems: A bipolar comparison of algal communities and habitats. *Nova Hedwigia, Beiheft, 123*, 173–197.

Mueller, D. R., Jeffries, M. O., & Vincent, W. F. (2003a). Ice shelf break-up and ecosystem loss in the Canadian High Arctic. *Eos, Transactions of the American Geophysical Union, 84*, 548,552.

Mueller, D. R., Vincent, W. F., & Jeffries, M. O. (2003b). Break-up of the largest Arctic ice shelf and associated loss of an epishelf lake. *Geophysical Research Letters, 30*, 2031.

Mueller, D. R., Vincent, W. F., Bonilla, S., & Laurion, I. (2005). Extremotrophs, extremophiles and broadband pigmentation strategies in a high Arctic ice shelf ecosystem. *FEMS Microbiology Ecology, 53*, 73–87.

Mueller, D. R., Vincent, W. F., & Jeffries, M. O. (2006). Environmental gradients, fragmented habitats and microbiota of a northern ice shelf cryoecosystem, Ellesmere Island, Canada. *Arctic, Antarctic, and Alpine Research, 38*, 593–607.

Mueller, D. R., Copland, L., Hamilton, A., & Stern, D. R. (2008). Examining Arctic ice shelves prior to 2008 breakup. *Eos, Transactions of the American Geophysical Union, 89*, 502–503.

Müller, T., Bleiß, W., Martin, C.-D., Rogaschewski, S., & Fuhr, G. (1998). Snow algae from Northwest Svalbard: Their identification, distribution, pigment and nutrient content. *Polar Biology, 20*, 14–23.

Narod, B. B., Clarke, G. K. C., & Prager, B. T. (1988). Airborne UHF radar sounding of glaciers and ice shelves, northern Ellesmere Island, Arctic Canada. *Canadian Journal of Earth Sciences, 25*, 95–105.

Oren, A. (2000). Salt and brines. In B. A. Whitton & M. Potts (Eds.), *The ecology of cyanobacteria: Their diversity in time and space* (p. 281–306). Dordrecht: Kluwer Academic Publishers.

Pearl, H. W., & Pinckney, J. L. (1996). A mini-review of microbial consortia: Their role in aquatic production and biogeochemical cycling. *Microbial Ecology, 31*, 225–247.

Pointing, S. B., Büdel, B., Convey, P., Gillman, L., Körner, C., Leuzinger, S., & Vincent, W. F. (2015). Biogeography of photoautotrophs in the high polar biome. *Frontiers in Plant Science, 6*, 692. doi:10.3389/fpls.2015.00692.

Price, P. B. (2007). Microbial life in glacial ice and implications for a cold origin of life. *FEMS Microbiology Ecology, 59*, 217–231.

Price, P. B., & Sowers, T. (2004). Temperature dependence of metabolic rates for microbial growth, maintenance, and survival. *Proceedings of the National Academy of Sciences of the United States of America, 101*, 4631–4636.

Priscu, J. C., & Christner, B. C. (2004). Earth's icy biosphere. In A. Bull (Ed.), *Microbial diversity and bioprospecting* (p. 130–145). Washington, DC: American Society for Microbiology.

Quesada, A., & Vincent, W. F. (1997). Strategies of adaptation by Antarctic cyanobacteria to ultraviolet radiation. *European Journal of Phycology, 32*, 335–342.

Quesada, A., Vincent, W. F., & Lean, D. R. S. (1999). Community and pigment structure of Arctic cyanobacterial assemblages: The occurrence and distribution of UV-absorbing compounds. *FEMS Microbiology Ecology, 28*, 315–323.

Rivkina, E. M., Friedmann, E. I., McKay, C. P., & Gilichinsky, D. A. (2000). Metabolic activity of permafrost bacteria below the freezing point. *Applied and Environmental Microbiology, 66*, 3230–3233.

Roos, J. C., & Vincent, W. F. (1998). Temperature dependence of UV radiation effects on Antarctic cyanobacteria. *Journal of Phycology, 34*, 118–125.

Schopf, J. W., & Walter, M. R. (1982). Origin and early evolution of cyanobacteria: The geological evidence. In G. Carr & B. A. Whitton (Eds.), *The biology of cyanobacteria* (p. 543–564). Oxford: Blackwell Scientific Publisher.

Schraeder, R. L. (1968). *Ablation of Ice Island ARLIS II, 1961.* M.Sc. Thesis, Department of Geology, University of Alaska, College, Fairbanks, pp. 59

Sjöling, S., & Cowan, D. A. (2003). High 16S rDNA bacterial diversity in glacial meltwater lake sediment, Bratina Island, Antarctica. *Extremophiles, 7*, 275–282.

Smith, D. D. (1961). Sequential development of surface morphology on Fletcher's Ice Island, T-3. In G. O. Raasch (Ed.), *Geology of the Arctic* (p. 896–914). Toronto: University of Toronto Press.

Smith, J. A., Hodgson, D., Bentley, M. J., Verleyen, E., Leng, M. J., & Roberts, S. J. (2006). Limnology of two Antarctic epishelf lakes and their potential to record periods of ice shelf loss. *Journal of Paleolimnology, 35*, 373–394.

Squier, A. H., Airs, R. L., Hodgson, D. A., & Keely, B. J. (2004). Atmospheric pressure chemical ionisation liquid chromatography/mass spectrometry of the ultraviolet screening pigment scytonemin: Characteristic fragmentations. *Rapid Communications in Mass Spectrometry, 18*, 2934–2938.

Stal, L. (2000). Cyanobacterial mats and stromatolites. In B. A. Whitton & M. Potts (Eds.), *The ecology of cyanobacteria: Their diversity in time and space* (p. 61–120). Dordrecht: Kluwer Academic Press.

Stal, L. (2003). Microphytobenthos, their extracellular polymeric substances, and the morphogenesis of intertidal sediments. *Geomicrobiology Journal, 20*, 463–478.

Tang, E. P. Y., Tremblay, R., & Vincent, W. F. (1997). Cyanobacteria dominance of polar freshwater ecosystems: Are high-latitude mat-formers adapted to low temperature? *Journal of Phycology, 33*, 171–181.

Thibault, D., Head, E. J. H., & Wheeler, P. A. (1999). Mesozooplankton in the Arctic Ocean in summer. *Deep Sea Research: Part I - Oceanographic Research Papers, 46*, 1391–1415.

Van Hove, P., Swadling, K., Gibson, J. A. E., Belzile, C., & Vincent, W. F. (2001). Farthest north lake and fjord populations of calanoid copepods *Limnocalanus macrurus* and *Drepanopus bungei* in the Canadian High Arctic. *Polar Biology, 24*, 303–307.

Van Hove, P., Belzile, C., Gibson, J. A. E., & Vincent, W. F. (2006). Coupled landscape-lake evolution in the coastal High Arctic. *Canadian Journal of Earth Sciences, 43*, 533–546.

Van Hove, P., Vincent, W. F., Galand, P. E., & Wilmotte, A. (2008). Abundance and diversity of picocyanobacteria in High Arctic lakes and fjords. *Algological Studies, 126*, 209–227.
Van Trappen, S., Mergaert, J., Van Eygen, S., Dawyndt, P., Cnockaert, M. C., & Swings, J. (2002). Diversity of 746 heterotrophic bacteria isolated from microbial mats from ten Antarctic lakes. *Systematic and Applied Microbiology, 25*, 603–610.
Varin, T., Lovejoy, C., Jungblut, A. D., Vincent, W. F., & Corbeil, J. (2010). Metagenomic profiling of Arctic microbial mat communities as nutrient scavenging and recycling systems. *Limnology and Oceanography, 55*, 1901–1911.
Varin, T., Lovejoy, C., Jungblut, A. D., Vincent, W. F., & Corbeil, J. (2012). Metagenomic analysis of stress genes in microbial mat communities from extreme Arctic and Antarctic environments. *Applied and Environmental Microbiology, 78*, 549–559.
Veillette, J., Mueller, D. R., Antoniades, D., & Vincent, W. F. (2008). Arctic epishelf lakes as sentinel ecosystems: Past, present and future. *Journal of Geophysical Research - Biogeosciences, 113*, G04014.
Veillette, J., Lovejoy, C., Potvin, M., Harding, T., Jungblut, A. D., Antoniades, D., Chénard, C., Suttle, C. A., & Vincent, W. F. (2011). Milne Fiord epishelf lake: A coastal Arctic ecosystem vulnerable to climate change. *Ecoscience, 18*, 304–316.
Vézina, S., & Vincent, W. F. (1997). Arctic cyanobacteria and limnological properties of their environment: Bylot Island, Northwest Territories, Canada (73°N, 80°W). *Polar Biology, 17*, 523–534.
Villeneuve, V., Vincent, W. F., & Komárek, J. (2001). Community structure and microhabitat characteristics of cyanobacterial mats in an extreme high Arctic environment: Ward Hunt Lake. *Nova Hedwigia, Beiheft, 123*, 199–224.
Vincent, W. F. (1988). *Microbial ecosystems of Antarctica* (304 pp). Cambridge: Cambridge University Press.
Vincent, W. F. (2000a). Evolutionary origins of Antarctic microbiota: Invasion, detection and endemism. *Antarctic Science, 12*, 374–386.
Vincent, W. F. (2000b). Cyanobacterial dominance in the polar regions. In B. A. Whitton & M. Potts (Eds.), *The ecology of cyanobacteria: Their diversity in time and space* (p. 321–340). Dordrecht: Kluwer Academic Press.
Vincent, W. F. (2007). Cold tolerance in cyanobacteria and life in the cryosphere. In J. Seckbach (Ed.), *Algae and cyanobacteria in extreme environments* (p. 287–301). Heidelberg: Springer.
Vincent, W. F. (2009). Cyanobacteria. In G. E. Likens (Ed.), *Encyclopedia of inland waters* (Vol. 3, p. 55–60). Oxford: Elsevier.
Vincent, W. F., & Howard-Williams, C. (1989). Microbial communities in southern Victoria Land streams (Antarctica). 2. The effects of low temperature. *Hydrobiologia, 172*, 39–49.
Vincent, W. F., & Howard-Williams, C. (2000). Life on snowball earth. *Science, 287*, 2421.
Vincent, W. F., & Neale, P. J. (2000). Mechanisms of UV damage to aquatic organisms. In S. J. de Mora, S. Demers, & M. Vernet (Eds.), *The effects of UV radiation in the marine environment* (p. 149–176). Cambridge: Cambridge University Press.
Vincent, W. F., Castenholz, R. W., Downes, M. T., & Howard-Williams, C. (1993). Antarctic cyanobacteria: Light, nutrients, and photosynthesis in the microbial mat environment. *Journal of Phycology, 29*, 745–755.
Vincent, W. F., Gibson, J. A., Pienitz, R., Villeneuve, V., Broady, P. A., Hamilton, P. B., & Howard-Williams, C. (2000). Ice shelf microbial ecosystems in the high Arctic and implications for life on snowball earth. *Naturwissenschaften, 87*, 137–141.
Vincent, W. F., Gibson, J. A. E., & Jeffries, M. O. (2001). Ice shelf collapse, climate change, and habitat loss in the Canadian High Arctic. *Polar Record, 37*, 133–142.
Vincent, W. F., Mueller, D. R., & Bonilla, S. (2004). Ecosystems on ice: The microbial ecology of Markham Ice Shelf in the high Arctic. *Cryobiology, 48*, 103–112.
Vincent, W. F., Whyte, L. G., Lovejoy, C., Greer, C. W., Laurion, I., Suttle, C. A., Corbeil, J., & Mueller, D. R. (2009). Arctic microbial ecosystems and impacts of extreme warming during the International Polar Year. *Polar Science, 3*, 171–180.

Webster-Brown, J. G., Hawes, I., Jungblut, A. D., Wood, S. A., & Christenson, H. K. (2015). The effects of entombment on water chemistry and bacterial assemblages in closed cryoconite holes on Antarctic glaciers. *FEMS Microbiology Ecology, 91*(12). doi:10.1093/femsec/fiv144.

Wheeler, P. A., Gosselin, N., Sherr, E., Thibault, D., Kirchman, D. L., Benner, R., & Whiteledge, T. E. (1996). Active cycling of organic carbon in the Central Arctic Ocean. *Nature, 380*, 697–699.

White, A., Mueller, D., & Copland, L. (2015). Reconstructing hydrographic change in Petersen Bay, Ellesmere Island, Canada, inferred from SAR imagery. *Remote Sensing of Environment, 165*, 1–13.

Yallop, M. L., Anesio, A. M., Perkins, R. G., Cook, J., Telling, J., Fagan, D., MacFarlane, J., Stibal, M., Barker, G., Bellas, C., Hodson, A., Tranter, M., Wadham, J., & Roberts, N. W. (2012). Photophysiology and albedo-changing potential of the ice algal communities on the surface of the Greenland Ice Sheet. *The ISME Journal, 6*, 2302–2313.

Zakhia, F., Jungblut, A. D., Taton, A., Vincent, W. F., & Wilmotte, A. (2008). Cyanobacteria in cold environments. In R. Margesin, F. Schinner, J. C. Marx, & C. Gerday (Eds.), *Psychrophiles: From biodiversity to biotechnology* (p. 121–135). Heidelberg: Springer.

Part III
Arctic Ice Shelf Calving Processes and Ice Islands

Chapter 10
Factors Contributing to Recent Arctic Ice Shelf Losses

Luke Copland, Colleen Mortimer, Adrienne White, Miriam Richer McCallum, and Derek Mueller

Abstract A review of historical literature and remote sensing imagery indicates that the ice shelves of northern Ellesmere Island have undergone losses during the 1930s/1940s to 1960s, and particularly since the start of the twenty-first century. These losses have occurred due to a variety of different mechanisms, some of which have resulted in long-term reductions in ice shelf thickness and stability (e.g., warming air temperatures, warming ocean temperatures, negative surface and basal mass balance, reductions in glacier inputs), while others have been more important in defining the exact time at which a pre-weakened ice shelf has undergone calving (e.g., presence of open water at ice shelf terminus, loss of adjacent multiyear landfast sea ice, reductions in nearby epishelf lake and fiord ice cover). While no single mechanism can be isolated, it is clear that they have all contributed to the marked recent losses of Arctic ice shelves, and that the outlook for the future survival of these features is poor.

Keywords Ice shelf • Calving • Multiyear landfast sea ice • Mass balance • Climate warming • Glaciers

---

L. Copland (✉) • A. White
Department of Geography, Environment and Geomatics, University of Ottawa, Ottawa, ON, Canada
e-mail: luke.copland@uottawa.ca; awhit059@uottawa.ca

C. Mortimer
Department of Earth and Atmospheric Sciences, University of Alberta, Edmonton, AB, Canada
e-mail: cmortime@ualberta.ca

M. Richer McCallum • D. Mueller
Department of Geography and Environmental Studies, Carleton University, Ottawa, ON, Canada
e-mail: miriam.richer-mccallum@canada.ca; derek.mueller@carleton.ca

L. Copland, D. Mueller (eds.), *Arctic Ice Shelves and Ice Islands*, Springer Polar Sciences, DOI 10.1007/978-94-024-1101-0_10

10.1 Introduction

There have been rapid changes to the Arctic cryosphere over the past decade, with increased melt rates on Canadian Arctic ice caps (Fisher et al. 2012; Sharp et al. 2011), a strongly negative mass balance for the Greenland Ice Sheet (Box 2013), an acceleration and increase in mass loss from many Greenland outlet glaciers (Rignot et al. 2011; Schrama and Wouters 2011), and rapid reductions in sea ice area, age and thickness (Kwok and Rothrock 2009; Yu et al. 2014; Simmonds 2015). Over this same period, there have also been dramatic changes to the ice shelves of northern Ellesmere Island, Nunavut, Canada, such as the complete loss of the 87 km^2 Ayles Ice Shelf in August 2005 (Copland et al. 2007), complete loss of the 50 km^2 Markham Ice Shelf in summer 2008 (Mueller et al. 2008), and ~90% loss of the ~180 km^2 Serson Ice Shelf between 2008 and 2011 (Table 10.1). The ice shelves in this region reduced in number from six to three between 2005 and 2012, with a decrease in area from 1043 km^2 at the start of summer 2005 to ~500 km^2 at the end of summer 2015 (Mueller et al. 2017). These ice shelf losses have produced a large number of ice islands that have typically drifted westward after formation into the

Table 10.1 Timing and size of major calving events from the Ellesmere Island ice shelves, 2000–2012

Ice Shelf	Date	Event	Area loss (km^2)	Area after loss (km^2)	Reference(s)
Ward Hunt	2000–2002	Calving	6	436	Mueller et al. 2003
	July–August 2008	Calving	42	394	Mueller et al. 2008
	August 2010	Calving	54	340	This study
	August 2011	Split in two	39	301	Mueller et al. 2017
	August 2012	Calving from western side	6	295	This study
Ayles	August 13, 2005	Complete loss	87	0	Copland et al. 2007
Markham	August 7–12, 2008	Complete loss	50	0	Mueller et al. 2008
Petersen	August 8–18, 2005	Calving	8.0	41	Pope et al. 2012; White 2012
	August 2008	Calving	9.0	32	White 2012
	Summer 2011	Calving	5.5	25	White 2012
	August 2012	Calving	5.5	19	White 2012
Serson	July 29–31, 2008	Calving	122	77	Mueller et al. 2008
	August 1–8, 2011	Calving	45	32	This study
Milne[a]	September 2009	–	–	205	Mortimer et al. 2012

Adapted from Copland (2009) and White (2012)

[a]Milne Ice Shelf did not experience any significant changes over the period 2000–2012 (Mortimer et al. 2012)

interior parts of the western Queen Elizabeth Islands or into the Beaufort Sea (Van Wychen and Copland 2017), where they may pose a significant risk to offshore infrastructure and shipping (Fuglem and Jordaan 2017).

To help understand the driving mechanisms behind recent ice shelf losses, it is useful to consider the location and timing of past calving events. In this chapter, calving events are defined as the partial or complete loss of an ice shelf, and can typically be identified easily in remote sensing imagery. They differ from breakup events, which represent the fracturing of an ice shelf, and which may occur in situ without the loss of ice shelf area. Due to the difficulty of detecting small-scale breakup events with remote sensing imagery, most of the discussion here is focused on calving.

Frequent repeat imaging using satellite sensors such as the Moderate Resolution Imaging Spectroradiometer (MODIS; ~250 m resolution), together with Synthetic Aperture Radar (SAR) wide swath images from sensors such as Radarsat and Envisat (~100 m resolution), has enabled detailed reconstructions of the timing and patterns of ice shelf calving events since the early 2000s. This imagery indicates that ice shelf losses have occurred from embayments across northern Ellesmere Island (Table 10.1; Fig. 10.1). Indeed, five out of the six ice shelves present at the start of the twenty-first century lost substantial area between 2000 and 2012. The only ice shelf which did not experience a significant change in area over this time was the Milne, which measured 206 km^2 in 2001 and 205 km^2 in 2009 (Mortimer et al. 2012).

Historical records from early explorers such as Lt. Aldrich in 1876 (Nares 1878) and Robert Peary in 1906 (Peary 1907) described a continual ice shelf fringe along the coast of northern Ellesmere Island, with an estimated area of 8900 km^2 at the start of the twentieth century (Vincent et al. 2001). For example, Nares (1878) described the ice shelves as permanent 'hummocky' ice with a rolling surface, and for the Ward Hunt Ice Shelf stated that '*The hummocks do not come in close to Ward Hunt Island… The line of hummocks is between five and six miles off, and does not seem to differ from those farther east.*'. Similarly, for Ayles Fiord, '*The actual line of hummocky ice is still about two miles from shore*'. As of 2015 this fringe is no longer continuous, and instead ice shelves are present in only three fiords (Petersen, Milne and Disraeli/Ward Hunt), totalling ~500 km^2 (see Mueller et al. 2017 for a full review). The twentieth century ice shelf losses were spatially extensive, both in terms of a reduction in the distance that ice shelves extended northwards into the Arctic Ocean, and longitudinally in terms of their distribution across northern Ellesmere Island.

Chronologically, the ice shelves of northern Ellesmere Island have lost area during two main periods: (1) the 1930s/1940s to 1960s (start date is poorly defined due to paucity of data); (2) since the start of the twenty-first century (particularly 2002–2012). During the earlier period, extensive ice islands were produced as the ice shelves calved, such as the loss of the >1300 km^2 ice shelf in Yelverton Bay between the mid-1930s and mid-1940s, which produced ice islands T-1, T-2 and T-3 (Koenig et al. 1952; Jeffries 1987; Pope et al. 2012). These ice islands were the largest ever observed in the Arctic Ocean, with Koenig et al. (1952) stating that T-1 had an area

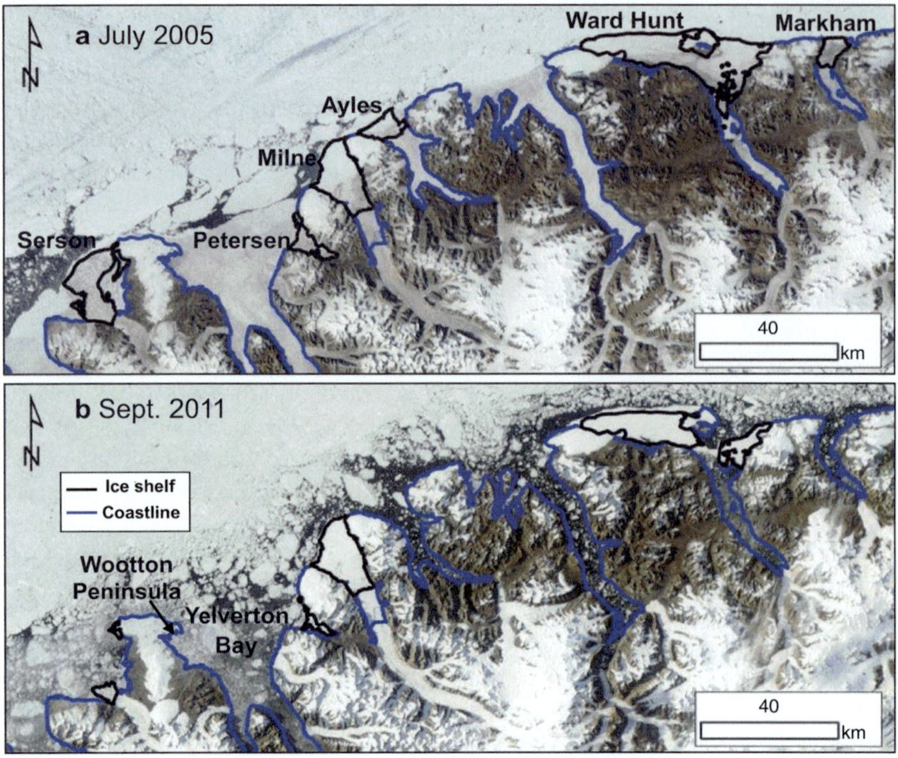

Fig. 10.1 Extent of ice shelves across northern Ellesmere Island in: (**a**) July 2005, and (**b**) September 2011. Losses between these dates have been widespread, with the complete loss of the Serson, Ayles and Markham ice shelves, and substantial reductions in area of the Ward Hunt and Petersen ice shelves (see Table 10.1 for details) (MODIS base imagery courtesy of the Rapid Response Project at NASA/GSFC)

of >200 square nautical miles (nm) (>687 km^2), while T-2 measured 17 × 18 nm and had an area of ~300 nm^2 (1030 km^2). Other evidence for ice shelf losses at this time comes from widespread observations of ice islands in aerial photography and from military flights, including the identification of >28 ice islands by Koenig et al. (1952) and an additional 31 ice islands by Montgomery (1952) (see Van Wychen and Copland 2017 for a full review). After the 1960s the ice shelves of northern Ellesmere Island were generally stable until the start of the twenty-first century, with the exception of relatively small losses from the Ward Hunt Ice Shelf of 35–40 km^2 from 1980 to 1982, and ~40 km^2 in 1982–1983 (Jeffries and Serson 1983). Since the start of the twenty-first century, ice shelf losses have been more extensive again, especially in the summers of 2005, 2008 and 2011, with the total remaining area in 2012 approximately half of the 1043 km^2 present in 2005 (Table 10.1).

Given the patterns described above, the large spatial extent of ice shelf losses implies that contributory factor(s) were likely acting at a regional, and not just local, scale. However, the contributing factor(s) did not occur at a constant rate as losses were dominant in the 1940s–1960s and since the start of the twenty-first century. With these constraints in mind, the following review addresses the main mechanisms that have likely been responsible for the ice shelf losses. This information is required to understand and predict the fate of the remaining Arctic ice shelves, and here we use a combination of field measurements, remote sensing observations and a review of existing literature to present conclusions about the dominant factors which have caused ice shelf losses. Most discussion focuses on calving events that have occurred over the period 2002–2012 due to the improved availability of data from this period compared to earlier times, although discussion of earlier losses is also included where relevant.

10.2 Climate

Major climate assessments have reported significant temperature increases in the Arctic over the past few decades, with the Arctic warming twice as fast as the rest of the planet since 1980, and warming mainly focused in the winter (AMAP 2012). At Eureka, the closest weather station to the Ellesmere ice shelves (~200 km south), Lesins et al. (2010) analyzed climate records collected between 1954 and 2007. The mean annual surface air temperature warmed 3.2°C since 1972, with summer exhibiting the least change of any season. There was also a 10% increase in precipitation since 1961, dominated by changes in the spring, summer and fall.

To understand changes occurring at the ice shelves, Copland et al. (2007) undertook NCEP/NCAR climate reanalysis of 1000 mbar daily air temperatures for the Ayles Ice Shelf. This indicated a mean annual warming of 0.37°C decade^{-1} between 1948 and 2006, for a total increase of ~2.1°C. This warming was not evenly distributed throughout the year, with particularly strong warming in the fall, winter and spring, but no significant trend in the summer. Mueller et al. (2009) also found similar patterns for regions adjacent to the Ward Hunt Ice Shelf, with mean annual warming of 0.48°C decade^{-1} between 1948 and 2007, and greatest increases in the fall (0.70°C decade^{-1}) and winter (0.68°C decade^{-1}), less in the spring (0.42°C decade^{-1}), and no statistically significant trend in the summer (0.08°C decade^{-1}). In terms of the energy available for ice shelf melting, this can be expressed in terms of positive degree days (PDDs), with Copland et al. (2007) finding that a threshold of >200 PDDs year^{-1} appeared to relate to periods of enhanced ice shelf calving, with this level reached during most years between 1948 and 1963, and since the mid-1990s. Copland et al. (2007) also found that there has been a strongly significant reduction in the number of freezing degree days (FDDs) since 1948, and particularly since the 1990s, with 2005 having the lowest on record (5472 FDDs year^{-1}, compared to the long-term average of 6370 FDDs year^{-1}). This lack of winter cold means that the ice shelves are less able to recover from summer melt, and likely experience less basal freeze-on than before.

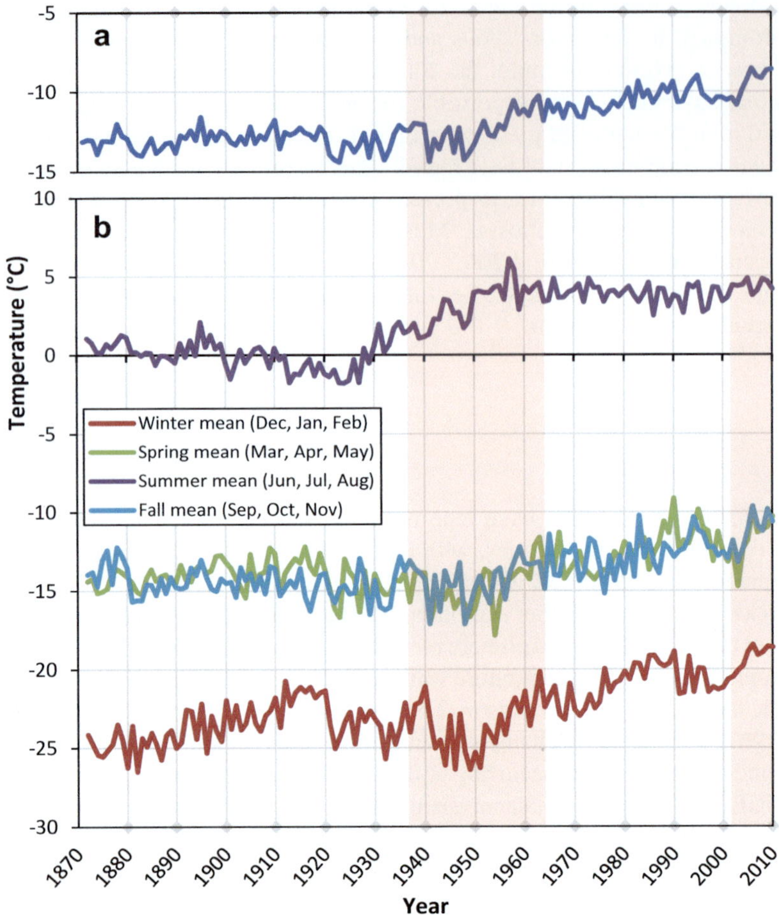

Fig. 10.2 Surface air temperature trends from 1870 to 2010 for northern Ellesmere Island (82.5°N, 82.0°W) derived from the Twentieth Century Reanalysis project (V2) (http://www.esrl.noaa.gov/psd/data/gridded/data.20thC_ReanV2.html): (**a**) Mean annual air temperatures; (**b**) Seasonal air temperatures. Main ice shelf calving periods indicated by *pink shading*

The completion of the Twentieth Century Reanalysis (V2) project (Compo et al. 2011) allows for the assessment of long-term temperature patterns for northern Ellesmere Island for the first time. Based on the grid cell centered at 82.5°N, 82.0°W (near the centre of the Petersen Ice Shelf), mean monthly air temperatures at the 1000 mbar level for the period January 1870 to December 2010 were retrieved from the NOAA Earth System Laboratory Website (http://www.esrl.noaa.gov/psd/data/gridded/data.20thC_ReanV2.html) (Fig. 10.2). In terms of mean annual temperatures, there was generally little change between 1870 and the late 1940s, but then a rapid warming occurred during the 1950s, followed by a more gradual warming from the 1960s to 1990s (Fig. 10.2a), consistent with the climate patterns described above. Temperatures have increased rapidly since 2005, with mean annual temperatures over 2006–2010 (−8.8°C) almost 5°C warmer than over the period 1946–1950 (−13.5°C). Broken down by season, summer temperatures warmed by >5°C between the early 1930s and late 1950s, but showed generally little variability before or after that (Fig. 10.2b). The 1930s–1950s summer warming preceded that observed in other seasons, with sustained winter warming only observed since the 1950s. Warming over the past decade has been primarily driven by increases in winter air temperatures, although fall and spring increases have also been apparent. Mean winter temperatures rose from −25.2°C between 1946 and 1950, to −18.7°C between 2006 and 2010. These warming patterns align well with the main periods of ice shelf losses in the 1940s–1960s and 2000s. It appears that summer warming was particularly important during the earlier period, while winter warming (and to a lesser extent, fall and spring warming) was more dominant over the past decade. However, White et al. (2015a) also demonstrated that recent calving events from the Petersen Ice Shelf (in 2005, 2008, 2011, and 2012) occurred during summers with record-breaking mean summer temperatures.

Given the above review it seems likely that there has been a connection between rising air temperatures and ice shelf losses over the past century or so. However, it is unclear as to which ice shelf physical characteristics have been most impacted by these temperature changes. This information is important to understand the exact causes of the losses and to understand how the ice shelves may evolve in the future. This is addressed in the remainder of this chapter.

10.3 Changes in Glacier Inputs

Canadian ice shelves are comprised of ice from several different sources, including meteoric ice that has formed from snow and rain deposition directly on their surface, marine ice formed in situ and added from lateral and basal freeze-on, and glacier ice derived from glacier flow into their margins (Mueller et al. 2006). Of the six ice shelves which were present in 2000, the Ayles, Milne, Petersen and Serson ice shelves received appreciable glacier input. For these ice shelves an important question is whether the glacier inputs have changed over time, as glaciers typically reduce in area, thickness and velocity in response to long term negative mass balance conditions (Thomson and Copland 2017). Information about the changes in glacier inputs for the Ayles, Milne and Petersen ice shelves is therefore reviewed below.

For the Ayles Ice Shelf, it is clear from air photos and satellite imagery that a major glacier which fed the rear of the ice shelf in 1959 had become completely disconnected by the early 2000s (Fig. 10.3). A report from an overflight in April 1966 (Hattersley-Smith 1967) suggests that this likely occurred in the mid-1960s, although the exact timing of this disconnection is unknown due to the lack of image data. It is likely that this glacier previously provided one of the biggest mass inputs to the ice shelf, and Copland et al. (2007) argued that the loss of this glacier input

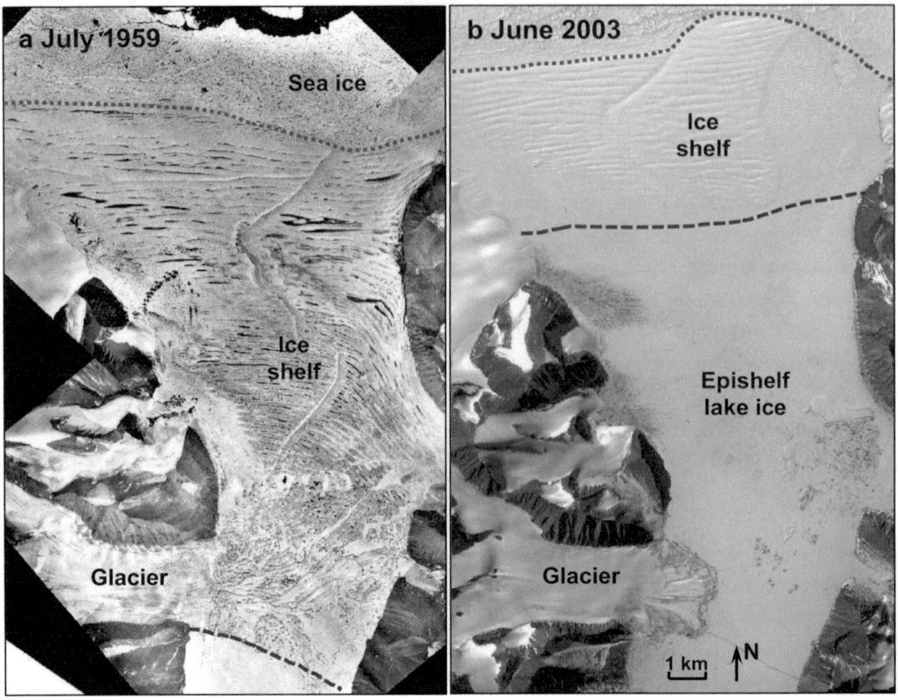

Fig. 10.3 Comparison of the Ayles Ice Shelf between: (**a**) July 1959 air photo (©Her Majesty the Queen in Right of Canada), and (**b**) June 2003 Advanced Spaceborne Thermal Emission and Reflection Radiometer (ASTER) satellite image. *Blue dotted line* indicates front of ice shelf, and *red dashed line* shows the rear. In the 1960s the front part of the ice shelf calved, and most of the remaining ice moved ~5 km northwards before refreezing in place at the head of the fiord (Jeffries 1986). Note how the main glacier input in 1959 had disconnected from the ice shelf by 2003; this disconnection likely occurred in the mid-1960s (Hattersley-Smith 1967)

likely contributed to the long-term negative mass balance of the Ayles Ice Shelf prior to its calving in August 2005.

Of all past and current Canadian ice shelves, the Milne Ice Shelf is the one that has received greatest glacier inputs. Mortimer (2011) used air photos and satellite imagery from 1950, 1959, 1984, 1993, 2001 and 2009 to identify and monitor changes in glacier inputs to the ice shelf over time. Five tributary glaciers terminated on the ice shelf in 1950, with Glaciers 1, 2 and 6 extending towards the ice shelf centre by at least 4.5, 2.5 and 1.5 km, respectively (Fig. 10.4a). By 1984, Glacier 1 terminated near the fiord sidewall, leaving a large remnant glacier tongue ~5 km long encompassed within the ice shelf. Air photo coverage for 1984 did not include the terminus of Glacier 3, but by 1993 it had retreated by ~3 km compared to 1959 (Fig. 10.4b), with its terminus separated from the ice shelf by an ice dammed lake. Further retreat of all tributary glaciers occurred between 1993 and 2009, with the glaciers nearest the rear of the ice shelf (Glaciers 3 and 4) retreating farthest, compared to those near the ice shelf front (Glaciers 1, 2 and 6). Field observations

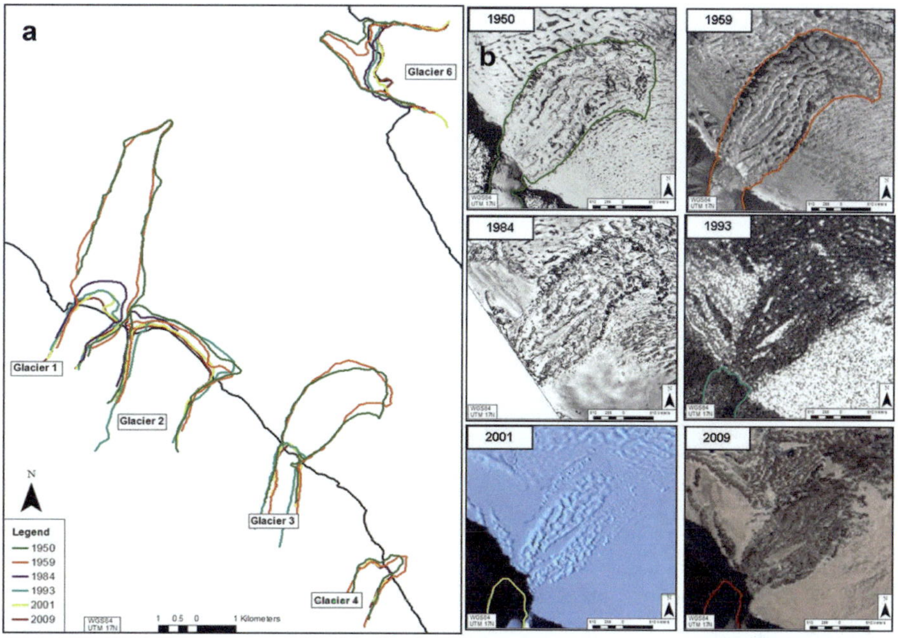

Fig. 10.4 (**a**) Changes in extent of the five main tributary glaciers flowing into the Milne Ice Shelf, 1950–2009; (**b**) Changes in the terminus position of Glacier 3 delineated from historical air photos (1950, 1959, 1984) and satellite imagery (1993: European Remote-Sensing Satellite-1 (ERS-1); 2001 and 2009: ASTER) (From Mortimer (2011))

in May 2009 indicated that Glacier 4 is now a hanging glacier, entirely disconnected from the ice shelf, whereas Glacier 1 still terminated on the ice shelf (Fig. 10.4a). Overall, all glaciers which provided direct input to the Milne Ice Shelf have retreated since 1950, with the largest retreat observed between 1959 and 1984. Mortimer et al. (2012) calculated the inflow from the last remaining tributary glacier in 2011 (Glacier 2; Fig. 10.4a) to be 0.048 m water equivalent (w.e.) year^{-1} averaged over the 2009 ice shelf area. This suggests that current glacier ice input compensates for <20% of the observed thinning rate of 0.26 m w.e. year^{-1} over the period 1981–2008/2009 (see Sect. 10.4 for details), and is unable to balance current mass losses. Glacier inputs also tend to be concentrated in ice tongues on the Milne Ice Shelf, meaning that many regions of the ice shelf receive effectively no glacier input to replace mass losses.

On the Petersen Ice Shelf, White et al. (2015a) recorded changes of the glaciers providing input to the ice shelf over the past 60 years. In 1959, three glaciers flowed into the ice shelf: one from the south (Fig. 10.5b), one from the northwest (not shown), and one from the north (not shown). These glaciers remained stable until the 2000s, with the northwest glacier advancing by ~250 m between 1959 and 1999. Major changes occurred to the southern glacier after the mid-2000s, with the glacier disconnecting from the ice shelf between 2007 and 2008 (Fig. 10.5c, d). Areas of ice

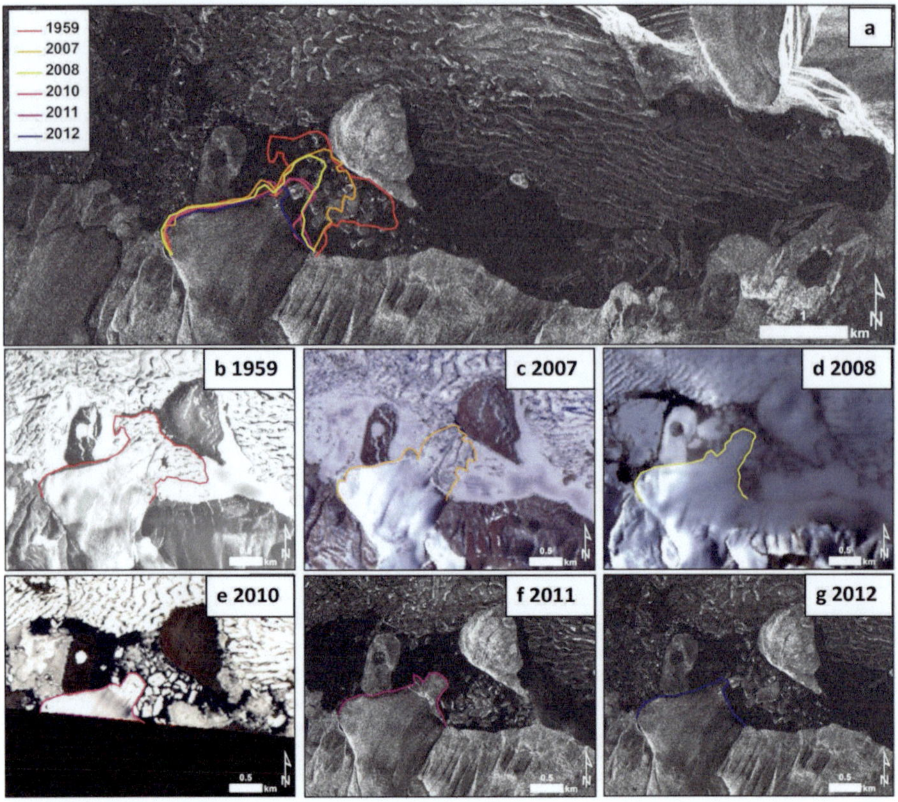

Fig. 10.5 (**a**) Temporal changes in the extent of the tributary glacier flowing from the southern coast into Petersen Bay, 1959–2012 (base image: Ultrafine Wide Radarsat-2 HH, February 3, 2012); (**b**) Aerial photography from August 13, 1959; (**c**) ASTER L1B scene acquired July 7, 2007; (**d**) ASTER L1B scene acquired August 22, 2008; (**e**) ASTER L1B scene acquired July 19, 2010; (**f**) Ultrafine Radarsat-2 HH scene acquired July 19, 2011; (**g**) Ultrafine Wide Radarsat-2 HH scene acquired February 3, 2012 (RADARSAT-2 Data and Products © MacDonald, Dettwiler and Associates Ltd. (2012), All Rights Reserved. From White (2012))

shelf which existed adjacent to this glacier in 1959 disintegrated, and produced new icebergs during an open water event in August 2008 (Fig. 10.5d). Further fracturing of the glacier terminus occurred between 2010 and 2012 during other open water events (Fig. 10.5e–g), with a large area of sea ice and icebergs now present between the glacier terminus and remaining ice shelf. The southern glacier therefore no longer provides any input to the ice shelf, although speckle tracking of Radarsat-2 scenes indicates that the northwestern glacier currently flows at ~20 m year^{-1} where it enters the ice shelf, and the northern glacier flows at ~7 m year^{-1} at the ice shelf boundary (White et al. 2015a). When combined with ice thickness measurements determined from ground-penetrating radar, White et al. (2015a) calculated that these glaciers currently provide a mass input of 0.07–0.12 m w.e. year^{-1} to the Petersen Ice Shelf when averaged over the February 2012 ice shelf area. However, as described above for the Milne Ice Shelf, the fact that glacier inputs are concentrated close to their source means that many parts of the ice shelf currently receive essentially no contribution from glacier ice.

10.4 In Situ Mass Balance

Given the strong evidence for climate warming and reductions in glacier input to the Ellesmere Island ice shelves over time, a related question is whether there have also been changes in their in situ mass balance. Long-term mass balance measurements on the ice shelves are few, but Braun (2017) provides a comprehensive review of measurements conducted on the Ward Hunt Ice Shelf and adjacent Ward Hunt Ice Rise since the 1950s (Fig. 10.1). The ice rise sits at a higher elevation than the adjacent ice shelf, is ~40–100 m thick, and is thought to have formed from the grounding of the ice shelf when it thickened over the past ~1500 years (Braun et al. 2004). Starting in 1959, >100 ablation stakes were installed on these ice masses, and measured annually until 1986 and infrequently since then. In 2002, a total of 14 new ablation stakes were installed on the ice rise, and in 2004 this network was expanded by the addition of a total of 30 stakes at 5 different locations on the Ward Hunt Ice Shelf (Braun et al. 2004; Mueller and Vincent 2006). These measurements indicate that winter snow accumulation remained relatively constant between 1952 and 2008, but that there was much more variability in summer ablation (Braun 2017). Negative mass balance years have dominated since the 1950s, particularly on the ice shelf compared to the ice rise, with cumulative surface mass losses on the ice shelf of ~6.3 m w.e., or a mean loss of 0.11 m w.e. $year^{-1}$, between 1952 and 2007. These losses are strongly influenced by particular years, however, such as a loss of 0.54 m w.e. in 2003 that accounted for half of total mass losses of 1.03 m w.e. between 1989 and 2003.

Mass balance measurements only provide information pertaining to mass changes at the ice shelf surface, whereas the effect of basal freeze-on or melting can also be important. Information concerning mass exchanges at the base of the Ward Hunt Ice Shelf suggest that basal freeze-on (accretion) was roughly equivalent to the negative surface mass balances observed between 1952 and 1982, leading to a stable ice thickness over this period (Braun 2017). Considerable thinning started in the early 1990s, however, with Vincent et al. (2001) and Mueller et al. (2003) using measurements of freeboard and changes in the depth of a freshwater layer trapped behind the ice shelf to indicate that the minimum ice shelf thickness reduced by up to ~50% (to an average of ~25 m) in the 1990s. Given that surface mass losses over this period were much less than the observed thinning, this suggests that basal losses have been much more significant than surface losses at the Ward Hunt Ice Shelf in the recent past (Braun 2017). Evidence for basal thinning is further supported by the fact that a major warming episode occurred in the Canadian portion of the Arctic Ocean in the 1990s that was much larger than any other event since the late 1940s (Gerdes et al. 2003). There is also evidence that another warm anomaly occurred in the Arctic Ocean around 2005–2007 (Beszczynska-Möller et al. 2012), which was even warmer than the event in the mid-1990s, and likely contributed to a 0.28–0.35 m loss in thickness of Arctic sea ice (Polyakov et al. 2010). These ocean warmings seem to mainly originate from an increase in heat flux through Fram Strait (Beszczynska-Möller et al. 2012). The impacts of long-term thinning became obvious when the central part of the Ward Hunt Ice Shelf fractured between 2000 and 2002 (Mueller et al. 2003), and then completely disintegrated in summer 2011, leaving the ice shelf separated into two individual parts for the first time in recorded history. There are no glacier inputs into the Ward Hunt Ice Shelf, so glacier changes can be ruled out as a cause of the observed losses.

On the Milne Ice Shelf, Mortimer et al. (2012) used 250 MHz ground-penetrating radar surveys in 2008/2009 to quantify changes in the thickness and volume of the ice shelf since measurements in 1981 by Prager (1983) and Narod et al. (1988). For direct line comparisons along a 7.5 km transect near the ice shelf front, there was a total average loss of 2.63 ± 2.47 m, equivalent to a mean specific mass balance of −0.085 ± 0.079 m w.e. year^{-1}. However, there was large spatial variability along the transect, with substantial thinning along the ~5 km seaward part of it (up to 9.69 m loss averaged over a 1 km distance), but some thickening in areas furthest inland (up to 5.01 m gain averaged over a 1 km distance). For the ice shelf as a whole, Mortimer et al. (2012) found that thinning dominated, with an average thinning of 8.1 ± 2.8 m (0.26 ± 0.09 m w.e. year^{-1}) over the period 1981–2008/2009, equating to a reduction in volume of 13% (1.5 ± 0.73 km^3 w.e.). There are no direct surface mass balance measurements for the Milne Ice Shelf over this period, but Mortimer et al. (2012) calculated that 73% of the measured thinning can be attributed to basal melt if it is assumed that the 1989–2003 surface mass loss rates of 0.07 m year^{-1} measured on the Ward Hunt Ice Shelf (Braun et al. 2004) can be applied to the Milne Ice Shelf.

The only other Canadian ice shelf with recent surface mass balance measurements is the Petersen Ice Shelf. The record is limited, with White (2012) reporting mean surface ablation of 1.18 m w.e. year^{-1} from measurements at two stakes over the period May 2011 to May 2012. Ice coring on and around the Petersen Ice Shelf provided no evidence for basal freeze-on in the recent past, while the glacier inputs described above (Sect. 10.3) compensated for <10% of this mass loss. It is difficult to draw conclusions about long-term trends based on a single year of mass balance data, but given the 2011 mean ice thickness of the Petersen Ice Shelf of 29 m, White et al. (2015a) predicted that it is unlikely to survive for more than a couple of decades given current mass balance conditions.

10.5 Sea Ice Changes

One of the most widely reported changes to the cryosphere over the past few decades has been the widespread loss of Arctic sea ice. Average Arctic sea ice extent is currently about 30% less than it was over the period 1979–2000 (AMAP 2012), with almost every September minimum extent of the past decade falling below the

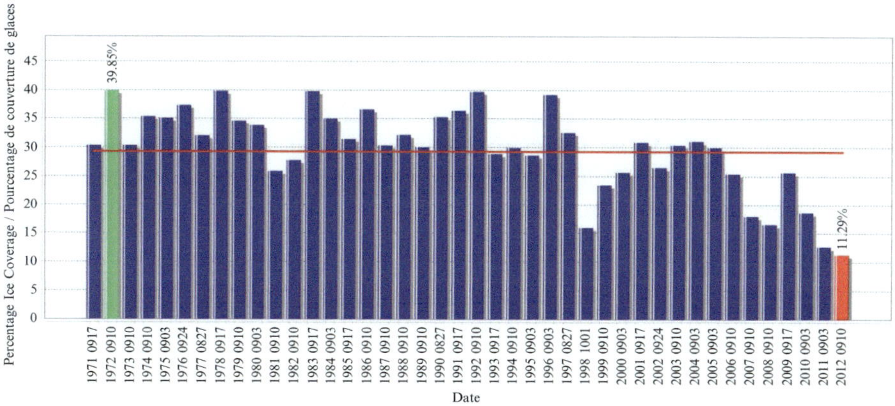

Fig. 10.6 Minimum sea ice coverage (%) over the Canadian Arctic, 1971–2012. *Red line* = 1981–2010 average (29.3%); *Green bar* = record maximum (1972, 39.9%); *Red bar* = record minimum (2012, 11.3%) (Source: Canadian Ice Service IceGraph v2.5 (http://iceweb1.cis.ec.gc.ca/IceGraph20))

long-term average. For example, the average September pan-Arctic minimum extent between 1981 and 2010 was 6.28×10^6 km^2, compared to the record minimum in September 2012 of 3.41×10^6 km^2 (http://nsidc.org/data/seaice_index/). The precise rate of Arctic sea ice decline is sensitive to the averaging period and method of calculation, but the most widely agreed upon rate is -11.5% decade^{-1} for September minimum extent between 1979 and 2012 (Wohlleben et al. 2013). In parallel with reductions in sea ice extent, there have also been marked reductions in average sea ice thickness and age. For example, Maslanik et al. (2011) reported a reduction in the proportion of multiyear sea ice in the Arctic from ~75% in the mid-1980s to ~45% in 2011. They also reported that the multiyear ice pack is becoming younger on average, with the proportion of ice >5 years old reducing from 50% to 10% over the same period. In terms of thickness, the mean winter thickness reduced from 3.64 m in 1980 to 1.89 m in winter 2008 for the central part of the Arctic Ocean (Kwok and Rothrock 2009).

Within the Canadian Arctic, recent sea ice losses have been dramatic. The minimum annual ice coverage derived from the Canadian Ice Service IceGraph tool has been markedly lower than the long-term average over the past decade (Fig. 10.6). In particular, 2012 reached a record low of 11.3% ice coverage in the Canadian Arctic, compared to the 1981–2010 normal minimum coverage of 29.3%. In terms of the significance of these changes for the Ellesmere Island ice shelves, it is clear that sea ice helps to stabilize ice shelves and tidewater glaciers (Reeh et al. 2001; Scambos et al. 2004; Williamson et al. 2008). The most detailed study to date of the loss of an Arctic ice shelf by Copland et al. (2007) found that there was a strong relationship between the timing of sea ice changes along the coastline of northern Ellesmere Island and the loss of the Ayles Ice Shelf. In particular, a semi-permanent fringe of multiyear landfast sea ice (MLSI) was lost from the front of the Ayles Ice Shelf in the ~2 weeks prior to its calving on August 13, 2005, accompanied by a period of open water along this coastline. A review of 179 SAR satellite scenes of this region from 1992 to 2005 by Copland et al. (2007) indicated that such sea ice conditions were unusual, with open water only observed in 7% of the SAR scenes and no evidence for loss of the MLSI fringe prior to August 2005. These observations align with the findings of Yu et al. (2014), who analyzed changes in the distribution of landfast sea ice across the Arctic over the period 1976–2007. They found significant reductions in the amount of winter landfast sea ice in almost all regions, with losses along the northern portion of the Canadian Arctic Archipelago (CAA) being among the largest of any location, changing at a rate of −19.7% decade^{-1}, compared to the Northern Hemisphere average of −6.7% decade^{-1}. This has been accompanied by a reduction in length of the landfast sea ice season for the northern CAA, with Yu et al. (2014) reporting change at a rate of −1.7 week decade^{-1} (−4.2% decade^{-1}) between 1977 and 2007. This aligns with an increase in the movement of multiyear sea ice from the Arctic Ocean to the CAA since 2005, most likely due to an increase in open water within the CAA which has provided more space for inflow to occur (Howell et al. 2013).

Copland et al. (2007) reported on the loss of >1000 km^2 of MLSI in Yelverton Bay in August 2005, both before and after the Ayles Ice Shelf losses. Pope et al. (2012) expanded on this study to reconstruct the long-term history of ice changes in Yelverton Bay and Inlet, and reported that MLSI losses also occurred in summer 2008 and summer 2010, with no multiyear ice cover left in this region at the end of summer 2010 for the first time in the historical record. Prior to 2005, this MLSI had been in place since at least 1950, and prior to the 1940s it appears that most of Yelverton Bay was occupied by an ice shelf. Analysis of satellite imagery and aerial photographs by White et al. (2015a) indicates that these MLSI losses were associated with the presence of open water adjacent to the Petersen Ice Shelf. At the same time as the MLSI losses, the Petersen Ice Shelf incurred losses of 8.0 km^2 (16%) in August 2005 and 9.0 km^2 in August 2008. Further open water conditions in Yelverton Bay in summer 2011 (Fig. 10.1b) and summer 2012 were also closely related to losses from the Petersen Ice Shelf in these years (Table 10.1). The 'Wootton Peninsula Ice Shelf' (unofficial name), on the west side of Yelverton Bay (Fig. 10.1), also lost 65.6% of its area between 2005 and 2009, with ~8 km^2 lost in August 2005 and ~8.4 km^2 lost in August 2008, during periods of open water adjacent to its terminus (Pope et al. 2012).

To understand broader relationships between the occurrence of open water and the loss of ice shelves, we developed an 'Open Water Index' to quantify the timing and relative extent of open water leads along the coastline of northern Ellesmere Island between 1997 and 2011 (Fig. 10.7). This index was calculated by multiplying the ice areas mapped in weekly Canadian Ice Service charts by the fraction of open water (in tenths) in each polygon that was considered to be a lead (<7/10 ice concentration) (Richer McCallum et al. 2014). The presence of leads was confirmed against MODIS satellite images, which were also used to more accurately define the start and end dates of open water events than was possible with the weekly ice charts. When the timing of ice shelf calving events is compared to the open water

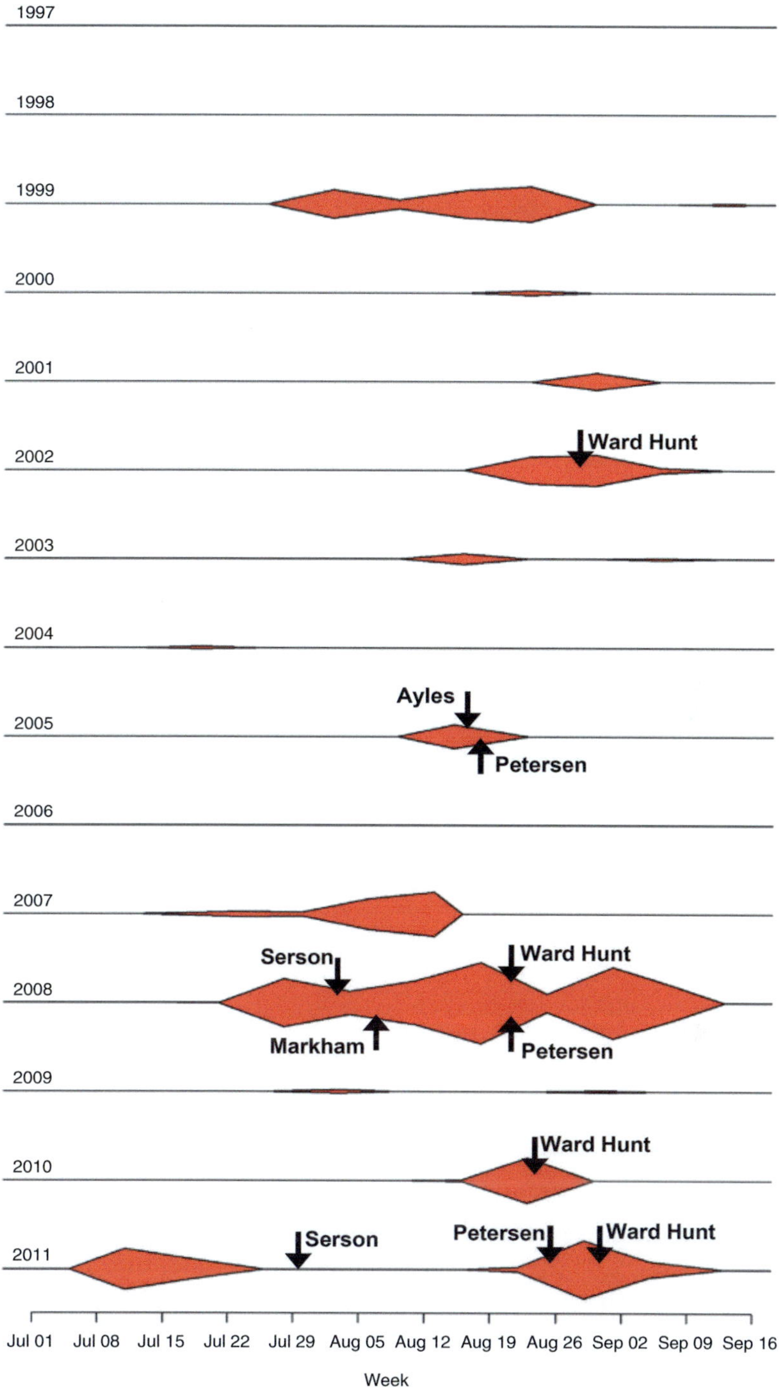

Fig. 10.7 Duration and relative size of open water events (*red shading*) along northern Ellesmere Island, 1997–2011, derived from an Open Water Index calculated from the Canadian Ice Service Digital Archive. In terms of area, the largest recorded event on August 18, 2008, had an Open Water Index of 8980 km^2. *Arrows* indicate the approximate date of major ice shelf calving events (see Table 10.1 for details)

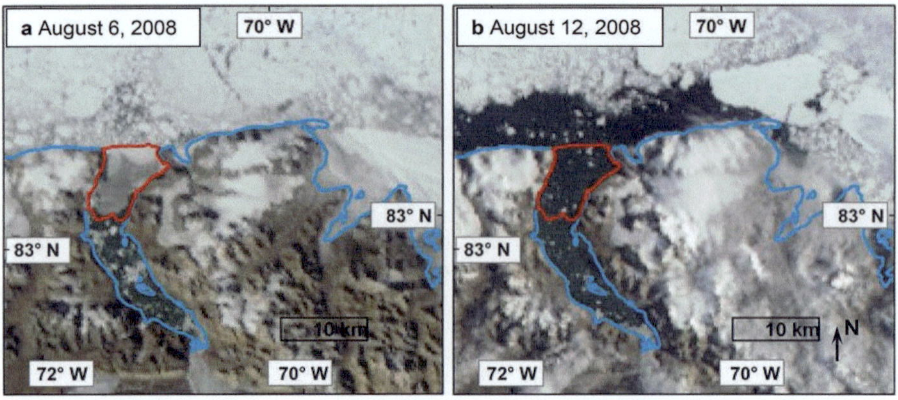

Fig. 10.8 MODIS satellite imagery showing the loss of the Markham Ice Shelf (outlined in *red*) during an open water event between: (**a**) August 6, 2008, and (**b**) August 12, 2008 (MODIS base imagery courtesy of the Rapid Response Project at NASA/GSFC)

index, it is clear that there is a strong relationship between the two (Fig. 10.7). Years with large and long-lived open water conditions along northern Ellesmere Island correspond to periods of extensive ice shelf calving events (e.g., in 2008), but there are few to no ice shelf losses in intervening years with little open water (e.g., 2006, 2009). For example, the entire Markham Ice Shelf was lost during an open water event in early August 2008 (Fig. 10.8), likely making this fiord ice-free for the first time in at least 3500 years (England et al. 2017). Similarly, 60% (122 km^2) of the Serson Ice Shelf calved away during a period of open water in late July 2008, with a further 45 km^2 of the ice shelf lost shortly after another open water event in early August 2011 (Fig. 10.7; Table 10.1).

10.6 Epishelf Lake Ice Changes

Most of the ice shelf calving events described above have occurred in relation to open water conditions at their seaward (outer) edge, such as the loss of the Ayles, Markham and Serson ice shelves. However, there are also several recent examples where open water conditions have existed in the fiords at the rear of ice shelves, resulting in the loss of extensive ice shelf areas from their landward margins. Unlike Antarctic ice shelves, which are typically fed by glaciers along their inner margin, many Arctic ice shelves have been historically bordered along their landward margins by perennially ice-covered epishelf lakes (Veillette et al. 2008, 2011; Jungblut et al. 2017). These epishelf lakes contain freshwater, which typically originates from snow and ice melt and runoff from the surrounding drainage basin (Hamilton 2016). This relatively buoyant freshwater forms a permanently stratified layer which overlies marine water below, and becomes trapped at the rear of the ice shelf. Veillette et al. (2008) suggested that a total of nine epishelf lakes were present along

Fig. 10.9 View of Disraeli Fiord and the rear of the Ward Hunt Ice Shelf, August 20, 2008. Note the extensive calving from the ice shelf into the area previously occupied by an epishelf lake (which drained in 2000–2002) (Photo used with permission of Denis Sarrazin, Université Laval)

northern Ellesmere Island in 1960. This number has been reducing over the past decade due to events such as the fracturing of the Ward Hunt Ice Shelf between 2000 and 2002, which resulted in the loss of the epishelf lake in adjacent Disraeli Fiord (Mueller et al. 2003). The freshwater layer in this epishelf lake was previously up to 43 m thick in the 1960s, 30 m thick in the 1980s and 28 m thick in 1999, with its thinning reflecting a reduction in the minimum draft of the Ward Hunt Ice Shelf over time. After the epishelf lake had drained from Disraeli Fiord it was replaced by seawater, which has been seasonally ice-free for many summers in the recent past. This has allowed extensive fracturing and iceberg calving to occur from the rear of the Ward Hunt Ice Shelf, such as that observed in summer 2008 (Fig. 10.9). Similarly, the majority of the Petersen Bay epishelf lake drained in August 2005, in close relation to the fracturing and calving of the Petersen Ice Shelf (White et al. 2015b). The epishelf lake partly reformed during 2005 and 2008, but there is little evidence that it has been present since 2009 when open water conditions have occurred at the rear of the ice shelf each summer. White et al. (2015a) provided evidence that substantial fracturing and calving from the rear of the Petersen Ice Shelf has also occurred during these times.

10.7 Other Factors: Tides and Seismic Activity

Past studies have suggested that other factors may also be important in the losses of Arctic ice shelves. For example, Holdsworth (1971) reported on the calving of the entire outer portion of the Ward Hunt Ice Shelf between August 1961 and April 1962, which reduced the ice shelf area at that time by almost half (Hattersley-Smith 1963). He argued that this calving event likely occurred in February/March 1962 during a period of abnormally high tides and shortly after a magnitude 3.4 earthquake that occurred ~50 km from the west end of the ice shelf. However, no direct observations were made of this calving event, so the timing of losses is ambiguous and it is difficult to envisage how calving could have occurred in midwinter when there is usually thick sea ice pushed up against the ice shelves. In a later study, Copland et al. (2007) found no evidence for a connection between tidal activity and the well-defined calving of the Ayles Ice Shelf on August 13, 2005. Northern Ellesmere Island has an unusually low tidal range, typically <0.20 m between high and low tides (http://www.tides.gc.ca), so tidal currents are likely relatively weak there. There is also no other evidence for a connection between seismic activity and the losses of an Arctic Ice Shelf, and it is questionable as to how much influence a relatively small (magnitude 3.4) earthquake would have on an ice shelf when any resultant waves would have been dampened by the extensive sea ice cover typically present in the winter. That is not to say that earthquakes are entirely insignificant for ice shelf calving events, as Brunt et al. (2011) reported on the importance of the 2011 Japanese Honshu earthquake and tsunami in causing calving from the Sulzberger Ice Shelf, Antarctica, >13,000 km away from the earthquake epicentre. However, this earthquake and tsunami were one of the largest in recorded history, and the Sulzberger ice shelf faces the open Southern Ocean, so does not provide a good analogue for the enclosed ocean conditions present on northern Ellesmere Island.

10.8 Discussion and Conclusions

It is evident that the Ellesmere Island ice shelves have been undergoing dramatic losses in the recent past. Long-term losses appear to be related to long-term changes in surface air temperature, with rapid warming in the 1940s–1960s (mainly in the summer) and in the 2000s (mainly in the winter) corresponding to periods of rapid ice shelf losses. There is also evidence that the Arctic Ocean has warmed recently, particularly in the 1990s (Gerdes et al. 2003) and the mid-2000s (Polyakov et al. 2010; Beszczynska-Möller et al. 2012). This long term warming has caused ice shelf thinning, both in terms of their in situ surface and basal mass balances, as well as in terms of reductions in mass inputs from tributary glaciers (e.g., Ayles, Milne and Petersen ice shelves; Figs. 10.3, 10.4, and 10.5). While these long term changes likely caused ice shelf weakening, they do not define the exact timing of major ice shelf calving events.

Frequent satellite observations of Arctic ice shelves over the past decade indicates that the precise timing of an ice shelf calving event is closely related to the presence of adjacent open water at the front and/or rear of the ice shelf (Fig. 10.7). Satellite imagery indicates that the presence of sea ice (particularly MLSI) plays a crucial control in stabilizing ice shelves, with few ice shelf calving events occurring when adjacent sea ice and/or epishelf lake ice is present. In the twentieth century the coastline of northern Ellesmere Island was typically fringed by a border of semi-permanent MLSI that remained in place for years to decades. However, this MLSI has been almost entirely lost since the start of the twenty-first century, removing the protection at the front of the ice shelves. There has been some debate as to exactly which physical effect causes an ice shelf to calve when exposed to open water, but suggestions have included increased exposure to wave action (Copland et al. 2007), offshore winds (Ahlnès and Sackinger 1988; Copland et al. 2007), changes in pack ice pressure and buttressing (Koenig et al. 1952), and collisions with mobile sea ice fragments. Further field measurements, remote sensing and modeling are needed to resolve the relative importance and interactions of these factors. There is little evidence to support the occurrence of earthquakes as a factor in the calving of Arctic ice shelves, while the role of tides is equivocal.

The mechanism of losses for Arctic ice shelves has been quite different from Antarctic ice shelves, particularly those on the Antarctic Peninsula. In particular, the unique ability for some Arctic ice shelves to calve from both their seaward and landward margins into the ocean and their own fiords, respectively, has likely accelerated their recent decline, particularly for the Ward Hunt and Petersen ice shelves since 2005. There are no known examples of Antarctic ice shelves that have calved from their landward margin into their own fiords as almost all Antarctic ice shelves are supplied by significant glacier inputs. The style of calving has also been quite different, with the Larsen A and B ice shelves on the Antarctic Peninsula disintegrating into thousands of parallel pieces due to an 'ice-shelf-fragment-capsize' mechanism driven by the penetration of water into surface crevasses (MacAyeal et al. 2003). The relative lack of glacier inputs to Arctic ice shelves means that crevasses on them are rare, and several of them have broken away as large single pieces in recent years (e.g., Ayles Ice Shelf: Copland et al. 2007; Markham Ice Shelf: Mueller et al. 2008). On the Antarctic Peninsula there have been dramatic increases in glacier motion and ice discharge in response to losses of adjacent ice shelves (De Angelis and Skvarca 2003; Berthier et al. 2012). However, the relative lack of glacier inputs to Arctic ice shelves, and the fact that these inputs are declining over time, also means that there has been little evidence of glacier changes in response to ice shelf losses on Ellesmere Island.

In conclusion, the outlook for survival of the remaining Ellesmere Island ice shelves is poor. There is no evidence for recent ice shelf regeneration, with current mass balances strongly negative due to surface melting, basal melting and reductions in inputs from tributary glaciers. Rapid reductions in Arctic ice shelf extent, age and thickness, and particularly losses of MLSI from the coastline of northern

Ellesmere Island, together with losses of epishelf lakes and their once permanent ice cover, have exposed the ice shelves to increasing periods of open water around their margins. This has led to ice shelf calving, which is likely to continue under current climate conditions and continuing reductions in Arctic sea ice.

Acknowledgements We thank the Natural Sciences and Engineering Research Council of Canada, Canada Foundation for Innovation, Ontario Research Fund, University of Ottawa, Polar Continental Shelf Program, ArcticNet, Northern Scientific Training Program, Canadian Space Agency, Alaska Satellite Facility and GLIMS project for support to complete this work. Twentieth Century Reanalysis V2 data kindly provided by the NOAA/OAR/ESRL PSD, Boulder, Colorado. RADARSAT is an official mark of the Canadian Space Agency. We thank the Nunavut Research Institute and communities of Resolute Bay and Grise Fiord for permission to undertake fieldwork on northern Ellesmere Island. We thank Dave Burgess and an anonymous reviewer for comments on the manuscript and Nicole Couture for coordinating the peer review process.

References

Ahlnès, K., & Sackinger, W. M. (1988). Offshore winds and pack ice movement episodes off Ellesmere Island. In W. M. Sackinger & M. O. Jeffries (Eds.), *Port and ocean engineering under Arctic conditions. Volume I* (p. 271–286). Fairbanks: Geophysical Institute, University of Alaska Fairbanks.

AMAP. (2012). *Arctic climate issues 2011: Changes in Arctic snow, water, ice and permafrost.* Oslo: Arctic Monitoring and Assessment Programme.

Berthier, E., Scambos, T. A., & Shuman, C. A. (2012). Mass loss of Larsen B tributary glaciers (Antarctic Peninsula) unabated since 2002. *Geophysical Research Letters, 39*, L13501. doi:10.1029/2012GL051755.

Beszczynska-Möller, A., Fahrbach, E., Schauer, U., & Hansen, E. (2012). Variability in Atlantic water temperature and transport at the entrance to the Arctic Ocean, 1997–2010. *ICES Journal of Marine Science, 69*, 852–863.

Box, J. E. (2013). Greenland Ice Sheet mass balance reconstruction. Part II: Surface mass balance (1840–2010). *Journal of Climate, 26*, 6974–6989. doi:10.1175/JCLI-D-12-00518.1.

Braun, C. (2017). The surface mass balance of the Ward Hunt Ice Shelf and Ward Hunt Ice Rise, Ellesmere Island, Nunavut, Canada. In L. Copland & D. Mueller (Eds.), *Arctic ice shelves and ice islands* (p. 149–183). Dordrecht: Springer. doi:10.1007/978-94-024-1101-0_6.

Braun, C., Hardy, D. R., Bradley, R. S., & Sahanatien, V. (2004). Surface mass balance of the Ward Hunt Ice Rise and Ice Shelf, Ellesmere Island, Nunavut, Canada. *Journal of Geophysical Research-Atmospheres, 109*(D22110). doi:10.1029/2004JD004560.

Brunt, K. M., Okal, E. A., & MacAyeal, D. R. (2011). Antarctic ice-shelf calving triggered by the Honshu (Japan) earthquake and tsunami, March 2011. *Journal of Glaciology, 57*(205), 785–788.

Compo, G. P., et al. (2011). The Twentieth Century Reanalysis project. *Quarterly Journal of the Royal Meteorological Society, 137*, 1–28. doi:10.1002/qj.776.

Copland, L. (2009). *Review of recent changes in Canadian Ice Shelves*. Report prepared for Canadian Ice Service, Environment Canada. Contract # KM149-08-2113, pp. 21.

Copland, L., Mueller, D., & Weir, L. (2007). Rapid loss of the Ayles Ice Shelf, Ellesmere Island, Canada. *Geophysical Research Letters, 34*, L21501. doi:10.1029/2007GL031809.

De Angelis, H., & Skvarca, P. (2003). Glacier surge after ice shelf collapse. *Science, 299*, 1560–1562.

England, J., Evans, D. A., & Lakeman, T. (2017). Holocene history of Arctic ice shelves. In L. Copland & D. Mueller (Eds.), *Arctic ice shelves and ice islands* (p. 185–205). Dordrecht: Springer. doi:10.1007/978-94-024-1101-0_7.

Fisher, D., Zheng, J., Burgess, D., Zdanowicz, C., Kinnard, C., Sharp, M., & Bourgeois, J. (2012). Recent melt rates of Canadian Arctic ice caps are the highest in four millennia. *Global and Planetary Change, 84–85*, 3–7. doi:10.1016/j.gloplacha.2011.06.005.

Fuglem, M., & Jordaan, I. (2017). Risk analysis and hazards of ice islands. In L. Copland & D. Mueller (Eds.), *Arctic ice shelves and ice islands* (p. 395–415). Dordrecht: Springer. doi:10.1007/978-94-024-1101-0_15.

Gerdes, R., Karcher, M. J., Kauker, F., & Schauer, U. (2003). Causes and development of repeated Arctic Ocean warming events. *Geophysical Research Letters, 30*(19). doi:10.1029/2003GL018080.

Hamilton, A. (2016). *Ice-Ocean interactions in Milne Fiord*. PhD Thesis. Department of Civil Engineering, University of British Columbia, Vancouver, Canada.

Hattersley-Smith, G. (1963). The Ward Hunt Ice Shelf: Recent changes of the ice front. *Journal of Glaciology, 4*(34), 415–424.

Hattersley-Smith, G. (1967). Note on ice shelves off the north coast of Ellesmere Island. *The Arctic Circular, XVII*, 13–14.

Holdsworth, G. (1971). Calving from Ward-Hunt Ice Shelf, 1961–1962. *Canadian Journal of Earth Sciences, 8*, 299–305.

Howell, S. E. L., Wohlleben, T., Dabboor, M., Derksen, C., Komarov, A., & Pizzolato, L. (2013). Recent changes in the exchange of sea ice between the Arctic Ocean and the Canadian Arctic Archipelago. *Journal of Geophysical Research Oceans, 118*(7), 3595–3607.

Jeffries, M. O. (1986). Ice island calvings and ice shelf changes, Milne Ice Shelf and Ayles Ice Shelf, Ellesmere Island, N.W.T. *Arctic, 39*, 15–19.

Jeffries, M. O. (1987). The growth, structure and disintegration of Arctic ice shelves. *Polar Record, 23*(147), 631–649.

Jeffries, M. O., & Serson, H. V. (1983). Recent changes at the front of Ward Hunt Ice Shelf, Ellesmere Island, N.W.T. *Arctic, 36*, 289–290.

Jungblut, A. D., Mueller, D., & Vincent, W. F. (2017). Arctic ice shelf ecosystems. In L. Copland & D. Mueller (Eds.), *Arctic ice shelves and ice islands* (p. 227–260). Dordrecht: Springer. doi:10.1007/978-94-024-1101-0_9.

Koenig, L. S., Greenaway, K. R., Dunbar, M., & Hattersley-Smith, G. (1952). Arctic ice islands. *Arctic, 5*, 67–103.

Kwok, R., & Rothrock, D. A. (2009). Decline in Arctic sea ice thickness from submarine and ICESat records: 1958–2008. *Geophysical Research Letters, 36*, L15501. doi:10.1029/2009GL039035.

Lesins, G., Duck, T. J., & Drummond, J. R. (2010). Climate trends at Eureka in the Canadian High Arctic. *Atmosphere-Ocean, 48*(2), 59–80.

MacAyeal, D. R., Scambos, T. A., Hulbe, C. L., & Fahnestock, M. A. (2003). Catastrophic ice-shelf break-up by an ice-shelf-fragment-capsize mechanism. *Journal of Glaciology, 49*, 22–36.

Maslanik, J., Stroeve, J., Fowler, C., & Emery, W. (2011). Distribution and trends in Arctic sea ice age through spring 2011. *Geophysical Research Letters, 38*, L13502. doi:10.1029/2011GL047735.

Montgomery, M. R. (1952). Further notes on ice islands in the Canadian Arctic. *Arctic, 5*, 183–187.

Mortimer, C. (2011). *Quantification of changes for the Milne Ice Shelf, Nunavut, Canada, 1950–2009*. MSc thesis, Department of Geography, University of Ottawa, Ottawa, Canada.

Mortimer, C., Copland, L., & Mueller, D. (2012). Volume and area changes of the Milne Ice Shelf, Ellesmere Island, Nunavut, Canada, since 1950. *Journal of Geophysical Research – Earth Surface, 117*, F04011. doi:10.1029/2011JF002074.

Mueller, D. R., & Vincent, W. F. (2006). Microbial habitat dynamics and ablation control on the Ward Hunt Ice Shelf. *Hydrological Processes, 20*, 857–876.

Mueller, D. R., Vincent, W. F., & Jeffries, M. O. (2003). Break-up of the largest Arctic ice shelf and associated loss of an epishelf lake. *Geophysical Research Letters, 30*(20), 2031. doi:10.1029/2003GL017931.

Mueller, D. R., Vincent, W. F., & Jeffries, M. O. (2006). Environmental gradients, fragmented habitats, and microbiota of a northern ice shelf cryoecosystem, Ellesmere Island, Canada. *Arctic, Antarctic, and Alpine Research, 38*(4), 593–607.

Mueller, D., Copland, L., Hamilton, A., & Stern, D. (2008). International Polar Year scientists and Canadian Rangers visit Arctic ice shelves just before massive loss in 2008. *EOS, Transactions, American Geophysical Union, 89*(49), 502–503.

Mueller, D., Van Hove, P., Antoniades, D., Jeffries, M. O., & Vincent, W. F. (2009). High Arctic lakes as sentinel ecosystems: Cascading regime shifts in climate, ice cover, and mixing. *Limnology and Oceanography, 54*(6, part 2), 2371–2385.

Mueller, D., Copland, L., & Jeffries, M. O. (2017). Changes in Canadian Arctic ice shelf extent since 1906. In L. Copland & D. Mueller (Eds.), *Arctic ice shelves and ice islands* (p. 109–148). Dordrecht: Springer. doi:10.1007/978-94-024-1101-0_5.

Nares, G. S. (1878). *Narrative of a voyage to the polar sea during 1875-6 in H.M. ships 'Alert' and 'Discovery'* (2 Vols.). London: Sampson Low, Marston, Searle and Rivington.

Narod, B. B., Clarke, G. K. C., & Prager, B. T. (1988). Airborne UHF sounding of glaciers and ice shelves, northern Ellesmere Island, Arctic Canada. *Canadian Journal of Earth Sciences, 25*, 95–105. doi:10.1139/e88-010.

Peary, R. E. (1907). *Nearest the pole*. London: Hutchinson.

Polyakov, I. V., et al. (2010). Arctic Ocean warming contributes to reduced polar ice cap. *Journal of Physical Oceanography, 40*, 2743–2756.

Pope, S., Copland, L., & Mueller, D. (2012). Loss of multiyear landfast sea ice from Yelverton Bay, Ellesmere Island, Nunavut, Canada. *Arctic, Antarctic, and Alpine Research, 44*(2), 210–221.

Prager, B. T. (1983). *Digital signal processing of UHF radio echo sounding data from northern Ellesmere Island*. MSc thesis, Department of Geophysics and Astronomy, University of British Columbia, Vancouver, Canada.

Reeh, N., Thomsen, H. H., Higgins, A. K., & Weidick, A. (2001). Sea ice and the stability of north and northeast Greenland floating glaciers. *Annals of Glaciology, 33*, 474–480.

Richer McCallum, M., Mueller, D. R., & Copland, L. (2014). *An assessment of the causes of open water leads and changes in sea ice conditions along northern Ellesmere Island, Canada.* International Symposium on Sea Ice in a Changing Environment, Hobart, Tasmania.

Rignot, E., Velicogna, I., van den Broeke, M. R., Monaghan, A., & Lenaerts, J. (2011). Acceleration of the contribution of the Greenland and Antarctic ice sheets to sea level rise. *Geophysical Research Letters, 38*, L05503. doi:10.1029/2011GL046583.

Scambos, T. A., Bohlander, J. A., Shuman, C. A., & Skvarca, P. (2004). Glacier acceleration and thinning after ice shelf collapse in the Larsen B embayment, Antarctica. *Geophysical Research Letters, 31*(L18402), 1–4.

Schrama, E. J. O., & Wouters, B. (2011). Revisiting Greenland ice sheet mass loss observed by GRACE. *Journal of Geophysical Research – Solid Earth, 116*, B02407. doi:10.1029/2009JB006847.

Sharp, M., Burgess, D. O., Cogley, J. G., Ecclestone, M., Labine, C., & Wolken, G. J. (2011). Extreme melt on Canada's Arctic ice caps in the 21st century. *Geophysical Research Letters, 38*, L11501. doi:10.1029/2011GL047381.

Simmonds, I. (2015). Comparing and contrasting the behaviour of Arctic and Antarctic sea ice over the 35 year period 1979–2013. *Annals of Glaciology, 56*(69), 18–28.

Thomson, L., & Copland, L. (2017). Multi-decadal reduction in glacier velocities and mechanisms driving deceleration at polythermal White Glacier, Arctic Canada. *Journal of Glaciology*. https://doi.org/10.1017/jog.2017.3.

Van Wychen, W., & Copland, L. (2017). Ice island drift mechanisms in the Canadian High Arctic. In L. Copland & D. Mueller (Eds.), *Arctic ice shelves and ice islands* (p. 287–316). Dordrecht: Springer. doi:10.1007/978-94-024-1101-0_11.

Veillette, J., Mueller, D. R., Antoniades, D., & Vincent, W. F. (2008). Arctic epishelf lakes as sentinel ecosystems: Past, present and future. *Journal of Geophysical Research – Biogeosciences, 113*, G04014.

Veillette, J., Lovejoy, C., Potvin, M., Harding, T., Jungblut, A. D., Antoniades, D., Chénard, C., Suttle, C. A., & Vincent, W. F. (2011). Milne Fiord epishelf lake: A coastal Arctic ecosystem vulnerable to climate change. *Écoscience, 18*(3), 304–316.

Vincent, W. F., Gibson, J. A. E., & Jeffries, M. O. (2001). Ice-shelf collapse, climate change, and habitat loss in the Canadian high Arctic. *Polar Record, 37*, 133–142.

White, A. (2012). *Dynamics and historical changes of the Petersen Ice Shelf and epishelf lake, Nunavut, Canada, since 1959*. MSc thesis, Department of Geography, University of Ottawa, Ottawa, Canada.

White, A., Copland, L., Mueller, D., & VanWychen, W. (2015a). Assessment of historical changes (1959–2012) and the causes of recent break-ups of the Petersen Ice Shelf, Nunavut, Canada. *Annals of Glaciology, 56*(69), 65–76. doi:10.3189/2015AoG69A687.

White, A., Mueller, D., & Copland, L. (2015b). Reconstructing hydrographic change in Petersen Bay, Ellesmere Island, Canada, inferred from SAR imagery. *Remote Sensing of Environment, 165*, 1–13. doi:10.1016/j.rse/2015.o4.017.

Williamson, S., Sharp, M., Dowdeswell, J., & Benham, T. (2008). Iceberg calving rates from northern Ellesmere Island ice caps, Canadian Arctic, 1999–2003. *Journal of Glaciology, 54*, 391–400.

Wohlleben, T., Tivy, A., Stroeve, J., Meier, W., Fetterer, F., Wang, J., & Assel, R. (2013). Computing and representing sea ice trends: Toward a community consensus. *EOS, Transactions,, American Geophysical Union, 94*(40), 352. doi:10.1002/2013EO400006.

Yu, Y., Stern, H., Fowler, C., Fetterer, F., & Maslanik, J. (2014). Interannual variability of Arctic landfast ice between 1976 and 2007. *Journal of Climate, 27*, 227–243. doi:10.1175/JCLI-D-13-00178.1.

Chapter 11
Ice Island Drift Mechanisms in the Canadian High Arctic

Wesley Van Wychen and Luke Copland

Abstract Ice islands are large tabular icebergs produced from the calving of Arctic ice shelves. The loss of ~8000 km^2 of ice shelves from the northern coast of Ellesmere Island over the past century has resulted in the production of numerous ice islands, with the first detected in the 1940s. Once calved, these ice islands take one of three routes: (1) they drift west and remain in the Arctic Ocean, where they can circulate for up to several decades; (2) they drift west and enter the interior islands of the Canadian Arctic Archipelago, where they disintegrate relatively rapidly; (3) in rare cases they drift east after calving and enter Nares Strait and then drift south along the east coast of Canada, reaching as far south as Labrador. Historically, the drift paths and disintegration patterns of ice islands were of military interest as they provided mobile platforms for the measurement of oceanographic and atmospheric properties, and they could potentially act as staging posts for Soviet or US access to the opposite side of the Arctic Ocean. Today, interest in ice islands primarily arises from the risks that can pose to shipping and offshore oil exploration, and their indication of the effects of climate change.

Keywords Ice island • Calving • Iceberg • Ice shelf • Ellesmere Island • Sea ice

11.1 Introduction

One of the dominant ways in which ice shelves lose mass is through a process termed calving, whereby the margins of an ice shelf detach and become freely floating pieces of ice. These calved pieces typically have a flat surface and are termed tabular icebergs when they originate from Antarctic ice shelves. These icebergs can reach very large sizes, with the largest on record, B-15, measuring 295 × 37 km and encompassing an area of ~11,000 km^2 when it broke away from the Ross Ice Shelf in March 2000 (Lazzara et al. 2008). In the Arctic these tabular icebergs are termed

W. Van Wychen (✉) • L. Copland
Department of Geography, Environment and Geomatics, University of Ottawa, Ottawa, ON, Canada
e-mail: wvanw046@uottawa.ca; luke.copland@uottawa.ca

L. Copland, D. Mueller (eds.), *Arctic Ice Shelves and Ice Islands*, Springer Polar Sciences, DOI 10.1007/978-94-024-1101-0_11

ice islands when they originate from ice shelves, with Koenig et al. (1952) describing the first use of the term in the late 1940s. In the Arctic, the largest recorded ice island had dimensions of 31.1 × 33.4 km (T-2; Jeffries 1992a).

The technical definition of an ice island varies slightly between sources. The operational manual MANICE (Canadian Ice Service 2005), used by operational sea ice forecasters, defines an ice island as

> *A large piece of floating ice protruding about 5 m above sea level, which has broken away from an Arctic ice shelf. They have a thickness of 30–50 m and an area of from a few thousand square metres to 500 sq. km or more. They are usually characterized by a regularly undulating surface giving a ribbed appearance from the air.*

Jeffries (1992a, p. 247) provides the following description:

> *Ice island: a tabular iceberg which has broken away or calved from an Arctic ice shelf. They have a gently undulating surface which gives them a ribbed appearance from the air. They initially show a minimum 2 m above sea level, but this decreases as they thin by melting during their drift. Their area and dimensions vary from as little as a few hundred square metres, tens of metres across, to a few hundred square kilometres, tens of kilometres across. The smallest ice islands may be created at the time of calving, or they may be a result of the disintegration of a larger ice island.*

Although there is no complete inventory of how many ice islands have existed, Jeffries (1992a) estimated that a total of ~600 occurred between 1946 and 1992. Further breakups of Arctic ice shelves since then have produced many more (Copland et al. 2007; Mueller et al. 2017). In this chapter we discuss the origin of ice islands, their physical characteristics and their previous drift patterns in the Canadian High Arctic. The drift patterns are of particular interest in relation to northern shipping and recent offshore oil development activities, as further discussed in Sect. 11.3. The mechanisms of ice shelf breakup that produce ice islands are addressed by Copland et al. (2017).

11.2 Origin and Identification of Ice Islands

Ice islands in the Arctic Ocean typically originate from the ice shelves found on the north coast of Ellesmere Island (Fig. 11.1). Radiocarbon dating of driftwood collected in the fiords behind these ice shelves suggests that they likely developed during a climatic deterioration approximately 3000–5500 years ago (England et al. 2008, 2017). The ice shelves have undergone extensive disintegration since this time, with a reduction in area from >8500 km^2 in 1906, to 1037 km^2 in 2003 and 535 km^2 in 2015 (Mueller et al. 2017). This has included the loss of the entire Ayles Ice Shelf in August 2005 and Markham Ice Shelf in August 2008.

Studies have also been conducted to determine if ice islands could originate from Greenland. In total, four Greenland glaciers were identified as possible sources of ice islands for the Arctic Ocean and Baffin Bay: Petermann, Ryder, C.H Ostenfeld and Hagen Brae (Higgins 1989). Additional ice island sources for East Greenland are the glacier-filled fiords of Nioghalvfjerdsfjorden and Jökelbugten (Higgins

Fig. 11.1 Map of current (outlines in *red*) and past (names in *brackets*) ice shelf locations across northern Ellesmere Island, as of July 2015 (MODIS base image from July 12, 2015, courtesy of the Rapid Response Project at NASA/GSFC)

1988). Upon study of the Greenland ice shelves and glaciers, Higgins (1989) found that they could produce ice islands that would be very difficult to distinguish from those calved from Ellesmere Island. Indeed, there have been many ice island sightings off the east coast of Canada over the past century, many of which likely originated from Greenland (e.g., Newell 1993; Peterson 2005; Hill et al. 2008). For example, Peterson (2005) describes the sighting of many ice islands up to 2.0 × 2.4 km in size off eastern Canada in 2001–2002 that originated from the Petermann Glacier, and Falkner et al. (2011) describe the calving of a 253 km^2 ice island from Petermann Glacier in August 2010. This ice island subsequently split into several smaller pieces as it drifted southwards, with Halliday et al. (2012) reporting average thicknesses of ~70 m on a ~72 km^2 piece when it was located off eastern Labrador in June 2011. This PII-A ice island had drifted >3000 km from its calving location in the space of only 10 months.

Ice shelves are widely described by explorers in early literature such as Peary (1907) and Storkerson (1921), with a full review provided by Koenig et al. (1952) and Jeffries (1987). Ice islands were not definitively identified until the 1940s, however, when post-WWII military exploration of the Arctic became more commonplace. The first ice island was identified on August 14, 1946, when the United States Air Force (USAF) found an ice island with an area of 500 km^2, later to be named T-1, approximately 500 km north of Barrow, Alaska (Jeffries 1992a). The second and third ice islands were located off northern Ellesmere and Axel Heiberg Islands

Fig. 11.2 Aerial view of an ice island in NW Yelverton Bay, July 13, 2015; likely origin is the Ward Hunt Ice Shelf. Note distinctive surface ridging. Ice island is ~0.9 km wide × 2.0 km long (Photo: Luke Copland)

in 1948 (T-4 and T-5). This was followed by the discovery of T-2 and T-3 in the Arctic Ocean in 1950 (Jeffries 1987).

The first ice islands were believed to have an ice shelf origin due to their characteristic regularly undulating (ribbed) surface topography (Fig. 11.2). This was later confirmed from further study of aerial photographs and ice samples (Koenig et al. 1952). The undulating surface topography found on Ellesmere Island ice shelves typically parallels the outer coast, but within fiords it often swings southwards and/or becomes disorganized (Hattersley-Smith 1957). Observations by the authors in July 2015 indicate that the undulations can reach amplitudes of >10 m between adjacent ridges and troughs at the southern end of the Milne Ice Shelf. This rolling surface topography is almost unique in the Northern Hemisphere, but is often observed on Antarctic Ice Shelves such as the McMurdo Ice Shelf. For the Ellesmere Island ice shelves, Hattersley-Smith (1957) argues the ridging originates from the effects of wind via the production of elongated snow dunes. Subsequent summer melting then tends to pool water in the ridges between dunes and enhance the melting in these areas, further enhancing the ridge and trough structure. Measurements by Crary (1960) indicated that prevailing winds along northern Ellesmere Island were parallel to the ridges and troughs, supporting this theory. However, further research is required to understand the exact processes occurring as observations of sand dunes indicates that they often occur at right-angles to the dominant wind direction (e.g., Barchan dunes; Tarbuck et al. 2009), and personal observations by the authors on the Milne Ice Shelf indicate that surface snow drifts frequently do not parallel the ridges and troughs. The summer meltwater ponds may therefore play a much more dominant role, as further discussed by Jeffries (2017) in this volume.

The thickness of an ice island is dependent on the thickness and structure of the ice shelf from which it originated. For example, the Milne Ice Shelf was found to average ~70 m depth in airborne radio-echo sounding measurements in 1981 (Narod et al. 1988), while a more recent survey by Mortimer et al. (2012) in 2008/2009 reported a mean thickness of 55 m, with a maximum thickness of 94 m. The Milne Ice Shelf appears to be unusually thick, however, with ice depths on most of the other Ellesmere Island ice shelves closer to 45 m or less (Jeffries 1986). For example, White et al. (2015) found a mean thickness of 28.5 m for the Petersen Ice Shelf. Copland et al. (2007) used radio-echo sounding to determine that the Ayles Ice Island was 42–45 m thick in May 2007, some 21 months after it calved. Crary (1958) measured ice thicknesses between 40 and 53 m at six points on T-3 in 1954, with an average of 48 m. It should also be noted that only a small proportion (typically 1/8th to 1/10th) of an ice island's total thickness is ever exposed above the ocean surface, in the same way as icebergs.

11.3 Significance

Knowledge of the occurrence, drift patterns and characteristics of ice islands has been important for past and current applications including oil development, shipping, and the establishment of platforms for scientific research and military operations. In addition, information on ice islands aids with the reconstruction of former ice shelf extents, provides a chronological record of the disintegration of ice shelves, and gives a basis for calculating the frequency of ice island calving events (Jeffries 1992b). These topics are explored more fully below.

11.3.1 Ice Island Drift and Oil Development

Research on the drift of ice islands has been of interest to oil developers in the Arctic since 1970, and as a consequence they have sponsored some of the related field research. For example, the Arctic Petroleum Operators Association conducted a 5-year program from 1972 to 1976 to catalogue the number and size of ice islands in the southern Beaufort Sea (Yan 1986). This was mainly motivated by the possibility of an ice island colliding with an oil drilling structure. Development of oil resources in the Beaufort Sea and off the North Slope of Alaska occurred on a relatively small scale with projects such as Amauligak on the Canadian Beaufort shelf in the mid-1980s (Kato and Kumakura 1996). There is currently a resurgence of interest in this region, however, as oil companies such as Shell and Conoco Philips have recently taken out new offshore oil leases there (Minerals Management Service 2008). These developments are close to the drift paths that previous ice islands have taken, meaning that a good understanding of these patterns is essential. Icebergs in the north Atlantic are typically towed or pushed out of the way with ocean-going tugs when they are on a potential collision course with oil rigs (Rudkin et al. 2005), but this solution would not be possible with most ice islands due to their large size (Fuglem and Jordaan 2017).

11.3.2 Ice Islands and Shipping

Ice islands currently pose little threat to shipping traffic in the Arctic given adequate reporting and surveillance coupled with the relatively small amount of ship traffic in northern waters. However, there have been significant recent increases in shipping in the Canadian Arctic (Pizzolato et al. 2014), which have been significantly correlated with reductions in sea ice in locations such as the Beaufort Sea and Western Parry Channel (Pizzolato et al. 2016), where previous ice islands have been observed. Recent rapid reductions in both average sea-ice extent and thickness has led to predictions that the frequency of navigable periods for trans-Arctic shipping will double by 2050 (Melia et al. 2016). Routes include the Northwest Passage, a waterway that connects the Atlantic and Pacific Oceans via passages in the Canadian Arctic Islands, and the Transpolar Route directly across the Arctic Ocean. These new routes could increase the potential for collisions with ice islands. In addition, the opening of more intra-island passageways in the Canadian Arctic due to thinning sea ice means that ice islands may drift into locations where they have not historically been found. For shipping, the largest risk arises not from large ice islands, which can be easily spotted, but from the many smaller icebergs that are produced when ice islands break up. These smaller ice fragments are still of considerable risk to ships despite their smaller size, but are very difficult to track individually and distinguish from surrounding sea ice because of their low profile.

11.3.3 Ice Islands as Scientific and Military Platforms

Due to their thickness, stability and residence times that can extend to decades, ice islands have previously been used as platforms for the establishment of semi-permanent bases. Historically, they have been used to study subjects such as

Table 11.1 Past ice island research stations

Name	Year established	Nationality	Duration (Years)
T-3 (Fletcher's)	May 1952	American	27
NP-6[a]	April 1956	Russian	3
ARLIS-II	1961	American	4
NP-19/WH-3	November 1962	Russian	4
NP-22	September 1973	Russian	9
NP-23	December 1975	Russian	3
NP-24	June 1978	Russian	2
Hobson's Choice	1983	Canadian	9

Adapted from Jeffries 1992a; Jeffries and Shaw 1993; Frolov et al. 2005
[a]Jeffries (1992a) states that NP-6 was established on an ice island, while Frolov et al. (2005) declares that NP-6 was established on an ice floe

bathymetry, ice strength and climate, with the first ice island research station established on T-3 in March 1952 by the USAF (Crary 1958, 1960; Table 11.1). This floating research station was operated for almost 27 years (Jeffries 1992a). A second research camp named ARLIS-II (Arctic Research Laboratory Island) was established in 1961 by the Naval Arctic Research Laboratory, approximately 200 km north of Barrow, Alaska (Jeffries 1992b). This station operated continuously for four years as it drifted across the Arctic Ocean. The first Canadian research base was established on Hobson's Choice Ice Island in 1983 (Foster and Marino 1986), with the camp operating yearly in the spring and summer until 1992 (Jeffries and Shaw 1993). There has also been extensive use of ice islands as drifting research stations by the Russians since their first discovery, as further described by Belkin and Kessel (2017).

While many scientific and meteorological measurements were made at the ice island camps, it is clear that they also had specific strategic importance for both the United States and the Soviet Union during the Cold War (Althoff 2017). Once it was proven that aircraft could safely land on ice islands and that camps could operate year-round on them, research focused on using ice islands as staging platforms for air strikes and detecting submarines (Althoff 2007, 2017). Indeed, the former Soviet Union took considerable interest in Arctic Ocean ice floes and ice islands at the same time as the Americans, possibly discovering T-1, T-2 and T-3 before the USAF (Zubov 1955). The Russians established their first drifting research station, NP-1 (North Pole 1), on an ice floe in 1937, and the success of this program proved the feasibility of conducting long-term Arctic monitoring directly from the ice (Frolov et al. 2005). Leary and LeShack (1996) provide an account of a secret military raid by the US Central Intelligence Agency on Russian drift station NP-8 in 1962 to determine the Soviet's progress in meteorology, oceanography and submarine detection. The Russian ice camps have continued to the present day, with station NP-41 (also referred to as North Pole-2015) operating between April 19, 2015 and mid-August 2015 (Barents Observer, '*Russian Arctic Scientists to be evacuated from ice floe*', August 5, 2015). Of the Russian drifting stations, five are believed to have been located on true ice islands (Table 11.1; Althoff 2017).

11.4 Sources of Evidence for Ice Island Drift Tracks

11.4.1 Aerial Reconnaissance

After T-1 was discovered by the USAF during a routine weather observing flight in 1946, aerial reconnaissance was the dominant method of detecting ice islands until the advent of synthetic aperture radar satellite imagery in the 1980s. The first systematic aerial photography of the Canadian High Arctic was undertaken by the USAF between 1946 and 1948 during Operation Polaris (Halliday 2008), which resulted in the discovery of ice islands T-4 and T-5 (Jeffries 1987). This operation used trimetrogon photography, which consisted of three cameras mounted along a line transverse to the flight direction, with one camera pointed vertically and the other two pointed sideways at a depression angle of ~30° from horizontal (Cogley and Adams 2000). The trimetrogon photography program was continued and expanded by the Royal Canadian Air Force (RCAF), which led to the discovery of an additional 28 ice islands in photos taken between 1947 and 1952 by Koenig et al. (1952), and a further 31 ice islands by Montgomery (1952). These were observed throughout the western and southern parts of the Canadian Arctic Archipelago, and ranged in size from ~0.5 to >12 km across. However, none were discovered along the eastern (Nares Strait/Lincoln Sea) side of Ellesmere Island at this time.

The first detailed National Topographic Survey maps of the ice shelves were produced from vertical aerial photography undertaken in 1959/1960 by the RCAF (Jeffries 1987). To this day, these maps and aerial photographs provide the most complete record of the Ellesmere Island ice shelves, and are typically used as the basis against which modern changes are compared (e.g., Mortimer et al. 2012; White et al. 2015). Public access to these photos, together with the earlier trimetrogon series, is available at the National Air Photo Library in Ottawa, Canada.

Operational detection and monitoring of sea ice conditions in the Canadian High Arctic was undertaken via aerial reconnaissance flights by the Polar Continental Shelf Project (PCSP) between 1961 and 1978, and by the Canadian Ice Service until the 1990s. During these flights, ice islands were often spotted (Lindsay et al. 1968) and their locations are preserved in the PCSP Ice Atlases (Lindsay 1975, 1977, 1982) and the Canadian Ice Service operational ice charts (since 1957) and Annual Ice Atlases (since 1986/1987). Today, aerial monitoring has largely been replaced with use of satellite imagery in the Canadian High Arctic, although the International Ice Patrol still undertakes extensive iceberg monitoring flights along the east coast of Canada. The International Ice Patrol was developed in 1914 after the sinking of the Titanic, and is operated by the US Coast Guard with funding from a variety of nations (http://www.navcen.uscg.gov/?pageName=IIPHome). Prior to World War II, iceberg detection was primarily undertaken from ships, but since then aerial reconnaissance from Hercules HC-130 aircraft has become the dominant survey method. This has produced an extensive record of iceberg occurrences along eastern Canada, which has included many reports of ice islands (Newell 1993; Peterson 2005; Hill et al. 2008).

11.4.2 Tracking Buoys

Once an ice island has been identified, tracking buoys are commonly deployed to provide convenient monitoring of its location. One of the early methods involved the placement of metal barrels in a recognizable fashion on the ice island surface, so that when aircraft flew overhead they could be easily recognized (Nutt 1966; Foster and Marino 1986). These metal barrels were also easily identified by on-board aircraft radar systems, allowing ice islands to be found and identified even when visual contact was not possible in poor weather conditions.

The first true beacons used on ice islands were deployed by the U.S. Navy (e.g., on WH-5; see Sect. 11.5.3), and provided a radio homing signal that aircraft could identify and track (Nutt 1966). The first satellite-based beacons utilized the ARGOS system developed in 1978, which computes location using the Doppler shift between signals. For example, early ARGOS buoys were used to track Hobson's Choice Ice Island in the 1980s (Sackinger et al. 1989, 1991), including the use of two buoys at either end of the island to track of its orientation (Foster and Marino 1986). At best, these signals could be used to locate a buoy to an accuracy of <250 m, but in modern buoys the addition of a GPS receiver provides positional accuracy to <10 m (Argos 2007). Current ARGOS data transmissions can also provide information on local environmental factors such as air pressures and temperatures, and can be made available in near real-time on the internet. Modern tracking beacons such as the Compact Air Launched Ice Beacon manufactured by MetOcean (Nova Scotia, Canada) are outfitted with parachutes to enable deployment directly from moving aircraft. GPS-based tracking beacons operating on other satellite systems, such as Iridium, are also now available.

11.4.3 Remote Sensing Platforms

Satellite observations provide an excellent means of finding and tracking ice islands, particularly via the use of Synthetic Aperture Radar (SAR). SAR has the advantage of being an active remote sensing technique, which means that the system sends out its own controlled signals which are detected back by the sensor. This enables illumination and observation of the Earth's surface regardless of cloud or light conditions, and has revolutionized the detection and monitoring of sea ice and ice island conditions, particularly during the polar night. The Canadian Ice Service first documented sea ice conditions with an aircraft-mounted Side-Looking Airborne Radar (SLAR) in 1978, and an aircraft-mounted SAR sensor in 1990. This was replaced with the use of RADARSAT-1 SAR satellite imagery in 1996, and most operational sea ice monitoring now relies on orbiting satellites with SAR capabilities such as RADARSAT-2, Sentinel-1 and ALOS-2.

Jeffries and Sackinger (1990) conducted a study of ice island detection in SAR imagery and determined three criteria that aid in their identification: (1) Size: ice islands are generally larger than surrounding pack ice floes; (2) Shape: ice islands are often longer and more angular than surrounding ice floes; (3) Returns: ice islands are typically distinctive in SAR images because:

(a) The radar returns are usually stronger than the surrounding pack ice, making ice islands appear brighter than the other features.
(b) Ice island edges appear bright in imagery due to their height above surrounding sea ice. The edges are also unusually long and straight.
(c) The regular ribbed-like texture of an ice island surface is very different from the more random patterning found on surrounding sea ice.

Thus, once an ice island is identified it can often be easily detected in SAR imagery by its characteristic size and shape, and because ice islands have a unique backscatter signature (bright tones that are in stark contrast to the dark tones of the surrounding pack ice). The signature of ice islands changes throughout the year, however, depending on the amount of melt taking place. Ice islands in imagery captured in late summer often appear darker because of the lower backscatter return due to a wet surface (Jeffries and Shaw 1993).

Passive satellite platforms (i.e., optical imagery) can also be used to find and track ice islands, although they are limited by their need for daylight and cloud-free conditions to observe the earth's surface. The launch of the Landsat program in 1972 provided the first readily-available high resolution satellite imagery of the Canadian High Arctic, although Landsat 1-3 could not obtain data above 79°N and were therefore unable to view the ice shelves directly. However, ice islands could be observed once they had drifted south of 79°N. Landsat 4 & 5 usually only obtained data up to ~80°N, although Jeffries et al. (1992) describe the first ever imaging of the ice shelves with Landsat 5 in summer 1990 via a special acquisition by the Alaska Satellite Facility. Today, Landsat 8 images up to ~82°N, which is still not far enough north to capture the northernmost ice shelves such as the Ward Hunt. However, special request SPOT imagery of the northernmost ice shelves has been occasionally collected from 1986 onwards.

The first regular imaging of the ice shelves didn't occur until the development of high resolution SAR and visible sensors in the mid- to late-1990s. In particular, the launch of the TERRA satellite in 1999 provided visible imagery of the ice shelves via: (a) the ASTER sensor, which captures high resolution (15 m) imagery but has a relatively infrequent repeat orbit of 16 days; and (b) the MODIS sensor, which images the Canadian High Arctic multiple times per day, but has a resolution of 250 m. The launch of a second MODIS sensor on the AQUA satellite in 2002 now provides repeat coverage of the Ellesmere Island ice shelves every ~90 min, and was used by Copland et al. (2007) to reconstruct the detailed dynamics of the Ayles Ice Shelf calving event in August 2005. The launch of the FORMOSAT-2 satellite in 2004 provided the first very high-resolution (2 m in panchromatic) imagery of the ice shelves on a daily repeat cycle, and was used by Mueller et al. (2008) to reconstruct the 2008 calving events from the Ward Hunt Ice Shelf.

11.4.4 Summary of Drift Track Evidence

In summary, there are a wide range of data sources available for reconstructing ice island drift tracks. Historical records, personal communications and aerial surveillance provide a good level of information about early ice islands. Earth observation satellites and tracking buoys provide reliable near real-time data for the location of recent and current ice islands. These data sources help to improve understanding of ice island motion and the factors driving it, and are used in the discussion below. Drift paths can be reconstructed with considerable accuracy when the ice islands are close to the Canadian Arctic Archipelago, but the record is more uncertain for older ice islands which were far out in the Arctic Ocean. This is simply because there was less surveillance undertaken in early years, and because positioning is much more difficult when there are no nearby land markers to use as reference points. Discussion here is limited to drift tracks of ice islands that originated from Canadian Arctic ice shelves.

11.5 Ice Island Drift Tracks

Information on the drift patterns of all the major ice islands that originated from the ice shelves of northern Ellesmere Island is discussed below and presented in Table 11.2. This list does not include all of the ice islands that have ever calved from these ice shelves, but provides information on the main ones and their drift paths. Where information is available, drift tracks are provided from calving to disintegration, and we consider the tracks to be generally representative of the paths taken by the wider ice island population.

Reconstructed drift patterns reveal that ice islands which calve from northern Ellesmere Island follow one of three general patterns:

1. *Arctic Ocean Drift*: in which ice islands stay within the Arctic Ocean and can complete several loops around the basin before disintegration and/or ejection via Fram Strait between Greenland and Svalbard.
2. *Archipelago Drift*: in which ice islands enter the intra-island channels of the Canadian Arctic Archipelago (CAA).
3. *Nares Strait Drift*: in which ice islands drift eastward after calving and enter Nares Strait, Baffin Bay and ultimately reach the east coast of Newfoundland and Labrador.

Table 11.2 List of all major ice islands previously tracked in the Canadian High Arctic

Name	Origin	Year of calving	Year of breakup	Location of breakup	Tracking method(s)	Duration (years)	Size	Refs
T-1	Ellesmere Island	1946[a]	1971	N coast Ellef Ringnes Island	Aerial photos, ice atlases	Approx. 25	~27 × 33 km	1,5
T-2	Ellesmere Island	1950[a]	1951	Atlantic Ocean	Aerial photos, ice atlases	Unknown	~31 × 33 km	1,9, 10
T-3 /Fletcher's	Ellesmere Island	1935	1984	Southern tip of Greenland	Aerial photos, research station, ice atlases	Approx. 40	~8 × 16.5 km	1,2, 11
ARLIS-II	Serson (Alfred Ernest) Ice Shelf	1955	1964	Exited into Greenland Sea	Aerial photos, research station, inferences from other ice islands	9	2.4 × 5.6 km	2,3
WH-1	Ward Hunt Ice Shelf	1961–1962	1968	NW coast Banks Island	Ice atlases, aerial reconnaissance	At least 3	73.5 km^2	1,2,5
WH-2	Ward Hunt Ice Shelf	1961–1962	1968[b]	N. Meighen Island	Ice atlases, aerial reconnaissance	At least 4	68 km^2	1,2,5
WH-3 /NP-19	Ward Hunt Ice Shelf	1961–1962	1966[b]	W coast Prince Patrick Island	Ice atlases, aerial reconnaissance	At least 3	8 × 14 km	1,2,5
WH-4	Ward Hunt Ice Shelf	1961–1962	1966[b]	W coast Prince Patrick Island	Ice Atlases, aerial reconnaissance	At least 3	93 km^2	1,2,5

WH-5	Ward Hunt Ice Shelf	1961–1962	1966	Grand Banks, Newfound-land	Aerial reconnaissance, tracking buoys, ice atlases	Approx. 4	~20 × 9 km	1,2,6
Hobson's Choice / Ice Island 3831	Ward Hunt Ice Shelf	1983	1992	Queens Channel	Satellite observations, tracking buoys, research station	Approx. 9	~10 × 5 km	2,4,10
Ayles Ice Island	Ayles Ice Shelf	August 2005	–	–	Tracking buoys, satellite observations, ice atlases	–	~15 × 6 km (66.4 km^2)	8

References: 1 = Lindsay et al. (1968), Lindsay (1975, 1977, 1982); 2 = Jeffries (1992a); 3 = Jeffries (1992b); 4 = Jeffries and Shaw (1993); 5 = Crary (1958); 6 = Nutt (1966); 7 = Hattersley-Smith (1963); 8 = Copland et al. (2007); 9 = Koenig et al. (1952); 10 = Yan (1986); 11= Wadhams (2000)

[a]Year of first observation, but calving date is unknown

[b]Year of last observation, but breakup date is unknown

11.5.1 Arctic Ocean Drift

After ice islands calve from northern Ellesmere Island, they typically begin a dominantly westward drift. This motion occurred for all of the ice islands reviewed here, except for WH-5 (Fig. 11.3, Table 11.2). Some of these ice islands ultimately end up drifting into the channels of the CAA (see Sect. 11.5.2), but many of them remain in the Arctic Ocean. These islands typically closely follow the north coast of the CAA until they reach the Beaufort Sea. Once there, they continue drifting along the coast of Alaska toward Russia, until they roughly reach Point Barrow, after which they begin to move almost straight towards the North Pole, carried by the Transpolar Drift ocean current. Once close to the North Pole, they either:

(a) Drift southward and meet the coastline of Ellesmere Island again, thus completing a full circuit around the Arctic Ocean
(b) Become caught in the Transpolar Drift, and exit the Arctic Ocean through Fram Strait between Greenland and Svalbard.

The typical clockwise motion around the Arctic Ocean is termed the 'Beaufort Gyre', and is driven by the dominant anti-cyclonic pressure systems that occur over this region (Fig. 11.4; Serreze et al. 1989). This flow dominates the motion of the sea ice pack in the Arctic Ocean, and explains why ice islands that calve from Ellesmere Island are usually immediately transported to the west. Occasionally, however, the anti-cyclonic system breaks down and is replaced by a persistent cyclonic system lasting for periods of up to several weeks (Serreze et al. 1989). This is relatively common in the late summer and early autumn, and is known as a 'Beaufort Gyre Reversal'. This can cause the pack ice to temporarily move counter-clockwise rather than clockwise.

The Transpolar Drift is another important aspect of sea ice motion in the Arctic Ocean (Fig. 11.4). This current results in an almost linear movement of sea ice from the Siberian Coast across the North Pole and out through Fram Strait (Serreze et al. 1989). This drift is responsible for exporting 20% of the Arctic Ice Pack annually, and can include the export of ice islands.

T-3 provides a good example of an ice island which followed the Arctic Ocean Drift route. This ice island was first discovered in 1950 in the northern Chukchi Sea at 75°24′N/173°00′W, although it is likely to have calved around 1935 (Li 1991; Wadhams 2000). From 1952 to 1979 it was used almost continuously as a research station, and thus its locations during that time are well known (Li 1991). During this period, it completed at least three circuits of the Beaufort Gyre and remained in the Arctic Ocean for at least 35 years (Fig. 11.3a; Sackinger et al. 1991). Finally, in 1983–1984 the ice island drifted out of the Arctic Ocean via the Transpolar Drift, and was spotted twice more off the southern tip of Greenland, presumably where it finally disintegrated. Li (1991) states that wind conditions near the North Pole define whether an ice island continues circulation in the Beaufort Gyre or whether it joins the Transpolar Drift. For T-3, these wind conditions only enabled it to join the Transpolar Drift on its fourth loop around the Arctic Ocean.

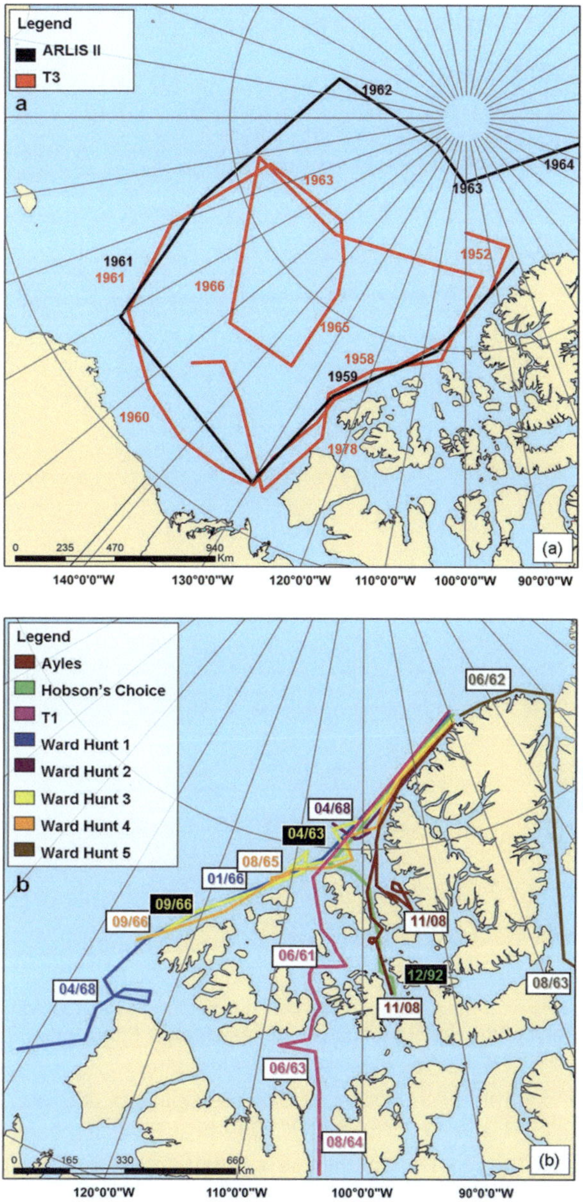

Fig. 11.3 (**a**) and (**b**) Reconstructed drift tracks for ice islands described in the text and Table 11.2

The drift of T-3 represents the longest recorded drift of an ice island, with a residence time in the Arctic Ocean of at least 35 years. It also illustrates the importance of the Beaufort Gyre in driving the drift of ice islands, as well as the role of the Transpolar Drift in ejecting ice islands from the Arctic Ocean. Ice Island ARLIS-II

Fig. 11.4 Surface currents in the Arctic Ocean and surrounding area (Reproduced from AMAP (1998), with permission of the Arctic Monitoring and Assessment Programme, Norway)

also provides an example of an Arctic Ocean Drift pattern that was similar to T-3 (Fig. 11.3a). ARLIS-II was first spotted in 1959 near Ellef Ringnes Island (Foster and Marino 1986), although its location was not regularly monitored until a research station was established on it in May 1961 when it was north of Point Barrow, Alaska (Tables 11.1, 11.2; Sackinger et al. 1991). It then took only three years to travel across the Arctic Ocean and be ejected by the Transpolar Drift through Fram Strait, without making a complete circuit in the Beaufort Gyre.

11.5.2 Archipelago Drift

There are many examples of ice islands that have drifted into the channels of the CAA, including T-1, Hobson's Choice and Ayles (Lindsay 1975, 1977; Sackinger et al. 1989; Jeffries and Shaw 1993; Copland et al. 2007; Fig. 11.3b). In addition, Koenig et al. (1952) and Montgomery (1952) found a total of 59 ice islands within the channels of the CAA in the late 1940s and early 1950s (as discussed in Sect. 11.4.1), and Sackinger et al. (1991) noted that in 1990 many smaller ice islands probably entered the Archipelago in addition to Hobson's Choice.

The drift of Hobson's Choice is perhaps the best documented of any ice island from calving to disintegration, in part helped by the presence of a Canadian research base on it (Foster and Marino 1986; Sackinger et al. 1989; Table 11.1), and also because it was tracked in some of the first SAR scenes of this region (Jeffries and Sackinger 1990). The ice island calved from Ward Hunt Ice Shelf in winter 1982 and was spotted nearby to the west for the first time in April 1983. ARGOS satellite tracking buoys were placed on the island in August 1983 (Sackinger et al. 1989), and by summer 1988 it had drifted a net distance of ~1000 km to the northern coast of Ellef Ringnes Island (Fig. 11.3b). This rate of movement was relatively slow, with the island experiencing frequent irregular motion and reversals along its travel path, particularly during the summer and falls of 1984–1986 when it became largely stalled at the northern edge of Axel Heiberg Island (Jeffries 1992a). Reconstructions by Jeffries and Shaw (1993) indicate that summer wind patterns over this period were often opposite to the normal direction of the Beaufort Gyre. This had the effect of both slowing the normal westward drift in this region and increasing the pack ice pressure against the northern coastline of the CAA. Further evidence for the compressed state of the pack ice during this period is provided by the fact that 8–12 smaller ice islands that broke off at a similar time to Hobson's Choice all stayed within a roughly 20 km radius of it (Jeffries and Sackinger 1990).

Sackinger et al. (1989) studied the motion of Hobson's Choice when it was located to the NW of Axel Heiberg Island during two periods of movement in May and June 1986, and found that its motion amongst the pack ice was generally preceded by an offshore surface wind and a wind speed of >6 m s^{-1}. Lu and Sackinger (1989) argued that the large NE motion of the ice island during these two events and three other events in summer 1986 was related to a 'mountain barrier effect', in which dominant westerly winds hit the mountains of Axel Heiberg Island and were deflected to the north by the Coriolis effect. They argued that this effect occurs up to 150 km away from the mountains, and that it should therefore be taken into account when predicting ice island movement.

In September and early October 1988, Hobson's Choice moved >100 km eastwards, towards the mouth of Peary Channel between Ellef Ringnes and Meighen Islands. Jeffries and Shaw (1993) state that this occurred during a reversal of the Beaufort Gyre, driven by a semi-stationary low pressure system over the Arctic Ocean. There was then a significant change in drift direction on October 10, 1988, when the ice island started moving southwards due to strong northerly wind flow off

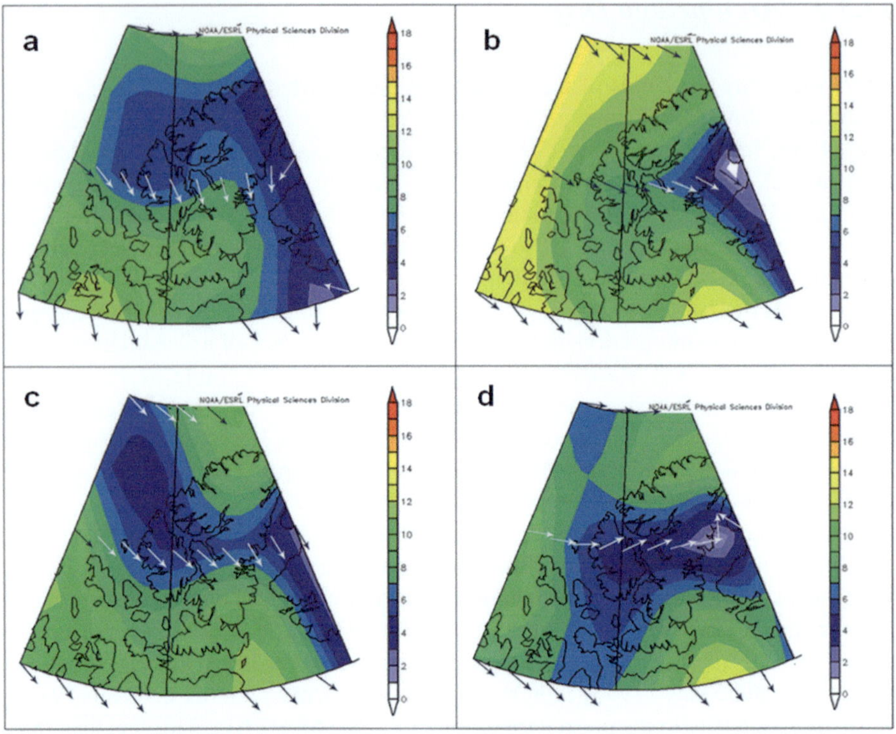

Fig. 11.5 NCEP reanalysis of surface wind conditions (in m s^{-1}) which allowed the drift of Hobson's Choice Ice Island into Peary Channel: (**a**) Oct. 9, 1988; (**b**) Oct. 10, 1988; (**c**) Oct. 11, 1988; (**d**) Oct. 12, 1988

the Arctic Ocean during October 9-13 (Jeffries and Shaw 1993). These conditions are confirmed by NCEP climate reanalyses which indicate that the wind strength averaged at least 9 m s^{-1} from October 9–12, 1988, with winds blowing directly into the opening of Peary Channel (Fig. 11.5). Winds alone are insufficient to explain the movement of Hobson's Choice, however, as the sea ice in the mouth of Peary Channel must have been mobile and in lower concentrations than normal (median 8–9/10 coverage) for this time. Once it had entered the mouth of the channel, it became stuck against the fast ice edge for almost a year until it fully entered it in August–September 1989 (Jeffries 1992a).

For the next couple of years Hobson's Choice generally moved short distances within Peary Channel, until August 1991 when it started moving rapidly southwards, passing into Queen's Channel (north of Cornwallis Island) and covering a total of 421 km by November 1991. Jeffries and Shaw (1993) analysed NOAA-AVHRR satellite imagery from this period, and found that there was extensive breakup of sea ice and open water in the channels where the ice island entered. This is confirmed in Canadian Ice Service charts from this time, which indicate clearing around Ellef Ringnes Island from August 22, 1991, until the ice pack consolidated again in mid-October 1991. In addition, air temperatures in summer 1991 were 2–4°C warmer than normal for this region. Thus, once sea ice conditions were open enough for movement to occur, winds further drove the ice island southwards. Indeed, Jeffries and Shaw (1993) highlight six periods of 4–7 days in autumn 1991 when winds were complimentary to the drift track. It is also likely that the drift was aided by southward flowing currents through Peary Channel (Fig. 11.4), although the strength of these currents is not quantified (Jeffries and Shaw 1993). Kwok (2006) showed that on average there is a net annual inflow of sea ice from the Arctic Ocean into the CAA via Peary and Sverdrup channels. However, this ice movement is typically restricted to a short period in the summer as fast ice covers the channels for most of the year (Dunbar 1973; Melling 2002).

Hobson's Choice ice island started disintegrating in late 1991, and was completely broken up by the end of 1992 (Jeffries and Shaw 1993). Up until summer 1991 it had been a stable platform strong enough to house a research base, so the question arises of why it became unstable so quickly. One possible answer is that the region where it broke up contains many polynyas and is an area of significant upwelling (Stirling 1997), so this warmer upwelling water likely increased basal melting. Alternatively, the widespread occurrence of summer surface meltwater ponds may have accelerated surface melt rates and exposed existing weaknesses in the ice structure via hydro-fracturing, as has been observed with the disintegration of some Antarctic Ice Shelves (MacAyeal et al. 2003)

The drift history of the Ayles Ice Island is similar to that of Hobson's Choice. This ice island first broke away from the Ayles Ice Shelf on August 13, 2005, and was initially 66.4 km^2 in area, with an additional 20.7 km^2 of smaller fragments lost at the same time (Copland et al. 2007). This event was unique as it was the first time that repeat satellite imagery (MODIS) recorded an Arctic ice shelf breaking apart in near real-time. When the ice island first calved it rapidly rotated away from the coast at a speed of ~1 km h^{-1} for the first 5 h (Copland et al. 2007). Some 5 days after it broke off it had drifted ~70 km to the west of the Ayles Ice Shelf, but a month later it had moved some 32 km back east. RADARSAT-1 imagery indicates that it then moved little until early 2007. A satellite tracking beacon was placed on the ice island in May 2007, and a second one was placed on the other half of it in April 2008 after it had broken apart in September 2007.

NCEP reanalysis reveals that the dominant winds in summer 2006 were from the southwest, along the coastline of Ellesmere Island, which would have acted against the westward drift of the Beaufort Gyre and worked to compress the pack ice along the coastline of Ellesmere Island. This is similar to the pattern that occurred with Hobson's Choice in 1984–1986 when it stalled along northern Axel Heiberg Island.

In 2007, the Ayles ice island moved ~55–75 km SW between mid-January and early February, but then moved little until mid-July 2007. In late July and in August 2007 it started moving SW again, with a total drift of ~220 km between August 6 and 30 (>9 km day^{-1}). NCEP reanalysis indicates a reduction in the prevailing onshore winds over northern Ellesmere Island during this time, with the resulting open coastal water allowing the rapid drift (Fig. 11.6a). In September 2007 the winds were stronger and more northerly, which pushed the pack ice back against the

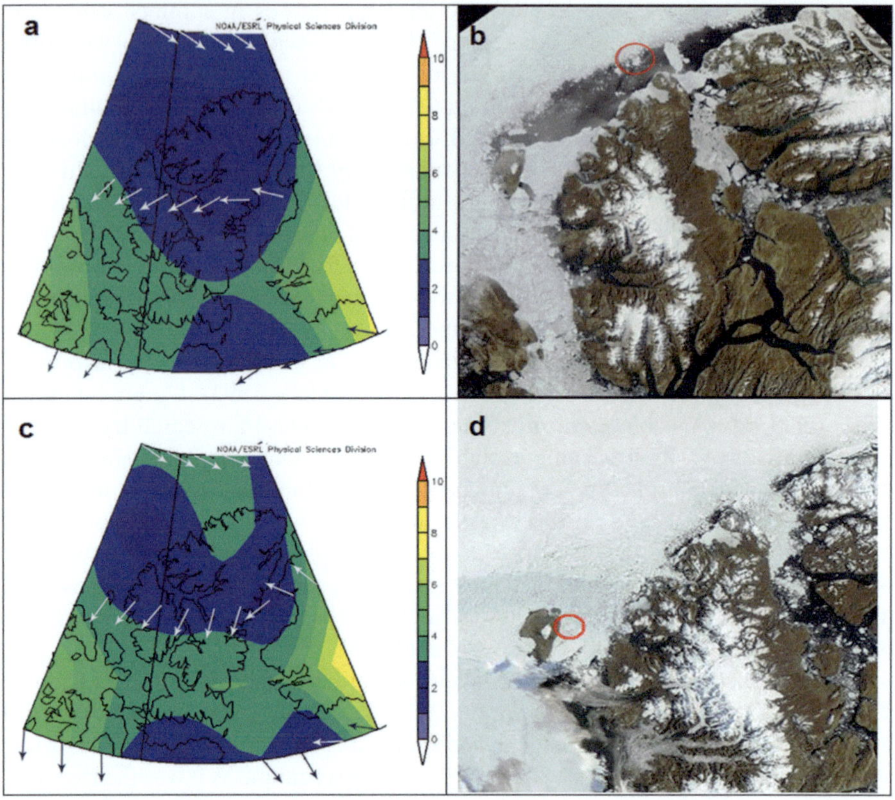

Fig. 11.6 Drift of the Ayles Ice Island: (**a**) Predominantly low surface winds (in m s^{-1}) over northern Ellesmere Island in August 2007; (**b**) Resulting low sea ice conditions along northern Ellesmere Island that allowed for the rapid westward drift of the ice island (marked by *red circle*; MODIS image from August 20, 2007); (**c**) Dominant onshore surface wind flow (in m s^{-1}), September 2007; (**d**) Resulting southerly motion of the ice island into Sverdrup Channel to the east of Meighen Island (MODIS image base from September 2, 2007, courtesy of the Rapid Response Project at NASA/GSFC)

coastline and allowed for the southward movement of the ice island into Sverdrup Channel between Meighen and Axel Heiberg islands (Fig. 11.6b), through the location where the Sverdrup Ice Plug has historically existed (Pope et al. 2017). On September 4, 2007, the Ayles Ice Island fragmented into two pieces on the east side of Meighen Island, possibly due to contact with the ocean floor given its proximity to the coast when it broke up (Fig. 11.7a, b).

During September 2007 the southward drift of the two pieces averaged ~3 km day^{-1}. Canadian Ice Service Ice Charts for Peary and Sverdrup channels showed that ice coverage was the lowest in the 40-year record for the week of September 10, 2007 (0.59 coverage versus 40-year median of 0.66), meaning that the ice islands were able to drift faster than usual at this time. However, regional temperatures dropped quickly at the end of September 2007 to about −15°C, which allowed the

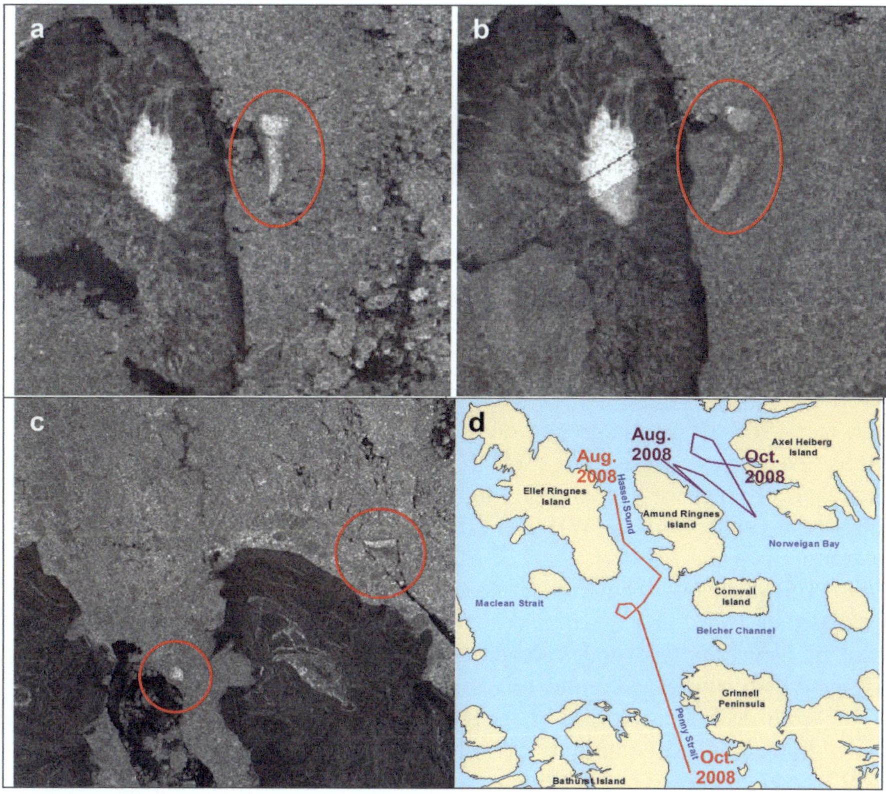

Fig. 11.7 RADARSAT-1 imagery of Ayles Ice Island: (**a**) On the east side of Meighen Island on September 3, 2007, the day prior to fracturing; (**b**) After fracturing on September 4, 2007; (**c**) The two pieces lodged in sea ice either side of Amund Ringnes Island on September 23, 2007, where they remained until early August 2008 (All imagery © Canadian Space Agency). (**d**) August to October 2008 tracks of the two ice island pieces

development of new sea ice which impeded the further drift of the fragments, and by late September 2007 they were locked in the sea ice either side of Amund Ringnes Island (Fig. 11.7c), ending any further drift until summer 2008.

Previous ice islands have also had documented drift through Sverdrup Channel. For example, 1946 aerial photographs reveal that an ice island had entered the channel, while in 1947 another ice island was located between Amund Ringnes and Haig-Thomas Islands (Montgomery 1952). Many of the ice islands that were sighted during this time moved little from one year to the next: Montgomery (1952) states that one ice island spotted in 1948 was found in the same location in 1950.

In 2008, the two pieces of the Ayles Ice Island started moving from their positions on either side of Amund Ringnes Island in the first week of August, once the sea ice started melting and breaking up in this area. The piece on the eastern side generally moved little throughout the summer, drifting in a north-south zigzag pattern between Amund Ringnes and Axel Heiberg islands (Fig. 11.7d). However, the piece on the western side moved much more, initially drifting northwards towards the northern coast of Amund Ringnes Island. Later, it drifted >250 km southwards between mid-August and early October to enter Queen's Channel to the NW of Cornwallis Island (Fig. 11.7d). RADARSAT-1 imagery indicates that this piece broke into three smaller fragments between October 2 and 3, 2008 at 76°14′N/97°12′W, at the same time that the southward motion was stopped by the formation of new sea ice in this region.

11.5.3 Nares Strait Drift

In winter 1961–1962, a large portion of the Ward Hunt Ice Shelf calved and produced several large ice islands (WH-1, WH-2, WH-3, WH-4 and WH-5, named from their original west to east calving locations; Hattersley-Smith 1963; Jeffries and Serson 1983). Of these, WH-1 to WH-4 drifted westwards, while WH-5 drifted eastward along the northern coast of Ellesmere Island (Nutt 1966; Lindsay et al. 1968). This is the only known example of a Canadian ice island which experienced prolonged eastward drift after breaking off, and the only one which made it to the coast of Newfoundland and Labrador. Hattersley-Smith (1963) argued that the divergent drift patterns were due to Ward Hunt Island acting as a barrier, blocking WH-5 from drifting westward under the influence of easterly and north-easterly winds. Rather, WH-5 could only be influenced by winds that would push it eastward, as were recorded in fall 1962 at Alert weather station (Nutt 1966).

Once at the entrance to the Lincoln Sea, it is likely that WH-5 was drawn into Nares Strait by the strong currents and large volumes of sea ice that pass through this channel each year. Sadler (1976) states that almost 15% of the total annual outflow from the Arctic Ocean occurs via this route. Studies of the Lincoln Sea and Nares Strait by Agnew (1999) and Dunbar (1973) found that ice drains annually from this region via Nares Strait, although the intensity of this drainage varies greatly between years and is highly dependent on the formation of ice bridges that block the outflow. Samelson et al. (2006) found that landfast sea ice consolidation usually occurs in the strait in mid- to late-January, with Kwok (2005) stating that sea ice drift typically ceases at this time and doesn't resume until the following July or August. Drift speeds were dependent on localized winds, coastal geography and ocean currents and could reach up to 20 km day^{-1} in September.

Many observations of the ice island were made in 1962 and 1963, and these indicated that it entered Nares Strait sometime between December 1962 and February 1963 (Nutt 1966). Upon entering the strait its motion accelerated to an average of >2.5 km day^{-1}, and as it drifted through the channel it rotated and occasionally became grounded and lodged. For example, its ends became stuck on Ellesmere and Hans islands between February 28 and July 14, 1963, which had the effect of blocking most of the sea ice export through Nares Strait at this time (Nutt 1966). By July 27, 1963, WH-5 had drifted at least 20 km south and broken into

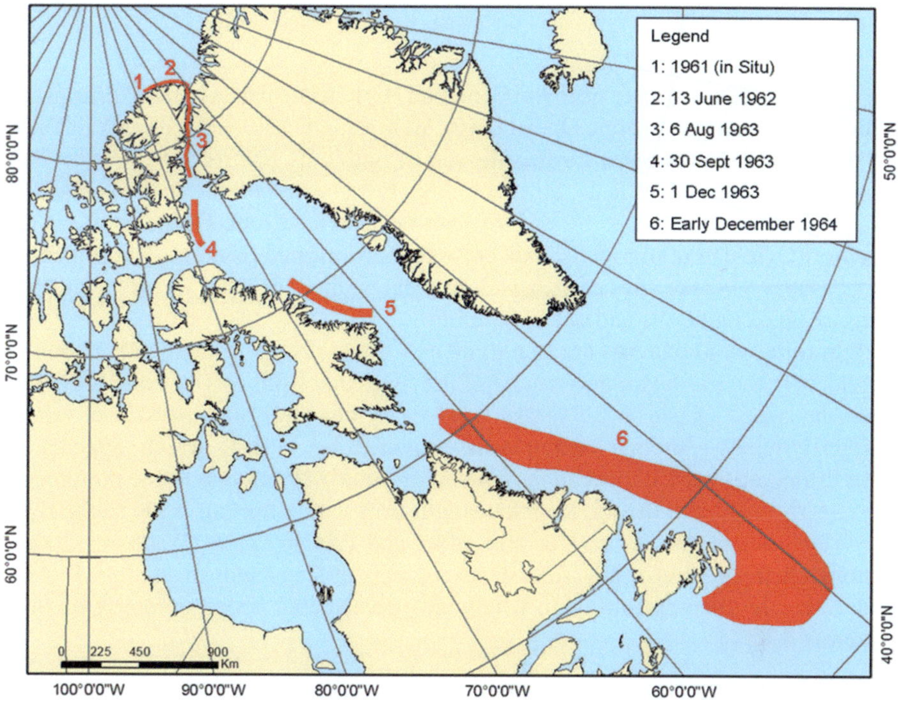

Fig. 11.8 Location of WH-5 ice island from calving to disintegration between 1961 and 1964 (Modified from Nutt (1966), with permission of the Arctic Institute of North America, Calgary)

three major pieces and many smaller pieces. Radio tracking beacons and oil drums were placed on two of the pieces in August 1963, which indicated that they were moving in a generally southerly direction (at speeds up to 16 km day^{-1} during strong wind events), although this motion was occasionally interrupted by stagnant or even slightly northward moving periods. Observations from the end of September onwards indicated that the pieces were now making rapid southerly progress, with aerial surveys in early 1964 finding sections scattered from the coast of Baffin Island in January to Labrador and the Grand Banks by April (Fig. 11.8). This rapid drift can be attributed to the Baffin and Labrador currents (Fig. 11.4; Nutt 1966), which are also effective at transporting icebergs from Greenland to this region. In total, it took <3 years from the time WH-5 calved to pieces of it reaching as far south as the Grand Banks, almost 5000 km away.

11.6 Summary

Ice islands which originate from the Canadian High Arctic can take three drift paths, which we have termed Arctic Ocean Drift, Archipelago Drift and Nares Strait Drift. Of these paths, Nares Strait Drift is the rarest, with only one recorded occurrence, while Arctic Ocean Drift is the most common. It is also important to note that either Archipelago Drift or Nares Strait Drift can occur after Arctic Ocean Drift, as the Beaufort Gyre can return ice islands close to their original location. The route that an ice island will take after calving is dependent on three primary factors: wind patterns, sea ice conditions and ocean currents.

Prevailing wind patterns are important as a direct driver of drift, with examples provided by the southward movement of the Ayles and Hobson's Choice ice islands from the Arctic Ocean into Sverdrup and Peary channels after periods of strong northerly winds. Another example is provided by WH-5, which appears to have drifted against local ocean currents after initial calving due to the dominance of westerly winds. In theory, ice islands will drift with a turning angle to the right of surface winds due to the Coriolis effect and Ekman spiral (Wadhams 2000), although there are few, if any, field measurements of ice islands to test this. In general, there are three reasons as to why winds have such a strong influence on drift patterns:

1. Ice islands have large surface areas and side profiles on which winds can act.
2. Ice islands can become entrained in surrounding pack ice, whose movement is strongly influenced by winds.
3. Winds can push pack ice away from shorelines creating open water areas in which ice islands can drift.

In relation to this final point, it is clear that wind has a strong influence on sea ice conditions, and that rapid ice island movements can typically only occur when there are low sea ice concentrations. This is particularly true in the summer, when reduced sea ice coverage makes the ice pack more susceptible to movement. Prevailing winds tend to push sea ice against the NW coastline of the CAA (Alt et al. 2006), which has the effect of closing all open water in this area. Consequently, for an ice island to drift freely along this coastline, a change in the prevailing wind patterns must occur to induce open water conditions. An example of this is provided by the motion of Hobson's Choice Ice Island close to Axel Heiberg Island, when the movement of pack ice away from the shoreline was followed by a rapid movement of the ice island parallel to the coast (Sackinger et al. 1989). Another example is provided by the Ayles Ice Island, which moved very little in summer 2006 when it was trapped in sea ice against the coast of Ellesmere Island, yet moved rapidly in August 2007 when a reduction or reversal in the prevailing onshore winds allowed sea ice to move away from the coastline (Fig. 11.6a, b). It is therefore clear that the effects of wind action on sea ice may be a more important factor for ice island motion than winds alone. Sea ice conditions also play a central role in determining ice island movement in the fall and winter, as drift significantly slows as new ice develops, and stops when the pack ice completely consolidates near a coastline. Consequently, many ice islands move little, if at all, during the winter months.

Ocean currents and dominant atmospheric circulation patterns influence ice island drift both on the very large scale (Beaufort Gyre, Transpolar Drift) and on the small scale (within the CAA). The initial westward drift experienced by all ice islands reviewed here (except WH-5) is caused by the Beaufort Gyre, which is driven by the anti-cyclonic winds common over this part of the Arctic Ocean. The Beaufort Gyre can trap and circulate ice islands in the Arctic Ocean for decades (e.g., T-3). For ice islands which follow this Arctic Ocean Drift path, they can only escape these circuits if wind conditions are favourable for them to enter the Transpolar Drift when they approach the North Pole (Li 1991). On the local scale, ocean currents in the CAA have been shown to strongly influence where ice islands drift once they have entered this region. For example, the rapid southerly movement of Ice Island WH-5 was influenced by the strong currents in Nares Strait, while the drift of Hobson's Choice Ice Island was strongly influenced by the currents through Peary and Queens Channels. In general, there is a net annual export of sea ice and ocean water from the Arctic Ocean through the CAA (Kwok 2005, 2006; Alt et al. 2006; Howell et al. 2013; Fig. 11.4).

In terms of the lifespan of an ice island, this is highly dependent on the drift pattern that it takes. Ice islands which experience Arctic Ocean Drift typically survive the longest (e.g., >35 years in the case of T-3), while those that undergo Archipelago or Nares Strait Drift may only survive for a few years (e.g., <5 years for WH-5). When drifting in the Arctic Ocean, summer surface melting accounts for approximately 1 m of surface ice loss per year, although this is somewhat dependent on latitude (Jeffries 1992a). For example, Smith (1961) found losses on T-3 of up to 1.35 m a^{-1}, and Schraeder (1968) measured surface melting of almost 1.0 m on ARLIS-II one summer. Jeffries et al. (1988) found that between 1952 and 1973, surface melting contributed to a steady reduction in the thickness of T-3 from ~60 to ~30 m (0.7 m $year^{-1}$). Surface melting does not necessarily equate to ice loss, however, as meltwater tends to pool in the troughs common on ice island surfaces, refreezing in the winter. This pooling can also increase surface melt rates by up to twice compared to bare ice, as lower albedo water absorbs more energy than reflective ice.

Ice island decay can also occur due to side and bottom melting from contact with warmer waters, particularly for ice islands that have drifted out of the Arctic Ocean via 'Archipelago Drift' or 'Nares Strait Drift'. This rate of melt is dependent on the temperature and salinity of the waters surrounding the ice island, as well as the relative velocity of the feature and near-surface waters (Wadhams 2000). This is because the melt from the ice island changes the temperature and salinity properties of the surrounding waters, meaning that the ice island can escape the influence of its own meltwater and lose mass more quickly if drift is fast, whereas melting is slowed if drift is slow and the ice island is surrounded by its own meltwater. There are no direct measurements of basal melt rates on ice islands, although the rapid disintegration of WH-5 once it drifted into relatively warm waters along Baffin Island suggests that it can be significant for some drift paths.

It may also be possible that ice islands can break apart due to a process termed 'break-up by flexure' (Wadhams 2000). In this scenario, ice islands can fracture due to the flex strain they experience when coming into contact with long period ocean swells and wave action. MacAyeal et al. (2006) argue that this was an important influence in the breakup of iceberg B-15 in Antarctica, although this mechanism would likely only be important for ice islands which have exited the Arctic Ocean due to the moderating influence of the sea ice pack on ocean swells. Flexure is also likely important when one part of an ice island becomes grounded, but the other part is still floating. In this case, a 'tide crack' would develop at the boundary between these areas and act as a weak point in the ice structure. Finally, ice island disintegration may also be caused by mechanical erosion via direct contact with the sea bed in shallow waters (Wadhams 2000).

11.7 Conclusions

Given the resurgence of interest in oil development in the Beaufort Sea, it is clear that we need a good understanding of ice island drift patterns to properly evaluate the risks of collision with these structures. The review here outlines how ice islands can take one of three different drift paths once they have calved from northern Ellesmere Island, with winds, sea ice conditions and ocean currents providing primary controlling factors. Past ice islands have predominantly stayed in the Arctic Ocean, although recent rapid reductions in sea ice may mean that future ice islands will find it easier to enter the channels of the CAA. This is a factor to be taken into account for the predicted increase in ship traffic through the NW Passage as it becomes increasingly ice-free in summer. The recent ice shelf breakup events along northern Ellesmere Island have produced more ice islands in the short term, although fewer will be present in the long term as the remaining ice shelves are completely lost.

Acknowledgments We would like to thank the Natural Sciences and Engineering Research Council of Canada, Canada Foundation for Innovation, Ontario Research Fund and the University of Ottawa for funding assistance. We would also like to thank the Canadian Ice Service, BBC News, Derek Mueller, Andrew Hamilton, Trudy Wohlleben, Bea Alt, Laurie Weir, Luc Desjardins, John Falkingham, Roger DeAbreu and Doug Bancroft for assistance with data collection and analysis. NCEP Reanalysis data provided by NOAA/OAR/ESRL PSD, Boulder, CO. Helpful reviews from Bea Alt and an anonymous reviewer are appreciated.

References

Agnew, T. A. (1999). Drainage of multiyear sea ice from the Lincoln Sea. In *Proceedings of the Conference on the Arctic Buoy Programme*, University of Washington, Seattle, Washington, 3–4 August 1998.

Alt, B., Wilson, K., & Carrieres, T. (2006). A case study of old-ice import and export through Peary and Sverdrup channels in the Canadian Arctic Archipelago: 1998–2005. *Annals of Glaciology, 44*, 329–338. doi:10.3189/172756406781811321.

Althoff, W. F. (2007). *Drift station: Arctic outposts of superpower science*. Washington, DC: Potomac Books.

Althoff, W. F. (2017). The military importance and use of ice islands during the Cold War. In L. Copland & D. Mueller (Eds.), *Arctic ice shelves and ice islands* (p. 343–366). Dordrecht: Springer. doi:10.1007/978-94-024-1101-0_13.

AMAP. (1998). *AMAP assessment report: Arctic pollution issues*. Arctic Monitoring and Assessment Programme (AMAP), Oslo, Norway, pp. 859.

Argos. (2007). *Argos user's manual 2007–2008*. Ramonville-Saint-Agne: Collecte Localisation Satellites.

Belkin, I. M., & Kessel, S. A. (2017). Russian drifting stations on Arctic ice islands. In L. Copland & D. Mueller (Eds.), *Arctic ice shelves and ice islands* (p. 367–393). Dordrecht: Springer. doi:10.1007/978-94-024-1101-0_14.

Canadian Ice Service. (2005). *MANICE: Manual of standard procedures for observing and reporting ice conditions* (Rev. 9th ed.). Ottawa: Environment Canada.

Cogley, J. G., & Adams, W. P. (2000). Remote-sensing resources for monitoring glacier fluctuations on Axel Heiberg Island. *Arctic, 53*, 248–259. doi:10.14430/arctic856.

Copland, L., Mueller, D. R., & Weir, L. (2007). Rapid loss of the Ayles Ice Shelf, Ellesmere Island, Canada. *Geophysical Research Letters, 34*, L21501. doi:10.1029/2007GL031809.

Copland, L., Mortimer, C., White, A., Richer McCallum, M., & Mueller, D. (2017). Factors contributing to recent Arctic ice shelf losses. In L. Copland & D. Mueller (Eds.), *Arctic ice shelves and ice islands* (p. 263–285). Dordrecht: Springer. doi:10.1007/978-94-024-1101-0_10.

Crary, A. P. (1958). Arctic ice island and ice shelf studies, Part I. *Arctic, 11*, 3–42. doi:10.14430/arctic3731.

Crary, A. P. (1960). Arctic ice island and ice shelf studies, Part II. *Arctic, 13*, 32–50. doi:10.14430/arctic3687.

Dunbar, M. (1973). Ice regime and ice transport in Nares Strait. *Arctic, 26*, 282–291. doi:10.14430/arctic2927.

England, J. H., Lakeman, T. R., Lemmen, D. S., Bednarski, J. M., Stewart, T. G., & Evans, D. J. A. (2008). A millennial-scale record of Arctic Ocean sea ice variability and the demise of the Ellesmere Island ice shelves. *Geophysical Research Letters, 35*, L19502. doi:10.1029/2008GL034470.

England, J., Evans, D. A., & Lakeman, T. (2017). Holocene history of Arctic ice shelves. In L. Copland & D. Mueller (Eds.), *Arctic ice shelves and ice islands* (p. 185–205). Dordrecht: Springer. doi:10.1007/978-94-024-1101-0_7.

Falkner, K. K., Melling, H., Munchow, A. M., Box, J. E., Wohlleben, T., Johnson, H., Gudmandsen, P., Samelson, R., Copland, L., Steffen, K., Rignot, E., & Higgins, A. K. (2011). Context for the recent massive Petermann Glacier calving event. *Eos, Transactions of the American Geophysical Union, 92*(114), 117–118.

Foster, M., & Marino, C. (1986). *The Polar Shelf: the saga of Canada's Arctic scientists*. Toronto: NC Press Limited.

Frolov, I. E., Gudkovich, Z. M., Radionov, V. F., Shirochkov, A. V., & Timokhov, L. A. (2005). *The Arctic Basin: Results from the Russian drifting stations*. Berlin: Springer.

Fuglem, M., & Jordaan, I. (2017). Risk analysis and hazards of ice islands. In L. Copland & D. Mueller (Eds.), *Arctic ice shelves and ice islands* (p. 395–415). Dordrecht: Springer. doi:10.1007/978-94-024-1101-0_15.

Halliday, H. A. (2008). Mapping in full flight: Air Force, Part 28. *Legion Magazine*. http://www.legionmagazine.com/en/index.php/2008/08/mapping-in-full-flight/

Halliday, E. J., King, A., Bobby, P., Copland, L., & Mueller, D. (2012). Petermann Ice Island 'A' survey results, offshore Labrador. In *Proceedings, Arctic Technology Conference*, Houston. doi: 10.4043/23714-MS.

Hattersley-Smith, G. (1957). The rolls on the Ellesmere ice shelf. *Arctic, 10*, 32–44. doi:10.14430/arctic3753.

Hattersley-Smith, G. (1963). The Ward Hunt Ice Shelf: Recent changes at the ice front. *Journal of Glaciology, 4*, 415–424. doi:10.3198/1963JoG4-34-415-424.

Higgins, A. K. (1988). Glacier velocities in North and North-East Greenland. *Grønlands Geologiske Undersøgelse Rapport, 140*, 102–105.

Higgins, A. K. (1989). North Greenland ice islands. *Polar Record, 25*, 207–212. doi:10.1017/S0032247400010809.

Hill, B. T., Ruffman, A., & Ivany, K. (2008, July 20–23). Historical data added to the Grand Banks iceberg database. In *Proceedings ICETECH 08 – 8th International Conference and Exhibition on Performance of Ships and Structures in Ice*, Banff, AB, Canada, Paper 109-R0.

Howell, E. L., Wohlleben, T., Dabboor, M., Derksen, C., Komarov, A., & Pizzolato, L. (2013). Recent changes in the exchange of sea ice between the Arctic Ocean and the Canadian Arctic Archipelago. *Journal of Geophysical Research, Oceans, 118*(7), 3595–3607. doi:10.1002/jgrc.20265.

Jeffries, M. O. (1986). Ice island calvings and ice shelf changes, Milne Ice Shelf, N.W.T. *Arctic, 39*, 15–19. doi:10.14430/arctic2039.

Jeffries, M. O. (1987). The growth, structure and disintegration of Arctic ice shelves. *Polar Record, 23*, 631–649. doi:10.1017/S0032247400008342.

Jeffries, M. O. (1992a). Arctic ice shelves and ice islands: Origin, growth and disintegration, physical characteristics, structural-stratigraphic variability and dynamics. *Reviews of Geophysics, 30*, 245–267. doi:10.1029/92RG00956.

Jeffries, M. O. (1992b). The source and calving of ice island ARLIS-II. *Polar Record, 28*, 137–144. doi:10.1017/S0032247400013437.

Jeffries, M. O. (2017). The Ellesmere ice shelves, Nunavut, Canada. In L. Copland & D. Mueller (Eds.), *Arctic ice shelves and ice islands* (p. 23–54). Dordrecht: Springer. doi:10.1007/978-94-024-1101-0_2.

Jeffries, M. O., & Sackinger, W. M. (1990). Ice island detection and characterization with airborne synthetic aperture radar. *Journal of Geophysical Research, 95*, 5371–5377. doi:10.1029/JC095iC04p05371.

Jeffries, M. O., & Serson, H. V. (1983). Recent changes at the front of Ward Hunt Ice Shelf, Ellesmere Island, N.W.T. *Arctic, 36*, 289–290. doi:10.14430/arctic2278.

Jeffries, M. O., & Shaw, A. M. (1993). The drift of ice islands from the Arctic Ocean into the channels of the Canadian Arctic Archipelago: The history of Hobson's Choice Ice Island. *Polar Record, 29*, 305–312. doi:10.1017/S0032247400023950.

Jeffries, M. O., Sackinger, W. M., & Shoemaker, H. D. (1988). Geometry and physical properties of ice islands. In W. M. Sackinger & M. O. Jeffries (Eds.), *Port and ocean engineering under Arctic conditions, Geophysical Institute* (Vol. 1, p. 69–83). Fairbanks: University of Alaska Fairbanks.

Jeffries, M. O., Reynolds, G. J., & Miller, J. M. (1992). First Landsat multi-spectral scanner images of the Canadian Arctic north of 80°N. *Polar Record, 28*(164), 1–6. doi:10.1017/S0032247400020192.

Kato, K., & Kumakura, Y. (1996). First year ice interactions on Molikpaq: Measurements and experiments. *Journal of Marine Science and Technology, 1*, 220–229. doi:10.1007/BF02390798.

Koenig, L. S., Greenaway, K. R., Dunbar, M., & Hattersley-Smith, G. (1952). Arctic ice islands. *Arctic, 5*, 67–103. doi:10.14430/arctic3901.

Kwok, R. (2005). Variability of Nares Strait ice flux. *Geophysical Research Letters, 32*, L24502. doi:10.1029/2005GL024768.

Kwok, R. (2006). Exchange of sea ice between the Arctic Ocean and the Canadian Arctic Archipelago. *Geophysical Research Letters, 33*, L16501. doi:10.1029/2006GL027094.

Lazzara, M. A., Jezek, K. C., Scambos, T. A., MacAyeal, D. R., & Van Der Veen, C. J. (2008). On the recent calving of icebergs from the Ross Ice Shelf. *Political Geography, 31*, 15–26. doi:10.1080/10889379909377676.

Leary, W. M., & LeShack, L. A. (1996). *Project Coldfeet: Secret mission to a Soviet ice station*. Annapolis: Naval Institute Press.
Li, F. C. (1991). *Simulation of the recurrence probability of ice islands in the Arctic Ocean*. Ph.D. dissertation, University of Alaska Fairbanks, Fairbanks, pp. 169
Lindsay, D. G. (1975). *Sea ice Atlas of Arctic Canada, 1961–1968* (pp. 213). Ottawa: Energy Mines and Resources Canada. doi:10.4095/299024.
Lindsay, D. G. (1977). *Sea-ice Atlas of Arctic Canada, 1969–1974* (pp. 219). Ottawa: Energy Mines and Resources Canada. doi:10.4095/298827.
Lindsay, D. G. (1982). *Sea-ice Atlas of Arctic Canada, 1975–1978* (pp. 139). Ottawa: Energy Mines and Resources Canada. doi:10.4095/298828.
Lindsay, D. G., Seifert, W., & Anderson, N. (1968). Ice islands, 1967. *Arctic, 21*, 103–106. doi:10.14430/arctic3255.
Lu, M., & Sackinger, W. M. (1989). The mountain barrier effect and modification of tabular iceberg motion in a coastal zone. *Journal of Coastal Research, 5*, 701–710.
MacAyeal, D. R., Scambos, T. A., Hulbe, C. L., & Fahnestock, M. A. (2003). Catastrophic ice-shelf break-up by an ice-shelf-fragment-capsize mechanism. *Journal of Glaciology, 49*, 22–36. doi:10.3189/172756503781830863.
MacAyeal, D. R., Okal, E. A., Aster, R. C., et al. (2006). Transoceanic wave propagation links iceberg calving margins of Antarctica with storms in tropics and Northern Hemisphere. *Geophysical Research Letters, 33*, L17502. doi:10.1029/2006GL027235.
Melia, N., Haines, K., & Hawkins, E. (2016). Sea ice decline and 21st century trans-Arctic shipping routes. *Geophysical Research Letters, 43*(18), 9720–9728. doi:10.1002/2016GL069315.
Melling, H. (2002). Sea ice of the northern Canadian Arctic Archipelago. *Journal of Geophysical Research, 107*, 3181. doi:10.1029/2001JC001102.
Minerals Management Service. (2008). Sale day statistics for Chukchi Sea sale 193. U.S. Department of the Interior. http://www.mms.gov/alaska/cproject/Chukchi193/193Saleday/Sale193SaleDayStats.htm. Accessed 12 Nov 2008.
Montgomery, M. R. (1952). Further notes on ice islands in the Canadian Arctic. *Arctic, 5*, 183–187. doi:10.14430/arctic3910.
Mortimer, C., Copland, L., & Mueller, D. (2012). Volume and area changes of the Milne Ice Shelf, Ellesmere Island, Nunavut, Canada, since 1950. *Journal of Geophysical Research – Earth Surface, 117*, F04011. doi:10.1029/2011JF002074.
Mueller, D., Copland, L., Hamilton, A., & Stern, D. (2008). International Polar Year scientists and Canadian Rangers visit Arctic ice shelves just before massive loss in 2008. *EOS. Transactions of the American Geophysical Union, 89*(49), 502–503. doi:10.1029/2008EO490002.
Mueller, D., Copland, L., & Jeffries, M. O. (2017). Changes in Canadian Arctic ice shelf extent since 1906. In L. Copland & D. Mueller (Eds.), *Arctic ice shelves and Ice Islands* (p. 109–148). Dordrecht: Springer. doi:10.1007/978-94-024-1101-0_5.
Narod, B. B., Clarke, G. K. C., & Prager, B. T. (1988). Airborne UHF sounding of glaciers and ice shelves, northern Ellesmere Island, Arctic Canada. *Canadian Journal of Earth Sciences, 25*, 95–105. doi:10.1139/e88-010.
Newell, J. P. (1993). Exceptionally large icebergs and ice islands in eastern Canadian waters: A review of sightings from 1900 to present. *Arctic, 46*, 205–211. doi:10.14430/arctic1345.
Nutt, D. C. (1966). The drift of Ice Island WH-5. *Arctic, 19*, 244–262. doi:10.14430/arctic3432.
Peary, R. E. (1907). *Nearest the Pole: A narrative of the Polar expedition of the Peary Arctic Club in the S. S. Roosevelt, 1905–1906*. New York: Doubleday, Page & Co.
Peterson, I. K. (2005). Large tabular icebergs and ice islands off eastern Canada in 2001–2003 and their probable source. In *Proceedings, 18th International Conference on Port and Ocean Engineering Under Arctic Conditions (POAC '05)* (Vol. 1, p. 143–152). Potsdam, Port and Ocean Engineering under Arctic Conditions.
Pizzolato, L., Howell, S. E. L., Derksen, C., Dawson, J., & Copland, L. (2014). Changing sea ice conditions and marine transportation activity in Canadian Arctic waters between 1990 and 2012. *Climatic Change, 123*(2), 161–173. doi:10.1007/s10584-013-1038-3.

Pizzolato, L., Howell, S., Dawson, J., Laliberte, F., & Copland, L. (2016). The influence of declining sea ice on shipping activity in the Canadian Arctic. *Geophysical Research Letters*, *43*(23), 12146–12158. doi:10.1002/2016GL071489.

Pope, S., Copland, L., & Alt, B. (2017). Recent changes in sea ice plugs along the northern Canadian Arctic Archipelago. In L. Copland & D. Mueller (Eds.), *Arctic ice shelves and ice islands* (p. 317–342). Dordrecht: Springer. doi:10.1007/978-94-024-1101-0_12.

Rudkin, P., Young, C., Barron Jr, P., & Timco, G. W. (2005). Analysis and results of 30 years of iceberg management. In *Proceedings, 18th International Conference on Port and Ocean Engineering Under Arctic Conditions(POAC '05)* (Vol. 2, p. 595–604). Potsdam, Port and Ocean Engineering under Arctic Conditions.

Sackinger, W. M., Jeffries, M. O., Tippens, H., Li, F., & Lu, M. (1989). Dynamics of ice-island motion near the coast of Axel Heiberg Island, Canadian High Arctic. *Annals of Glaciology, 12*, 152–156.

Sackinger, W. M., Jeffries, M. O., Li, F., & Lu, M. (1991). *Ice island creation, drift, recurrences, mechanical properties, and interactions with Arctic offshore oil production structures*. U.S. Department of Energy Final Report DOE/MC/25027-3112, pp. 34.

Sadler, H. E. (1976). Water, heat and salt transports through Nares Strait, Ellesmere Island. *Journal of the Fisheries Research Board of Canada, 33*, 2286–2295.

Samelson, R. M., Agnew, T., Melling, H., & Munchow, A. (2006). Evidence of atmospheric control of sea-ice motion through Nares Strait. *Geophysical Research Letters, 33*, L02506. doi:10.1029/2005GL025016.

Schraeder, R. L. (1968). *Ablation of ice island ARLIS-II, 1961*. M.Sc. Thesis, University of Alaska, Fairbanks, pp. 59.

Serreze, M. C., Barry, R. G., & McLaren, A. S. (1989). Seasonal variations in sea ice motion and effects on sea ice concentrations in the Canada Basin. *Journal of Geophysical Research, 94*(C8), 10955–10970. doi:10.1029/JC094iC08p10955.

Smith, D. D. (1961). Sequential development of surface morphology on Fletcher's Ice Island, T-3. In G. O. Raasch (Ed.), *Geology of the Arctic* (Vol. 2, p. 896–914). Toronto: University of Toronto Press.

Stirling, I. (1997). The importance of polynyas, ice edges, and leads to marine mammals and birds. *Journal of Marine Systems, 10*, 9–21. doi:10.1016/S0924-7963(96)00054-1.

Storkerson, S. T. (1921). Drifting in the Beaufort Sea. In V. Stefanson (Ed.), *The friendly Arctic* (1st ed., p. 784). New York: MacMillan.

Tarbuck, E. J., Lutgens, F. K., Tsujita, C. J., & Hicock, S. R. (2009). *Earth: An introduction to physical geology* (2nd Canadian ed.). Toronto: Pearson Education Canada.

Wadhams, P. (2000). *Ice in the ocean*. London: Gordon and Breach Science Publishers.

White, A., Copland, L., Mueller, D., & Van Wychen, W. (2015). Assessment of historical changes (1959–2012) and the causes of recent break-ups of the Petersen Ice Shelf, Nunavut, Canada. *Annals of Glaciology, 56*(69), 65–76. doi:10.3189/2015AoG69A687.

Yan, M. H. (1986). *The relationship between ice island movement and weather conditions*. MSc. Thesis, University of Alaska Fairbanks, Fairbanks, pp. 91.

Zubov, N. N. (1955). *Arctic ice-islands and how they drift, Translation T176R* (p. 1–10). Ottawa: Defence Research Board.

Chapter 12
Recent Changes in Sea Ice Plugs Along the Northern Canadian Arctic Archipelago

Sierra Pope, Luke Copland, and Bea Alt

Abstract For most of the twentieth century, multiyear landfast sea ice (MLSI) existed in semi-permanent plugs across Nansen Sound and Sverdrup Channel in the northern Queen Elizabeth Islands (QEI), Canada. Between 1961 and 2004, these ice plugs only experienced simultaneous break-ups in 1962 and 1998. However, break-ups of both ice plugs have occurred in 9 out of the 12 years since 2005, indicating that these features are not reforming. The history of these plugs is reviewed using Canadian Ice Service ice charts, satellite imagery and a literature review. The weather systems associated with plug break-up events are related to the synoptic patterns defined by Alt (Atmos-Ocean 3:181–199, 1979). Most break-ups occur during Type III synoptic conditions, when a low centers over the Asian side of the Arctic Ocean and a warm pressure ridge develops over the QEI, creating warm temperatures, clear skies, and frequent wind reversals. Ice plug break-ups are also associated with reductions in sea ice concentration along the northwest coast of Ellesmere Island. The removal of these MLSI plugs in recent years aligns with ice shelf losses and reductions in age and thickness of sea ice in the Canadian Arctic Archipelago, with implications for ice import and export through these channels and the response of Arctic sea ice to a changing climate.

Keywords Ice plug • Sea ice • Multiyear landfast sea ice • Synoptics • Canadian Arctic Archipelago • Queen Elizabeth Islands

S. Pope (✉) • L. Copland
Department of Geography, Environment and Geomatics, University of Ottawa, Ottawa, ON, Canada
e-mail: sierra.g.pope@gmail.com; luke.copland@uottawa.ca

B. Alt
Balanced Environments Associates, Carlsbad Springs, Ottawa, ON, Canada
e-mail: bea.alt@sympatico.ca

L. Copland, D. Mueller (eds.), *Arctic Ice Shelves and Ice Islands*, Springer Polar Sciences, DOI 10.1007/978-94-024-1101-0_12

12.1 Introduction

There is strong evidence that Arctic sea ice extent is decreasing, with large area losses over the past decade compared to the long-term mean (Simmonds 2015), and a decrease in annual mean sea ice thickness over the central Arctic Basin from 3.59 m in 1975 to 1.25 m in 2012 (Lindsay and Schweiger 2015). Additionally, there is evidence of other cryospheric changes in the Arctic such as strongly negative mass balance on the Greenland Ice Sheet and the acceleration of outlet glaciers (Khan et al. 2015), break-up and mass loss from the Ellesmere Island ice shelves (Copland et al. 2007; Mueller et al. 2008, 2017; White et al. 2015), and widespread thawing of permafrost (Grosse et al. 2016).

This chapter reviews recent changes in ice plugs, which are semi-permanent sea ice features that form across channels in the Queen Elizabeth Islands (QEI) in the Canadian High Arctic, north of 74.5°N. In the past they remained in place for decades, but they have largely failed to reform since major break-up events in 1998. Changes to these multiyear landfast sea ice (MLSI) features have occurred in tandem with recent ice shelf losses in the Canadian Arctic Archipelago (CAA); this chapter addresses the history of these MLSI features and the mechanisms surrounding their break-ups.

The recent response of Arctic sea ice to changing climate conditions assists in putting ice plug changes in context. Observations since the 1950s indicate accelerating reductions in Arctic sea ice extent up to the present day, with the biggest losses in September. For example, reductions in September sea ice extent occurred at a rate of 6.9% decade^{-1} between 1979 and 2000, compared to a rate of 24.3% decade^{-1} between 2001 and 2013 (Meier et al. 2014). The increasing melt of first year ice, formed during the previous winter, is coupled with a decrease in area and thickness of floating multiyear ice, sea ice which has survived at least one melting season (Maslanik et al. 2007). An increased percentage of the total Arctic sea ice cover is therefore now first year ice, representing a significantly thinner and weaker barrier between ocean and atmosphere. For example, at the end of summer 2011 only 25% of the Arctic sea ice cover was >2 years old, compared to almost 60% during the 1980s (Stroeve et al. 2012). Landfast sea ice has also been forming later in the fall and breaking up earlier in the spring in many regions across the CAA since the 1980s (Galley et al. 2012). The break-up and subsequent melting of the thick, multiyear ice plugs and their replacement by thinner, annually-melting first year ice discussed here exemplifies the limited recovery of the old sea ice regime in the CAA since 1998.

12.1.1 Definition and Significance

Atmospheric circulation patterns in the Canadian High Arctic typically result in thick, multiyear sea ice being pushed against the northwest edge of the QEI (Agnew et al. 2001; Jeffers et al. 2001). This process contributes to the formation of MLSI, described further by Jeffries (2002) and Pope et al. (2012). This multiyear sea ice forms semi-permanent blockages, known as ice plugs, at the head of channels between islands. The presence of these decades-old plugs can act as effective barriers against movement of pack ice from the Arctic Ocean into the interior channels of the QEI (Jeffers et al. 2001; Agnew et al. 2001; Kwok 2006).

There is some disagreement about the terminology of sea ice blockages in these regions, so here we follow the definitions of Alt and Lindsay (2005). Ice arches or ice bridges are defined as structurally controlled, short-lived summer features that occur between narrow points of land surrounding a channel, sound, or between islands (e.g. Nares Strait; Fig. 12.1). Ice barriers are generally poorly defined, but are taken here to consist of large areas of consolidated ice at the border between the Arctic Ocean and the channels and seas of the QEI (e.g. Peary Channel; Fig. 12.1). Finally, ice plugs are small areas of consolidated perennial sea ice formed between narrow points of land, and are defined largely by their historical longevity and presence over decades. In the Canadian High Arctic, there are only two ice plugs known to have existed over the past century: the Nansen and Sverdrup (Fig. 12.1b). These are the focus of this chapter.

The Nansen Sound and Sverdrup Channel ice plugs have been observed as perennial ice features since the first exploration of the Canadian High Arctic in the early twentieth century. Peary (1907) described the northern coastline of Ellesmere Island as being fringed by a permanent body of ice. This fringe continued across Nansen Sound, creating a blockage now known as the Nansen Ice Plug (Vincent et al. 2001). Ice plugs form due to the in situ growth and agglomeration of sea ice in narrow channels in the coldest, most northerly parts of the QEI, where nearby land pins them in place and summer melting is insufficient to remove them (Jeffers et al. 2001; Serson 1972). Their formation is further supported by the fact that the Beaufort Gyre typically pushes thick, old sea ice towards the northwest coast of the QEI. This has the effect of providing a near-continual supply of additional ice and protects their northern margins from wave action. Once formed, the sea ice within ice plugs typically remains landfast and builds over time via snow accumulation and basal freeze-on of sea water.

These ice plugs are subject to in situ fracture and melting during the summer, but have retained their old MLSI for much of recorded history in this region. While each of the ice plugs has experienced occasional individual break-ups, the particularly warm summer of 1998 saw the break-up of both ice plugs for the first time since 1962 (Fig. 12.2) (Alt et al. 2006; Jeffers et al. 2001). Since 1998 the MLSI ice plugs have largely failed to reform, and in this chapter we describe the climate and synoptic conditions which contributed to both earlier break-up events and recent changes.

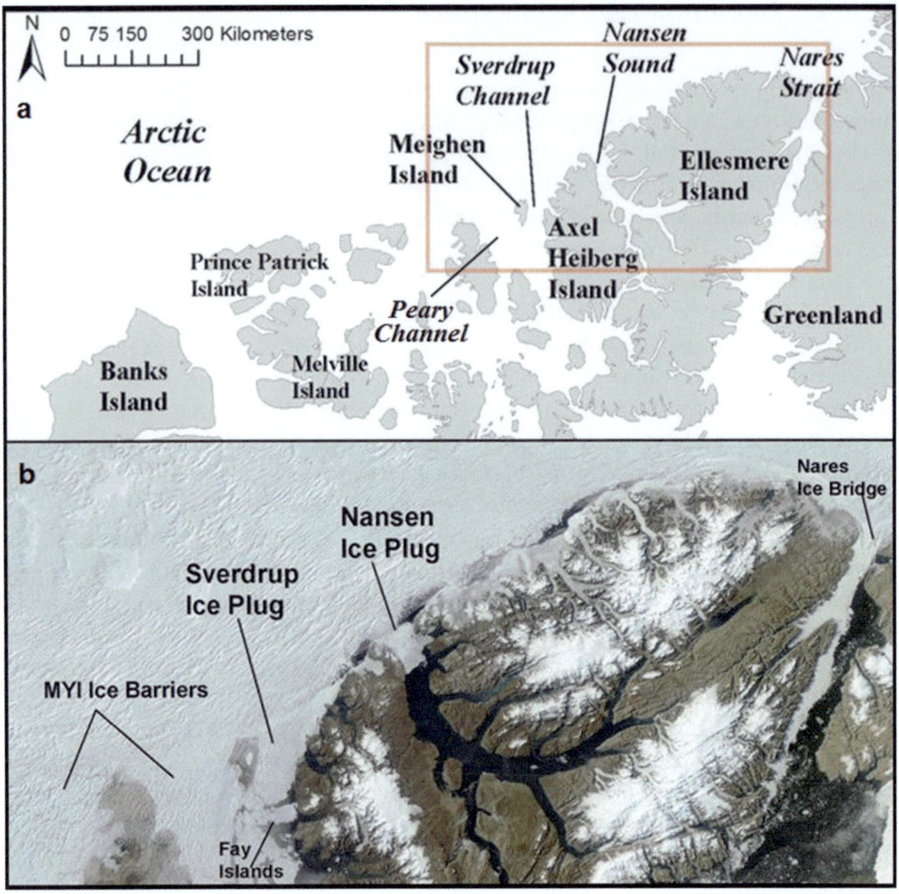

Fig. 12.1 (**a**) Location map of northern Queen Elizabeth Islands. (**b**) Location of major ice blockages on MODIS TERRA image, August 13, 2005 (imagery courtesy of MODIS Rapid Response Project, NASA/GSFC and University of Maryland, http://rapidfire.sci.gsfc.nasa.gov)

12.2 Regional Background

Sea ice atlases published by the Polar Continental Shelf Project (PCSP) provide data on ice plug conditions between 1961 and 1978 (Lindsay 1975, 1977, 1982). Ice charts produced by the Canadian Ice Service (CIS) provide coverage of the ice plugs from 1960 to present, and weekly regional ice charts for the past several decades are available online as part of the CIS Digital Archive (http://iceweb1.cis.ec.gc.ca/Archive/). Since 2000, during summer cloud-free conditions, daily or better 250 m resolution optical satellite imagery of the ice plugs is available from the Moderate Resolution Imaging Spectroradiometer (MODIS), which is flown on both the TERRA and AQUA satellites.

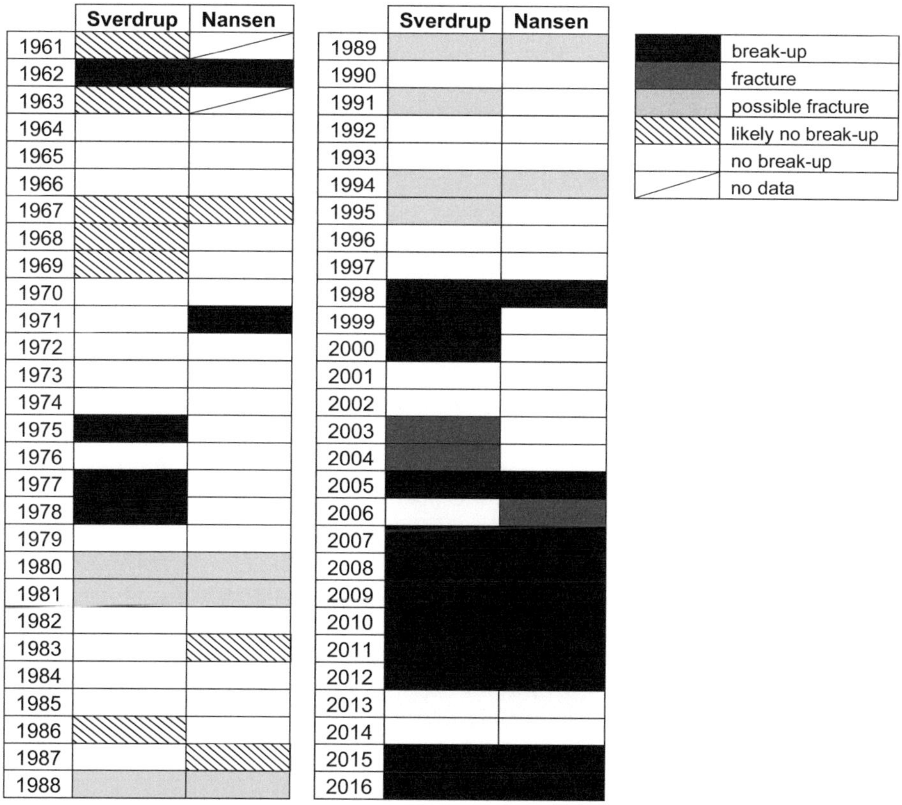

Fig. 12.2 Recorded break-up and fracture events of the Sverdrup and Nansen ice plugs, 1961–2016. Years without complete melt season coverage of the plug areas are analyzed for possible break-ups via ice codes and concentrations in ice charts at the end of the season or the behavior of the plug ice the following spring. Instances where in situ fractures cannot be ruled out due to lack of coverage or generalized ice codes are listed as 'possible fracture,' and instances with no evidence of break-up but incomplete coverage are listed as 'likely no break-up'

Data sources prior to 1960 are sparse and consist largely of visual observations associated with expeditions to the northern QEI (Peary 1907; Sverdrup et al. 1904; Stefansson 1938). A comprehensive survey of sea ice in the QEI was completed by Black (1962), who observed the fracturing of the Nansen and Sverdrup ice plugs and a subsequent movement of polar-basin ice into the interior passages of the QEI. Melling (2002) completed an overview of the region's sea ice conditions using the CIS Digital Archive from 1970 to 1999 and ice thickness measurements. His study indicated that landfast ice at 10/10ths concentration covers the Sverdrup Basin for over half the year, with 9/10ths ice presence during the summer months and a minimum ice extent in early September. The semi-permanent Sverdrup and Nansen ice plugs were a noted exception to this summer disintegration, remaining fast in most years of the study (Melling 2002). The Climate Change Action Fund (CCAF)

Summer 1998 Project team completed a comprehensive review of cryospheric variability in the Canadian Arctic over the anomalously warm summer of 1998. This included analysis of the 1998 Sverdrup and Nansen ice plug break-ups, as well as comparisons of regional temperature, wind and sea ice coverage patterns surrounding these events (Alt et al. 2001; CCAF Summer 1998 Project Team 2001; Jeffers 2001; Jeffers and McCourt 2001).

Annual sea ice flux near Sverdrup Ice Plug was estimated using AMSR-E and scatterometer image data at $2 \pm 6 \times 10^3$ km^2 a^{-1} from the Arctic Ocean in to the northern QEI between 1997 and 2002 (QEI-North Flux Gate in Kwok 2006). Agnew et al. (2008) calculated no net ice flux between 2002 and 2007 for the same region, noting lower annual variability in motion than the westerly regions of the northern QEI. The limited movement through this northern gate, which demonstrates the landfast nature of the sea ice in this region, suggested limited winter ice movement even without the presence of the northern plugs. However, decreasing sea ice extent and increasing sea ice movement in response to extremely warm summers suggests that the areas of mobile sea ice and the open-water limit will continue to move northward (Jeffers et al. 2001; Agnew et al. 2008). Regional studies of multiyear sea ice activity from 1968 to 2006 by Howell et al. (2008) indicate that since 2000, the CAA has experienced increases in dynamic import of multiyear ice from the Arctic Ocean. This has continued more recently, with an increase in Arctic Ocean multiyear ice inflow over the period 2005–2012 attributed to increased open water in the CAA that has provided more leeway for ice import to occur (Howell et al. 2013). These patterns and trends are consistent with evidence of weakening multiyear ice barriers in the QEI (Alt et al. 2006), and the increasing import of multiyear ice. Old sea ice import and export in Sverdrup Channel was assessed by Alt et al. (2006) following the 1998 plug break-up, who concluded that significant loss of old ice occurred there via melt and its movement out of the channel. Limited import of old Arctic sea ice into the channel was observed in 1999 and 2005, and significant through-flow of old ice was observed in 2000 (Alt et al. 2006).

The location of the QEI at a critical boundary between the ice-covered Arctic Ocean and open-water Baffin Bay creates difficulty in isolating a distinct system of synoptics to characterize weather behaviour in this region (Alt 1987). Early mass balance studies in the northern QEI were completed by Koerner (1977, 1979), and indicated low accumulation in the colder, dry interior of Ellesmere and Axel Heiberg islands and higher accumulation rates along the Arctic Ocean between Ellesmere Island's northwest coast and Meighen Island (Fig. 12.1). The relationships between regional synoptics and mass balance were studied by Alt (1979, 1987), who developed a categorization of summer atmospheric circulation controls on Meighen Ice Cap and other QEI ice caps and glaciers. These synoptics were revisited by Gardner and Sharp (2007) and Bezeau et al. (2015) in an assessment of High Arctic glacier mass balance, and are further discussed in Sect. 12.4 to provide a context in which the region's MLSI break-up events can be better understood.

12.2.1 Nansen Ice Plug

The Nansen Ice Plug occurs at the northern edge of Nansen Sound, between Ellesmere and Axel Heiberg islands (Fig. 12.1). All available evidence points to the fact that it existed as a near-permanent feature for most of the twentieth century. For example, while crossing the northern section of Nansen Sound in 1906, Peary noted that the ice in the sound appeared contiguous with the glacial fringe (ice shelf) along Ellesmere Island's northwest coast (Peary 1907). Visual observations of new pressure ice and open water north of Axel Heiberg Island from a 1932 expedition suggest minor fracturing in the plug at that time (Krüger Search Expeditions 1934), but that the blockage was still in place. The first formal sea ice surveys of this area also noted its presence (Black 1962), and the plug appears on all CIS ice charts from the beginning of the regional record in 1961 (Alt et al. 2001). The first detailed studies of this plug were completed by Serson in 1970–1971, who estimated an ice thickness of >6 m at the plug's northern edge and 10 m near the southern edge. The rolling ice surface featured hummocks ranging in height from 1 to 3 m (Serson 1972). Hummock heights increased across the plug, from the east side with 60 cm of snow cover to the wind-polished ice on the west side (Serson 1972). The plug was observed later by Jeffries et al. (1992), who concluded that the ice mass in Nansen Sound did not fit the characteristics of either an ice shelf or a large sea ice block, but instead consisted of an "*agglomeration of ice of different ages and thicknesses... subject to rearrangement of its component floes*". This complicated floe matrix indicates that the plug may, at times, be unstable and subject to rearrangement or disintegration (Jeffries et al. 1992).

12.2.2 Sverdrup Ice Plug

The Sverdrup Ice Plug occurs at the northern edge of Sverdrup Channel, between Axel Heiberg and Meighen Islands (Fig. 12.1). In this region, thick multiyear sea ice that is pushed by the polar ice pack from the northwest in the summer and fall meets a ridged mix of first year, second year and multiyear ice to the south (Melling 2002). The resulting MLSI plug creates a semi-permanent feature that is clearly identified in early sea ice maps (Black 1962; Alt et al. 2001). Sverdrup et al. (1904) described the ice in the channel as "*going up and down in waves*", indicating the rolling surface typical of old MLSI. The curved northern boundary of this ice plug was observed in 1916 by Stefansson (1938), and appears to be the same in aerial photography from 1950 to 1959 (Serson 1974). Spot depth measurements were conducted on the ice plug in the early 1970s by Serson (1974), who identified ice up to 6 m thick near its northern margin on the Axel Heiberg Island side. The ice surface was rolling, paralleling the dominant northwest winds and featured steep-sided hummocks and melt pools extending downwind and undercutting the hummock sides (Serson 1974). The position of Sverdrup Ice Plug makes it the more vulnerable of the two plug features and thus more frequently broken (Fig. 12.2). This occurs because open water areas to the north and south of Sverdrup Plug, and frequent lead development north of Meighen Island, leave it less protected from wind and ocean currents than the Nansen Plug. In addition, the Fay Islands, located directly to the south of the Sverdrup Plug (Fig. 12.1b), are the source of frequent sea ice fracturing.

12.3 Chronology of Recent Ice Plug Break-Ups and Activity

Previous observations of the behaviour of the ice plugs has been undertaken via ground surveys (since 1900), aerial reconnaissance (1960s–1980s) and satellite image analysis (since the 1990s) (Alt et al. 2006; Black 1962; Jeffers et al. 2001; Jeffries et al. 1992; Serson 1972, 1974). In this study, we use previously published papers and reports, ice charts from the CIS and PCSP, and satellite imagery to quantify the presence/absence and area changes of the Nansen and Sverdrup ice plugs between 1961 and 2016 (Fig. 12.2). Prior to 1998 the plugs were semi-permanent features, with only limited fracturing occurring in most years. Simultaneous break-ups of both plugs occurred only twice in the last half of the twentieth century, but have occurred nine times since the start of the twenty-first century. The chronology and detailed characteristics of these events is presented below.

12.3.1 1962 Ice Plug Events

The first recorded break-up of either of the ice plugs occurred in summer 1962, when they both broke up during an extremely warm period characterized by high temperatures at all climate stations in the QEI. For example, mean summer (June–July–August) temperatures at Eureka were 5.0°C, compared to the 1961–1970 mean of 3.4°C (Fig. 12.3). This summer featured the highest percentage of open water in channels of the QEI from 1961 until the extreme light ice year of 1981 (Jeffers et al. 2001). The Ward Hunt Ice Shelf also experienced a significant calving event in 1961/1962, forming several ice islands with a total area of 596 km^2 (Hattersley-Smith 1963). Koerner (1979) states that 1962 was the most negative mass balance year over the period 1960–1977 for regional glaciers, noting that significant melt was so widespread at high elevations over the ice caps of Devon, Ellesmere and Axel Heiberg islands that the 1962 melt layer is used as a reference horizon in ice core stratigraphy.

Black (1962) observed the break-ups of both ice plugs in 1962, noting fracturing around Meighen Island beginning in early August during dominantly NW, N and NE winds at 10–14.7 knots (5.1–7.6 m s^{-1}). The ice atlas images for August and September 1962 demonstrate the temporal progression of the Nansen and Sverdrup ice plug break-up events (Fig. 12.4; Lindsay 1975). The August 4–5, 1962, image

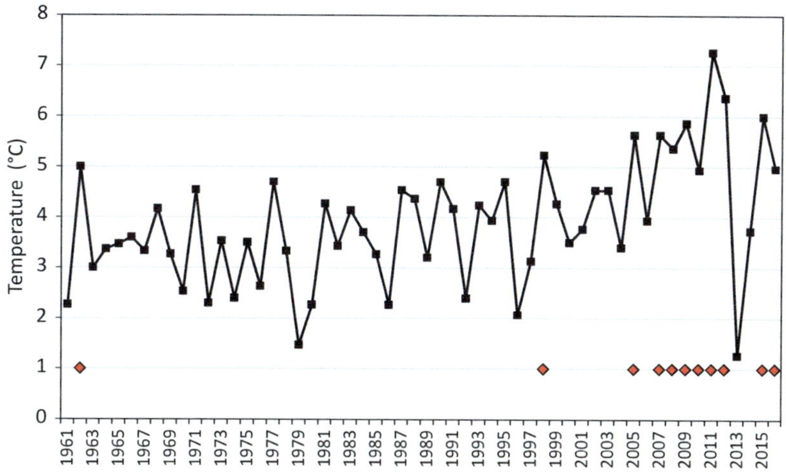

Fig. 12.3 Mean June–July–August temperatures at Eureka weather station, 1961–2016. *Red diamonds* indicate years when both Nansen and Sverdrup ice plugs broke up (Data source: Environment Canada (2016))

shows solid ice in both Nansen Sound and Sverdrup Channel that had not yet experienced any fracturing that summer (Fig. 12.4a). In contrast, the August 26 to September 5, 1962, composite chart (the time of maximum open water for the season), shows a broken Nansen Ice Plug with a large, refrozen plug fragment present at the channel's northern edge and a weakened, pre-break-up Sverdrup Ice Plug (Fig. 12.4b; Lindsay 1975). There are no ice charts of the plug area after September 5, 1962, but late September ice charts from the region directly south of Meighen Island show 8/10 old ice and 1/10 young ice, suggesting that the Sverdrup Plug area had fractured. Moreover, direct observation by Black (1962) confirms that the Sverdrup Ice Plug experienced complete break-up in the 1962 melt season.

12.3.2 1963–1997 Ice Plug Events

In 1963, both the Nansen and Sverdrup ice plugs appear to have been re-established based on spring map and air photo observations. Between 1963 and 1997, occasional individual break-ups and fractures of the plugs were observed (Fig. 12.2). The Nansen Plug remained stable until 1971; on August 11 of that year an area of open water was observed in Nansen Sound, and by September 6, 1971, the plug had completely broken up (Lindsay 1977). First-year ice in Nansen Sound on ice charts from spring 1972 provides evidence of the previous year's break-up (Lindsay 1977). Despite open water to the south of the Sverdrup Ice Plug and along Meighen Island, the Sverdrup Plug remained intact in 1971 (Lindsay 1977). The Sverdrup Plug experienced a break-up in 1975, which was observed September 21, 1975, on an ice

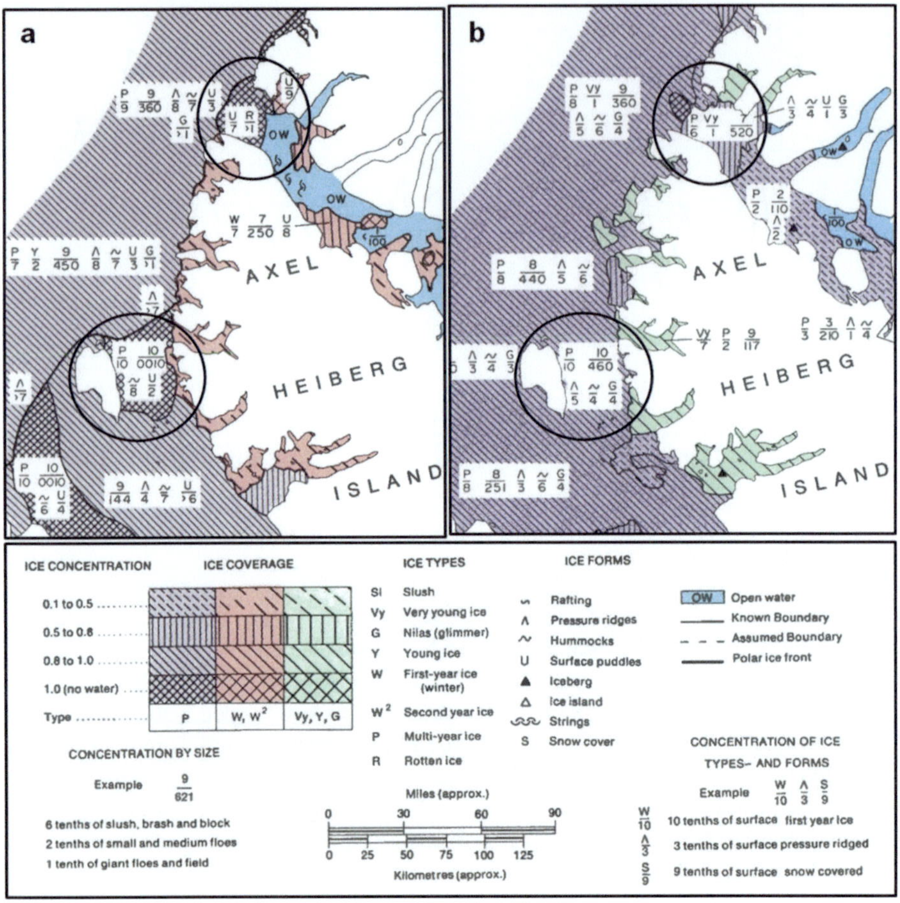

Fig. 12.4 Polar Continental Shelf Project Ice Atlas images of the Nansen and Sverdrup ice plugs in summer 1962. Ice plug locations are *circled*. (**a**) August 4–5, 1962; (**b**) August 26–September 5, 1962 (From Lindsay (1975), reproduced with permission of Natural Resources Canada)

charting flight, while the Nansen Plug remained solid throughout 1975. A flight directly over Meighen Island for the September 30–October 9, 1977, sea ice atlas charts observed a solid (10/10) Nansen Plug and a late season Sverdrup Plug fracture and break-up in 1977 (Lindsay 1982). Regional ice charts indicate that in the following year the weak first-year ice in Sverdrup Channel broke, but the Nansen Plug remained fast.

In 1981 there was extensive open water in the QEI, as evidenced in the sea ice coverage data provided by the CIS (Fig. 12.5a). However, there was no identified break-up of the Nansen or Sverdrup plugs in this year despite the warm summer temperatures and high percentage of open water. Generalized ice chart coverage in August and early September 1981 suggests that there may have been limited fracturing in the plug areas, but the September 8 regional chart indicates 10/10 ice in

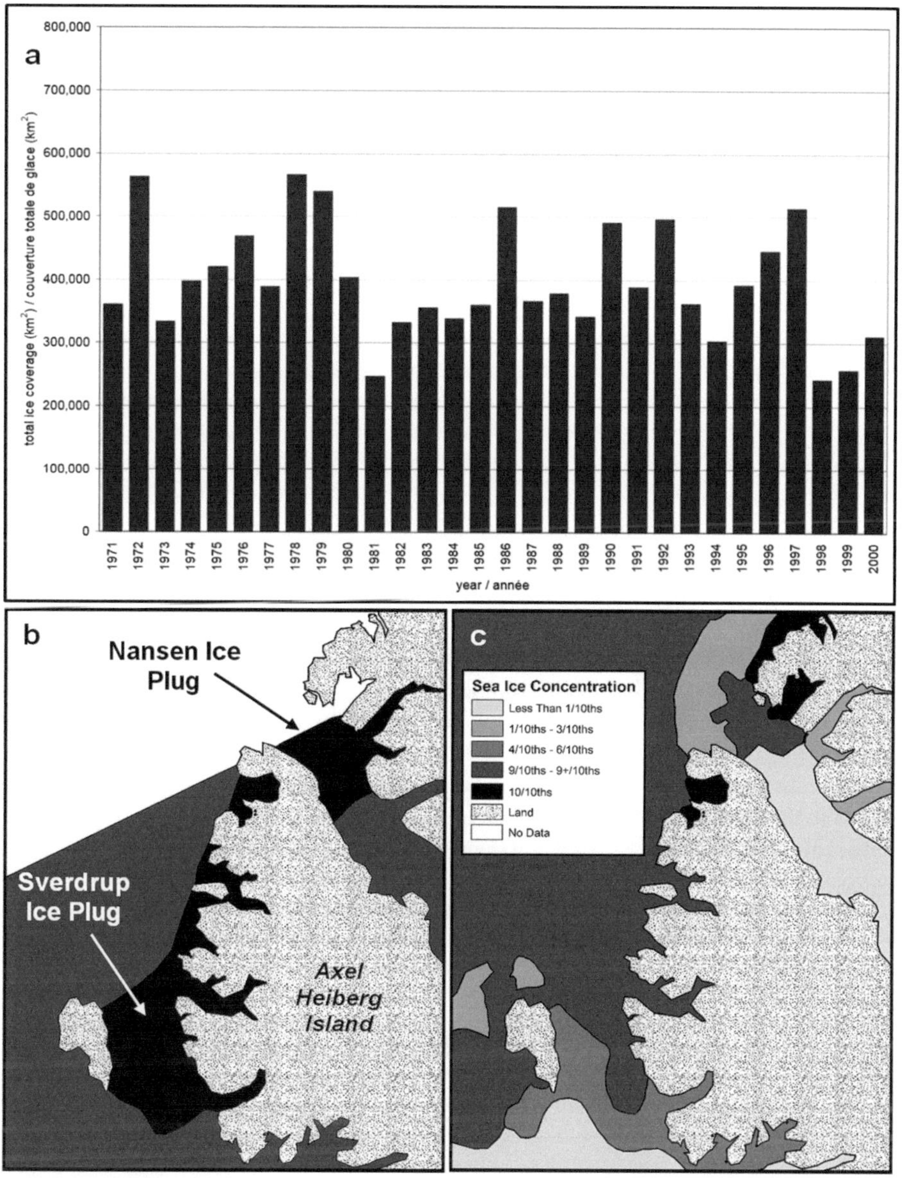

Fig. 12.5 Regional ice coverage indices: (**a**) Total ice coverage graph on September 10, 1971–2000, for the Canadian Arctic Archipelago; (**b**) Median sea ice concentration around ice plugs on September 24, 1971–2000; (**c**) Concentration of sea ice around ice plugs on September 28, 1998 (From Canadian Ice Service (2002), reproduced with the permission of Environment Canada)

clearly outlined plug areas. Daily temperature and wind data indicates that temperatures dropped below freezing in the last week of August and winds shifted from S/SW to SE during the second half of the month. This change in temperature and wind direction could have limited any possible early plug fracturing from becoming full break-ups during 1981.

12.3.3 1998 Ice Plug Events

In 1998 the melt season began early, characterized by a high pressure ridge, creating warm southerly flow, increased solar radiation due to clear skies, and above-normal spring temperatures. This continued with high summer air temperatures, high melt percentages of first and multiyear sea ice, and a long melt season (Figs. 12.3, 12.6a; Jeffers et al. 2001). Based on an analysis of CIS Regional Ice Analysis Charts and RADARSAT ScanSAR Wide imagery, Jeffers et al. (2001) found that the sea ice was significantly weakened during this summer, with a notable maximum of open water and the latest maximum open water date (September 28) in the 38-year period since 1961. Areas of light ice conditions in the QEI (less than 1/10 sea ice) were more prevalent in 1998 (Fig. 12.5c) than in the years preceding it (Fig. 12.5b), and the larger areas of open water and greater dispersal of sea ice exacerbated the ice decay.

Between August 17 and 24, 1998, the Sverdrup Plug fractured and a large piece of the Nansen Plug broke off as a single ice mass. The large areas of open water to the south of the plugs facilitated the southward movement and melt of some of the plug fragments (Jeffers et al. 2001); however, some large fragments survived to return to the plug area as the wind direction shifted. New ice growth in the fractured plug region began between August 31 and September 7, 1998, with a RADARSAT-1 image from September 21 (Fig. 12.7a) indicating a portion of landfast ice which had remained intact along the margin of Nansen Sound, and the formation of new nilas and grey ice south of this location. Jeffers and McCourt (2001) attribute this new ice formation to decreasing maximum air temperatures measured at Eureka (Fig. 12.6a), and to light and variable winds near the channel.

A contributing factor that made 1998 an unusual year was the fact that between September 21 and 27, 1998, a low-pressure system brought warmer than average temperatures and strong southerly winds to the area. A surface low pressure system then moved northeast through the region behind an upper warm trough and brought gale force winds (40 knots, 20.5 m s^{-1}) to Sverdrup Channel and winds >20 knots (10.3 m s^{-1}) to Nansen Sound (Alt et al. 2006). These factors facilitated the final break-up of the Nansen Ice Plug and entirely dislodged the Sverdrup Ice Plug, melting and fracturing the new ice which had formed in the previous month (CCAF Summer 1998 Project Team 2001) and completing the first simultaneous plug break-up since 1962.

The impact of these strong winds is indicated in RADARSAT-1 images from September 21 and 28, 1998, which show the highly fractured ice pack at this time (Fig. 12.7). In the September 28 image (Fig. 12.7b), pieces of the broken Nansen

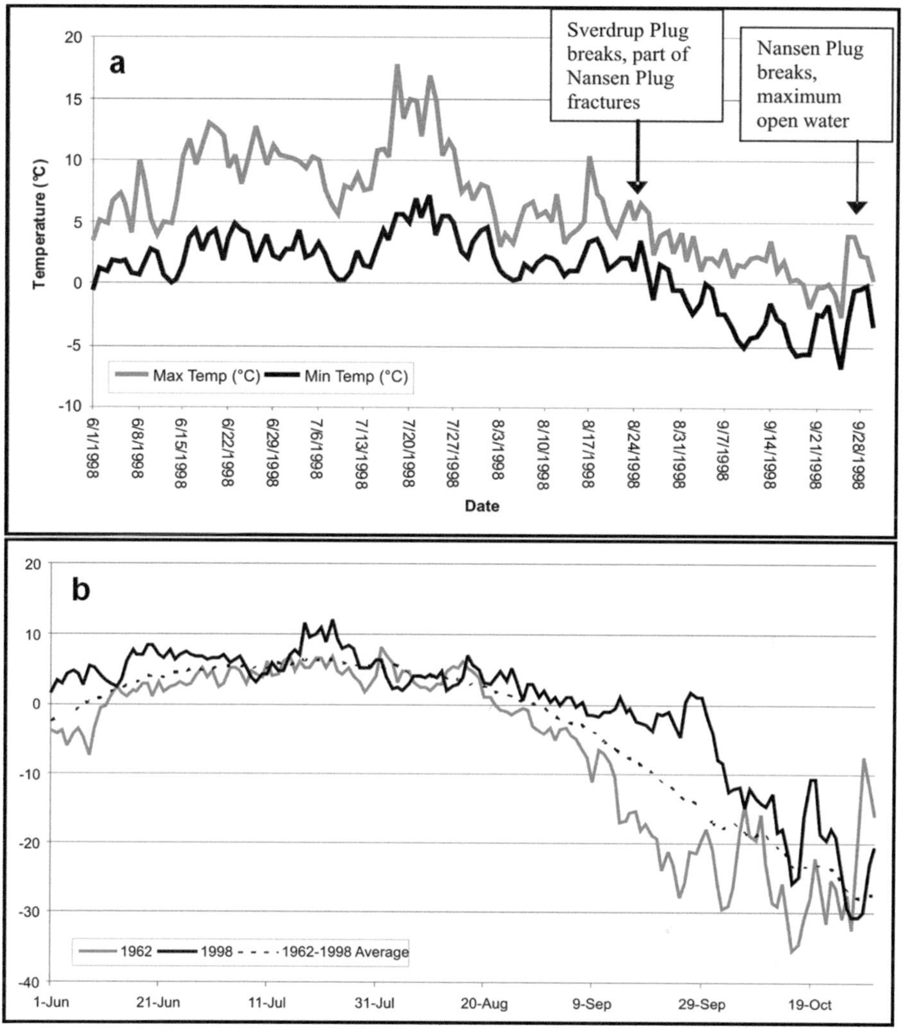

Fig. 12.6 Summer conditions at Eureka weather station: (**a**) Summer 1998 daily maximum and minimum air temperatures. (**b**) Daily June–October air temperatures for 1962, 1998 and long-term mean (Data source: Environment Canada (2016))

Plug are visible north and east from its original position, and the area to the south of Nansen Sound is largely clear of ice. Sverdrup Channel is clear of ice to the south, with open water as far north as Meighen Island. The remnants of the Sverdrup Plug are not identifiable; Jeffers and McCourt (2001) suggest that they may have melted completely or drifted and were incorporated into the Arctic Ocean ice pack, while Alt et al. (2006) suggest they returned to the plug area when the winds shifted to northwesterly behind the storm track.

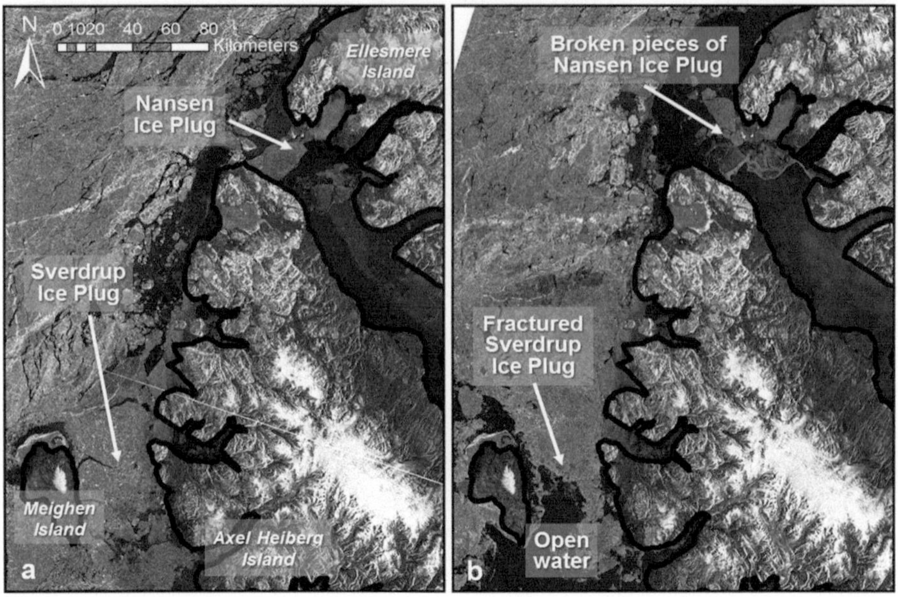

Fig. 12.7 RADARSAT-1 images of 1998 plug break-ups: (**a**) September 21, 1998; (**b**) September 28, 1998 (Radarsat images copyright Canadian Space Agency)

12.3.4 1999–2005 Ice Plug Events

The response of the Nansen and Sverdrup ice plugs to the extreme summer of 1998 continued well into the following year. Alt et al. (2006) present an overview of the movement of ice in Sverdrup and Peary channels between the 1998 extreme break-up events and 2005 (Fig. 12.2). The reformed ice in southern Nansen Sound broke again in the summer of 1999, but an absence of strong winds allowed it to stabilize as an in situ blockage. Wilson et al. (2004) note that reformation of the Nansen Plug appeared to consist of a different type of ice than had been previously found there, forming from in situ landfast ice, as opposed to the Arctic Ocean multiyear ice floes combined with first year ice that dominated before. The Sverdrup Plug fractured in both summer 1999 and 2000, which allowed ice to move from Sverdrup Channel north to the Arctic Ocean. The freeze-up in Sverdrup Channel occurred over 3 weeks earlier in 2000 than in 1999, due to a circulation pattern that enhanced the flow of cold air from the north.

Summer 2001 was characterized by a relatively cold summer season (Fig. 12.3) with heavy old-ice presence and early freeze-up, featuring the first summer without a Sverdrup Plug break-up since 1997 (Alt et al. 2006). The Sverdrup Plug remained stable in 2002, but in 2003 an early melt season and complete first-year ice loss lead to its fracturing, although the ice remained in place. The plug also fractured in situ in 2004, following a short melt season and very limited ice loss, and in late August 2005 it broke up once again. The Nansen Plug also broke up in August 2005, after maintaining relative stability since its 1998 break-up. The simultaneous break-up of the Nansen and Sverdrup plugs in 2005 was not matched by the light sea ice conditions of the previous break-up events (Alt et al. 2006), although many other ice losses also occurred along northern Ellesmere Island that summer. For example, almost the entire 87.1 km^2 Ayles Ice Shelf was lost during a strong offshore wind event on August 13, 2005, together with the removal of 330 km^2 of young MLSI and 690 km^2 of 58+ year old MLSI from Yelverton Bay (Copland et al. 2007; Pope et al. 2012; Mueller et al. 2017). Summer temperatures at Eureka were also the warmest on record since 1961 (Fig. 12.3).

The frequent fracturing of the Sverdrup Plug and the 2005 simultaneous break-up event suggest that a solid recovery of the plugs did not occur after the extreme summer of 1998. Alt et al. (2006) and Wilson et al. (2004) used CIS digital ice charts from 1997 to 2003 to track ice movement and gains/losses of multiyear ice from Sverdrup and Peary channels, and found that the removal of the ice blockages has allowed for increased movement southward of old ice through the QEI.

12.3.5 2006–2016 Ice Plug Events

CIS weekly ice charts and MODIS satellite imagery were used to determine the activity of the ice plugs between 2006 and 2016 (Fig. 12.2). Summer 2006 featured less ice loss and earlier regional freeze-up than in summer 2005. The Sverdrup Plug, now consisting of first year and refrozen pack ice, remained solid, and the Nansen Plug fractured in situ in late August 2006. Ice charts from August 28, 2006, indicated 9+/10 ice in the Nansen Plug area, with areas of 1/10–2/10 ice alongside both plugs.

The Sverdrup Plug fractured in early August 2007 and broke away completely between August 20 and 27, 2007, allowing for the movement of pack ice and a piece of the Ayles Ice Island into the southern part of Sverdrup Channel (Van Wychen and Copland 2017). The Nansen Plug also broke up in mid-August 2007.

CIS ice charts from 2008 indicated reductions in sea ice concentration along the northwestern coast of Ellesmere Island and throughout the study region compared to the two previous years, and show fracture and break-up of both ice plugs between July 31 and August 4, 2008. MODIS images show the process of break-up and export of this ice (Fig. 12.8). The break-up of the plugs occurred during a warm summer (Fig. 12.3) and period of very strong wind and wind reversals; marine wind prognosis charts indicate strong southeasterly winds at up to 35 knots (18 m s^{-1}) moving directly parallel with both Sverdrup Channel and Nansen Sound on August 2, 2008 (M. Schaffer, pers. comm. 2009). The Sverdrup Plug area remained filled with fractured ice through the second week of August 2008, when northerly winds dominated the region, but by August 17, 2008, both plugs had completely cleared. Many ice shelf losses also occurred in summer 2008, such as the complete loss of the Serson Ice Shelf in July 2008 (Mueller et al. 2008) and Markham Ice Shelf on August 6, 2008 (Mueller et al. 2017).

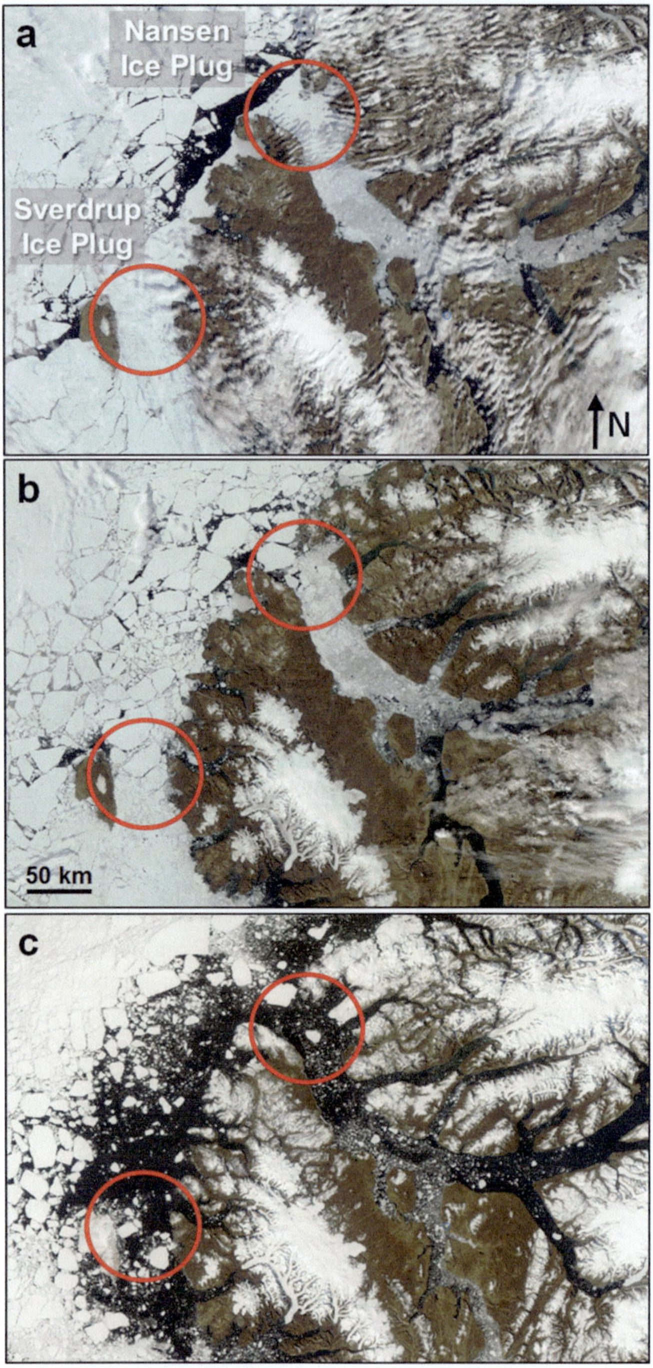

Fig. 12.8 MODIS imagery of the 2008 ice plug break-ups (ice plugs indicated by *red circles*). (**a**) July 31, 2008: both plugs appear stable; (**b**) August 5, 2008: both plugs fractured, ice movement northwards into the Arctic Ocean; (**c**) August 29, 2008: both Nansen Sound and Sverdrup Channel are largely ice-free, with southerly movement of pack ice into Eureka Sound and through Sverdrup Channel (Imagery courtesy of MODIS Rapid Response Project, NASA/GSFC and University of Maryland, http://rapidfire.sci.gsfc.nasa.gov)

Since 2008, the complete loss of the Nansen and Sverdrup ice plugs has occurred in every summer (typically between early August and mid-September), with the exceptions of 2013 and 2014. Summer 2013 was very cold across the CAA, with the coldest mean summer temperature at Eureka over the entire period of record (Fig. 12.3) and the first positive mass balance recorded at White Glacier (~150 km SE of Sverdrup Ice Plug) for a decade (Thomson et al. 2017). Timelapse cameras at Milne and Petersen ice shelves showing a snowline that remained near sea level for the entire summer (Copland, unpublished data). The cold conditions resulted in the formation of landfast sea ice along northern Ellesmere Island that was reminiscent of conditions last seen in the 1990s, and the associated formation of stable Nansen and Sverdrup ice plugs. Temperatures in summer 2014 were still below the decadal average (Fig. 12.3), and the stable ice plugs that formed the previous summer remained locked in place throughout 2014. However, summer 2015 saw the return of the warm and sunny conditions more typical of the past decade, resulting in fracturing of the ice plugs in the first week of August and their complete loss by the first week of September. In 2016 MODIS imagery shows the complete loss of both ice plugs by August 11.

12.4 Discussion

Between 1961 and 2016, our analysis indicates that both the Sverdrup and Nansen ice plugs broke up in 1962, 1998, 2005, and every year since 2007 with the exception of 2013 and 2014 (Fig. 12.2). These events occurred during record warm summers at Eureka, typically when the mean June-July-August temperature was >5°C (Fig. 12.3). Many of the break-ups also occurred during strong regional wind events with a southerly component, making it clear that the plug break-ups are strongly related to regional weather patterns.

12.4.1 Weather Systems Affecting the Canadian Arctic Archipelago

The Nansen and Sverdrup ice plugs respond to weather systems which have been described in detail by Alt (1979, 1987), Gardner and Sharp (2007) and Bezeau et al. (2015). Alt (1979, 1987) developed a classification system of three distinct synoptic systems for the Canadian High Arctic based on the summer position and characteristics of the 500 mbar circumpolar vortex, a mid-tropospheric cyclonic system with counter-clockwise winds moving around a cold polar air mass (Table 12.1). Further discussion and examples of each synoptic type can be found in Alt (1979, 1987). Following this classification system, Gardner and Sharp (2007) and Bezeau et al. (2015) described the relationship between Arctic circumpolar vortex variability and

Table 12.1 Synoptic type classifications (after Alt 1987)

Synoptic classification	Name	Characteristics	Placements
Type I	Polar Ocean Circulation	Baffin low, Arctic Ocean high	Vortex in North American sector
Type IIs	Cyclonic System Circulation with Snow	Tracking cyclone with snow	Vortex moving across Queen Elizabeth Islands (QEI) from central Polar Ocean
Type IIr	Cyclonic System Circulation with Rain	Tracking cyclone with rain	Vortex in Beaufort Sea sector
Type III	Island Circulation	Anticyclone	Ridge in QEI; vortex in Eurasian sector

glacier mass balance in the Canadian High Arctic. As regional weather patterns associated with these synoptics have distinct influences on the behaviour and survival of sea ice in this region, the synoptic types are useful in explaining the break-up of the ice plugs.

Synoptic Type I of Alt (1979, 1987), Polar Ocean Circulation, describes conditions dominated at the surface by a high pressure system in the Arctic Ocean to the west of Meighen Island and a low pressure system over Baffin Bay. Low stratus or fog is carried from the Arctic Ocean pack ice into the northern QEI by northerly flow during this circulation type, with the resulting precipitation generally limited to trace amounts of drizzle, freezing drizzle or snow. The upper cold low in the Baffin Bay – Hudson Bay area leads to a predominance of storm tracks from the north continental coast of mainland Canada to Baffin Bay (Alt 1979). Type I scenarios are used by Alt (1979, 1987) to explain periods of low melting degree days on Meighen Ice Cap, when there is a "*suppression of ablation*". Type I features the lowest occurrence of melting degree days of the three scenarios. When combined with frequent northerly and northwesterly wind flow, limited wind reversals, and frequent fog blocking solar radiation, this reduces the potential for Type I systems to cause multiyear sea ice melt and break-up in the northern Queen Elizabeth Islands.

Synoptic Type II, termed Cyclonic System Circulation, occurs when the 500 mbar cold low is positioned over the Beaufort Sea region. The frontal systems between the radiatively heated Siberian and Alaskan landmasses and the colder polar ocean pick up moisture over the peripheral ice-free ocean areas and move around the upper cold low (Alt 1979). This results in tracking surface lows and developed frontal zones along the northern edge of the CAA, low temperatures and significantly high precipitation. The differences in precipitation type produce a secondary classification into Type IIs (tracking cyclone with snow) and Type IIr (tracking cyclone with rain). Type II systems feature strong winds and potential wind direction reversals, but do not feature the high temperatures and solar radiation which would facilitate ice weakening and MLSI break-up.

Synoptic Type III, termed Island Circulation, occurs when a low at the 500 mbar level is positioned on the Asian side of the Arctic Ocean and a ridge develops at all levels over the eastern CAA and Greenland. A warm ridge in eastern North America can dominate air flow, causing the western Arctic to experience higher temperatures, lighter winds and increased solar radiation reaching the ground and ice surface. Gardner and Sharp's (2007) classification system, using the circumpolar vortex position and strength, present a corresponding synoptic condition to Alt's Type III, which they refer to as Type I-B. This situation features a vortex centered in the Eastern Hemisphere that is either weak or strong with no trough reaching into the Canadian High Arctic, and is associated with warm air temperatures and extremely negative regional glacier mass balances (Gardner and Sharp 2007). Similar conditions are also described by Bezeau et al. (2015). Synoptic Type III frequently develops only partially, but even this is enough to produce clear, warm weather and southerly wind flow over Meighen Ice Cap. Alt's (1979) data for this synoptic type features significantly more melting degree-days over the measured period than the other two synoptic conditions. Additionally, Type III is identified as bringing storms northward and creating southerly and easterly winds, tracking from the land surfaces rather than the Arctic Ocean. Intrusion of well-developed mid-latitude frontal systems are also observed in Type III cases, leading to strong winds in the QEI and rapid wind direction reversals. Preconditioning of the ice by high air temperatures and disturbance by strong, changeable winds strongly impact the ability of local freeze-up to maintain or rebuild ice.

12.4.2 Synoptic Controls on Ice Plug Break-Up Events

The synoptic situations described above indicate that the position of low pressure systems (and the Polar Vortex) is of vital importance to local climate for the ice plugs, influencing both temperature and wind patterns and snow and ice changes in this region. The high summer temperatures associated with Type III synoptic patterns increase the likelihood of ice plug break-up, as demonstrated by the occurrence of warm temperatures at Eureka during years of known ice plug losses (Fig. 12.3). The dynamic system of lows in the CAA means that there is potential for significant wind direction reversals. These wind shifts are necessary for the melt and ice removal mechanisms in the northern QEI, with warm southerly winds causing melt and northerlies transporting the melted and separated ice southward away from its origin.

Summer 1962 is identified via several climatic variables as a clear example of a dominantly Type III synoptic condition. Alt (1987) notes that a ridge extending from Hudson Bay to the QEI in early June 1962 led to near-melting temperatures in the region. This was followed by well above normal seasonal temperatures at most stations in mid-July, corresponding to significant anticyclonic circulation, and well above normal temperatures at the remaining stations in late July, during the passage of the warm frontal system (Alt 1987). These conditions combined to produce

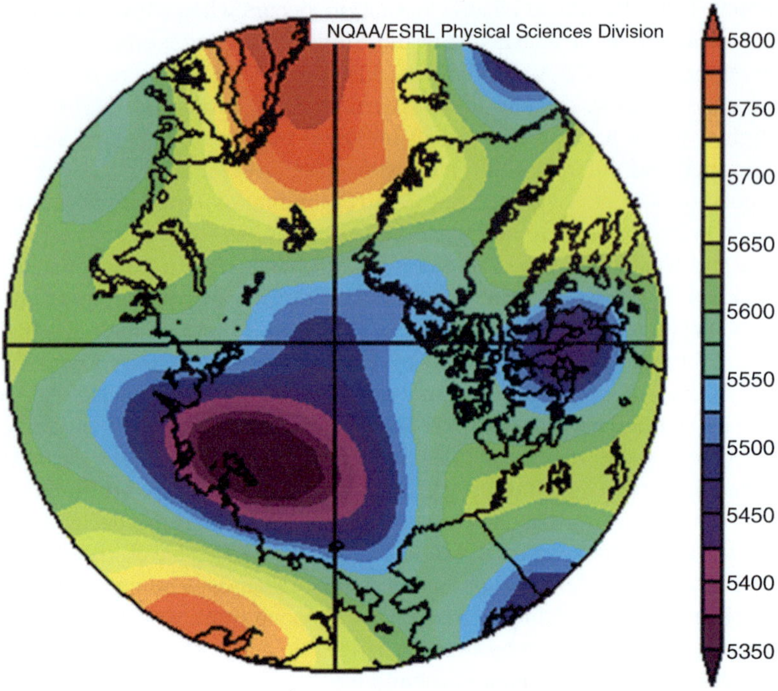

Fig. 12.9 Example of Type III Synoptic conditions: NCEP/NCAR reanalysis of 500 mbar geopotential height from July 28, 2008, just before the vortex begins to move into the Beaufort sector of the Arctic Ocean and fractures appear in Nansen and Sverdrup ice plugs. Note the position of the low on the Asian side of the Arctic Ocean, and the development of a ridge in the Eastern Queen Elizabeth Islands

strongly negative mass balance conditions across the CAA. Similarly, the high summer temperatures and winds in summers 1998 and 2008 provided ideal conditions for the final fracture and removal of the Nansen and Sverdrup ice plugs. Gardner and Sharp (2007) describe a weak, Eastern Hemisphere centered 500 mbar Arctic circumpolar vortex in July 1998 (categorized as their Type I-B condition), which created warm July surface air temperatures and a very negative regional mass balance that summer. Conditions just before the simultaneous plug break-up in 2008, illustrated in a National Centers for Environmental Prediction/National Center for Atmospheric Research (NCEP/NCAR) reanalysis for July 28, 2008, exemplify a Type III synoptic condition (Fig. 12.9), which contributed to the strong southeasterly winds associated with the 2008 break-up events. Bezeau et al. (2015) describe how similar conditions occurred during every summer between 2007 and 2012, with frequent summer anticyclonic circulation over the CAA resulting in strongly negative mass balance conditions on glaciers and ice caps. This period exactly overlaps with the longest period of annual break-ups of both the Nansen and Sverdrup ice plugs observed in the historical record.

12.4.3 Comparisons Between Nansen and Sverdrup Ice Plug Break-Up Events

As discussed above, well above normal summer temperatures and sea ice melt occurred during all of the years in which both the Nansen and Sverdrup ice plugs broke. However, there is a distinction of scale between the two twentieth century removals and the frequent break-ups since the start of the twenty-first century. The older 1962 and 1998 break-up events occurred as anomalous removals of decades-old MLSI plugs within a context of infrequent break-up events and long periods of stability. In contrast, the more recent events consist of the almost annual fracture and break-up of first year ice and young multiyear ice (Fig. 12.2). This facilitates a classification of the break-ups into two distinct groups: early (1962 and 1998) and recent (since 2005).

In the early period, Alt et al. (2001) indicate a similarity between the July mean 500 mbar pattern in 1962, noted by both Alt (1987) and Gardner and Sharp (2007) as a significant negative mass balance and open water situation, and the 500 mbar pattern for spring 1998. Jeffers (2001) notes that the temperature patterns preceding the 1962 and 1998 break-up events were very similar, although melt in the region of the plugs was much higher following the 1998 break-ups than the 1962 events. In 1962, the Sverdrup Ice Plug broke, but the ice did not clear into the Arctic Ocean in the same way that it did in 1998. In summer 1962, the synoptics featured strong southerly flow and a freeze-up beginning 3–4 weeks earlier than in 1998. This explains the moderate post-break-up melting in 1962, whereas in 1998 the open water season lasted until late September and late summer storms delayed the fall freeze-up (Alt et al. 2001). Daily mean temperatures at Eureka showed a steady decrease in late August and September 1962, whereas they remained close to 0°C through the end of September 1998 (Fig. 12.6b). The extension of warm temperatures into fall 1998 is the most significant difference between the 1962 and 1998 situations, and these increased temperatures impacted the timing of freeze-up around the ice plugs.

In the recent period, ice plug break-up events have occurred in young multiyear sea ice (2005 and 2007) or in first year ice (2008–2012, 2016). There is a strong connection between sea ice age and thickness, which in turn determines its ability to withstand melt and wind forcing (Maslanik et al. 2007). This is evident in the timing of break-ups at the start of this period: the 2005 and 2007 break-ups occurred in mid/late August, whereas the 2008 break-ups of first year ice occurred at the beginning of August. Daily mean temperatures from Eureka show very little difference between years which featured complete plug break-ups (e.g. 2005, 2007, 2008) and years with no break-ups or with single plug fractures (e.g. 2004, 2006). A more important variable associated with these recent simultaneous break-up events is the presence of a lead of low sea ice concentration along the northern edges of the plugs, occurring as the pack ice is pushed away from the land. CIS ice charts indicate that prior to the 2005 break-ups, a lead with 5/10 sea ice concentration developed along the coast across all of Nansen Sound and extended into Sverdrup Channel.

In 2007, sea ice at 1/10 concentration bordered the northern coast in the 2 weeks preceding the plug break-ups. Similarly, in 2008, 2/10 concentration sea ice was present along the plugs' northern borders 1 week prior to their break-ups, and similar conditions were observed in 2009–2012 and 2015–2016. This area of low-concentration ice (lead) was not present in 2004 or 2006; these years feature sea ice at 9+/10 concentration flush against the fast coastal ice. This suggests that the concentration of sea ice present between the Arctic Ocean and the plugs' northern edges is an important factor in the survival or break-up of ice in these waterways. Mahoney et al. (2007) suggest that the presence of a lead at the landfast ice edge reduces the stability of landfast sea ice via wave action or solar radiation absorption by the adjacent open water.

12.5 Conclusions

From the review presented here, it is clear that the creation of ice plugs is gradual and takes many years, requiring summer climate conditions of cool temperatures and frequent northwesterly winds. The plugs have historically resisted break-up due to their thick ice and because of the presence of anchoring land on either side. The break-up events, on the other hand, have occurred rapidly during warm summers characterized by mean temperatures >5°C at Eureka, an early melt season and long warm periods sufficient to produce significant surface melting and a weakened plug structure. These conditions are typically associated with Alt's (1979, 1987) Type III synoptic system when a warm ridge dominates eastern North America, bringing southerly winds, warm temperatures, and high wind speeds with reversals in direction. For the complete break-up of both MLSI plugs to occur, as happened in 1962 and 1998, there must also be weakening of the surrounding sea ice. For example, Jeffers (2001) discusses the importance of high proportions of open water to the south of the ice plugs, which weakens the protective matrix around them, adding tension to the system and facilitating greater fracture and movement of the ice plug pieces.

An important second phase of the break-up process is the requirement for strong storm systems to move through the region, which fracture the pre-weakened ice plug. This was particularly evident in the case of the 1998 plug losses, when a low pressure system and high winds in mid to late August created a forcing strong enough to cause a distinct fracture in the Sverdrup Plug and complete breakage of the Nansen Plug. Continued warm weather into fall 1998 facilitated open water conditions northward into Sverdrup Channel and the removal of the last piece of the Nansen Plug. Higher summer temperatures in recent decades have created a backdrop of less ice and more snow-free land and open water; this results in storms that draw in moisture, track along the land edges, and behave more distinctly like southerly storms. For example, Gardner and Sharp (2007) note that their Type I-B synoptic condition, associated with Alt's Type III classification (1979) and exemplified by the 1962 and 1998 weather systems, created a temperature system in the High Arctic which was "*almost thermally homogeneous with continental North America*".

Following the 1998 break-ups, the patterns of late fall sea ice buildup appear to have changed. Since 1998 there has not been a classifiable 'extreme heavy ice year' in the northern QEI. This is in contrast to the 1962 break-ups and the extreme light ice year of 1981, after which there were heavy ice years recorded within 5 years, which aided in the recovery of the plugs and local MLSI. The plug break-up events of 1962 and 1998, occurring in conditions of thick MLSI, represent responses to anomalous years of high average summer temperatures, high proportions of open water, and extreme southerly wind events and wind reversals. The recent lack of old ice recovery in the northern QEI means that the plug break-ups and fractures since 1998 have occurred in conditions of young MLSI or first year ice. Significant inter-annual variability in ice conditions within the 1961–2016 measurement period and the break-ups since 2005 indicate a changing sea ice regime around the plugs. The two-part plug break-up process can now be placed within the context of the younger, thinner Arctic ice regime, suggesting limited or no plug recovery in the future.

One of the major potential impacts of recent ice plug losses is the increased ability of old sea ice to move from the Arctic Ocean in to the CAA. For example, Black (1962) observed that the 1962 Sverdrup Plug clearing allowed for a significant movement of multiyear ice from the Arctic Ocean into the QEI between late 1962 and fall 1963. Simultaneously there was a northward movement of the open-water limit, which was largely attributed to the extreme summer air temperatures contributing to sea-surface temperature increases (Jeffers et al. 2001). More recently, Howell et al. (2009, 2013) found an increase in dynamic import of multiyear ice from the Arctic Ocean into the QEI.

The removal of the Nansen Sound and Sverdrup Channel ice plugs represents only part of the destabilization of previously landfast sea ice in the northern CAA. The break-ups of these ice plugs have occurred in tandem with the removal of significant areas of ice shelf and MLSI from embayments along northern Ellesmere Island (Mueller et al. 2017). The loss of landfast sea ice in this region represents an increase of open water and the removal of an important buffer against further old ice export and ice shelf break-up.

Acknowledgments The authors would like to thank the Natural Sciences and Engineering Research Council of Canada, Canada Foundation for Innovation, Ontario Research Fund and the University of Ottawa for funding contributions. Helpful comments from two anonymous reviewers and Derek Mueller are appreciated. We are very grateful to Ed Hudson for wind data and Matt Arkett for RADARSAT-2 images.

References

Agnew, T. A., Alt, B. T., De Abreu, R., & Jeffers, S. (2001, May 14–18). The loss of decades old sea ice plugs in the Canadian Arctic Islands. In *Extended Abstracts: Polar Meteorology and Oceanography Conference*, American Meteorological Society, San Diego.

Agnew, T., Lambe, A., & Long, D. (2008). Estimating sea ice area flux across the Canadian Arctic Archipelago using enhanced AMSR-E. *Journal of Geophysical Research, 113*, C10011. doi:10.1029/2007JC004582.

Alt, B. T. (1979). Investigation of summer synoptic climate controls on the mass balance of Meighen Ice Cap. *Atmosphere-Ocean, 3*, 181–199. doi:10.1080/07055900.1979.9649060.

Alt, B. T. (1987). Developing synoptic analogs for extreme mass balance conditions on Queen Elizabeth Island ice caps. *Journal of Applied Meteorology, 26*, 1605–1623. doi:10.1175/1520-0450.

Alt, B., & Lindsay, D. (2005). *Preliminary investigation of unique features of the Canadian Arctic Archipelago sea ice conditions as they relate to short range forecast modeling with emphasis on differences from the East Coast*. Contract Report. Prepared by Balanced Environments Associates for Canadian Ice Service, Ottawa.

Alt, B. T., Jeffers, S., McCourt, S., & Chagnon, R. (2001). Spatial and temporal studies of maximum open water in the Queen Elizabeth Islands. In CCAF Summer 1998 Project Team (Eds.), *The state of the Arctic cryosphere during the extreme warm summer of 1998: Documenting cryospheric variability in the Canadian Arctic*. CCAF final report, Waterloo.

Alt, B., Wilson, K., & Carrieres, T. (2006). A case study of old-ice import and export through Peary and Sverdrup channels in the Canadian Arctic Archipelago: 1998–2005. *Annals of Glaciology, 44*, 329–338. doi:10.3189/172756406781811321.

Bezeau, P., Sharp, M., & Gascon, G. (2015). Variability in summer anticyclonic circulation over the Canadian Arctic Archipelago and west Greenland in the late 20th/early 21st centuries and its effect on glacier mass balance. *International Journal of Climatology, 35*, 540–557. doi:10.1002/joc.4000.

Black, W. A. (1962). *Sea-ice survey, Queen Elizabeth Islands region, summer 1962, Geographical paper no. 39*. Ottawa: Department of Mines and Technical Surveys.

Canadian Ice Service (CIS). (2002). *Sea ice climatic atlas: Northern Canadian waters 1971–2000*. Ottawa: Environment Canada.

Climate Change Action Fund (CCAF) Summer 1998 Project Team. (2001). Brief executive summary. In CCAF Summer 1998 Project Team (Eds.), *The state of the Arctic cryosphere during the extreme warm summer of 1998: Documenting cryospheric variability in the Canadian Arctic*. CCAF final report, Waterloo.

Copland, L., Mueller, D., & Weir, L. (2007). Rapid loss of the Ayles Ice Shelf, Ellesmere Island, Canada. *Geophysical Research Letters, 34*, L21501. doi:10.1029/2007GL031809.

Environment Canada. (2016). Historical climate data online: monthly data. http://climate.weather.gc.ca/historical_data/search_historic_data_e.html. Accessed 17 Dec 2016.

Galley, R. J., Else, B. G. T., Howell, S. E. L., Lukovich, J. V., & Barber, D. G. (2012). Landfast sea ice conditions in the Canadian Arctic: 1983–2009. *Arctic, 65*(2), 133–144.

Gardner, A. S., & Sharp, M. (2007). Influence of the Arctic circumpolar vortex on the mass balance of Canadian high Arctic glaciers. *Journal of Climate, 20*, 4586–4598. doi:10.1175/JCLI4268.1.

Grosse, G., Goetz, S., McGuire, A. D., Romanovsky, V. E., & Schuur, E. A. G. (2016). Changing permafrost in a warming world and feedbacks to the Earth system. *Environmental Research Letters, 11*(4), 040201. doi:10.1088/1748-9326/11/4/040201.

Hattersley-Smith, G. (1963). The Ward Hunt Ice Shelf: Recent changes at the ice front. *Journal of Glaciology, 4*, 415–424. doi:10.3198/1963JoG4-34-415-424.

Howell, S. E. L., Tivy, A., Yackel, J., & McCourt, S. (2008). Multi-year sea ice conditions in the western Canadian Arctic Archipelago region of the Northwest Passage: 1968–2006. *Atmosphere-Ocean, 46*(2), 229–242. doi:10.3137/ao.460203.

Howell, S. E. L., Duguay, C. R., & Markus, T. (2009). Sea ice conditions and melt season duration variability within the Canadian Arctic Archipelago: 1979–2008. *Geophysical Research Letters, 36*, L10502. doi:10.1029/2009GL037681.

Howell, S. E. L., Wohlleben, T., Dabboor, M., Derksen, C., Komarov, A., & Pizzolato, L. (2013). Recent changes in the exchange of sea ice between the Arctic Ocean and the Canadian Arctic

Archipelago. *Journal of Geophysical Research, Oceans, 118*(7), 3595–3607. doi:10.1002/jgrc.20265.

Jeffers, S. (2001) Ice conditions in the Queen Elizabeth Islands in 1998 (1962, 1981): A detailed analysis. In CCAF Summer 1998 Project Team (Eds.), *The state of the Arctic cryosphere during the extreme warm summer of 1998: Documenting cryospheric variability in the Canadian Arctic*. CCAF final report, Waterloo.

Jeffers, S., & McCourt, S. (2001). Winds and the Nansen Sound and Sverdrup Channel sea ice plugs. In CCAF Summer 1998 Project Team (Eds.), *The state of the Arctic cryosphere during the extreme warm summer of 1998: Documenting cryospheric variability in the Canadian Arctic*. CCAF final report, Waterloo.

Jeffers, S., Agnew, T. A., Alt, B. T., De Abreu, R., & McCourt, S. (2001). Investigating the anomalous sea ice conditions in the Canadian High Arctic (Queen Elizabeth Islands) during the summer of 1998. *Annals of Glaciology, 33*, 507–512. doi:10.3189/172756401781818761.

Jeffries, M. O. (2002). Ellesmere Island ice shelves and ice islands. In R. S. Williams & J. G. Ferrigno (Eds.), *Satellite image atlas of glaciers of the world: North America* (p. J147–J164). Washington: United States Geological Survey.

Jeffries, M. O., Reynolds, G. J., & Miller, J. M. (1992). First Landsat multi-spectral scanner images of the Canadian Arctic north of 80°N. *Polar Record, 28*, 1–6. doi:10.1017/S0032247400020192.

Khan, S. A., Aschwanden, A., Bjork, A., Wahr, J., Kjeldsen, K. K., & Kjaer, K. H. (2015). Greenland ice sheet mass balance: A review. *Reports on Progress in Physics, 78*(4), 046801. doi:10.1088/0034-4885/78/4/046801.

Koerner, R. M. (1977). Ice thickness measurements and their implications with respect to past and present ice volumes in the Canadian High Arctic ice caps. *Canadian Journal of Earth Sciences, 14*, 2697–2705. doi:10.1139/e77-237.

Koerner, R. M. (1979). Accumulation, ablation and oxygen isotope variations on the Queen Elizabeth Islands ice caps, Canada. *Journal of Glaciology, 22*, 25–41. doi:10.3198/1979JoG22-86-25-41.

Krüger Search Expeditions. (1934). Krüger search expeditions, 1932. *Polar Record, 1*(8), 121–129. doi:10.1017/S0032247400031181.

Kwok, R. (2006). Exchange of sea ice between the Arctic Ocean and the Canadian Arctic Archipelago. *Geophysical Research Letters, 33*, L16501. doi:10.1029/2006GL027094.

Lindsay, D. G. (1975). *Sea ice atlas of Arctic Canada, 1961–1968* (pp. 213). Ottawa: Energy Mines and Resources Canada. doi:10.4095/299024.

Lindsay, D. G. (1977). *Sea-ice atlas of Arctic Canada, 1969–1974* (pp. 219). Ottawa: Energy Mines and Resources Canada. doi:10.4095/298827.

Lindsay, D. G. (1982). *Sea-ice atlas of Arctic Canada, 1975–1978* (pp. 139). Ottawa: Energy Mines and Resources Canada. doi:10.4095/298828.

Lindsay, R., & Schweiger, A. (2015). Arctic sea ice thickness loss determined using subsurface, aircraft, and satellite observations. *The Cryosphere, 9*, 269–283. doi:10.5194/tc-9-269-2015.

Mahoney, A., Eicken, H., Gaylord, A. G., & Shapiro, L. (2007). Alaska landfast sea ice: Links with bathymetry and atmospheric circulation. *Journal of Geophysical Research, 112*(C2), C02001. doi:10.1029/2006JC003559.

Maslanik, J. A., Fowler, C., Stroeve, J., Drobot, S., Zwally, J., Yi, D., & Emery, W. (2007). A younger, thinner Arctic ice cover: Increased potential for rapid, extensive sea-ice loss. *Geophysical Research Letters, 34*, L24501. doi:10.1029/2007GL032043.

Meier, W. N., Hovelsrud, G. K., van Oort, B. E. H., Key, J. R., Kovacs, K. M., Michel, C., Haas, C., Granskog, M. A., Gerland, S., Perovich, D. K., Makshtas, A., & Reist, J. D. (2014). Arctic sea ice in transformation: A review of recent observed changes and impacts on biology and human activity. *Reviews of Geophysics, 52*(3), 185–217. doi:10.1002/2013RG000431.

Melling, H. (2002). Sea ice of the northern Canadian Arctic Archipelago. *Journal of Geophysical Research, 107*(C11), 3181. doi:10.1029/2001JC001102.

Mueller, D. R., Copland, L., Hamilton, A., & Stern, D. R. (2008). Examining Arctic ice shelves prior to 2008 breakup. *EOS. Transactions of the American Geophysical Union, 89*, 502–503. doi:10.1029/2008EO490002.

Mueller, D., Copland, L., & Jeffries, M. O. (2017). Changes in Canadian Arctic ice shelf extent since 1906. In L. Copland & D. Mueller (Eds.), *Arctic ice shelves and ice islands* (p. 109–148). Dordrecht: Springer. doi:10.1007/978-94-024-1101-0_5.

Peary, R. E. (1907). *Nearest the Pole: A narrative of the polar expedition of the Peary Arctic Club in the S.S. Roosevelt, 1905–1906*. New York: Doubleday Page & Company.

Pope, S., Copland, L., & Mueller, D. (2012). Loss of multiyear landfast sea ice from Yelverton Bay, Ellesmere Island, Nunavut, Canada. *Arctic, Antarctic, and Alpine Research, 44*(2), 210–221.

Serson, H. V. (1972). *Investigation of a plug of multi-year ice in the mouth of Nansen Sound, Technical note 72-6*. Ottawa: Defence Research Establishment.

Serson, H. V. (1974). *Sverdrup Channel, Technical note 74-10*. Ottawa: Defence Research Establishment.

Simmonds, I. (2015). Comparing and contrasting the behaviour of Arctic and Antarctic sea ice over the 35 year period 1979–2013. *Annals of Glaciology, 56*(69), 18–28.

Sverdrup, O. N., Bay, E., Schei, P., et al. (1904). *New land; four years in the Arctic regions*. London: Longmans Green and Co.

Stefansson, V. (1938). *The problem of Meighen Island*. New York: The Explorers Club.

Stroeve, J. C., Serreze, M. C., Kay, J. E., Holland, M. M., Meier, W. N., & Barrett, A. P. (2012). The Arctic's rapidly shrinking sea ice cover: A research synthesis. *Climatic Change, 110*, 1005–1027. doi:10.1007/s10584-011-0101-1.

Thomson, L., Zemp, M., Copland, L., Cogley, G., & Ecclestone, M. (2017). Comparison of geodetic and glaciological mass budgets for White Glacier, Axel Heiberg Island, Canada. *Journal of Glaciology, 63*(237), 55–66. doi:10.1017/jog.2016.112.

Van Wychen, W., & Copland, L. (2017). Ice island drift mechanisms in the Canadian High Arctic. In L. Copland & D. Mueller (Eds.), *Arctic ice shelves and ice islands* (p. 287–316). Dordrecht: Springer. doi:10.1007/978-94-024-1101-0_11.

Vincent, W. F., Gibson, J. A. E., & Jeffries, M. O. (2001). Ice-shelf collapse, climate change, and habitat loss in the Canadian High Arctic. *Polar Record, 37*, 133–142. doi:10.1017/S0032247400026954.

White, A., Copland, L., Mueller, D., & VanWychen, W. (2015). Assessment of historical changes (1959–2012) and the causes of recent break-ups of the Petersen Ice Shelf, Nunavut, Canada. *Annals of Glaciology, 56*(69), 65–76. doi:10.3189/2015AoG69A687.

Wilson, K. J., Falkingham, J., Melling, H., & De Abreu, R. (2004, September 20–24). Shipping in the Canadian Arctic: Other possible climate change scenarios. In *Proceedings from IEEE International Geoscience and Remote Sensing Symposium*, Anchorage, Alaska. doi:10.1109/IGARSS.2004.1370699.

Chapter 13
The Military Importance and Use of Ice Islands During the Cold War

William F. Althoff

Abstract The Arctic Ocean, because it is ice-covered, facilitates research difficult to conduct in the open seas. The polar pack and, upon discovery, ice islands offer natural (if problematic) platforms. When first detected, 'floating islands' aroused immediate interest by the U.S. Air Force then the U.S. Navy. Large, deep-draft masses, they resist sea ice pressure and breakup; as platforms, ice islands confer long-term occupation. Superpower rivalry propelled Cold War science. The circumpolar North had become a theater of operations, an exposed flank. Field programs multiplied on both sides of the Central Arctic, as did the number of ice-based outposts for research. Air-deployed drifting stations are a Soviet logistic invention; from 1954 to 1991, the USSR sustained a continuous presence on drifting ice islands. As U.S. Air Force concern for the Arctic eased, the under-ice capabilities of nuclear submarines intensified and programs to understand (and exploit) the Polar Basin were developed: oceanography, geophysics, underwater acoustics, sea-ice physics, meteorology, marine biology. The ballistic-missile submarine introduced anti-submarine warfare into Arctic waters, further stimulating research pertaining to the floating ice cover and Arctic Ocean acoustics.

Aviation helped open the Arctic to systematic observation. To date, Russia has deployed five drifting stations onto ice islands, the United States two, and Canada one. These were tiny communities, with deep isolation at close quarters. On sea ice or ice island, the potential for emergency haunted every camp manager. Success depended on the cooperation of researchers with logistics-support staff. In the post-Cold War era there are different challenges in the Arctic: climate change, some now argue, poses a threat to national security.

Keywords Drifting station • Ice island • Severnyy Polyus • T-3 • ARLIS II • Hobson's Choice

W.F. Althoff (✉)
Geologist/Naval Historian, Whitehouse Station, New Jersey, USA
e-mail: skyships@juno.com

L. Copland, D. Mueller (eds.), *Arctic Ice Shelves and Ice Islands*, Springer Polar Sciences, DOI 10.1007/978-94-024-1101-0_13

13.1 Historical Backdrop

The sciences of the Earth are heavily observational, that is, each relies on fieldwork and ever-improving data sets. To penetrate an ocean, one needs platforms. The Arctic Ocean is relatively unknown; before vessels could reach the central polar basin, the only alternatives for the explorer-scientist were on-foot traverses or frozen-in ships, which, one hoped, would be shifted poleward.

The Arctic Ocean and its peripheral seas have known European penetration since the sixteenth century. Systematic observation, however, dates from Fridtjof Nansen and the three-year drift of his ice-ship Fram (1893–1896). First to demonstrate the drift technique, Nansen's science expedition opened the modern era in the North. Twentieth-century technology, such as the radio, icebreaker, nuclear power, aviation, satellites and telecommunications, have since opened circumpolar realms to survey, basic research and long-term environmental monitoring.

Russians have long occupied their northlands. Intensive development efforts, however, date from the Soviet era. The absence of observations between Siberia and North America complicated the analysis of air-ice-ocean processes along the Northern Sea Route (Northeast Passage). In the 1930s, the notion arose of reviving Nansen's concept of a North Pole science station: deploying quasi-permanent camps onto the ice itself, to record synoptic observations.

Sea ice is an ensemble of floes, not a continuous sheet. Cracks open at any time. Fracturing can threaten or terminate an ice-based camp if safety is compromised or the runway for airlift support drops below minimums. First detected in 1946–47, 'floating islands' aroused immediate military interest. Ice islands are exceptionally large, deep-draft masses; owing to their size and thickness they resist sea ice pressure and breakup. In short, as platforms they offer longevity. The Cold War posed alarm and threat. If landings proved practicable, ice islands might be used for air operations and air defence, as a base for submarines or a site for early-warning radar or air-sea rescue. For research, their ice might support a weather base or a geophysical laboratory.

Drifting stations deployed by air are a Soviet logistic invention: on the eve of the Second World War, the Soviet Union airlifted a field party onto the pack mere miles off the geographic North Pole with a mission to conduct basic observational work. That expedition verified the feasibility of ice-based camps for reconnaissance and research, a template for subsequent stations in its Severnyy Polyus ('North Pole') series.

Countless air, surface and submarine-based missions (and satellites) have probed the Arctic Ocean and fringing landmasses, most funded by the military. Environmental factors vital to understanding this theater of operations (e.g., underwater acoustics) are outlined in the next section; programs and techniques for working atop ice follow. Ice-based stations deployed by Soviet and American institutions are then profiled, as is the Canadian experience with ice islands in the Arctic Ocean.

13.2 Physiographic Factors

13.2.1 Baseline Environmental Data/On-Ice Scientific Programs

'The Arctic Ocean, because it is ice covered, provides several opportunities for observations which are difficult or impossible to make in an open ocean' (Crary and Goldstein 1957). Seismic detectors, for example, can be arrayed on the pack ice for ocean bathymetric, seismic reflection, and seismic refraction profiles along the drift path. Commencing in 1950, reconnaissance research exploiting ice islands embraced an array of meteorological, oceanographic, and geophysical studies:

- Physical and chemical oceanography, including seismic depth soundings
- Marine geology and sub-bottom geophysics, such as seismic reflections
- Regional gravity and magnetic surveys
- Ice-movement studies
- Physical properties of sea ice
- Marine biology
- Meteorological programs, including upper-air studies and synoptic observations for correlation with other investigations
- Arctic Ocean acoustics
- Ice morphology

As well, basin climatology and paleo-circulation were studied by means of bottom-sediment grab and core sampling.

13.2.2 Mean Surface Circulation in the Arctic Basin

Before earth-orbiting satellites, to study the ocean one went to sea. Early explorers used ice-strengthened-ships, which were frozen-in and shifted with the polar pack. It was realized that the polar pack and, once discovered, ice islands are useful platforms offering a natural (if somewhat problematic) foundation.

The notion of air-supported encampments exploiting moving ice was pioneered by the Soviet Union. The Arctic and Antarctic Research Institute in Leningrad (now St. Petersburg) was, and remains, the agency primarily responsible for polar scientific services and research. Forecasting weather and ice conditions along the Northern Sea Route is a primary mission. In May 1937, the Institute installed a manned station on sea ice. A pioneer expedition, the 274-day, nearly 3000 km odyssey of the North Pole Drifting Observatory Station (later SP-1) confirmed the feasibility of modern ice-rafted bases for research. The United States, for its part, would not exploit Arctic ice until 1950.

The Soviets, meantime, had confirmed the presence of in-flowing Atlantic Water beneath the cold, relatively low-salinity surface layer in the Arctic Ocean. Defined by winds and surface currents, ice motion (hence drift paths) are predictable only in a general sense. Two patterns predominate; the Beaufort Gyre is a clockwise motion requiring about ten years for one orbit, with its mean center off Canada-Alaska, and the Transpolar Drift Stream is a large-scale, faster-moving motion of ice away from the Siberian coast and across the pole, debouching through Fram Strait, the basin's high-energy connection to (and exchange with) the world ocean. As well, the region is a major site of global deep-water formation.

For its Severnyy Polyus (or SP) series of drifting stations, the Arctic and Antarctic Research Institute often deployed off Eastern Siberia, a focal area for extended drifts (as long as three years) riding the outflow of cold water and multiyear ice exported from the Arctic Ocean into the high North Atlantic.

13.2.3 Underwater Acoustics

Antisubmarine warfare was the U.S. Navy's top priority in the days of the Soviet submarine threat. Under ice, information is transferred acoustically (and the pack itself is an active source of background noise). In the maritime Arctic battle space, acoustic energy is a tool of detection; accordingly, propagation and ambient-noise studies have probably attended every U.S. ice-based camp.

The Arctic cruises by Nautilus (SSN-571) and Skate (SSN-578) in 1958 were followed by decades of research to understand (and exploit) the sea-ice cover. Detection and tracking are essential for military control in the Arctic Ocean and its approaches, which form prospective patrol areas for attack and missile-carrying submarines (SSBNs). In 1958, the operational unknowns relative to all-season, under-ice capability were formidable, as Leary (1999) notes:

> *...investigating in-ice surfacing techniques; developing and evaluating procedures for coordinated operations by nuclear submarines; testing torpedoes; providing oceanographic and hydrographic data; testing communications and sound propagation; evaluating the protection afforded by the ice pack; and investigating the problems of submarine versus submarine warfare in ice areas.*

The physics of ice-acoustic interaction is challenging: a geometrically complex, dynamic solid-plate boundary floating on an upward refracting fluid. Performance of high-frequency systems is fundamental to sensors related to weapons (torpedoes and mines). The propagation of low-frequency acoustic waves applies to long-range detection and tracking. Using explosive sources, long-range propagation experiments were conducted between stations. And, often, ice-based experts worked a 'live' submarine target maneuvering nearby, beneath the sea ice canopy:

> *The unique environmental feature of the Arctic—the one that effectively precludes extrapolation of generalized acoustics theory, models, and data from the more thoroughly researched open ocean areas—is the ice canopy. Its presence grossly affects the two parameters, sound propagation and noise, that are of prime importance to the ultimate users of acoustics knowledge, the sonar designers and operators.* (Untersteiner et al. 1976)

The technological adjunct of underwater acoustics is sonar engineering. Ice-detecting sensors reduce the danger of collision in shallow water and allow tactical use of the canopy itself. In sum, the Navy began a long-term research program in Arctic underwater acoustics along with studies of other environmental factors that affect naval operations in this unique theater. The ballistic-missile submarine introduced the field of antisubmarine warfare into Arctic waters, further stimulating work in sea-ice physics.

13.3 Prelude to the International Geophysical Year: The Decade 1946–1956

The Second World War had interrupted study of the Soviet Arctic and, as well, ended pure northern science. The circumpolar North had become strategic space, a theater of operations. A peripheral realm hitherto on the outskirts of military notice, by 1945 'north' to policymakers embodied an exposed flank, a zone of competing interests – a direction. In short, the new superpowers had an immense geostrategic stake in the region.

According to Ottawa's Joint Intelligence Bureau (1962), in addition to economic development in the Soviet North, advances in the defence field:

> *...have centred around the use of the area as a forward base for possible air operations directed against North America, as a base for naval operations and for air defence facilities; around the Northern Sea Route as a route for the transfer of naval vessels from the West to the Far East and around the use of the Novaya Zemlya area for the testing of nuclear weapons.*

After 1945, operations accelerated on and over as well as under Arctic ice. Obsessed with secrecy, Soviet activities were conducted with no fanfare, inciting no alarm in the West. Their program was not driven by military concerns, although, certainly, such applications would be made. What was sought, instead, was a scientific understanding of the maritime Arctic, a realm the Soviets deemed theirs.

Field programs multiplied on both sides of the Central Arctic, as did the number of scientists deployed. Unknown to the West, the Soviets revived the drift technique pioneered in 1937, air-deploying SP-2 in April 1950 off eastern Siberia. Their second drifting station, its mission was to improve the accuracy of ice and weather forecasts along the Northern Sea Route, by taking measurements over a full annual cycle. According to Treshnikov (1956):

> *The observations of this station notably increased our knowledge of the relief of the eastern part of the Arctic Basin, of the currents, the ice drift, and the heat balance of the ice, and, most important of all, the station systematically reported the weather and ice movements to the Arctic weather bureaus, which could use this data to give more reliable information on the weather and ice drift along the Northern Sea Route.*

Weather reconnaissance flights by United States aircraft from Alaska to the geographic North Pole had begun in 1946. That August, a U.S. Air Force (USAF)

WB-50 (shortened version of the B-29) reported the detection, by radar, of an enormous mass. Designated Target One or T-1, this object was detected at different positions on subsequent missions. In 1950, radar and visual searches found another two 'ice islands', T-2 and T-3. By then, the USAF and Navy were advocating landings onto sea ice, for basic research. During 1951–52, both services airlifted field teams into the Beaufort Sea, to record geophysical and oceanographic data. These were spot observations; no camps were set. In March 1952, the first ice-island landing (onto T-3) was logged by the Alaskan Air Command. The Air Force decision to remain and occupy would result in more than two decades of geological, geophysical, meteorological, oceanographic, biological, acoustic and other research (see Sect. 13.6).

The Arctic Institute in Leningrad, for its part, would all but run a shuttle to seaward, using two complimentary programs of data collection. In addition to its spring re-supply of the semi-permanent drifting stations, each Sever ('North') air expedition established a temporary camp; in effect, an ice-borne airport. From these, mobile scientific detachments 'jumped' to pre-selected touchdowns. Unlike the SP stations, these were brief tarries: spot soundings; measurements of temperature, ice thickness and snow cover; a sampling of the bottom; magnetic and astronomical as well as meteorological observations. Large tracts were thus covered. In contrast, the slow, erratic course of drifting stations granted an intensive but narrow picture of the area through which they moved.

In April 1954, the next in the series of Soviet drifting stations (SP-3 and SP-4) were declared operational. Their science roster included oceanography, meteorology, glaciology and geomagnetism. The Soviets would maintain a continuous drift-presence to 1991, with a policy of keeping two SP stations in simultaneous operation. If evacuation became necessary, aircraft sought a suitable multiyear floe to raft a replacement. And if the ice outlasted its one-year shift of personnel, the men were relieved. Concurrently, 'jumping' detachments worked sectors of interest, taking spot measurements at selected coordinates. In 1954 alone, these specialized field operations flew 250,000 miles over the central Arctic and made hundreds of landings without mishap. According to Fletcher (1968):

> *It should be kept in mind that the Soviet post-war activity was unknown to the West. It was conducted clandestinely and no announcements were made until mid-1954. So the interest building up on the American side was completely independent. Believe me, had we been able to show our authorities the information... it would not have been so difficult to get authorization and support for Arctic Basin investigation.*

On T-3, a four-man party completed 5 months of observations in September 1955. The ice island was not reoccupied until March 1957, when a USAF survey party arrived for construction of a new base: International Geophysical Year Arctic Ocean Station 'B' or BRAVO.

13.4 International Geophysical Year (IGY)

The IGY of 1957–1958 held a strong polar research component. When 1956 opened, two Soviet stations were operational: SP-4 and SP-5. That April, SP-6 (1956–59) was installed north of Wrangel Island. Measurements derived from SP-6 and its sister, SP-7 (1957–59) would constitute Moscow's contributions to the IGY program for the Arctic Ocean. Sustained by airlift, SP-4 and SP-5 were relieved and re-supplied that spring, each for the second time. Operational during March-June 1957, the Sever-10 high latitude air-expedition closed down SP-4, relieved SP-6, established SP-7 and set up automatic weather stations and radio beacons for tracking ice movement.

To counter Soviet initiatives, no U.S. station was operational. For the IGY, the U.S. Department of Defense was directed to provide the logistic support necessary to field-deploy scientific teams. U.S. planning called for two ice camps in the Arctic Ocean. Staging from Alaska, the Air Force would support ice-based IGY outposts hosting military personnel and civilian investigators from various universities and institutions:

> *The government officials who supplied the money, while not indifferent to pure scientific discovery, expected the new knowledge would have civilian and military applications. The American and Soviet governments and their allies further hoped to win practical advantages in their Cold War competition. Under the banner of the IGY they could collect global geophysical data of potential military value. Along the way each nation hoped to gather intelligence about its rivals, and meanwhile enhance its own prestige.* (Weart 2003)

The new IGY BRAVO base on Ice Island T-3 hosted its inaugural researchers in May 1957. Their field program concerned gravity, seismology, aurora and airglow phenomena, geomagnetism, oceanography, ocean currents, navigation, ice drift and meteorology. Field parties would deploy to T-3 throughout the Geophysical 'Year' (extended into 1960). The second camp was Station ALPHA, the first U.S. sea-ice camp. In September 1958, the Navy and USAF concluded an agreement whereby, after the IGY, the Office of Naval Research (ONR) would provide support for continued research on ALPHA while the USAF assumed responsibility for T-3. When progressive fracturing and ridging obliged abandonment of ALPHA in November 1958 (after a 19 month occupation), a replacement camp was established in May 1959, to continue the program. But severe pressure and fracturing required an emergency air-evacuation of CHARLIE in January 1960. The U.S. IGY effort in the Arctic Ocean was reduced to a single station, on T-3.

A ten-man party deployed onto T-3 in summer 1959 and conducted (for example) gravity and magnetic work, geophysics (seismic, resistivity), oceanography, ice-movement studies, ice morphology and micrometeorology measurements. '*Those [IGY stations] were rich sources of information*' a geophysicist recalls, '*because almost anything you discovered up there was new at the time*' (Hunkins, pers. comm., 21 May 1991)

When the IGY concluded, the Department of Defense authorized a continuation of the Arctic Ocean drift-station program (Fig. 13.1). Most of the IGY participants

Fig. 13.1 Ice Island T-3, summer 1960. The platform supported more than two decades (1952–1973) of geological, meteorological, oceanographic, biological, and acoustics research as the island orbited off Alaska/Canada, in the Beaufort Gyre. Note the icebreaker (USS Burton Island) nearing rendezvous in the distance (Source: Dr. Gerry H. Cabaniss)

continued their respective programs in cooperation with ONR, Department of the Navy or with the Geophysics Research Directorate of the Air Force.

13.5 Soviet Ice-Island Drifting Stations

The Soviet Union operated more drifting stations over a longer span of years than any other nation. Commencing in 1954, the Arctic Institute sustained a continuous on-ice presence. SP-4, for example, was the first in continual operation for three years. Deployed in April 1956 as part of USSR contributions to the IGY, SP-6 exploited an ice island, a first for the Soviet Union. Three years later, following the annual (spring) relief, SP-6 was evacuated in 1959, with Sever-12 logging nine flights to complete the closeout. So as to satisfy IGY commitments, SP-7 had been deployed in 1957 about midway between Bering Strait and the North Pole.

Established in April 1959, SP-8 lasted to March 1962, a total of 1057 drift-days. Its floe fractured several times, obliging staff to relocate both camp and airstrip. Two months after the evacuation, the Central Intelligence Agency clandestinely inspected the abandoned outpost, to assess Soviet acoustical research. Findings proved suggestive though inconclusive. Still, analysts from the Office of Naval Intelligence concluded:

> *In spite of the austere, unsanitary, and politically pressured conditions under which the Soviet scientists lived, all evidence indicates that they were engaged in a highly developed, successful, and extremely useful program of science.* (Leary and LeShack 1996)

Eleven drift expeditions were established during the 1960s: SP-9 through SP-19. Except for '18' (briefly) and '19,' each exploited sea ice. Nikolai Dimitrievich Vinogradov was a station leader at both SP-22 and SP-31. Which ice is the platform of choice, he was asked. '*An ice island is preferable, of course*' Vinogradov replied, smiling. '*Certainly the work is much easier…than on a thin ice floe*' (Vinogradov, pers. comm., 21 May 1992).

SP-19 logged a full transect of the Arctic Basin between November 1969 and April 1973, with three shifts of personnel. When SP-19 floated north of Greenland on approach to Fram Strait, an air-evacuation was ordered. SP-22, SP-23, and SP-24 each exploited an ice island. Deployed by icebreaker, SP-22 logged an unusually productive occupation (1973–1982). In May 1978 it escaped the Beaufort Gyre into the Transpolar Drift. Arcing past the geographic North Pole in a near-transect of the Arctic Ocean basin, SP-22 was occupied by nine shifts of personnel and hosted two Sever expeditions. The last aircraft supporting the operation lifted clear in April 1982. Prior to air-evacuation, '22' incited a media and diplomatic dustup when, in 1977, it penetrated the Canadian Arctic sector and waters claimed sovereign by Ottawa. What was Moscow doing 'on our side of the Arctic?' (see Sect. 13.11).

As well, SP-23 (1975–78) and SP-24 (1978–80) transected the Arctic Ocean, nearing West Greenland and final evacuation, a testament to the enduring nature of freshwater ice shelf ice. The former had been deployed north of Wrangel Island, a focal area for extended drift expeditions. SP-24 also sustained a fertile run, hosting three shifts of personnel and basing Sever air expeditions in 1979, 1980 and 1981.

The mid-1970s found the Kremlin wary of Pentagon plans in the 'Arctic operational zone.' According to the Soviet Military Encyclopedia (1976):

> *The military leadership of the U.S. assigns great strategic importance to the Arctic. In accordance with the development by the Pentagon of a 'polar strategy', U.S. Arctic research work mainly services military purposes. The influence of climatic and hydrographic conditions on the operation of forces and on the possibility of using weapons and military equipment in the Arctic is being studied….The main efforts of the USSR in the Arctic are directed toward continued research and economic utilization for peaceful purposes.*

Typical staffs at Soviet stations included two meteorologists, two or three oceanographers, four aerologists, three to four geophysicists, two electrical engineers, and one or two mechanics. A physician and cook accompanied them. Weather data were radioed daily and letters and scientific reports describing results dispatched to shore. P.A. Gordienko, a seasoned station head, outlined the responsibilities of on-ice staffs for Western readers:

> *Since the station operates all day every day and supply problems limit the size of the crew, a good deal of versatility is expected from each man. The magnetologist usually serves as astronomer, the physician doubles as housekeeper, the meteorologists readings of the direct-heating power of the rays of the sun and the fliers assist the hydrologists and aerologists. All hands take turns at kitchen duty….And all lend a hand in hacking a runway out of the ice* (Gordienko 1961).

Off northern Alaska, the third shift manning SP-31 was emergency-evacuated in July 1991 because of floe breakup. This was the last station of the Soviet era. A twelve-year hiatus ensued; not until 2003 did the Institute in Leningrad (now St. Petersburg) install another drifting station, establishing SP-32.

In recent years, petrodollars have revived Russian polar programs. In the post-Soviet era, SP-33 was deployed in 2004, SP-34 in 2005, SP-35 in 2007 and SP-36 in 2008. SP-37 was evacuated at the end of May 2010 and SP-38 was deployed off Wrangel Island 5 months later.

13.6 Ice Island T-3 (United States)

Scientific and military interest in ice islands was incited in 1946 when a very large (24 by 30 km) tabular mass was detected by the USAF via radar. Classified 'secret,' Target X (later T-1) caused considerable excitement: such an enormous feature was unknown for the Arctic Ocean. Speculation as to its possible source and probable future soared. '*For the next three years, the existence of T-1 remained a military secret, and strategists dreamed of making a landing field on it. Apart from its possible military importance, valuable scientific work could be carried out on it if a base could be established there*' (Miller 1956). Repeat radar sightings during reconnaissance flights had realized a specific rate of movement. And low-level air passes confirmed a remarkable tableland of ice.

In April 1947, an unusual rippled pattern off Borden Island was photographed during a joint USAF/Royal Canadian Air Force (RCAF) flight. Squadron-Leader Keith R. Greenaway, RCAF, later recalled the sighting:

> *The undulating surface of [ice identified in 1950 as] T-3 was immediately associated by us with that of the ice shelf along the north coast of Ellesmere Island. We had no idea of the mechanics of the formation of the ice shelf, or undulations....At the time of discovery, we speculated that landings could be made on T-3 without undue risk* (Greenaway, pers. comm., 20 Nov. 1991).

Joint Canadian-U.S. expeditions (1953–1954) would investigate the Ward Hunt Ice Shelf, the presumed parent ice mass for T-3. This work laid the foundation for research programs concerning 'floating ice-island sources' and, as well, ice-shelf formation and their geologic history. The first USAF orders to establish a manned on-ice project date from June 1950:

> *The possibility of increased air traffic over the Arctic Ocean requires a thorough knowledge of weather conditions and communications problems peculiar to polar flying operations. Information concerning the logistical support of a manned weather station on the Polar Ice Pack and the capabilities and limitations of rescue operations is essential for successful air operations in this area....* (Wise 1978)

In March 1951, an experimental pack-ice camp was annihilated within 3 weeks. And so the next deployment exploited an ice island. In March 1952, the United States established an experimental outpost on T-3. A party under Colonel Joseph

O. Fletcher, USAF, remained to set up a temporary base. Its status exploratory, the inaugural camp was necessarily small and experimental. Still:

> *T-3 also provided the 58th Reconnaissance Squadron with its only emergency landing strip in the Arctic. With improved radio facilities, runway lights and a hardened runway, T-3 could be a significant target for emergency landings and navigation during the winter* (Ibid.).

The decision to remain marked the beginning of systematic at-sea geological, geophysical, meteorological, oceanographic and biological research. Logistics furnished by the Alaskan Air Command, Air Weather Service personnel and scientists of the Air Research and Development Command, Cambridge Research Center, initiated meteorological and other studies. Hostage to the Beaufort Gyre, the occupation of T-3 would persevere to 1974. Meantime, because of proximity to Alert, on Ellesmere Island, the first evacuation was logged in May 1954 to Thule Air Force Base, 901 km away. According to the *Boston Sunday Globe* on 22 May:

> *Although many details still are secret, it can be disclosed that during the occupation period, scientific information of immense value, both in the practical military sense and in the realm of pure science, was obtained on T-3....Since the North Polar basin rapidly is gaining importance as a strategic crossroads for air power and probably for atomic submarine operation, any additions to United States knowledge of the area are more than welcome in headquarters of all the armed services.*

As Air Force concern for the Arctic eased, that of the Navy Department intensified, due in large measure to the projected use of the Polar Basin and adjoining seas by nuclear submarines. But how does one carry science and research to an ice-congested ocean where the ONR had obvious requirements? Dr. Max Britton headed the Arctic Research Program, ONR:

> *The obvious answer was to utilize the ice itself as the Soviet Union had long been doing, and as the U.S. Air Force was doing....We in ONR were determined to get into the drifting station business.* (Britton 1968)

When it became practicable to operate from the ice, Britton was pleased to note, '*work at sea could really progress*'. In April 1962, the Alaskan Air Command officially transferred T-3 to the Navy and its Arctic Research Laboratory (ARL). Established in 1947, ARL became a major field logistic support activity for research in the western Arctic (see Sect. 13.7).

T-3 hosted field parties from a variety of disciplines and agencies. Acoustic measurements (for instance) were initiated by the Underwater Sound Laboratory (USL). In 1958, the nuclear-submarine *Skate* had surfaced at ALPHA. On winter cruise in February 1960, *Sargo* (SSN-583) logged a rendezvous with T-3, a USL electronics team on the ice listening to and recording the boat. Later, a small satellite camp to T-3 was set up for underwater low-frequency work. Sited off-island, so as to exclude acoustic and electromagnetic interferences, it was manned for experiments with submarines on an all-season basis. During 1968, representatives from universities in California, New York and Washington State exploited the ice island, among them meteorologists, seismologists, oceanographers, geologists and biologists. With all personnel off by 1974, T-3 drifted unmanned (Shoemaker, pers. comm., 31 March

2002). On 20 August 1983, the ice island's last confirmed position was 83.1°N, 4.2°W, on approach to Fram Strait and export into the high North Atlantic.

13.7 ARLIS II (United States)

Located near Point Barrow, the ARL was a year-round basic-research facility funded by the ONR, Department of the Navy. Its role was to provide logistic support for contracted research groups. In 1959, the laboratory acquired the ability to conduct research offshore, on the ice, a capability conferred by aircraft.

Following the IGY, ARL (renamed Naval Arctic Research Laboratory) adopted a new model of ice-based station: 'small and simple' camps supported by aircraft exploiting unprepared ice. That is, drifting stations independent of fixed and maintained airstrips. One result was the ARLIS series (Arctic Research Laboratory Ice Station). Excepting ARLIS II, each was a small, low-noise camp devoted largely to underwater acoustics research.

ARLIS I (1960) had been the first station established by ARL independent of the Air Force. After its evacuation, a reconnaissance flight seeking a suitable floe for the next temporary camp sighted an ice island. Measuring 5.2 × 3.2 km, this shelf-ice fragment was littered with glacial debris (Fig. 13.2). In May 1961, ARLIS II was established as an ONR/NARL station.

> *Although its location was farther south than desired, the relative permanence and stability of an ice island far outweighed the southerly locale. The aircraft circled and inspected the island. The first Cessna [180] landed on the island and the second landed on the sea ice, on a refrozen lead that cut deeply into the island, to investigate possible landing sites for the R4D [Navy version of the DC-3]. After three trial landings by the Cessnas, the R4D was waved in.* (Schindler 1968)

The decision was taken to occupy. An airlift ensued: 28 flights by an R4D aircraft, 41 by two Cessna 180s. The first prefabricated hut was up by the evening of the second day (23 May), and the first scientific measurement taken by 1 June. And so the United States boasted two active ice-island stations. That summer had found T-3 grounded northwest of Point Barrow, on the continental shelf. The Air Force elected to evacuate. This occurred a few months too soon; during routine re-supply to ARLIS II in February 1962, T-3 (reduced in size) was rediscovered. Useful again, the T-3 camp was rehabilitated and a new chapter opened: that April, the Alaskan Air Command transferred T-3 to the Navy. Through ONR, the Navy would sustain the two bases; operated as a logistics pair, T-3 and ARLIS II supported concurrent oceanographic, meteorological and acoustic programs. During the next four years (1962–65), moreover, as ARLIS II was shifted toward Greenland on the Transpolar Drift Stream, T-3 served as an alternate for flights between Barrow and a receding outpost.

En route to Fram Strait, ARLIS II had reached latitude 88°N, longitude 117°W when 1963 ended. The scientific roster that year consisted of marine biology,

Fig. 13.2 Rock debris on Ice Island ARLIS II, June 1964. Note prefabricated huts and impounded meltwater in background. Following the IGY, a new approach to ice-based research was initiated by the Office of Naval Research: 'small and simple' floe camps of the ARLIS series (Source: Dr. Kenneth L. Hunkins)

gravity, geomagnetism, seismology, surface and micro-meteorology, snow, ice, and oceanographic studies. Orbiting in the Beaufort Gyre, T-3 that December was at latitude 81°N, longitude 147°W. That summer, an ocean-bottom thermal program had begun (with cores) and measurements of bottom currents initiated (Arctic Research Laboratory 1964).

Ice islands are undesirable for acoustics work due to their rarity, thickness and dissimilarity from sea ice. For access to the ocean, hydro stations were required on floe ice near an ice island edge. Further, the main camp itself interfered due to background acoustic interferences:

> *While permanent and safe from breakup, experience at T-3 indicated that [big camps] were hardly ideal sites for low-frequency acoustics research. Forty kW [kilowatt] generators going full time, road graders on the runway at all hours, uncleared personnel, electromagnetic radiations from strong HF [high-frequency] radios and aircraft beacons – made an acoustician's life a hard one. Small, quiet camps were obviously needed.* (Buck 2003)

Large hydrophone arrays were installed, with the ice providing a platform for suspending them. During spring 1964, for instance, T-3 and ARLIS II (along with a small, short-term floe camp) were exploited for long-range sound-propagation experiments. And when ARLIS II was evacuated in May 1965, this Navy program continued from T-3. Multiyear ice limited icebreaker penetrations for re-supply. Consequently, although their cargo spaces were limited, aircraft were the primary means used to deploy personnel and materials (particularly fuel) onto the shifting ice.

The drift track of ARLIS II arced fully across the basin, from the Beaufort Sea sector (73°N) to the geographic Pole (88°39′N) and thence Fram Strait. The strait is a zone of violent ice action; there surface currents increase on approach to the exit. Concluding four years of occupation, abandonment was logged on 11 May 1965. The ice island had hosted 14 separate research projects (most shifted to T-3), 118 scientific personnel, 132 support staff as well as transients and aircraft crewmen (Schindler 1968).

13.8 Ice Island WH-5

Five large ice islands broke away from the Ward Hunt Ice Shelf between August 1961 and April 1962 (Hattersley-Smith 1963). The largest of these was WH-5. Four drifted westward. WH-5, however, moved east, towards the Lincoln Sea; it rounded Ellesmere Island, passed down Robeson Channel though Hall Basin into Kennedy Channel (Van Wychen and Copland 2017). There it lodged, an ice-plug barrier to floe ice drifting south. The resulting open water incited a study of the fragment and its local influence, taking advantage of the rare absence of ice in a sector normally closed to shipping. Among the outcomes were the first direct current measurements in Smith Sound, cross-sections across Kane Basin and other observations. This work was sponsored by the Arctic Institute of North America in collaboration with the Woods Hole Oceanographic Institute (Massachusetts), and the U.S. Navy and Air Force (Arctic Research Laboratory 1964).

13.9 Hobson's Choice Ice Island (Canada)

In April 1983, newly calved ice islands were detected adjacent to the Ward Hunt Ice Shelf. (Jeffries and Shaw 1993). Their drift was tracked by Ottawa. The Polar Continental Shelf Project is Canada's Arctic logistics agency. 'Polar Shelf' (as it is popularly called) supports studies by government departments and universities in all fields of science. In 1984, the agency initiated construction of a research station on the largest shelf fragment for geological, geophysical, meteorological and oceanographic studies of the Arctic Ocean and the Canadian polar margin:

> *The ice island will offer an excellent opportunity to add significantly to our knowledge of oceanic conditions off Canada's northern coast. The major advantage of the ice island over ice camps on sea ice is that studies can be done year round. The proposed research programs will allow us to answer basic questions regarding the oceanography of the Arctic Ocean in support of wide ranging national and international climate studies; understanding the seasonal variability of oceanographic phenomena in our most northern ocean; and assessing the structure and function of the Arctic marine system as a basis for predicting the*

Fig. 13.3 The ice-island platform known as 'Hobson's Choice', June 1987. Occupied by Canada, its floating outpost is barely discernible (*circle*). Note how ice-island topography contrasts with the enclosing pack ice (Source: George D. Hobson)

> *potential impact of hydrocarbon exploration, exploitation or transport in the Arctic.* (Department of Fisheries and Oceans 1985)

Northern sovereignty is a persistent Canadian theme. As well as probing the nation's 'geological roots', Hobson's Choice Ice Island (named for then Director of Polar Shelf) tendered a platform for securing sovereign offshore limits (Fig. 13.3). And hung over an ice island edge, at least one Defence Department hydrophone would be listening for submarines. Washington's possible military role in the northlands was, again, becoming a concern among Canadians:

> *The increasing military use of the Arctic region both by the U.S. and Soviet forces has raised the possibility of it becoming a site of superpower confrontation. Canada's geographic position between the two superpowers, and our collective security agreements, of course make Canadian involvement in any superpower conflict unavoidable, and recognition of this reality has prompted renewed interest in the strategic importance of the Canadian Arctic and of the adjoining oceans and airspace.* (Haydon 1988)

Spring 1985 brought upgrades to Canada's new floating outpost. An evolving 'Main Street' had sleeping cabins, an office, navigation quarters, a generator hut, a workshop, and a combined kitchen and dining area. This last area was a godsend to the cook/chef, who'd been persuaded to return with a promise of modern workspaces. The latest ice-island facility to probe the Arctic Basin was open for business (Fig. 13.4).

Fig. 13.4 Diesel shack for the Hobson's Choice Ice Island facility, 1990. The hut housed a 30-kilowatt generator (Source: William Althoff)

A mechanic, cook, and station manager were the minimum support staff. The 1985 field season (spring-summer) was a pacesetter, with investigations by several federal and private organizations. Research included structural geology, oceanography and a refraction survey. As well, gravity and bathymetric data were taken beneath the seismic profiles. Water-column and bottom samples were retrieved. Geophysical investigations continued, including mapping bottom topography and sub-bottom stratigraphy. Continuous depth profiles of seismic reflection data were also recorded. The field season for 1986 resembled that of the year before, except that a geothermal project and several heat-flow stations were added to the station's suite of programs.

At times crowded, the facility at its peak could support 35 people in 14 wooden buildings and six large Parcols (tents), with a half-dozen sleeping cabins provided for the care and keep of researchers (Fig. 13.5). The field season commenced in February and extended into September, after which the base floated unoccupied through boreal dark. When the occupation for 1986 closed out, Polar Shelf could boast two seasons of fruitful research. That November, base manager Michael Schmidt tendered a prospectus:

> *The ice island has to date provided an excellent opportunity to develop the technology and means to operate a remote scientific research station in the Arctic Ocean. If, as envisioned, the island drifts around this vast ocean, the time, effort and money devoted at this time will be repaid by an efficient and safe environment in which to conduct scientific experiments. This last season has provided a solid foundation upon which to build.* (Polar Continental Shelf Project 1986a)

Fig. 13.5 Canadian field camp on Hobson's Choice Ice Island, August 11, 1987. Note the meltwater impounded by the island's ridge-and-trough topography (Source: George D. Hobson)

Polar logistics are costly. The 1986 budget for Polar Shelf was $5.1 million, of which approximately $2 million was devoted to ice island work. Economizing to maintain only what was essential, George H. Hobson cautioned his managers:

> *Unused buildings should be closed down to save fuel. I remind you that fuel costs at least $14.00 per gallon and is not that easily transported to the island. Eight people should not occupy 4 buildings. I would strongly suggest that 8 people occupy 2 buildings, thereby shutting down and saving fuel on the other 2.* (Polar Continental Shelf Project 1986b)

Shifted out of the Gyre in 1989, into Peary Channel, the ice-island platform was shunted further south during 1991, into the inter-island channels of the Arctic Archipelago. Useless, scientifically, the station was dismantled and salvaged:

> *The Canadian ice island has proved to be a valuable stable platform from which a broad spectrum of scientific studies can be carried out to obtain new data on a previously unknown part of the Arctic Ocean.* (Hobson 1989)

13.10 Life-on-Ice

Reaction to extreme geographic isolation is not wholly predictable. Most on-ice personnel adapted well. A few, though, sank into depression within days (sequestration for Soviet staff was 12 months). These were tiny communities: deep isolation at close quarters. Communal living under outpost conditions stokes inter-personal

Fig. 13.6 The remains of a Jamesway shelter after a fire on ice island T-3, spring 1960. On the ice, the potential for accident and emergency haunted every camp manager (Source: Dr. Gerry H. Cabaniss)

irritations that can sour into resentments. Camp leaders therefore were alert to the abrupt welling up of trouble.

The potential for accident and emergency, compounded by the presence of guns and explosives (for seismic work), haunted every camp manager. Shelter is a survival commodity, so fire presented an exceptional hazard (Fig. 13.6). The Alaskan Air Command records a further threat:

> *Between 1 July 1959 and July 1960, the activity on Bravo [T-3] was uneventfully dull and drab with nothing to break up the white monotony of everyday jobs and chores. However, when there was an exciting moment it usually became a matter of survival as Ice Station Alpha had proven. In October 1959 such a moment occurred when a giant polar bear attacked the campsite.* (Wise 1978)

Success depended on the cooperation of researchers with logistics-support staff. It still does. In those years, spring meant rotation of personnel, and delivery of all possible freight before runways softened. On Arctic ice, June through September is a time of thaw, ablation and an inescapable wet. Meltwater impounds, structures pedestal due to differential melting, dark objects (e.g., tools, even a drop of engine oil) sink from view, leaving craters (Fig. 13.7). Except for paradrop re-supply (Fig. 13.8a), air operations ceased. The men were uniquely isolated:

> *During the summer, when the only contact with others is via radio, morale reaches its lowest ebb. Feelings of isolation, helplessness, and hopelessness become predominant. The long hours of daylight make sleeping difficult, and time passes slowly. Personnel problems*

Fig. 13.7 End of the 1958 melt season: station trailers at IGY BRAVO (ice island T-3). Seasonal ablation and melt were major ice-island headaches, closing runways and pedestalling shelters due to shading of the underlying surface (Source: Dr. Gerry H. Cabaniss)

build up and tempers become short. The arrival of the first plane in the fall becomes the most discussed item of daily business. (Johnston 1968)

Soviet and U.S. field parties relied almost entirely on airlift support: survival and research equipment, vehicles, fuel, freight, rations, personnel. Runway construction and maintenance proved to be a major chore, particularly in the 1950s and 1960s, when U.S. stations relied on large aircraft (e.g., the C-124), long airstrips and the heavy equipment needed to groom them (Fig. 13.8b).

On T-3, routine observations passed into the hands of contractors: employees of federal agencies (e.g., U.S. Weather Bureau), civilian institutions and private companies under contract to the ONR. In July 1970, a heated discussion ended with a rifle being discharged, killing the station manager. A call went out; heard by ham radio operators, the laboratory at Barrow was alerted and, in turn, the head of the Arctic Program at ONR. What actions could, and should, be taken? An investigation was imperative. And both the body and suspect had to be returned to the States. But the T-3 runway was closed to fixed-wing aircraft, and would remain so until late September or October. The wait option was rejected in favour of immediate action: the suspect and the deceased had to be retrieved. 'We could only speculate on the possibility of further trouble', the Chief of Naval Research would recall. He faced a further complication:

The question of jurisdiction became apparent immediately. The laws of the sea are clear for jurisdiction over ships, boats, rafts, and debris, but are undefined for ice islands. If an ice island has the same status as a ship, clearly the United States had jurisdiction. (Holmquist 1972)

Fig. 13.8 (**a**) A melt-season airdrop is made to the Ice Island T-3 station. (**b**) Unloading fuel from a C-124 on T-3. Runways were a major chore when U.S. stations relied on large aircraft, long strips and the heavy equipment needed to maintain them. Except for the occasional icebreaker, airlift resupply sustained Soviet and American drifting stations, delivering survival and research equipment, vehicles, fuel, rations, freight and personnel (Source: Dr. Kenneth L. Hunkins)

T-3 floated 600 miles from the nearest U.S. airfield. Uncertain of the potential legal implications, it was decided that the suspect not touch down in Canada (the only other country that could logically have claimed jurisdiction). Hence Alert, on Ellesmere Island, was ruled out for the return flight from T-3. Courtesy of the USAF, an air-refueled H-3 type helicopter airlifted an investigative party (and a replacement for the deceased) to the ice island then flew the suspect nonstop to Thule and thence to the state of Virginia, for trial.

13.11 East-West Intersections

The conduct of northern science in that era was rooted in a rivalry shaped, steered and dominated by superpower tensions. Scientists tend not to share the enmity of their governments, with reason trumping rhetoric. On the working level, a bond of respect and camaraderie united field parties. '*We just don't get much Cold War here*', an American in Alaska observed, '*either between amateur radio operators or between scientists*'. Each side experienced flyovers, such as three inspection passes over ARLIS II by a Soviet aircraft in 1961. As well, the occasional drop-in to the ice camp of the other took place.

In May 1962, an aircraft assigned to NARL out of Barrow sighted Severnyy Polyus-11. Neither side in this campaign for data had yet visited the other. Granted permission, the American pilots landed. Despite fragile diplomatic relations ashore, on the ice genuine emotion ensued:

> *Because of the language barrier, the amenities were limited to smiles, handshakes, and taking photographs of all present, but this in no way impaired the cordiality of both parties. After [a somewhat awkward] 15 min. on the station the plane continued on its way to ARLIS II.* (Schindler 1968)

Canadian personnel logged a May Day visit to SP-25 (1981–84) in 1983. Polar Shelf's George Hobson (of Hobson's Choice) recalled the experience:

> *We had wonderful treatment.... We were there just as they were changing personnel. About a week later we were on television in Leningrad. [The utilitarian AN-2s], huge things, on wooden skies [had a] bloody big padlock on the doors. And the station manager had the keys. Pilots had to turn in the keys. Because with the range in the tanks that they had, they could very well escape to North America.* (Hobson, pers. comm., 25 April 1993)

In March 1977, a Twin Otter aircraft landed at SP-22. Earlier, Canadian surveillance and fighter aircraft had checked out the Soviet ice island station. None had touched down, until this plane. CBC (Canadian Broadcasting Corporation) reporter Jim Bitterman deplaned with a cameraman. Why? The news media had been promoting that Cold War paradigm of a Soviet military threat, in this instance an 'ice-island threat'. Hitherto, Soviet drifting stations had tended to shift away from North America. Following deployment in 1973, SP-22 instead had orbited off Alaska-Canada; 42 months later, the island was tracking westward about 300 km out, in waters claimed by Ottawa.

Introductions settled, permission was requested to inspect and to film. Led by Nikolai Vinogradov, head of the station, a tour was granted. Along with laboratories, radio shack and other facilities, a series of vertical shafts proved intriguing. Via ladder, the CBC investigated. '*The main interest of Canadians*', Vinogradov laughed, recalling the drop-in, '*was accumulated around these pits*'. Suspected missile silos, the pits were active. Electric heaters produced liquid water for the station. When the ice had been drilled to about thirty feet, the heater was extracted, the next shaft begun. The resulting 'silos', Bitterman discovered, were in fact refrigerator-storage for perishables. Oiled by vodka, cognac and dinner, East-West talks proved cordial. One myth regarding this offshore Soviet 'threat' was agreeably dispelled.

In 1988, inside the Soviet sector, a Mi-8 twin-turbo Aeroflot helicopter pounded in to CRYSTAL, an acoustics U.S. floe station. The craft circled then landed. Twenty-eight men disembarked, nearly the entire third shift for SP-28 (1986–89). While brief, the drop-in proved warm, with 'smiles all around'. With permission from Washington granted, a reciprocal visit was made to SP-28, whose position that April was above 89°N and mere miles from CRYSTAL. The Americans were well received. '*They were very gracious and extremely friendly*', an ONR officer remembered.

That year also, a Soviet TU-95 bomber pounded past APLIS 88 (Arctic Polar Research Lab Ice Station), a U.S. high-frequency research camp floating west of the Date Line, that is, in Soviet sovereign waters. U.S. Navy scientist Robert Francois was on the ice.

> *I could clearly make out the dual, counter-rotating propellers on the front of the four engine nacelles. They made an enormous racket as their four blades crossed each other...At a slightly higher altitude came the U.S. escort [and tanker], flying a zigzag course to keep its net forward progress at the same rate as the [TU-95] Bear...Only one pass was made.... Two days later, we had a second visit...* (Francois, pers. comm., 25 April 1993)

13.12 Post-Cold War

Beyond the ground-based net, the Arctic is largely air-dependent for transportation and logistical support. People as well as freight are moved almost invariably by air (heavy or outsized cargo, notably fuel, is reserved for seaborne arrival, exploiting spring breakup). The sole means of regular and reliable access to most communities, weather stations, bases and bivouacs is by air. Offshore, aircraft have landed in almost every part of the Arctic Ocean.

From 1952 to 2008, Russia's Arctic and Antarctic Research Institute deployed five stations onto ice islands, the United States two, and Canada one. During much of that span, the Soviets held overwhelming leadership in Arctic science (Leary 1999). Today, research and observation in the Arctic remain high among Russian priorities (Russia, for example, was a proponent and active participant in the International Polar Year of 2008). And near both poles, intensive ice-based research continues.

The exigencies of Cold War competition vanished in 1991. Its legacies contribute still. Archives of air, ice and ocean observations (e.g., submarine sonar profiles of ice thicknesses) hold immense utility for paleoclimate modelers. Circumpolar weather stations add regularly to the international data network. In a warming world, the Arctic Ocean and its peripheries are the focus of atmospheric, biological, ecological, geological and oceanographic research designed to understand the climate system and predict its future evolution. Teams monitor indices to change, such as sea-ice extent and thickness, mountain glaciers and ice fields, permafrost and circumpolar biota. Paralleling global trends, negative mass balances for glaciers and sea ice are confirmed. Climate change, many now argue, poses a threat to national security.

References

Arctic Research Laboratory. (1964). Work of Arctic Research Laboratory, Point Barrow, Alaska, 1963. *Polar Record, 12*(77), 181–182.

Britton, M. E. (1968). Administrative background of the developing program. In J. E. Sater (Ed.), *Arctic drifting stations: A report on activities supported by the Office of Naval Research* (p. 29–35). Washington, DC: Arctic Institute of North America.

Buck, B. M. (2003). *Accomplishments in Arctic underwater surveillance acoustics, 1960–1988.* Unpublished 11-page paper, courtesy Buck [Special Projects Officer, Undersea Branch, ONR/Polar Research Lab].

Crary, A. P., & Goldstein, N. (1957). Geophysical studies in the Arctic Ocean. *Deep Sea Research, 4*, 185–201.

Department of Fisheries and Oceans. (1985). *A unique opportunity: An oceanographic program for the Canadian Arctic ice island.* Internal Report, Marine Sciences and Information Directorate, Department of Fisheries and Oceans, Ottawa.

Fletcher, J. O. (1968). Origin and early utilization of aircraft-supported drifting stations. In J. E. Sater (Ed.), *Arctic drifting stations: A report on activities supported by the Office of Naval Research* (p. 1–13). Washington, DC: Arctic Institute of North America.

Gordienko, P. A. (1961). The Arctic Ocean. *Scientific American, 204*(5), 88–106.

Hattersley-Smith, G. (1963). The Ward Hunt Ice Shelf: Recent changes at the ice front. *Journal of Glaciology, 4*, 415–424.

Haydon, P. T. (1988). The strategic importance of the Arctic: Understanding the military issues. *Canadian Defence Quarterly, 17*(4), 27–34.

Hobson, G. (1989). Ice island field station: New features of Canadian polar margin. *Eos, Transactions American Geophysical Union, 70*(37), 833–839.

Holmquist, C. D. (1972). The T-3 Incident. *U.S. Naval Institute Proceedings, 98*, 44–53.

Jeffries, M. O., & Shaw, M. (1993). The drift of ice islands from the Arctic Ocean into the channels of the Canadian Arctic Archipelago: The history of Hobson's Choice Ice Island. *Polar Record, 29*, 305–312.

Johnston Jr., C. B. (1968). Problems working on a drifting station. In J. E. Sater (Ed.), *Arctic drifting stations: A report on activities supported by the Office of Naval Research* (p. 91–99). Washington, DC: Arctic Institute of North America.

Joint Intelligence Bureau, Ottawa. (1962). The Soviet North: Economic aspects. National Archives of Canada, Record Group 25, Vol. 3306, file 9059-D-40, part 1.

Leary, W. M. (1999). *Under ice: Waldo Lyon and the development of the Arctic submarine.* College Station: Texas A&M University Press.

Leary, W. M., & LeShack, L. A. (1996). *Project COLDFEET: Secret mission to a Soviet ice station*. Annapolis: Naval Institute Press.
Miller, M. M. (1956). Floating islands. *Natural History Magazine, 65*, 233–239, 274, 276.
Polar Continental Shelf Project. (1986a). Ice island 1986 status report (memorandum, November 1986), Ottawa, Canada.
Polar Continental Shelf Project. (1986b). Some guidelines and information for ice island group managers (memorandum, 5 March 1986), Ottawa, Canada.
Schindler, J. F. (1968). The impact of ice islands – The story of ARLIS II and Fletcher's Ice Island, T-3, since 1962. In J. E. Sater (Ed.), *Arctic drifting stations: A report on activities supported by the Office of Naval Research* (p. 49–78). Washington, DC: Arctic Institute of North America.
Soviet Military Encyclopedia. (1976). *Arktika, Vol. 1, A171*. Moscow: Voyennoye Izdatel'stuo. Translated by: CIS Multilingual Section, National Defence Headquarters, Ottawa.
Treshnikov, A. P. (1956). The Soviet drifting station SP-3, 1954–55. *Polar Record, 8*(54), 222–229. doi:10.1017/S0032247400048993.
Untersteiner, N., Hunkins, K. L., & Buck, B. M. (1976). Arctic science: Current knowledge and future thrust. In E. I. Salkovitz (Ed.), *Science, technology, and the modern Navy, thirtieth anniversary* (p. 1946–1976). Arlington: Office of Naval Research.
Van Wychen, W., & Copland, L. (2017). Ice island drift mechanisms in the Canadian High Arctic. In L. Copland & D. Mueller (Eds.), *Arctic ice shelves and ice islands* (p. 287–316). Dordrecht: Springer. doi:10.1007/978-94-024-1101-0_11.
Weart, S. R. (2003). *The discovery of global warming*. Cambridge, MA: Harvard University Press.
Wise, M. L. (1978). *Ice islands of the Arctic: Alaskan Air Command's Arctic experience, AAC historical monograph, Elmendorf Airforce Base* (pp. 118). Alaska: Alaskan Air Command.

Chapter 14
Russian Drifting Stations on Arctic Ice Islands

Igor M. Belkin and Sergey A. Kessel

Abstract A summary of Russian discoveries of Arctic ice islands – peculiar tabular icebergs – is presented, complete with a chronological account of drifting stations installed on ice islands. Of 40 'North Pole' drifting stations established from 1937 through 2013, six were set up on five ice islands: North Pole-6, 18/19 (same ice island), 22, 23, and 24. These ice islands served as reliable long-term research platforms as evidenced by the extensive bibliography of scientific publications based on observations made from manned ice island stations. Studies were conducted of structure and morphology of ice islands; under-ice biota; deep Arctic Ocean benthos; meteorology and climate; and oceanography. Biological collections from these ice islands are still being analyzed.

Keywords Ice island • Arctic Ocean • Ice shelf • Canada Basin • North Pole drifting stations • NP-22

14.1 Introduction

The first observations of ice islands by Russians might have been made centuries ago. In the sixteenth to seventeenth centuries, Russian fishermen, hunters, traders, and surveyors began to explore Russia's northern seas. In the early sixteenth century, the first ethnic Russian group to colonize the coasts of the White and Barents seas, the so-called *pomors* began hunting walrus and seals off Novaya Zemlya, and, probably, Svalbard (Yurchenko 2005). Gradually, Russians advanced eastward from the Barents Sea, exploring the Kara Sea, Laptev Sea, East Siberian Sea, and the Chukchi Sea, up to Bering Strait.

While exploring the newly discovered coasts and coastal seas, they sighted what appeared to be distant offshore lands whose reality seemed unquestionable at the time. Over the next centuries, many expeditions tried in vain to reach these lands,

I.M. Belkin (✉)
Graduate School of Oceanography, University of Rhode Island, Narragansett, RI, USA
e-mail: igormbelkin@gmail.com

S.A. Kessel
Arctic and Antarctic Research Institute (retired), St. Petersburg, Russia

L. Copland, D. Mueller (eds.), *Arctic Ice Shelves and Ice Islands*, Springer Polar Sciences, DOI 10.1007/978-94-024-1101-0_14

hence dubbed "phantom lands". These "phantom lands" are now believed to have been made largely of ground ice which were subsequently destroyed by thermal erosion (Gavrilov et al. 2003; Günther et al. 2015).

The number of such sightings grew with time until 1946 when, independently, the U.S./Canadian and Soviet polar aviators on ice reconnaissance flights into the Arctic Ocean discovered unusual tabular icebergs with a characteristically corrugated (undulating) upper surface that have become known as ice islands (Koenig et al. 1952; Burkhanov 1954). The first-ever manned drifting station on an ice island was established by the U.S. in 1952. By that time, the Russians already had extensive experience with a series of long-term drifting stations known as 'North Pole' (NP), which were set-up on ice floes. The first-ever drifting station NP-1 operated for 9 months in 1937, followed after World War II by NP-2, 3, 4, and 5 (1950–1956). It was a matter of time for Soviet sea ice reconnaissance to find an ice island suitable for a manned drifting station, which happened to be NP-6. Since then, the Soviets set up five more stations on four ice islands: NP-18/19 (same island), 22, 23, and 24.

This chapter is a brief account of Russian discoveries of Arctic ice islands and a summary of scientific studies conducted from the North Pole drifting stations NP-6, 18, 19, 22, 23, and 24. The references include mostly English-language publications that could serve as entry-points to the relevant Russian-language literature. Two Russian-language books by Kessel (2005) and Dremlyug and Kessel (2007) are dedicated exclusively to ice islands and their history, while the English-language monograph by Frolov et al. (2005) and the Russian-language compendium by Kornilov et al. (2010) sum up multi-disciplinary studies conducted from all manned drifting polar stations established by the USSR/Russian Federation in the Arctic Ocean.

14.2 Early Discoveries of Arctic Ice Islands

14.2.1 Ferdinand Wrangell: Possible Discovery of Ice Islands in 1821

In 1820–1824, Lieutenant Ferdinand Petrovich Wrangell led a Russian expedition along the Arctic coast of Northeastern Siberia, conducting offshore surveys east of the Kolyma River mouth. Wrangell's observations were first published by Professor Friedrich Parrot, to whom Wrangell gave his materials (Parrot 1826; Wrangel 1826). Later, slightly different versions of Wrangell's full account were published in French, German, English, and Russian in 1839–1841 (e.g., Wrangell 1841) so that this account became well-known among scientists and polar explorers as evidenced by Charles Darwin's citation of Wrangell's book (Darwin 1873).

The most complete version of Wrangell's account contains tantalizing descriptions of peculiar ice formations; some of them might have been ice islands or their fragments. On April 4, 1821, traveling along the coast of the East Siberian Sea, the expedition traversed huge conical ice hills (up to 27 m high), with long valleys in-between; these hills were utterly different from the previously encountered immense "winter hummocks" (up to 24 m in height). The conical hills consisted of dark ice, described as "smooth and even, its colour varying from whitish grey to black: it had a perfectly fresh taste, and was large-grained and opaque" (Wrangell 1841, p. 142). These ice formations might have been remnants of grounded ice islands. Based on the above description, Wrangell might reasonably be credited as a possible discoverer of the Arctic ice islands even though in the original English-language account (Wrangell 1841) the term ice-island is only used once to describe a common ice floe, upon which Wrangell and his people drifted across a coastal polynya, and not a true ice island defined in this chapter as a tabular iceberg with a corrugated (undulating) surface.

14.2.2 The Modern Era: Arctic Aviation and Ice Islands

The most credible observation of an ice island before the advent of polar aviation was made by the Russian schooner *Krestianka* in summer 1934 in the Chukchi Sea. There the crew saw an island-like feature, described it, determined its coordinates, and reported their findings by radio (Dremlyug and Kessel 2007, p. 9). Based on the radio report, Soviet hydrographers put Krestianka Island on a map. Tragically, on return passage *Krestianka* and all aboard perished in October 1934 during a storm in the Okhotsk Sea (Bollinger 2003, p. 58). Yet, Soviet hydrographers were confident enough to send the expedition vessel *Smol'nyi* to the Chukchi Sea in 1943 to search for Krestianka Island. Even though search conditions were good, with almost no sea ice and good visibility, the island was not found and was removed from nautical charts. If Krestianka Island indeed was a drifting ice island, then its disappearance from the original location can be easily explained by its drift.

From mid-September until October 12, 1937, the personnel of the Soviet polar station on Henrietta Island (77°06′N, 156°30′E; the northernmost island of De Long Islands), repeatedly observed what seemed to be an ice island to the northeast of Henrietta Island. Drawings were made of the ice island (Burkhanov 1954, his Fig. 3), with two characteristic hills. Later, a large tabular iceberg with two hills was observed in the same area by pilot I.I. Cherevichny and navigator V.I. Akkuratov; the iceberg was slowly moving to the north, then northwest. This "phantom land" was alternatively named "Mukhanov Land" (after the station leader) or "Polyarniks Land" (Burkhanov 1954, his Figs. 1 and 3).

In the late 1930s before World War II, numerous sightings of drifting ice islands were reported by Soviet ice reconnaissance north of Kotelny Island (part of the New Siberian Islands or Novosibirskie ostrova) and east of Severnaya Zemlya (Burkhanov 1954). In March 1941, a large ice island was observed at 74°N north of Bear Islands (Medvezhyi Islands) in the East Siberian Sea. This ice island was quite different from the aforementioned ice islands observed north of Henrietta Island and northwest of Kotelny Island. It had undulating surface, with several frozen streams and lakes clearly visible; even the most experienced Russian ice observers could hardly tell it from Arctic tundra. Later this ice island was observed drifting to the north and was named Andreev Land after the Russian explorer Stepan Andreev who reported seeing a land north of Bear Islands in 1764 (Burkhanov 1954).

One of the most experienced Russian ice specialists, oceanographer P.A. Gordienko reported close encounters with – and even visiting – tabular icebergs (possibly, ice islands) during ice reconnaissance on icebreaker *Anastas Mikoyan* in the late 1930s and through the 1940s in the East Siberian and Chukchi seas (Burkhanov 1954). One of these ice islands spotted north of Wrangell Island was surveyed on August 23, 1947 by a boat team from *Mikoyan* led by P.A. Gordienko. After approaching the ice island, oceanographer A.L. Sokolov made close-up drawings of the ice island's walls with well-defined horizontal layers, 2–20 cm thick (published by Kessel 2005, his Fig. 6). A brief account of this survey appeared in *Polar Record* in May 1955 (Anonymous 1955), apparently based on Admiral Vasily Burkhanov's first public presentation of Soviet discoveries of ice islands (delivered on February 18, 1954) and his subsequent article (Burkhanov 1954) described below.

Another ice island encountered by icebreaker *Mikoyan* on September 21, 1948 in the central East Siberian Sea was examined by P.A. Gordienko and the ship's captain who boarded the ice island. According to Gordienko's account, the ice island's upper surface featured gentle hills up to 6 m high, with frozen streams in-between; the ice island's walls had bluish color and consisted of exceptionally strong ice that the icebreaker could not crush (Burkhanov 1954).

During ice reconnaissance in De Long Strait in 1939–1942 P.A. Gordienko repeatedly observed strange cube-shaped ice blocks, up to 20 m across, and up to 7 m higher than the surrounding ice floes. These blocks consisted of monolith ice of peculiar light-to-deep-blue colour. As determined by subsequent chemical analysis, the ice was absolutely fresh. Similar ice blocks were also observed by N.A. Volkov in 1935–1937 in the Bering Strait (Burkhanov 1954).

The advent of polar aviation accelerated the Arctic Ocean exploration. In 1946–1950, extensive airborne ice reconnaissance by the U.S./Canada and USSR led to independent, practically simultaneous discoveries of three large ice islands. These discoveries remained classified for several years in all three countries involved. In November 1950, the U.S./Canadian discoveries and observations of these ice islands (named T-1, T-2, and T-3) were made public at the First Alaskan Science Conference in Washington, D.C. (Fletcher 1950; Koenig et al. 1952). On February 18, 1954 Admiral Vasily Burkhanov made presentation at Moscow Branch of Geographical Society of the USSR, in which he revealed the hitherto classified Soviet observations of the same ice islands (T-1, T-2, and T-3) that have been independently discovered and reported by Soviet sea ice reconnaissance pilots; therefore, Burkhanov referred to these ice islands using the respective Soviet pilot's last name as "Kotov Island" (T-1), "Mazuruk Island" (T-2), and "Perov Island" (T-3) (Fig. 14.1, reproduced from Burkhanov 1954, Fig. 6). This presentation was publicized by Russian newspapers, whose detailed accounts were summed up by *Polar Record* (Anonymous 1955). Numerous Russian publications on ice islands and their putative associations

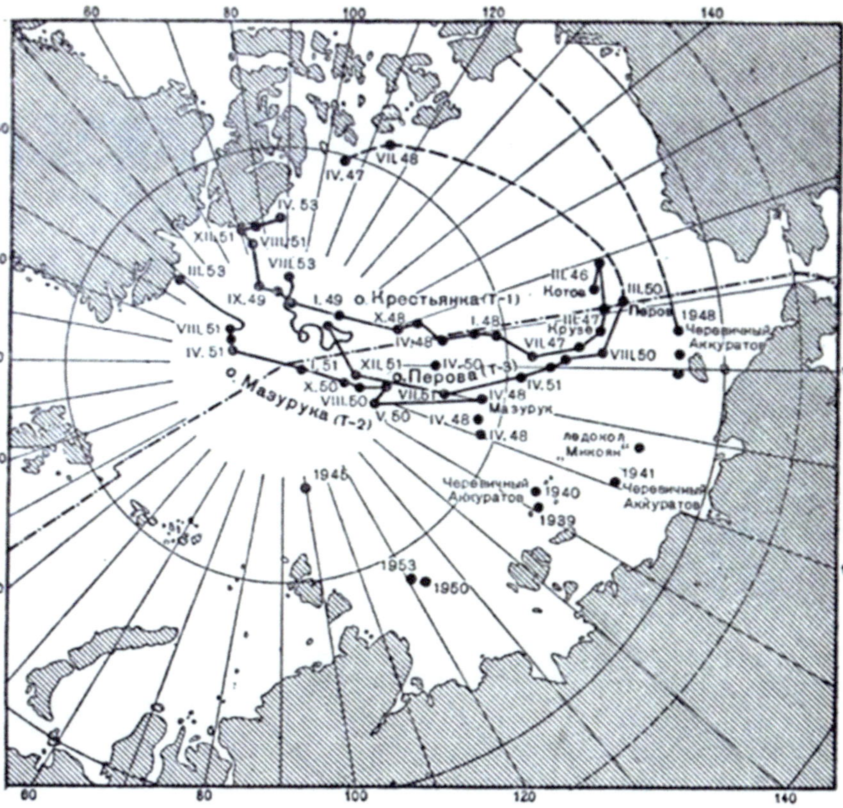

Fig. 14.1 Sightings and drift of ice islands reported by Russians in 1939–1953. Transliterations and translations of Russian names, with comments: "о. Крестьянка (Т-1)" ("ostrov Krestianka/Krestyanka"), Krestianka Island (also known as Kotov Island) identified with T-1; "о. Мазурука (Т-2)" ("ostrov Mazuruka"), Mazuruk Island identified with T-2; "о. Перова (Т-3)" ("ostrov Perova"), Perov Island identified with T-3; "Черевичный Аккуратов", sightings of ice islands made by pilot Cherevichny and navigator Akkuratov in 1939, 1940, 1941, and 1948; "Крузе" (Kruze), sighting of Krestianka Island by pilot L.K. Kruze; "ледокол Микоян" ("ledokol Mikoyan"), sighting of an ice island from icebreaker Mikoyan; also shown are three Russian sightings of ice islands between 100°E and 123°E made in 1945, 1950, and 1953 (observers unknown). I.I. Cherevichny, I.S. Kotov, L.K. Kruze, I.P. Mazuruk, and V.M. Perov were among the most experienced pilots in the Soviet sea ice reconnaissance, while V.I. Akkuratov was a famed navigator (Reproduced from Burkhanov (1954, Fig. 6) with permission)

with "phantom lands" followed; some articles were promptly translated into English (e.g., Zubov 1955).

Beginning in spring 1946, Soviet sea ice observers on reconnaissance flights across the Chukchi Sea northeast of Wrangell Island, at 75°N–76°N, reported numerous sightings of ice islands, whose height and appearance were markedly different from the surrounding ice floes. In March 1946, pilot I.S. Kotov spotted what he believed was Krestianka Island northeast of Wrangell Island at 76°N, 165°W. It

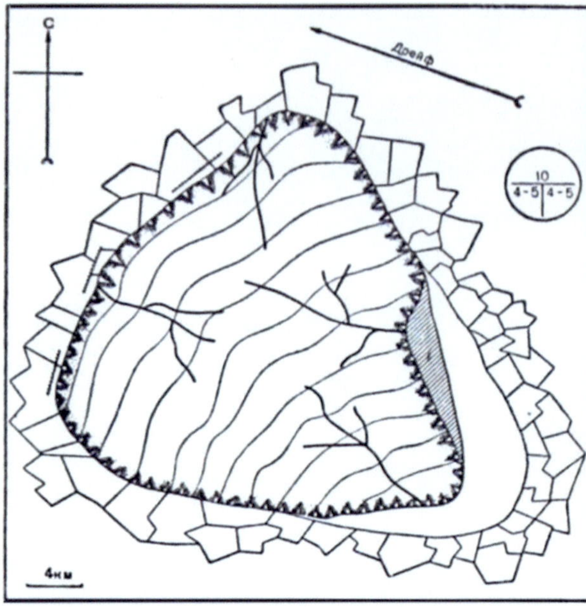

Fig. 14.2 Drawing of Kotov Ice Island made by navigator V.I. Akkuratov on March 19, 1947. The 4-km scale bar in the *bottom left corner* is for the island size only, not for the wave pattern, which is shown highly schematically, with the wave length (crest-to-crest distance) exaggerated many times (see text; also cf. Koenig et al. 1952, Fig. 6). North is at the *top*. Dendritic patterns are frozen streams that drain meltwater in summer. The hatched area east of the ice island is fast ice. The ice island was surrounded by ice floes except for open water (*white*) in its wake as the ice island was drifting WNW. The drift direction is shown with *arrow* at the top, where "Дрейф" means "Drift" (Reproduced from Burkhanov (1954, Fig. 7) with permission)

measured 25 km × 30 km, with an area of 520 km^2. In retrospect, this identification of Kotov Island with Krestianka Island seems highly dubious given the time span of 12 years between these sightings of drifting ice islands in 1934 and 1946. On March 19, 1947 pilot L.K. Kruze spotted the same ice island that pilot I.S. Kotov saw in March 1946 but it was now at 76°N, 173.5°W – 200 km west of the original sighting. Evidently, this was a drifting ice island. A drawing made during that flight by navigator V.I. Akkuratov and first published by Burkhanov (1954, his Fig. 7) shows a heart-shaped feature with a characteristic undulating surface pattern (Fig. 14.2). U.S. Air Force pilots independently discovered this same ice island, later termed T-1, measuring 28 km × 33 km and in a shape resembling an arrowhead or a chicken's heart, on August 14, 1946 at 76°15′N, 160°15′W (Koenig et al. 1952). Burkhanov (1954) identified T-1 with Kotov Island (Fig. 14.1). It must be noted here that Akkuratov's sketch has grossly exaggerated the wavelength of the corrugated surface of this ice island. Indeed, his sketch shows 11 waves (hence, an average crest-to-crest distance of 3 km), while a low-altitude photo of the same island (T-1) taken on August 1, 1951 (Koenig et al. 1952, their Fig. 6) shows >40 parallel

Table 14.1 Russian ice island sighting dates, dimensions and other characteristics

Ice island	First and last sighting (month/year)	Dimensions (km)	Thickness (m)	Mean distance (m) between waves (ridges) on the upper surface
NP-6	04/1956 – 09/1959	13.8 × 8.3	6–12	No waves
NP-18/19	04/1968 – 04/1973	12.6 × 6.7	32–36	210
NP-22	04/1973 – 04/1982	4.8 × 2.0	22–30	220
NP-23	08/1975 – 11/1978	7.5 × 3.0	6–18	170
NP-24	03/1978 – 11/1980	16.0 × 6.5	15–40	280

Data from Grishchenko and Simonov (1985, their Table 1), Dremlyug and Kessel (2007, foldout table pp. 100–101), with corrections, and from Sychev (1961, 1962) for NP-6

troughs filled with meltwater. As we know today, the wave-like pattern of most ice islands and ice shelves of the Arctic Ocean has a length scale of a few hundred metres, between 170 and 280 m according to Grishchenko and Simonov (1985, their Table 1; Table 14.1).

In April 1948, at 82°N, 170°E, pilot I.P. Mazuruk observed an ice island with a characteristic undulating surface measuring 28 km × 32 km. In May 1950, Soviet pilots reported to have seen the same island at 87°N, 155°E (Dremlyug and Kessel 2007, pp. 28–29; Burkhanov 1954, his Fig. 6; Fig. 14.1). On July 21 1950, U.S. pilots independently discovered and reported this same ice island measuring 31 km × 33 km at 86°40′N, 167°00′E, and termed it T-2 (Koenig et al. 1952). Burkhanov (1954) identified T-2 with Mazuruk Island (Fig. 14.1).

In March 1950, pilot V.M. Perov discovered an ice island with an area of 100 km^2 northeast of Herald Island at 74.5°N, 169°W (Burkhanov 1954; Fig. 14.1). The U.S. Air Force pilots independently discovered this same ice island on July 29, 1950 at 75.4°N, 173°W (Koenig et al. 1952). It measured 16.6 km × 8.3 km, had a kidney shape, and was termed T-3. Later, it was realized that U.S. pilots had observed T-3 much earlier, on April 24 and 27, 1947 north of Ellef Ringnes Island and on July 9, 1948 off the west coast of Prince Patrick Island (Koenig et al. 1952). Burkhavov (1954) identified T-3 with Perov Island (Fig. 14.1).

Burkhanov's identification of Russian vs. U.S./Canadian ice islands remains uncontested to date, even though no comparative study of Russian and U.S./ Canadian discoveries of these islands has ever been published in the scientific literature.

Numerous Russian publications on ice islands were summed up by Kessel (2005) and Dremlyug and Kessel (2007), who also compiled sightings of small ice islands in the East Siberian and Chukchi seas reported by Soviet airborne ice reconnaissance and observers from naval icebreakers in the 1930s–1940s. No Soviet reports of ice islands observations exist between 1950 and April 1956, when an ice island was discovered on which NP-6, the first Soviet ice island station, was established, thereby opening an era of Soviet/Russian scientific exploration and exploitation of ice islands, described in the next section.

14.3 North Pole Manned Drift Stations on Ice Islands

In 1956–1980, six Russian manned stations were set up on five ice islands that drifted in the Beaufort Gyre and Transpolar Drift (Fig. 14.3).

NP-6 In April 1956, at 74°24′N, 177°10′W, pilot V.I. Maslennikov discovered an ice island measuring 13.8 km × 8.3 km. The first Soviet station on an ice island, North Pole-6 (NP-6), was soon established and was occupied for 3 years (Sychev 1961, 1962) (Fig. 14.4).

Relative to other ice islands, this ice island was rather thin, with its thickness ranging from 6–12 m (Sychev 1961, 1962). Despite being relatively thin, this ice

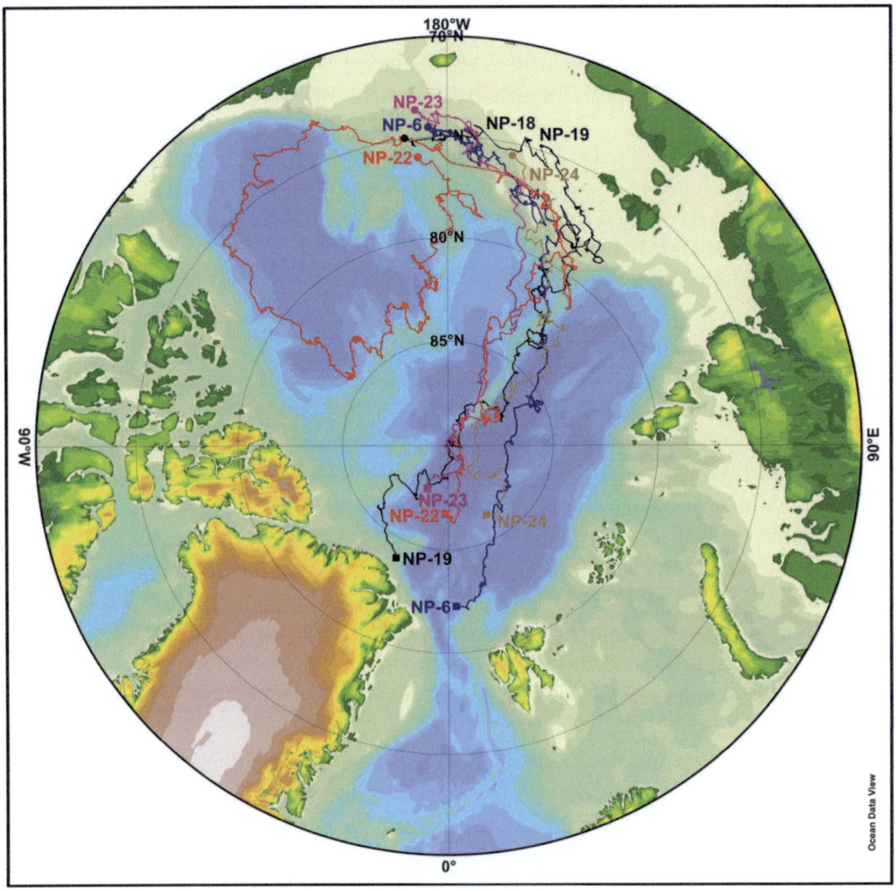

Fig. 14.3 Drift tracks of Russian North Pole stations established on ice islands. Drift data from the Arctic Climatology Project (Environmental Working Group 2000a) are mapped with the Ocean Data View software (Schlitzer 2016). The beginning and end of each track are marked with *circles* and *squares*, respectively. NP-19 was set up on the same ice island as NP-18 (both shown in *black*), with a 6-month break in-between (*black dashed line*)

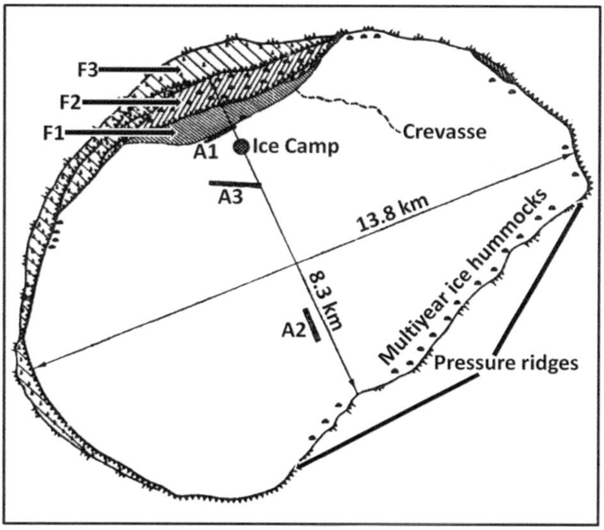

Fig. 14.4 Ice island NP-6. Acronyms: *A1* airstrip in April–May 1957, *A2*, airstrip in June–August 1957, *A3* airstrip since September 1957, *F1* multiyear fast ice with hummocky surface, *F2* multiyear fast year with ice hummocks smoothed, *F3* 2-to-3-year-old fast ice with old and young ice hummocks (Adapted from Kessel (2005, Fig. 20, after Sychev 1962))

island survived intact for the entire duration of the manned station NP-6, from April 1956 up to September 1959. Ice studies by Cherepanov (1964) revealed a very peculiar structure and properties of this ice island that set it apart from other ice islands. In particular, the crystalline structure of NP-6 markedly differed from that of other ice islands studied before and also of the ice shelf of northern Ellesmere Island. Therefore, Cherepanov (1964) stressed that NP-6 is not a typical ice island.

NP-18 On April 19, 1968 a roughly rectangular ice island was found at 71°36′N, 163°00′W, measuring 13.6 km × 7.4 km (Dremlyug and Kessel 2007, foldout table between pp. 100 and 101; yet 12.6 km × 6.7 km according to Grishchenko and Simonov 1985, their Table 1), with an area of 82 km^2 and an estimated thickness of 35 m. The drifting station NP-18 set up on this ice island became operational on October 9, 1968. On May 9, 1969 the NP-18 station was relocated to an ice floe 230 km northeast of the ice island (Kornilov et al. 2010, p. 245), but the ice island was later used as a platform for the next station, NP-19; therefore, we refer to this ice island as NP-18/19 (Fig. 14.5).

NP-19 This station was set up on the same ice island as NP-18 and began regular observations on November 7, 1969. In January 1970, the ice island grounded on shoals off De Long Islands and lost 19 km^2 or ~20% of its area. The station survived the break-up. NP-19 closed on April 14, 1973. The drift of NP-19 is described by Chilingarov et al. (1972, 1986, 2014).

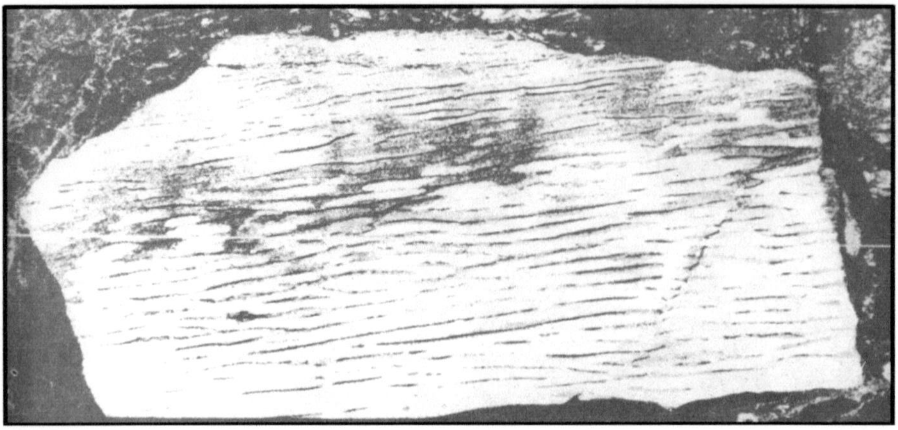

Fig. 14.5 Ice island NP-18/19. Aerial photo from an altitude of 10,000 m on April 19, 1968, when this ice island was discovered by the Russian sea ice reconnaissance (Adapted from Kessel (2005, Fig. 23, after Chilingarov et al. 1972, photo on p. 5))

NP-22 This ice island was discovered on April 6, 1973 (Kessel 2005, his Table 1 on pp. 29–32). The drifting station NP-22 (Figs. 14.6 and 14.7) was manned without interruptions for 8½ years, from September 13, 1973 until April 8, 1982 (Romanov et al. 1997) – the longest continuously occupied drifting station ever. After completing a full circle around the Beaufort Gyre, NP-22 escaped from the gyre into the Transpolar Drift. Owing to the island's longevity, successive NP-22 crews expanded and upgraded NP-22 facilities and conducted diverse studies of the ice island and its environment.

NP-23 In August 1975, an ice island measuring 7.5 km × 3.0 km was found northeast of Wrangell Island, at 72°45′N, 176°05′W (Kessel 2005, his Table 1 on pp. 29–32). It was chosen as a platform for NP-23 (Fig. 14.8) that remained in operation from December 5, 1975 until November 1, 1978. The ice island's thickness varied spatially from 6–18 m and averaged 15 m. In summer 1977, the island lost 1.5–2.0 m of ice from its upper surface because of intense ablation (Kessel 2005, pp. 62–63).

NP-24 This station was set up from icebreaker *Sibir* on June 23, 1978 at 76°45′N, 163°00′E on an ice island found on March 9, 1978 at 75°12′N, 172°00′E (Kessel 2005, p. 63; Fig. 14.9). This ice island was first observed in the northern Chukchi Sea in 1977: visually in August, then with airborne side-looking radar in December (Kornilov et al. 2010, p. 391). The ice island dimensions were 17 km × 6 km (Kessel 2005; 16.0 km × 6.5 km according to Grishchenko and Simonov 1985, their Table 1). Its thickness varied spatially from 15–40 m (Kessel 2005). NP-24 was occupied until November 19, 1980 (Romanov et al. 1997).

NP-24 was the last Russian station thus far established on an ice island. A pause in the disintegration of the Ward Hunt Ice Shelf – the main supplier of ice islands in

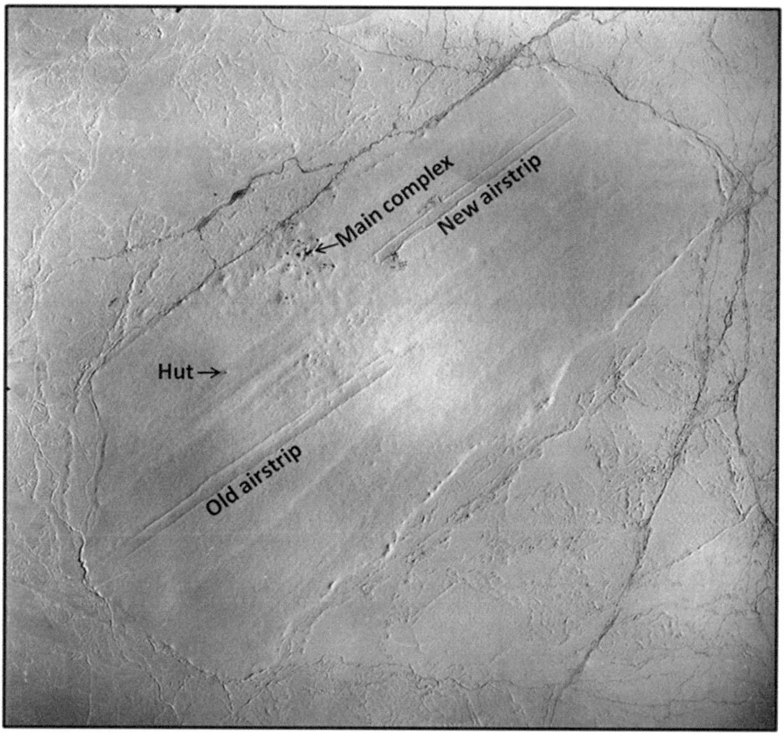

Fig. 14.6 Aerial photo (nadir view) of ice island NP-22 in 1979 (Adapted from Kessel (2005, Fig. 28, from V.K. Yakimyuk's personal archive)) The ice island dimensions are 4.8 km × 2.0 km (Grishchenko and Simonov 1985; Table 14.1). The main complex of several interconnected prefabricated huts is in the *centre* of the ice camp (for its close-up photo see Fig. 14.7) surrounded by individual huts. Photos of the stand-alone hut on the *left* taken before and after the 1977 melt season are shown in Figs. 14.10 and 14.11 respectively. Air strips were prepared each season anew as previous air strips were destroyed by summer melt. Faint surface manifestations of older air strips are still visible in this photo

the Arctic (Jeffries 1992) – was the main reason for the absence of ice-island-based stations.

A summary of key parameters of the Russian ice islands (Table 14.1) is based mostly on Grishchenko and Simonov (1985, their Table 1) and Dremlyug and Kessel (2007, foldout table between pp. 100 and 101), with corrections, while data for NP-6 are from Sychev (1961, 1962); included are only those ice islands on which North Pole manned drifting stations were set up. Dimensions and thickness were usually measured early during the occupation of each ice island. In case of NP-22, its attrition was studied over several years (Grishchenko 1980; Grishchenko and Simonov 1985). Thickness is observed to decrease with time due to the upper surface ablation and the bottom surface scouring by currents. For example, NP-22 thinned by ~1 m $year^{-1}$ while occupied.

Fig. 14.7 Main complex of several interconnected prefabricated huts on drifting station NP-22 in May 1977. Developed by Russian designers Kanaki and Ovchinnikov, these comfortable mobile huts were officially termed PDKO after them, with *PD* standing for *polyarny dom* (in Russian) or polar house. The complex included mess room, kitchen, food storage, recreation room, and sauna. The sign reads: "NP-22 Vladivostok Square" after icebreaker *Vladivostok*, which together with M/V *Captain Kondratiev* delivered supplies, including the huts, for the first party of 10 in September 1973. The huts were assembled onboard before the ice camp was set up on the ice island (Photo by Igor Belkin)

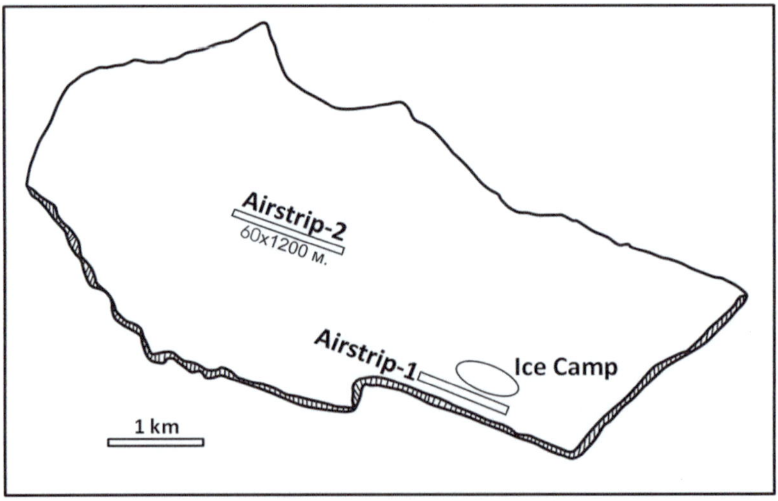

Fig. 14.8 A sketch of Ice Island NP-23 (Adapted from Kessel (2005, Fig. 34))

Fig. 14.9 Ice Island NP-24 (nadir view with airborne side-looking radar). The ice island dimensions are 16.0 km × 6.5 km (Grishchenko and Simonov 1985; Table 14.1) (Adapted from Kessel (2005, Fig. 35))

A comparison between ice island stations and other stations set up on ice floes (Romanov et al. 1997; Frolov et al. 2005, pp. 29–33, their Tables 1.1 and 1.2; Kornilov et al. 2010) illustrates the substantial benefits conferred by ice islands. First, the longevity of ice islands far exceeds that of ice floes. Therefore, as observational platforms, ice islands are more useful than ice floes. Second, ice islands are solid enough to withstand landings of relatively large aircraft, thus allowing heavy and bulky scientific equipment to be delivered. On the balance, the ice islands turned out to be superior platforms as evidenced by the number and duration of studies conducted from these islands and briefly described in the next section.

14.4 Scientific Observations from the Russian Ice Islands

Scientific investigations conducted from the Russian drifting stations are described by Frolov et al. (2005) and Kornilov et al. (2010), including all six stations established on five ice islands: NP-6, NP-18/19, NP-22, NP-23, and NP-24. Many investigations conducted from ice islands were similar to the studies conducted from drifting stations on ice floes, especially if the emphasis was on standard meteorological and oceanographic observations. There were, however, studies for which ice islands played a key role, e.g., the biology of the under-ice and benthic realms,

Beaufort Gyre oceanography, and others. Finally, glaciological studies of the ice islands were quite specific – if not unique – since the ice islands' structure and morphology differ from that of ice floes. Below we review a few fields, where the ice islands played the most important role.

14.4.1 Ice Island Structure and Morphology

Stratigraphic studies are especially important in helping to determine ice island source, thereby providing clues to its drift pattern after calving. Three ice islands – NP-6, NP-18/19, and NP-22 – were studied by ice specialists, using various means, including ice coring and scuba diving (Cherepanov 1964; Legen'kov 1973; Legen'kov and Chugui 1973; Legen'kov et al. 1974; Grishchenko 1980; Grishchenko and Simonov 1985). The most extensive observations of ice island structure and morphology were conducted from NP-22, owing largely to its longevity.

NP-6 The first Russian studies of ice island stratigraphy and crystallography were conducted by N.V. Cherepanov using data collected on NP-6, where six pits were dug in the 10–12 m thick ice island (Cherepanov 1964). This study yielded the first-ever data on the structure of very thick sea ice growing up to 12 m thick under natural conditions. The ice island's structure was quite peculiar and different from that of previously studied ice islands and ice shelves. Therefore, Cherepanov (1964) stressed that this ice island is a rare ice formation in the Arctic. At the same time, he noted that such ice formations are encountered in the Arctic quite often, featuring a flat even upper surface, a rather low freeboard of 0.5–1.0 m, a moderate size of less than 0.5–1.0 km, and lack of terrestrial debris. The crystalline structure of NP-6 was unusual for multi-year sea ice in that ice crystals were uniformly shaped and their optical axes were uniformly oriented (aligned). Below the top surface 30–50 cm layer with numerous inclusions (sediment debris), the ice was completely uniform down to 8.5 m depth, with a distinct fibrous (sponge-like) structure of drainage channels traceable all the way to the bottom surface. The ice salinity increased from 0 at the upper surface to 3 ppt at 2 m depth and remained constant between 2 and 6 m, then decreased to 2.5 ppt at 8.5 m depth. The vertically-averaged salinity was 2.26 ppt, similar to that of multi-year ice, e.g., the NP-4 ice floe (2.2 ppt). Based on ice texture, crystalline structure, salinity and density, Cherepanov (1964) concluded that the NP-6 ice island was of marine origin.

NP-18/19 The morphology, stratigraphy, temperature regime, and thermally induced stresses and deformations of this ice island were documented by Legen'kov (1973), Legen'kov and Chugui (1973), and Legen'kov et al. (1974). The ice island was composed of two very different strata (Legen'kov et al. 1974): the upper 0–16 m stratum consisted of fresh ice while the stratum between 16 and 32.5 m (bottom surface) consisted of saline ice whose salinity was approximately 8 ppt. According to Kessel (2005, p. 53), stratigraphic studies conducted by another group of researchers revealed three strata: 0–17 m, glacier ice (which could be 'iced firn' - a meteoric ice type that forms from snow and rainfall (in situ accumulation) discussed in the early Ellesmere ice shelf literature; for a review see Jeffries(1992)); 17–25 m, infiltration-congelation ice (from sea water and wet snow); and 25–36 m, congelation ice (from sea water). The close-up fixed-location stratigraphic inspection of a 1-m thick stratum between 5.25 and 6.25 m depth from the upper surface of the ice island revealed a series of fine layers, 10–14 cm thick (average, 12 cm), apparently formed by annual accumulation; thus, the maximum thickness of 36 m corresponds to 300 years of accumulation (Legen'kov and Chugui 1973).

The upper surface of the ice island featured parallel undulations, with a crest-to-crest distance of 200–300 m (Legen'kov and Chugui 1973), with an average distance of 210 m (Grishchenko and Simonov 1985, their Table 1). The summer ablation was most pronounced (up to 50 cm) on the island's topographic highs, thereby gradually smoothing its undulating relief (Kessel 2005, p. 53).

Annual variations of ice temperature decreased with depth: While at the upper surface the ice temperature ranged from −29°C in February to 0°C in August, at 13 m depth the magnitude of seasonal variations was just 0.8°C, decreasing to 0°C at 32.5 m depth (near the bottom surface). Below 20 m, ice temperature was nearly constant throughout the year, increasing with depth from −7°C at 20 m to −1.8°C at 32.5 m (bottom surface) (Legen'kov et al. 1974).

NP-22 Morphological and structural characteristics of this ice island and its attrition were studied by Grishchenko and Simonov (1985). In 1974, the ice island's parameters were: maximum length 4.8 km; maximum width 2.0 km; mean freeboard 3.4 m; mean draft 25 m; mean thickness 29 m; mean distance between ridges 220 m. During the summer of 1974 (June 18–September 10), the surface ablation/melt averaged 40–45 cm across the ice island, whereas the surrounding ice floes lost 26–30 cm to surface melting. The bottom ablation/melting continued even after September 10.

During the anomalously warm melting season of 1977 Igor Belkin and Igor Afanasyev surveyed the entire freeboard (approximately 12 km in length) of NP-22 and counted 16 streams draining meltwater into the ocean. As a result of the exceptionally sunny and warm late spring and early summer of 1977, NP-22 has lost to ablation about 1 m of ice from the ice island's upper surface because of intense surface melt (Figs. 14.10 and 14.11). This is comparable to the loss of 1.5–2 m of ice observed during that same season from another Russian ice island, NP-23 (see below).

Scuba divers surveyed NP-22 from September 1974 until April 1975 (Grishchenko 1980; Grishchenko and Simonov 1985). They found that ice accretion on the ice island's sidewalls was only observed near the surface; the accretion rapidly decreased with depth and vanished at 5 m depth. Below 5 m, both along the island's sidewalls and on its bottom surface, ice erosion by currents dominated in the ice island's ablation. Ice loss due to bottom ablation (scouring by currents) was 47 cm year^{-1}. Thus, the total decrease of the ice island thickness due to the combined effects of surface melt and bottom erosion was ~1 m year^{-1}. The ablation rate at the bottom and sidewalls correlated strongly with mean monthly water temperature in

Fig. 14.10 Triple-hut complex housing a laboratory and living quarters for three people on drifting station NP-22. This photo was taken before the 1977 melt season (compare with Fig. 14.11 taken after the melt season). Geophysicist Alexander Baranov is standing in front of the entrance, accompanied by Toros (the dog) (Photo by Igor Belkin)

Fig. 14.11 The same triple-hut complex as in Fig. 14.10 but after the 1977 melt season (Photo by Igor Belkin)

the upper 0–25 m layer and with mean monthly drift speed of the ice island (Grishchenko and Simonov 1985, their Fig. 3).

Throughout its thickness, this ice island consisted of fresh glacier ice. The ice island's stratigraphic structure included two layers of fine-grained mineral inclusions of aeolian origin at depths of 10 and 23 m below the ice island's upper surface. Annual layers were pronounced throughout the entire thickness of NP-22. In the upper 10-m stratum, the annual layers consisted of blue ice (10–12 cm) separated by whitish ice (1–3 cm; its colour ascribed to numerous air bubbles); hence, the annual accumulation was approximately 13 cm, similar to 12 cm on NP-19 (Legen'kov and Chugui 1973). Below the top 10-m stratum, the annual layers were thicker, composed of 20–30 cm of blue ice and 5 cm of whitish ice.

14.4.2 Marine Biology

Biological studies conducted from ice islands included phytoplankton, zooplankton, benthos, nekton, and cryobiota (Frolov et al. 2005; Kornilov et al. 2010). Historically, first investigations based on observations and collections from drifting ice stations, including ice islands, focused on zooplankton (Brodsky and Pavshtiks 1976) and fishes, particularly gadoids (Andriashev et al. 1980). Multidisciplinary studies based on collections from Russian manned drifting stations, including those set up on ice islands, comprised the first-ever monograph on the biology of the central Arctic Basin (Vinogradov and Melnikov 1980a, b). Extensive observations on the Arctic sea ice ecosystem, including the NP-22 and NP-23 ice islands, were made by Melnikov (1989).

During the drift of NP-22 in 1976–1977 across the hitherto unexplored northern and eastern parts of the Canada Basin, Afanasyev (1978) made 50 hydrobiological benthic stations over the Canadian Abyssal Plain, in a water depth between 2600 and 3550 m. According to Afanasyev (1978), only 13 deep stations (>3000 m) were made in the Arctic Basin prior to 1977. Thus, the biological collections from NP-22 in 1976–1977 resulted in a step-like four-fold increase of the Arctic Ocean deep water data base. The Afanasyev dataset is still the only one yet collected in the eastern Canada Basin, available from the Census of Marine Life's Arctic Ocean Diversity Project database, http://www.arcodiv.org/Database/Plankton_datasets.html.

Bluhm et al. (2011) pointed out that the largest data gaps in sampling Arctic deep-sea (>3000 m) invertebrate benthos exist in the eastern Canada Basin. It is precisely the eastern side of the Canada Basin that NP-22 drifted across, enabling sampling of the hitherto unexplored region (Afanasyev 1978) and allowing the biology of the Canada Basin to be comprehensively studied for the first time (Vinogradov and Melnikov 1980a, b). Biological collections made in the 1970s from NP-22 and NP-23 were promptly described in a series of papers (Afanasyev and Filatova 1980 (benthic fauna); Averincev 1980 (polychaetes); Belyaeva 1980 (phytoplankton); Geinrikh et al. 1980 (copepods); Jirkov 1980 (polychaetes);

Kamenskaya 1980 (amphipods); Kondakov et al. 1980 (octopodes); Kosobokova 1980 (copepods); Moskalev 1980 (benthic fauna); Pasternak 1980 (sea pens); Pavshtiks 1980 (zooplankton); Tsynovsky 1980a, b (fishes); Tsynovsky and Melnikov 1980 (snailfish); Vinogradov and Melnikov 1980b (pelagic ecosystems); Zezina 1980 (brachiopods); Levenstein 1981 (polychaetes), and Kosobokova 1982 (zooplankton)). Most of these papers comprised the Russian-language monograph "Biology of the Central Arctic Ocean" (Vinogradov and Melnikov 1980a, b); some of these papers have been translated into English (see References).

Collections made from NP-22 and other ice islands were later used in numerous studies; a partial chronological list of selected papers (with just a few representative papers by the same author) includes Margulis 1982 (siphonophores); Kussakin 1983 (isopods); Markhaseva 1984 (calanoid copepods); Vasilenko 1988 (cumaceans); Kruglikova 1989 (radiolarians); Petryashov 1989 (mysids); Stepanjants 1989 (hydrozoans); Vassilenko 1989 (cumaceans); Petryashov 1993 (mysids); Malyutina and Kussakin 1995a, b, 1996 (isopods); Markhaseva 1996 (calanoid copepods); Kosobokova and Hirche 2000 (zooplankton); Jirkov 2001 (polychaetes); Sirenko 2001 (invertebrates); Bjørklund and Kruglikova 2003 (radiolarians); Buzhinskaja 2004 (polychaetes); Petryashov 2004 (mysids); Sirenko et al. 2004 (deep-water communities); Rogacheva 2007 (holothurians); Salazar-Vallejo et al. 2007 (polychaetes); Kruglikova et al. 2009 (radiolarians); Sanamyan et al. 2009, 2016 (sea anemones); Kosobokova et al. 2011 (zooplankton); Jirkov and Leontovich 2012 (polychaetes); Zasko and Kosobokova 2014 (plankton radiolarians) and a special volume on invertebrates (Gebruk et al. 2014). Massive sampling of bottom fauna from the previously unexplored abyssal depths of the Canada Basin contributed greatly to zoogeographical studies of the Arctic Basin (Vinogradova 1997). A comprehensive up-to-date bibliography of biological studies based on collections from ice islands is yet to be compiled.

14.4.3 Oceanography and Meteorology

The drift of NP-22 supported sustained oceanographic and meteorological observations in the Canada Basin and northern Beaufort Gyre. Other ice islands (NP-6, 18/19, 23, and 24) drifted mostly in the southern Beaufort Gyre or in the Transpolar Drift. The data collected from these stations, *inter alia*, shed a new light on the distribution of the Atlantic and Pacific waters in the central Arctic Basin (Rusanov et al. 1979; Frolov et al. 2005). Meteorological, climatological, and radiosonde observations from these ice islands filled data gaps in the most remote parts of the Arctic, thereby enhancing climatological summaries of various observables (e.g., Kahl et al. 1999; Warren et al. 1999). Oceanographic and meteorological observations from the Russian stations on ice islands have comprised an important part of the database used in the atlases published under the aegis of the Arctic Climatology Project (Environmental Working Group 1997, 2000a, b). At the same time, the original oceanographic data collected from these ice islands have not made it yet to the

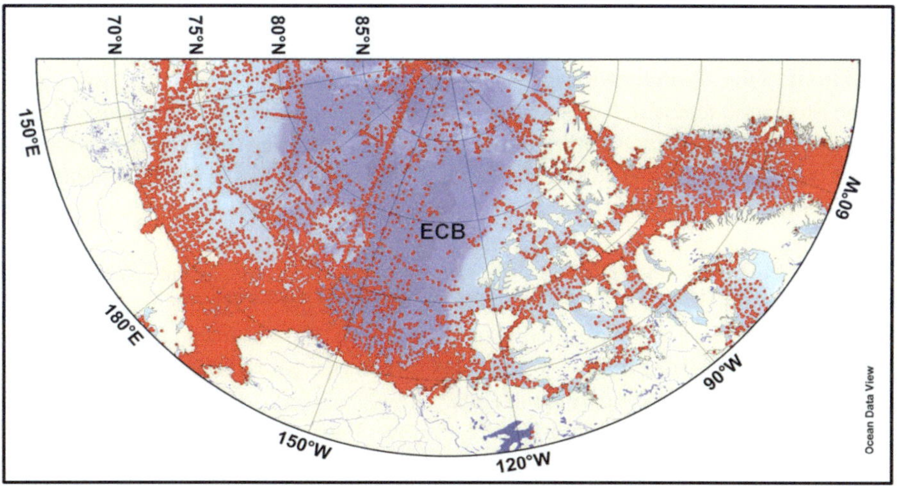

Fig. 14.12 Oceanographic data distribution map illustrating data paucity in the eastern Canada Basin (*ECB*) and over the nearby continental shelves of the Canadian Arctic Archipelago. Shown are all oceanographic stations with temperature and salinity measurements (both Nansen bottle data and CTD) available at the National Oceanographic Data Center (NODC; presently National Centers for Environmental Information, NCEI/NOAA) as part of the World Ocean Database 2013 (Boyer et al. 2013) and mapped with the Ocean Data View software (Schlitzer 2016)

publicly available World Ocean Database (WOD), which for the last half-century has been maintained by the National Oceanographic Data Center of the National Oceanic and Atmospheric Administration (now – since 2015 – part of the National Centers for Environmental Information, NCEI). As a result, the northern and eastern parts of the Canada Basin feature the sparsest data coverage in the entire Arctic Ocean (Fig. 14.12).

The legacy collections and data sets from ice islands also serve as valuable benchmarks, against which future changes in the Arctic could be evaluated. Comparisons with observations from NP-22 made in the 1970s and early 1980s revealed drastic changes of Arctic ecosystems, including those of sea ice; some of the ecosystem changes are deemed to have resulted from the long-term decrease of multi-year fraction of sea ice cover and its replacement with seasonal ice (Melnikov 2005; Melnikov and Semenova 2013). Future changes in the warming Arctic Ocean (Wassmann et al. 2011; Arrigo 2014) are expected to include range expansion of various species, including benthos (Renaud et al. 2015).

14.5 Conclusions

With the world's longest coastline facing the Arctic Ocean between 30°E and 170°E, Russia was traditionally at the forefront of Arctic exploration, including the search for "phantom lands" that have been sighted numerous times over centuries. There is little doubt that most sightings were of real objects (islands or "lands") yet all attempts to reach these lands turned out to be fruitless, thus giving rise to the term "phantom lands." Some of these "phantom lands" were likely islands partly composed of ground ice, which explains their eventual disappearance as a result of thermal erosion due to climate amelioration and mechanical erosion due to actions of waves and sea ice (Gavrilov et al. 2003; Günther et al. 2015). Other "phantom lands" were probably drifting (or temporary grounded) icebergs, including ice islands.

One of the first explorers of the Siberian coast, Ferdinand Petrovich Wrangell led an expedition in 1820–1824 (Wrangell 1826, 1841; Parrot 1826) that made observations of strange ice formations that were drastically different from regular ice floes, hummocks, or pressure ridges. In retrospect, these ice formations could have been remnants of what we now call ice islands.

With the advent of polar aviation in the twentieth century and ever-increasing shipping activity along the Northern Sea Route, pilots and navigators of the Soviet Ice Reconnaissance Service reported numerous sightings of ice islands in the Soviet sector of the Arctic Ocean and beyond, although those reports have not been made publicly available. The growing importance of the Arctic has justified the establishment of manned drifting stations on ice floes (Frolov et al. 2005; Kornilov et al. 2010). Five large ice islands discovered by Russians since 1946 were chosen as platforms for drifting stations North Pole-6, 18/19, 22, 23, and 24 that operated much longer than other stations thanks to the superior thickness (between 10 and 40 m) of these ice islands and their near indestructibility. Indeed, these stations combined served 23 years as reliable platforms and were the objects of various studies. Of particular importance was the 8½-year trajectory of NP-22 – the longest continuously occupied drifting station ever – that facilitated studies of physical and biological oceanography of the Beaufort Gyre and Canada Basin.

Until this day the Canada Basin remains poorly explored, especially its northern and eastern parts. Notwithstanding the ongoing warming in the Arctic and the attendant long-term attrition (shrinking and thinning) of the Arctic ice cover, the multiyear pack ice in the eastern part of the Canada Basin still presents a formidable problem even for modern icebreakers. Similarly, research submarines would typically circumvent this area since the thick ice pack would prevent emergency surfacing while deep keels of ice hummocks would present a constant hazard. The Submarine Arctic Science Program (SCICEX) scientific cruises of U.S. Navy's nuclear submarines in 1993–2003 carefully avoided the eastern Canada Basin (SCICEX Science Advisory Committee 2010).

The data paucity in the Canada Basin is particularly notable with regard to biological collections and sediment sampling (e.g., Bjørklund and Kruglikova 2003; Jirkov and Leontovich 2012; Hunt et al. 2014; Xiao et al. 2014; Zasko and Kosobokova 2014; Bluhm et al. 2015; Wassmann et al. 2015). Therefore, the legacy collections from Russian ice islands, especially NP-22, remain indispensable. Numerous new species of pelagic and benthic fauna were discovered in the Canada Basin thanks to the collections from NP-22 and other ice islands. More species likely await discovery in these collections.

Acknowledgments The splendid record of the Russian manned drifting stations on Arctic ice islands owes everything to the courage, determination, skill, and vision of Russian *polyarniks* – specialists in all things polar – seafarers, airmen, ice observers, and scientists – who devoted their lives to Arctic exploration and research. We are deeply indebted to all of them and particularly to the crew members of the North Pole drifting stations who worked under the most trying environmental conditions to collect unique scientific data. Logistical support provided by the Arctic and Antarctic Research Institute (AARI), St. Petersburg, during our work in the Arctic is greatly appreciated. We are especially thankful to Martin Jeffries for suggesting this chapter and to Derek Mueller and Luke Copland, editors, for encouraging us. The manuscript has been substantially improved thanks to numerous edits and comments by William Althoff, Derek Mueller, Daphne Johnson, Nikolai Kornilov, Zalman Gudkovich, Eduard Sarukhanian, Yuri Gorbunov, Vladimir Radionov, Vasily Smolyanitsky, Andrey Glazovsky, Igor Melnikov, Ksenia Kosobokova, and Yuri Yakovlev, while Alexander Merkulov, Igor Jirkov, Elena Markhaseva, and Yuri Rudyakov provided hard-to-get Russian papers. Maps in Figs. 14.3 and 14.12 were created by Daphne Johnson using the Ocean Data View software (Schlitzer 2016). Permission to reproduce figures from the "Problems of Geography" series was kindly granted by its Editor-in-Chief, Vladimir Kotlyakov. The series is freely available online courtesy of the Russian Geographical Society. For 6 months in 1977, Igor Belkin worked together with Igor Afanasyev on the best drifting station of all time – North Pole-22 – under the cheerful leadership of Igor Simonov. This chapter is dedicated to his memory, as well as the memory of Valentin Dremlyug, who inspired Sergey Kessel's interest in ice islands.

References

Afanasyev, I. F. (1978). Studies of the deep-sea benthic fauna of the central part of the Arctic Ocean. *Oceanol, 18*(5), 621. Originally published in Okeanologiya 18(5), 950–951. [in Russian].

Afanasyev, I. F., & Filatova, Z. A. (1980). To the investigation of the deep-sea benthic fauna of the Canada Basin of the Arctic Ocean. In M. E. Vinogradov & I. A. Melnikov (Eds.), *Biology of the central Arctic Basin* (p. 219–229). Moscow: Nauka. [in Russian].

Andriashev, A. P., Mukhomediyarov, B. F., & Pavshtiks, E. A. (1980). On great amounts of cryopelagic gadid fishes (*Boreogadus saida* and *Arctogadus glacialis*) in near-pole Arctic regions. In M. E. Vinogradov & I. A. Melnikov (Eds.), *Biology of the central Arctic Basin* (p. 196–211). Moscow: Nauka. [in Russian].

Anonymous. (1955). Soviet reports of "ice islands" in the Arctic Ocean. *Polar Record, 7*(50), 416–417. doi:10.1017/S0032247400046556.

Arrigo, K. R. (2014). Sea ice ecosystems. *Annual Review of Marine Science, 6*, 439–467. doi:10.1146/annurev-marine-010213-135103.

Averincev, V. G. (1980). *Chauvinelia arctica*, sp. n. (Acrocirridae, Polychaeta) from the Canada Basin. *Explorations of the Fauna of the Seas, 25*(33), 57–62. [in Russian].

Belyaeva, T. V. (1980). Phytoplankton of the drift area of North Pole-22 station. In M. E. Vinogradov & I. A. Melnikov (Eds.), *Biology of the central Arctic Basin* (p. 133–142). Moscow: Nauka. [in Russian].

Bjørklund, K. R., & Kruglikova, S. B. (2003). Polycystine radiolarians in surface sediments in the Arctic Ocean basins and marginal seas. *Marine Micropaleontology, 49*(3), 231–273. doi:10.1016/S0377-8398(03)00036-7.

Bluhm, B. A., Ambrose Jr., W. G., Bergmann, M., Clough, L. M., Gebruk, A. V., Hasemann, C., Iken, K., Klages, M., MacDonald, I. R., Renaud, P. E., Schewe, I., Soltwedel, T., & Wlodarska-Kowalczuk, M. (2011). Diversity of the Arctic deep-sea benthos. *Marine Biodiversity, 41*(1), 87–107. doi:10.1007/s12526-010-0078-4.

Bluhm, B. A., Kosobokova, K. N., & Carmack, E. C. (2015). A tale of two basins: An integrated physical and biological perspective of the deep Arctic Ocean. *Progress in Oceanography, 139*, 89–121. doi:10.1016/j.pocean.2015.07.011.

Bollinger, M. J. (2003). *Stalin's slave ships: Kolyma, the gulag fleet, and the role of the west*. Westport: Greenwood Publishing Group.

Boyer, T. P., Antonov, J. I., Baranova, O. K., Coleman, C., Garcia, H. E., Grodsky, A., Johnson, D. R., Locarnini, R. A., Mishonov, A. V., O'Brien, T. D., Paver, C. R., Reagan, J. R., Seidov, D., Smolyar, I. V., & Zweng, M. M. (2013). *World Ocean Database 2013, NOAA Atlas NESDIS 72*. Silver Spring: National Oceanic and Atmospheric Administration (NOAA).

Brodsky, K. A., & Pavshtiks, E. A. (1976). Plankton of the central part of the Arctic Basin (based on collections of the North Pole drifting stations). *Problems of Geography* [Voprosy Geografii], *101*, 48–157. [in Russian]. http://lib.rgo.ru/reader/flipping/Resource-4322/vg_101/index.html. English translation: *Polar Geography, 1*(2), 143–161. doi:10.1080/10889377709388621.

Burkhanov, V. F. (1954). Origin of ice islands in the Arctic. *Problems of Geography* [Voprosy Geografii], *36*, 3–29. [in Russian]. http://lib.rgo.ru/reader/flipping/Resource-4190/002_R/index.html

Buzhinskaja, G. N. (2004). Two new genera of the pelagic family Yndolaciidae (Polychaeta) from the Arctic Ocean with an addition to the description of *Yndolacia lopadorrhynchoides* Støp-Bowitz. *Sarsia, 89*(5), 338–345. doi:10.1080/00364820410002604.

Cherepanov, N. V. (1964). Structure of sea ice of great thickness. *Proceedings of the Arctic and Antarctic Research Institute, 267*, 13–18. [in Russian]. Translation T 448 R, Directorate of Scientific Information Services, Defence Research Board Canada, January 1966.

Chilingarov, A., Yevseyev, M., & Sarukhanyan, E. (1972). *Life on an ice island* [Pod Nogami Ostrov Ledyanoi] (pp. 160). Moscow: Molodaya Gvardiya. [in Russian]. English translation: Chilingarov, A., Sarukhanyan, E., & Yevseyev, M. (1975). *Life on an ice island* (NTIS Product No. ADA018072, Cold Regions Research and Engineering Laboratory, pp. 208). Hanover: U.S. Army Corps of Engineers.

Chilingarov, A., Yevseyev, M., & Sarukhanyan, E. (1986). *Life on an ice island* [Pod Nogami Ostrov Ledyanoi] (2nd ed.). Leningrad: Gidrometeoizdat. [in Russian].

Chilingarov, A. N., Yevseyev, M. P., & Sarukhanyan, E. I. (2014). *Life on an ice island* [Pod Nogami Ostrov Ledyanoi] (3rd ed.). Moscow: Paulsen.

Darwin, C. (1873). Origin of certain instincts. *Nature, 7*(179), 417–418. doi:10.1038/007417a0.

Dremlyug, V. V., & Kessel, S. A. (2007). *Enigmatic Arctic Islands* [Ostrova – Zagadki Arktiki]. St Petersburg: GeoGraf Press. [in Russian].

Environmental Working Group. (1997). In L. Timokhov & F. Tanis (Eds.), *Environmental Working Group Joint U.S.-Russian atlas of the Arctic Ocean*. Boulder: National Snow and Ice Data Center. doi:10.7265/N5H12ZX4.

Environmental Working Group. (2000a). In F. Tanis & V. Smolyanitsky (Eds.), *Environmental Working Group Joint U.S.-Russian Arctic sea ice atlas*. Boulder: National Snow and Ice Data Center. doi:10.7265/N5C82766.

Environmental Working Group. (2000b). In F. Fetterer & V. F. Radionov (Eds.), *Environmental Working Group Arctic meteorology and climate atlas*. Boulder: National Snow and Ice Data Center. doi:10.7265/N5MS3QNJ.

Fletcher, J. O. (1950). Floating ice islands in the Arctic Ocean. *Tellus, 2*(4), 323–324. doi:10.1111/j.2153-3490.1950.tb00346.x.

Frolov, I. E., Gudkovich, Z. M., Radionov, V. F., Shiroshkov, A. V., & Timokhov, L. A. (2005). *The Arctic Basin: Results from the Russian drifting stations*. Berlin: Springer-Praxis.

Gavrilov, A. V., Romanovskii, N. N., Romanovsky, V. E., Hubberten, H. W., & Tumskoy, V. E. (2003). Reconstruction of ice complex remnants on the eastern Siberian Arctic Shelf. *Permafrost and Periglacial Processes, 14*(2), 187–198. doi:10.1002/ppp.450.

Gebruk, A. V., Thiel, H., & Thurston, M. (Eds.). (2014). Deep-sea fauna of European seas: An annotated species check-list of benthic invertebrates living deeper than 2000 m in the seas bordering Europe: Introduction. *Invertebrate Zoology, 11*(1), 1–2.

Geinrikh, A. K., Kosobokova, K. N., & Rudyakov, Y. A. (1980). Seasonal variations in vertical distribution of some prolific copepods of the Arctic Basin. In M. E. Vinogradov & I. A. Melnikov (Eds.), *Biology of the central Arctic Basin* (p. 155–166). Nauka: Moscow. [in Russian]. Translated as: *Can Transl Fish Aquat Sci, 4925*. Translation Bureau, Department of the Secretary of State of Canada, Ottawa, 1983. http://www.dfo-mpo.gc.ca/Library/19332.pdf

Grishchenko, V. D. (1980). Surface morphology of ice in the Arctic Basin. In M. E. Vinogradov & I. A. Melnikov (Eds.), *Biology of the central Arctic Basin* (p. 33–55). Moscow: Nauka. [in Russian].

Grishchenko, V. D., & Simonov, I. M. (1985). Morphological and structural features of the drifting ice island NP-22. *Problems of the Arctic and Antarctic, 59*, 60–68. [in Russian].

Günther, F., Overduin, P. P., Yakshina, I. A., Opel, T., Baranskaya, A. V., & Grigoriev, M. N. (2015). Observing Muostakh disappear: Permafrost thaw subsidence and erosion of a ground-ice-rich island in response to Arctic summer warming and sea ice reduction. *The Cryosphere, 9*(1), 151–178. doi:10.5194/tc-9-151-2015.

Hunt, B. P. V., Nelson, R. J., Williams, B., McLaughlin, F. A., Young, K. V., Brown, K. A., Vagle, S., & Carmack, E. C. (2014). Zooplankton community structure and dynamics in the Arctic Canada Basin during a period of intense environmental change (2004–2009). *Journal of Geophysical Research, 119*(C4), 2518–2538. doi:10.1002/2013JC009156.

Jeffries, M. O. (1992). Arctic ice shelves and ice islands: Origins, growth and disintegration, physical characteristics, structural-stratigraphic variability, and dynamics. *Reviews of Geophysics, 30*(3), 245–267. doi:10.1029/92RG00956.

Jirkov, I. A. (1980). Toward the Polychaeta fauna of the abyssal of the Canada Basin. In M. E. Vinogradov & I. A. Melnikov (Eds.), *Biology of the central Arctic Basin* (p. 229–236). Moscow: Nauka. [in Russian].

Jirkov, I. A. (2001). *Polychaeta of the Arctic Ocean* (p. 633). Moscow: Yanus-K. [in Russian]. http://www.twirpx.com/file/1941820/

Jirkov, I. A., & Leontovich, M. K. (2012). Biogeography of Polychaeta of the Eurasian North Polar Basin. *Invertebrate Zoology, 9*(1), 41–51. http://www.nature.air.ru/invertebrates

Kahl, J. D. W., Zaitseva, N. A., Khattatov, V., Schnell, R. C., Bacon, D. M., Bacon, J., Radionov, V., & Serreze, M. C. (1999). Radiosonde observations from the former Soviet "North Pole" series of drifting ice stations, 1954–90. *Bulletin of the American Meteorological Society, 80*(10), 2019–2026. doi:10.1175/1520-0477.

Kamenskaya, O. E. (1980). Deep-sea amphipods (Amphipoda, Gammaridea) from collections made during the expedition on the drifting station "North Pole-22". In M. E. Vinogradov & I. A. Melnikov (Eds.), *Biology of the central Arctic Basin* (p. 241–251). Moscow: Nauka. [in Russian].

Kessel, S. A. (2005). *Arctic ice islands* [Ledyanye Ostrova Arktiki] (pp. 153). St Petersburg: Otechestvo Foundation [in Russian].

Koenig, L. S., Greenaway, K. R., Dunbar, M., & Hattersley-Smith, G. (1952). Arctic ice islands. *Arctic, 5*(2), 66–103. doi:10.14430/arctic3901.

Kondakov, N. N., Moskalev, L. I., & Nesis, K. N. (1980). *Benthoctopus sibiricus* Løyning, an endemic octopod of the Eastern Arctic. In A. P. Kuznetsov (Ed.), *Ecological investigations of the shelf* (p. 42–56). Moscow: P.P. Shirshov Institute of Oceanology, USSR Academy of Sciences. [in Russian, with English summary].

Kornilov, N. A., Kessel, S. A., Sokolov, V. T., & Merkulov, A. A. (2010). *Russian scientific studies on the Arctic drifting ice* [Rossijskie issledovaniya na drejfuyushchih l'dah Arktiki]. St. Petersburg: Arctic and Antarctic Research Institute [in Russian].

Kosobokova, K. N. (1980). Seasonal variations in the vertical distribution and age composition of *Microcalanus pygmaeus, Oithona similis, Oncaea borealis* and *O. notopus* populations in the central Arctic Basin. In M. E. Vinogradov & I. A. Melnikov (Eds.), *Biology of the central Arctic Basin* (p. 167–182). Moscow: Nauka. [in Russian]. Translated as: *Can Transl Fish Aquat Sci, 4926*. Translation Bureau, Department of the Secretary of State of Canada, Ottawa, 1983. http://www.dfo-mpo.gc.ca/Library/61095.pdf

Kosobokova, K. N. (1982). Composition and distribution of the biomass of zooplankton in the central Arctic Basin. *Oceanology, 22*(6), 744–750.

Kosobokova, K., & Hirche, H. J. (2000). Zooplankton distribution across the Lomonosov Ridge, Arctic Ocean: Species inventory, biomass and vertical structure. *Deep-Sea Research Part I, 47*(11), 2029–2060. doi:10.1016/S0967-0637(00)00015-7.

Kosobokova, K. N., Hopcroft, R. R., & Hirche, H. J. (2011). Patterns of zooplankton diversity through the depths of the Arctic's central basins. *Marine Biodiversity, 41*(1), 29–50. doi:10.1007/s12526-010-0057-9.

Kruglikova, S. B. (1989). Arctic Ocean radiolarians. In Y. Herman (Ed.), *The Arctic seas: Climatology, oceanography, geology and biology* (p. 461–480). New York: Van Nostrand Reinhold.

Kruglikova, S. B., Bjørklund, K. R., Hammer, O., & Anderson, O. R. (2009). Endemism and speciation in the polycystine radiolarian genus Actinomma in the Arctic Ocean: Description of two new species *Actinomma georgii* n. sp and *A. turidae* n. sp. *Marine Micropaleontology, 72*(1–2), 26–48. doi:10.1016/j.marmicro.2009.02.004.

Kussakin, O. G. (1983). The first new isopod species from the abyssal of the Canada Basin of the Arctic Ocean. *Marine Biology* [Biologiya morya], *3*(1), 13–17. [in Russian, with English summary].

Legen'kov, A. P. (1973). Thermal stresses and deformations of the drifting ice island North Pole-19. *Oceanology, 13*(6), 804–808.

Legen'kov, A. P., & Chugui, I. V. (1973). Morphological measurements of the drifting ice island North Pole-19. *Problems of the Arctic and Antarctic,* 42, 44–48. [in Russian]. Translated as: Results of morphometric measurements of the ice island of the Severnyi Polyus-19 drifting station. In A. F. Treshnikov (Editor-in-Chief), *Problems of the Arctic and Antarctic, 42*, 53–60. Translated and published for the Office of Polar Programs, NSF by Amerind Publishing in collaboration with Oxonian Press, New Delhi, India, 1976.

Legen'kov, A. P., Chuguy, I. V., & Yeremin, N. N. (1974). Temperature measurements of the drifting ice island North Pole-19. *Oceanology, 14*(4), 495–498. doi:10.1109/OCEANS.1974.1161384.

Levenstein, R. Y. (1981). Some peculiarities of the distribution of the family Polynoidae from the Canada Basin of the Arctic Ocean. *Proceedings of the Institute of Oceanology* [Trudy Instituta Okeanologii], *115*, 26–36. [in Russian, with English summary].

Malyutina, M. V., & Kussakin, O. G. (1995a). Addition to the Polar Sea bathyal and abyssal Isopoda (Crustacea). Part 1. Anthuridea, Valvifera, Asellota (Ischnomesidae, Macrostylidae, Nannoniscidae). *Zoosystematica Rossica, 4*(1), 49–62.

Malyutina, M. V., & Kussakin, O. G. (1995b). Addition to the Polar Sea bathyal and abyssal Isopoda (Crustacea, Malacostraca). Part 2. Asellota: Desmosomatidae. *Zoosystematica Rossica, 4*(2), 239–260.

Malyutina, M. V., & Kussakin, O. G. (1996). Addition to the Polar Sea bathyal and abyssal Isopoda (Crustacea, Malacostraca). Part 3. Asellota: Munnopsidae. *Zoosystematica Rossica, 5*(1), 13–27.

Margulis, R. Y. (1982). A new siphonophore *Rudjakovia plicata* gen. n., sp. n., (Coelenterata, Hydrozoa) from the Polar Basin and some notes on other siphonophores. *Zoological Journal* [Zoologichesky Zhurnal], *61*(3), 440–444. [in Russian].

Markhaseva, E. L. (1984). Aetideidae copepods (Copepoda, Calanoida) of the eastern sector of the central Arctic Basin. *Oceanology, 24*(3), 391–393.

Markhaseva, E. L. (1996). Calanoid copepods of the family Aetideidae of the World Ocean. *Proceedings of the Zoological Institute of the Russian Academy of Sciences, 268*, 1–331.

Melnikov, I. A. (1989). *The ecosystem of the Arctic Sea ice* [Ekosistema arkticheskogo morskogo l'da] (pp. 191). Moscow: Nauka. [in Russian]. (Translated as: Melnikov, I. A. (1997). *The Arctic Sea ice ecosystem* (pp. 221). Amsterdam: Gordon and Breach Science Publishers.

Melnikov, I. A. (2005). Sea ice–upper ocean ecosystems and global changes in the Arctic. *Russian Journal of Marine Biology, 31*(Suppl. 1), S1–S8. doi:10.1007/s11179-006-0010-8.

Melnikov, I. A., & Semenova, T. N. (2013). Cryopelagic fauna of the present sea ice cover of the central Arctic Basin. *Problems of the Arctic and Antarctic, 98*, 14–25. [in Russian]. http://www.aari.ru/misc/publicat/paa/PAA-98/PAA-98-02-(14-25).pdf

Moskalev, L. I. (1980). Benthic fauna of the outer shelf of the Chukchi Sea. In A. P. Kuznetsov (Ed.), *Ecological investigations of the shelf* (p. 73–79). Moscow: P.P. Shirshov Institute of Oceanology, USSR Academy of Sciences. [in Russian, with English summary].

Parrot, G. F. (1826). Bemerkungen des Herausgebers über die Nordlichte. In F. von Wrangel (Ed.), *Physikalische Beobachtungen des Capitain-Lieutenant Baron von Wrangel während seiner Reisen auf dem Eismeere in den Jahren 1821, 1822 und 1823* (p. 61–99). Berlin: Reimer.

Pasternak, F. A. (1980). Pennatularia *Umbellula encrinus* (L.) from the Canada Basin of the Arctic Ocean. In M. E. Vinogradov & I. A. Melnikov (Eds.), *Biology of the central Arctic Basin* (p. 236–239). Moscow: Nauka. [in Russian].

Pavshtiks, E. A. (1980). Some patterns in the life of the plankton of the central Arctic Basin. In M. E. Vinogradov & I. A. Melnikov (Eds.), *Biology of the central Arctic Basin* (p. 142–154). Moscow: Nauka. [in Russian]. Translated as: *Can Transl Fish Aquat Sci, 4917*. Translation Bureau, Department of the Secretary of State of Canada, Ottawa, 1983. http://www.dfo-mpo.gc.ca/Library/64564.pdf

Petryashov, V. V. (1989). Arctic Ocean Mysids (Crustacea, Mysidacea): Evolution, composition, and distribution. In Y. Herman (Ed.), *The Arctic seas: Climatology, oceanography, geology and biology* (p. 373–396). New York: Van Nostrand Reinhold.

Petryashov, V. V. (1993). Deep-water Mysids (Crustacea, Mysidacea) of the Arctic Basin. *Explorations of the Fauna of the Seas, 45*(53), 70–89. [in Russian].

Petryashov, V. V. (2004). Mysids (Crustacea, Mysidacea) of the Eurasian sub-basin of the Arctic Basin and adjacent seas: Barents, Kara, and Laptev seas. *Explorations of the Fauna of the Seas, 54*(62), 124–145. [in Russian].

Renaud, P. E., Sejr, M. K., Bluhm, B. A., Sirenko, B., & Ellingsen, I. H. (2015). The future of Arctic benthos: Expansion, invasion, and biodiversity. *Progress in Oceanography, 139*, 244–257. doi:10.1016/j.pocean.2015.07.007.

Rogacheva, A. V. (2007). Revision of the Arctic group of species of the family Elpidiidae (Elasipodida, Holothuroidea). *Marine Biology Research, 3*(6), 367–396. doi:10.1080/17451000701781880.

Romanov, I. P., Konstantinov, Y. B., & Kornilov, N. A. (1997). *"North Pole" drifting stations (1937–1991)* [Dreyfuyushchie Stantcii "Severnyy Polyus" (1937–1991 gg.)] (2nd ed., p. 213). St Petersburg: Gidrometeoizdat. [in Russian]. English translation: https://web.archive.org/web/20070930224455/http://www.aari.nw.ru/projects/Atlas/Meteorology/HTML/HISTORY/COLLECTION/NPhistory.htm

Rusanov, V. P., Yakovlev, N. I., & Buynevich, A. G. (1979). Chemical oceanography of the Arctic Ocean. *Proceedings of the Arctic and Antarctic Research Institute* [Trudy AANII], *355*, 1–144. [in Russian].

Salazar-Vallejo, S. I., Gillet, P., & Carrera-Parra, L. F. (2007). Revision of *Chauvinelia*, redescriptions of *Flabelliseta incrusta*, and *Helmetophorus rankini*, and their recognition as acrocirrids (Polychaeta: Acrocirridae). *Journal of the Marine Biological Association of the United Kingdom, 87*(2), 465–477. doi:10.1017/S0025315407054501.

Sanamyan, N. P., Cherniaev, E. S., & Sanamyan, K. E. (2009). *Bathyphellia margaritacea* (Cnidaria: Actiniaria): The most northern species of the world. *Polar Biology, 32*(8), 1245–1250. doi:10.1007/s00300-009-0685-3.

Sanamyan, N. P., Sanamyan, K. E., & Grebelnyi, S. D. (2016). Two poorly known Arctic sea anemones, *Cactosoma abyssorum* and *Halcampa arctica* (Actiniaria: Halcampidae). *Invertebrate Zoology, 13*(1), 1–14. doi:10.15298/invertzool.13.1.01.

Schlitzer, R. (2016). *Ocean Data View*. http://odv.awi.de. Accessed 1 Sept 2016.

SCICEX Science Advisory Committee. (2010). *SCICEX (submarine Arctic science program) phase II science plan, part I: Technical guidance for planning science accommodation missions* (p. 76). Arlington: US Arctic Research Commission. http://purl.fdlp.gov/GPO/gpo13693

Sirenko, B. I. (Ed.). (2001). List of species of free-living invertebrates of Eurasian Arctic seas and adjacent deep waters. *Explorations of the Fauna of the Seas, 51*(59), 1–132.

Sirenko, B., Denisenko, S., Deubel, H., & Rachor, E. (2004). Deep water communities of the Laptev Sea and adjacent parts of Arctic Ocean. *Explorations of the Fauna of the Seas, 54*(62), 28–73.

Stepanjants, S. D. (1989). Hydrozoa of the Eurasian Arctic seas. In Y. Herman (Ed.), *The Arctic seas: Climatology, oceanography, geology and biology* (p. 397–430). New York: Van Nostrand Reinhold.

Sychev, K. A. (1961). *On drifting ice island* (pp. 112). Moscow: Morskoi Transport. [in Russian].

Sychev, K. A. (1962). On drifting ice island. In V. F. Burkhanov (Ed.), *On Ice Island* (pp. 7–52). Moscow: Geografgiz. [in Russian].

Tsynovsky, V. D. (1980a). Fishes collected from the drifting station "North Pole 22" in the winters of 1978–79 and 1979–80. In N. N. Parin (Ed.), *Fishes of the open ocean* (p. 110–112). Moscow: P.P. Shirshov Institute of Oceanology, USSR Academy of Sciences. [in Russian].

Tsynovsky, V. D. (1980b). On the ichthyofauna of deep-sea basins of the Central Arctic Ocean. In M. E. Vinogradov & I. A. Melnikov (Eds.), *Biology of the central Arctic Basin* (p. 214–218). Moscow: Nauka. [in Russian].

Tsynovsky, V. D., & Melnikov, I. A. (1980). About the finding *Liparis koefoedi* (Liparidae, Osteichthyes) in waters of the central Arctic Basin. In M. E. Vinogradov & I. A. Melnikov (Eds.), *Biology of the central Arctic Basin* (p. 211–214). Moscow: Nauka. [in Russian].

Vasilenko, S. V. (1988). A new deep-sea species *Hemilamprops canadensis sp. n.* (Crustacea: Cumacea) from the Canada Basin of the Arctic Ocean. *Zoological Journal* [Zoologichesky Zhurnal], *67*(6), 945–949. [in Russian].

Vassilenko, S. V. (1989). Arctic Ocean Cumacea. In Y. Herman (Ed.), *The Arctic seas: Climatology, oceanography, geology and biology* (p. 431–444). New York: Van Nostrand Reinhold.

Vinogradova, N. G. (1997). Zoogeography of the abyssal and hadal zones. *Advances in Marine Biology, 32*, 325–387. doi:10.1016/S0065-2881(08)60019-X.

Vinogradov, M. E., & Melnikov, I. A. (Eds.). (1980a). *Biology of the central Arctic Basin* (pp. 260). Moscow: Nauka. [in Russian].

Vinogradov, M. E., & Melnikov, I. A. (1980b). The study of the pelagic ecosystem of the central Arctic Basin. In M. E. Vinogradov & I. A. Melnikov (Eds.), *Biology of the central Arctic Basin* (p. 5–14). Moscow: Nauka. [in Russian]. Translated as: *Can Transl Fish Aquat Sci, 4098*. Translation Bureau, Department of the Secretary of State of Canada, Ottawa, 1983. http://www.dfo-mpo.gc.ca/Library/14842.pdf

von Wrangel, F. (1826). Physikalische Beobachtungen. In F. von Wrangel (Ed.), *Physikalische Beobachtungen des Capitain-Lieutenant Baron von Wrangel während seiner Reisen auf dem Eismeere in den Jahren 1821, 1822 und 1823* (p. 11–31). Berlin: Reimer.

Warren, S. G., Rigor, I. G., Untersteiner, N., Radionov, V. F., Bryazgin, N. N., Aleksandrov, Y. I., & Colony, R. (1999). Snow depth on Arctic sea ice. *Journal of Climate, 12*(6), 1814–1829. doi:10.1175/1520-0442.

Wassmann, P., Duarte, C. M., Agusti, S., & Sejr, M. K. (2011). Footprints of climate change in the Arctic marine ecosystem. *Global Change Biology, 17*(2), 1235–1249. doi:10.1111/j.1365-2486.2010.02311.x.

Wassmann, P., Kosobokova, K. N., Slagstad, D., Drinkwater, K. F., Hopecroft, R. R., Moore, S. E., Ellingsen, I., Nelson, R. J., Carmack, E., Popova, E., & Berge, J. (2015). The contiguous domains of Arctic Ocean advection: Trails of life and death. *Progress in Oceanography, 139*, 42–65. doi:10.1016/j.pocean.2015.06.011.

Wrangell, F. P. (1841). *Narrative of an expedition to the Polar Sea in the years 1820, 1821, 1822 and 1823 commanded by lieutenant, now Admiral Ferdinand Wrangell, of the Russian imperial navy* (pp. 302). New York: Harper and Brothers. Reprinted in 2008 as: Wrangel, F. P. (1842) *Narrative of an expedition to the Polar Sea in the years 1820–23* (pp. 306). Whitefish: Kessinger Publishing.

Xiao, W. S., Wang, R. J., Polyak, L., Astakhov, A., & Cheng, X. R. (2014). Stable oxygen and carbon isotopes in planktonic foraminifera *Neogloboquadrina pachyderma* in the Arctic Ocean: An overview of published and new surface-sediment data. *Marine Geology, 352*, 397–408. doi:10.1016/j.margeo.2014.03.024.

Yurchenko, A. (2005). Pomor. In M. Nuttall (Ed.), *Encyclopedia of the Arctic* (p. 1680–1682). New York: Routledge.

Zasko, D. N., & Kosobokova, K. N. (2014). Radiolarians in plankton of the Arctic Basin: Species composition and distribution. *Zoological Journal* [Zoologichesky Zhurnal], *93*(9), 1057–1069. [in Russian].

Zezina, O. N. (1980). On a deep sea find of brachiopods in the Arctic Basin. In M. E. Vinogradov & I. A. Melnikov (Eds.), *The biology of the central Arctic Basin* (p. 240–241). Moscow: Nauka. [in Russian].

Zubov, N. N. (1955). Arctic ice islands and how they drift. *Priroda, 2*, 7–45. [in Russian]. Translated as: T-176R. Ottawa: Defence Research Board, 1955, p. 1–10. http://www.dtic.mil/dtic/tr/fulltext/u2/682865.pdf

Chapter 15
Risk Analysis and Hazards of Ice Islands

Mark Fuglem and Ian Jordaan

Abstract When considering man-made structures for offshore Arctic regions where ice islands may transit, the probability of encountering different sizes of ice islands and ice island fragments needs to be estimated. This can be difficult to determine in near-shore regions where the occurrence of ice islands is infrequent, yet if there is an occurrence and the ice island grounds in shallow water, it could break into a number of smaller ice island fragments. Probabilistic design methods can be used to determine if consideration of ice island impact in structural design criteria is required and to choose appropriate levels of ice strengthening. This chapter includes a brief description of the design issues involved, available approaches and areas where additional information would be useful.

Keywords Ice islands • Probabilistic design • Ice loads • Gouges • Offshore structures

15.1 Introduction

When considering man-made structures for offshore Arctic regions where ice islands may transit, the probability of encountering different sizes of ice islands and ice island fragments needs to be estimated. Structures could include offshore oil and gas exploration and production platforms, subsea systems, and pipelines as well as other facilities. Design and operational considerations may include seasonal versus full-time operations, structure shape and ice strengthening as well as ice management.

Ice islands break off ice shelves along northern Ellesmere Island and from northern Greenland (Fig. 15.1). The majority from Ellesmere Island move west and south, a significant proportion of which get grounded in the Arctic Archipelago, while the rest move further west into the Beaufort Gyre. Ice islands may circle

M. Fuglem (✉)
C-CORE, St. John's, NL, Canada
e-mail: Mark.Fuglem@c-core.ca

I. Jordaan
Ian Jordaan and Associates Inc., St. John's, NL, Canada

L. Copland, D. Mueller (eds.), *Arctic Ice Shelves and Ice Islands*, Springer Polar Sciences, DOI 10.1007/978-94-024-1101-0_15

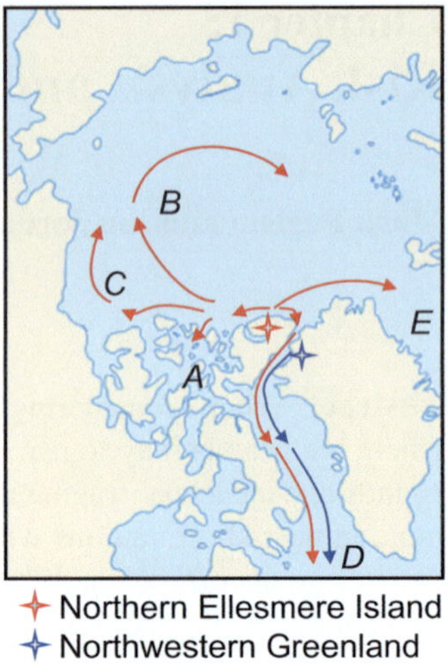

Fig. 15.1 Possible ice island trajectories from Northern Ellesmere Island (*red cross*) and Northwestern Greenland (*blue cross*): into the Canadian Arctic Archipelago (*A*); into the Beaufort Gyre (*B*); along the Beaufort and Chukchi coasts (*C*); through Nares Strait into Baffin Bay and south (*D*); through Fram Strait into the North Atlantic (*E*)

within the gyre several times before finally exiting into neighbouring regions such as the North Atlantic (Van Wychen and Copland 2017). Ice islands from Greenland and rarely from Ellesmere Island move down through Nares Strait reaching Baffin Bay or further south. Ice islands driven into shallow waters may ground or create gouges (scours) in the seabed. Changes in driving forces, ice island mass and shape, and water level can cause grounded ice islands to reposition themselves, scour further or move into deeper water. Sackinger et al. (1988) indicate that breakup is much more likely where islands ground in shallower water than in the Beaufort Gyre.

Ice islands present a challenge for offshore structural designers. Because of their immense size, it is generally not practical to design for loads resulting from contact with ice islands and larger fragments. At the same time, the probability of encountering an ice island is very small for regions distant from the ice shelves where the ice islands calve. In probabilistic structural design, some level of risk is considered acceptable. If the probability of encountering an ice island (or ice island fragment) is extremely small, structural loads resulting from impact may not need to be explicitly considered. Where Arctic structures are strengthened for loads resulting from interactions with first and multi-year sea ice, they may be capable of withstanding impacts from some ice island fragments as well. If the associated probability of failure is still too high, additional structural strengthening may be required. The probability of impact can be reduced through ice management or, in the case of floating systems, through avoidance by moving off site when detected ice islands approach. When considering management and avoidance, potential difficulties in conducting operations in pack ice or severe weather should be taken into account. Management will not be possible for large features and in heavy pack ice conditions.

There are several gravity based structure (GBS) designs that may resist horizontal loading resulting from impact with ice islands; the cylindrical GBS (Fig. 15.2a) is relatively large at the waterline and will therefore be subject to high wave loads,

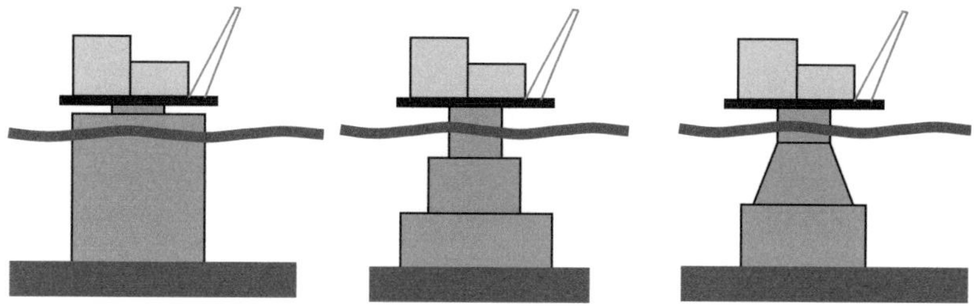

Fig. 15.2 Gravity-Based Structure (GBS) configurations: (**a**) the Cylindrical GBS configured for large wave loads (due to large width at waterline) and relatively large ice loads; (**b**) the Stepped GBS is configured for significantly reduced wave loads and reduced ice loads as a result of smaller probability of impact, reduced contact area with penetration and transfer of initial ice mass kinetic energy into rotation; (**c**) the Sloped GBS is configured for significantly reduced wave loads and reduced ice loads as a result of smaller probability of impact

the stepped GBS (Fig. 15.2b) has reduced size at the waterline to reduce wave loads, and the sloped GBS (Fig. 15.2c) has a sloped face at the waterline that will deflect impacting ice upward, reducing the horizontal force and inducing a downward force that counteracts the overturning moment and provides additional resistance to sliding.

Methods for protecting subsea equipment such as wellheads and manifolds from ice island contact include protective structures, built to withstand maximum impact load, or placement of the equipment in glory hole excavations to avoid contact (Fig. 15.3). Consideration must be given to the possibility of soil being pushed into the glory hole or the ice feature becoming trapped in the glory hole.

Burial depths of Arctic pipelines are chosen to achieve specified safety targets considering frequency of gouges resulting from ice features and associated gouge depths and widths as well as sub-gouge deformations (Fig. 15.4). Gouges will generally occur on seabed slopes where ice features impact the seabed. Gouge depth will depend on the strength of the ice feature keel, the strength of the soil and the wind, current and sea ice driving force pushing the ice island. Ice islands will be subject to large driving forces and are unlikely to rotate significantly, so they could generate very large gouge depths and widths. On the other hand, the probability of an ice island gouging over a pipeline may be extremely small, so it may not contribute significantly to the overall risk. Smaller ice island fragments may have shapes that are more unstable and therefore rotate upward upon contact with the seabed, reducing the gouge force (though they might scour further upslope). When considering pipeline burial depth, the effect of soil deformation below the gouge is important. Soil deformation will increase with gouge width and decrease with soil strength. Consideration should be given to the diameter of the pipe or pipeline bundle, and the backfill material used.

If possible, structures and pipelines could be located where bathymetry does not allow access by larger ice islands. The use of man-made berms to stop large ice islands from impacting a structure or buried pipeline might also be considered. In determining

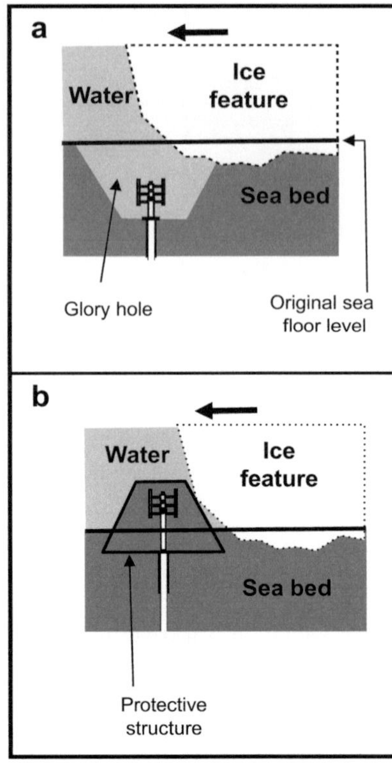

Fig. 15.3 Options for protection of subsea equipment: (**a**) glory hole, where subsea equipment is placed below the maximum gouge depth; (**b**) protective structure which can withstand maximum impact load

requirements for ice management, avoidance and ice strengthening, it is necessary to consider the probability of encountering ice islands and fragments, the effectiveness of management and avoidance systems, and the distribution of structural loading when interactions occur. A number of notable studies on the design requirements for ice islands have been published. Dunwoody (1983) briefly discusses application of the probabilistic methodology to design and presents methods for determining ice island encounter rates and their associated size and velocity distributions. Dunwoody considered ice island kinetic energies rather than the resulting loads. Refinements that could be made include more explicit consideration of the number and size of ice islands and fragments in general and the number and behaviour of grounded ice islands in shallow areas. Sackinger et al. (1985) present useful information on ice island calving, sighting and drift velocity and also discuss possible defensive measures.

15.2 Probabilistic Design

The International Organisation for Standardisation (ISO) Standard for Arctic offshore structures (ISO 19906) is a code that encourages reliability-based design (ISO 2010). Safety objectives and target reliability levels are given for:

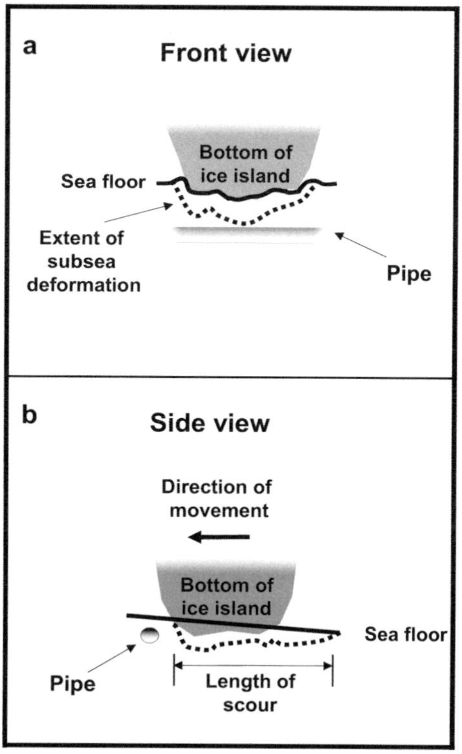

Fig. 15.4 Illustration of potential gouge depth, length and subgouge deformation for an ice island impacting a slope. (**a**) Front view. (**b**) Profile view

(i) three life safety categories: a manned but not to be evacuated facility (S1), manned facility with provision for safe evacuation prior to extreme environmental events (S2), and an unmanned facility with occasional visits for maintenance or inspection (S3); and
(ii) three failure consequence categories, referenced as high (C1), medium (C2) and low (C3).

The selection of a consequence category includes life safety (to persons onboard or near the facility), potential environmental loss and potential economic loss. Currently proposed target reliabilities for different combinations of life safety and consequence are summarized in Table 15.1.

ISO 19906 specifies annual target exceedance probabilities that loads should not exceed. In addition, load factors (Table 15.2) for design loads based on different load event types are provided. The Extreme-level Ice Event (ELIE) and Abnormal-level Ice Event (ALIE) are defined at annual exceedance probabilities of 10^{-2} and 10^{-4}, or return periods of 100 years and 10,000 years, respectively (ISO 2010). Calibration of the ISO 19906 Standard is aimed at ensuring that the factored design load (the product of a load associated with an ELIE or ALIE event times the load factor) in combination with reserve capacity in the factored structural resistance (the product of the structural resistance times a load resistance factor) results in the required safety level (ISO 2010).

Table 15.1 Annual reliability[a] targets based on safety class and consequence of failure

Safety class	Consequence of failure		
	C1 – High	C2 – Medium	C3 – Low
S1 – Manned and non-evacuated	$1–10^{-5}$	$1–10^{-5}$	$1–10^{-5}$
S2 – Manned and evacuated	$1–10^{-5}$	$1–10^{-4}$	$1–10^{-4}$
S3 – Unmanned	$1–10^{-5}$	$1–10^{-4}$	$1–10^{-3}$

[a]Annual Reliability = 1 – Annual Probability of Failure

Table 15.2 Target safety level and load factor for the S1 safety class and C1 consequence class

Ice load event type	Annual exceedance probability P_E	Load factor
Extreme-level ice event (ELIE) – frequent environmental events	10^{-2}	1.35
Abnormal-level ice event (ALIE) – rare environmental events	10^{-4}	1.0

To design against an ALIE, the annual exceedance probability is specified as 10^{-4} and the load factor is one. To design against an ELIE, the designer must consider the load at a 10^{-2} annual exceedance probability with a load factor of 1.35. It is of note that rare events with annual probability of occurrence less than 1-in-100 will have a zero load at the 10^{-2} level. The ISO Standard also requires consideration of combined loads, such as wave loads in combination with ice loads (ISO 2010). Jordaan (2005) provides theoretical background for the application of probabilistic methods to engineering design, including the analysis of extremes (here, the maximum ice load on a structure). A number of applied methods can be used to determine the structural loads associated with the specified probabilities of exceedance. These include Monte-Carlo simulation, FORM (first-order reliability method), SORM (second-order reliability method) and importance sampling. Each has relative advantages and disadvantages relating to execution time, simplicity, and robustness. Monte-Carlo simulation is often used because it is very robust and relatively simple to implement, although it can require long computer run times if complex models are used. A flow chart of the steps involved in setting up a Monte Carlo simulation (Fig. 15.5) includes the following steps:

(i) for each year simulated, the number of encounters is randomly generated and for each encounter, the associated load value is determined using appropriate ice-structure interaction models with randomly simulated input values;
(ii) the maximum load during the year is determined;
(iii) the annual maximum loads for all years simulated are sorted and ranked; and
(iv) the value associated with the specified probabilities of exceedance are determined.

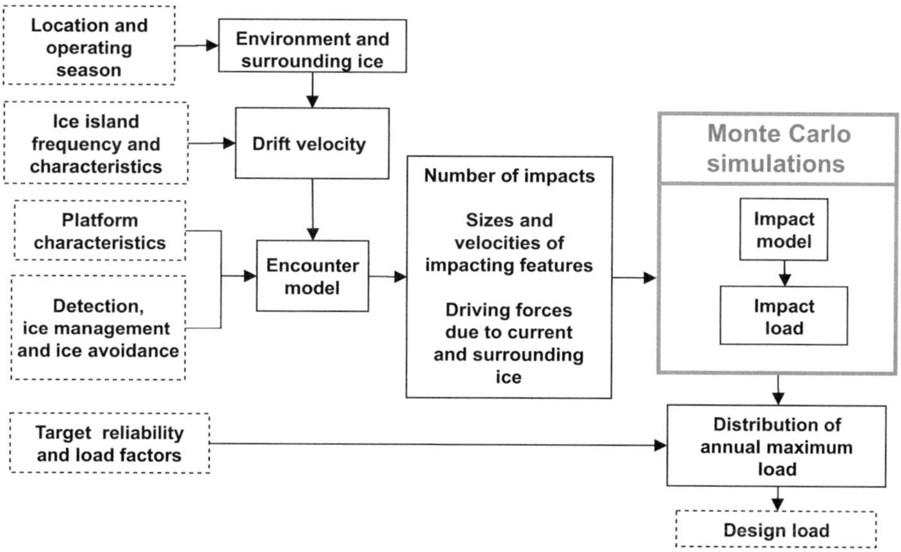

Fig. 15.5 Probabilistic framework for determining design load using a Monte Carlo approach

The distributions of ice island mass and shape as well as environmental conditions (wind, waves and surrounding ice movements) can be used to estimate the ice island velocities (zero velocities associated with grounded ice islands should be excluded), from which encounter rates can be determined. If ice management or avoidance strategies are in place, the number of encounters may be reduced and the distributions of size and shape of impacting ice islands and associated environmental conditions changed. For example, small ice fragments may be difficult to detect in some situations; ice islands and large ice island fragments may be impossible to tow; and towing may not be feasible in moderate to heavy ice conditions. In determining impact forces on a structure, consideration must be given to the size and shape of the impacting feature, the impact offset relative to the centers of the structure and ice island, the ice island velocity, the added mass at impact associated with the water around the ice island, and 'driving' forces on the ice island during the impact due to winds, currents and especially, where present, surrounding sea ice. These driving forces will vary considerably between winter, when surrounding first and multi-year sea ice will often result in significant forces, and summer, when first-year sea ice is absent.

Design loads are determined from the distribution of annual maximum impact loads (Fig. 15.6). The Extreme-level Ice Event (ELIE) corresponds to a probability of exceedance of 10^{-2} and load factor of 1.35, whereas the Abnormal-level Ice Event (ALIE) corresponds to a probability of exceedance of 10^{-4} and a load factor of 1.0. Figure 15.7 illustrates the effect of encounter rate on the maximum annual structural load distribution. Figure 15.7a shows a possible load distribution given that impact has occurred. Figure 15.7b shows the resulting annual maximum load distributions

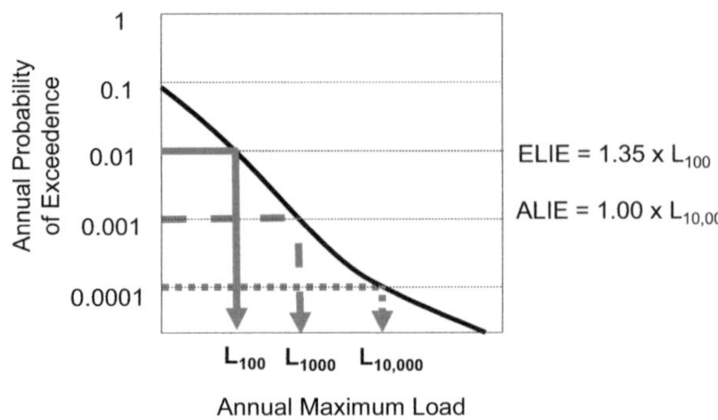

Fig. 15.6 Calculation of design loads given annual maximum load distribution

associated with encounter return periods of 1, 10 and 100 years. It is of note that for rare events such as ice island impacts, if the encounter rate corresponds to a return period greater than the structural load return period, the associated maximum load is zero, for example the 100-year load in Fig. 15.7b.

15.3 Encounter Rates

In the case of vertically sided structure, an encounter is defined as an event in which an ice island or fragment will impact a structure if no management or avoidance steps are taken. Expected encounter rates can be estimated using straightforward geometric considerations. Additional steps are required for more complex geometries and for facilities that are submerged or below the seabed. In the later cases, the draft and shape of the ice keel and heave motions or depth and width of gouge will need to be considered. Equations have also been developed for the probability of an ice feature passing over a linear feature such as a pipeline.

Geometric arguments (Dunwoody 1983; Nessim et al. 1986) are used to determine the rate with which ice features (ice islands and their fragments) encounter a cylindrical structure, based on the areal density, average projected width and average velocity of the ice features. A single ice feature randomly placed in a reference region of area A containing the structure is considered. In a small interval of time Δt, the ice feature can be considered to pass over or sweep out an area equal to $V\Delta t \times (W_S + W_I)$, where V is the ice feature velocity, W_S is the effective width of the structure and W_I is the effective width of the ice feature (Fig. 15.8). If the center of the structure is within the swept out area, an encounter will occur. The probability of encounter is therefore $V\Delta t \times (W_S + W_I)/A$. Integrating over time and over the distribution of ice feature sizes and velocities, one obtains the average number of encounters η during the duration of the period being investigated, T_D (typically 1 year):

$$\eta = \rho(\bar{W}_I + W_S)\bar{V}_I T_D \tag{15.1}$$

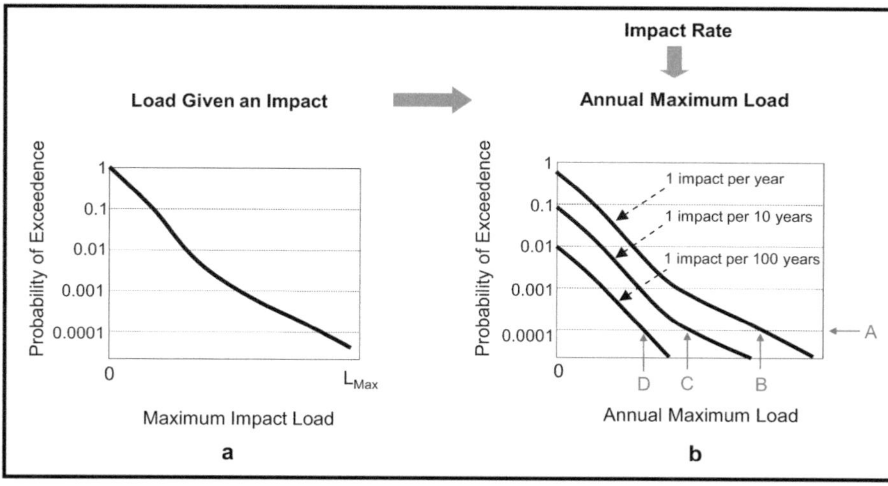

Fig. 15.7 Effect of encounter rate on specified annual maximum load. A line defines the 10,000 year values (*arrow A*) as well as values given for 1, 10 and 100 impacts per year (*arrows B, C* and *D*, respectively)

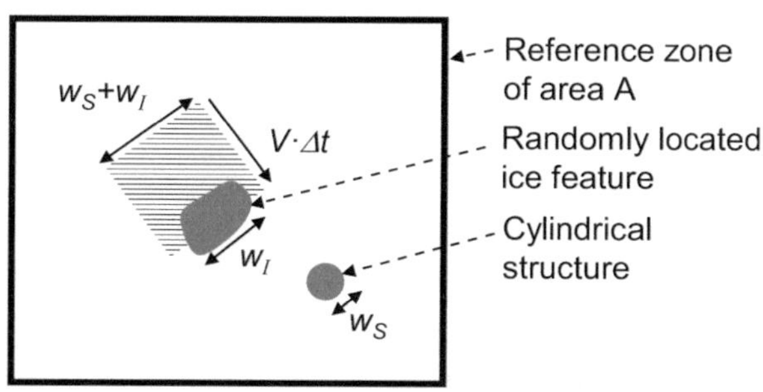

Fig. 15.8 Illustration of basis for method for determining encounter rates

where ρ is the average areal density of ice features (per m^2), $\overline{W}_I$ is the average projected width (m) of the ice features, W_S is the structure diameter (m), $\overline{V}_I$ is the average instantaneous drift speed (m s^{-1}).

The model approach corresponding to Eq. 15.1 avoids both the problem of determining the number of ice islands entering a region (the flux) and explicit modeling of ice island trajectories. Areal density is defined as the average number of ice islands per unit area at any random instant in time. This is a relatively easy parameter to obtain, although care is required to account for periods of zero velocity and possible long term trends in the numbers of ice islands. The estimated average value over the life of the structure should be considered.

In the case of ice islands, the areal density will be correlated to previous periods of major calving. Recently, fairly large portions of the Ellesmere Island ice shelves and Petermann Glacier have calved (Mueller et al. 2017; Münchow et al. 2014). As a result, the probability of encountering an ice island over the next decade could be relatively high. Calving rates might then be expected to decrease on Ellesmere Island since so little of the ice shelves remain, but the calving of ice tongues, such as Petermann Glacier, may remain elevated.

The effective width of an ice island is defined as the width perpendicular to the direction of motion of that part of the ice island that could impact the structure (Fig. 15.9). For a structure of constant width from all directions over the whole water column (such as a cylindrical GBS), the effective width is defined by the

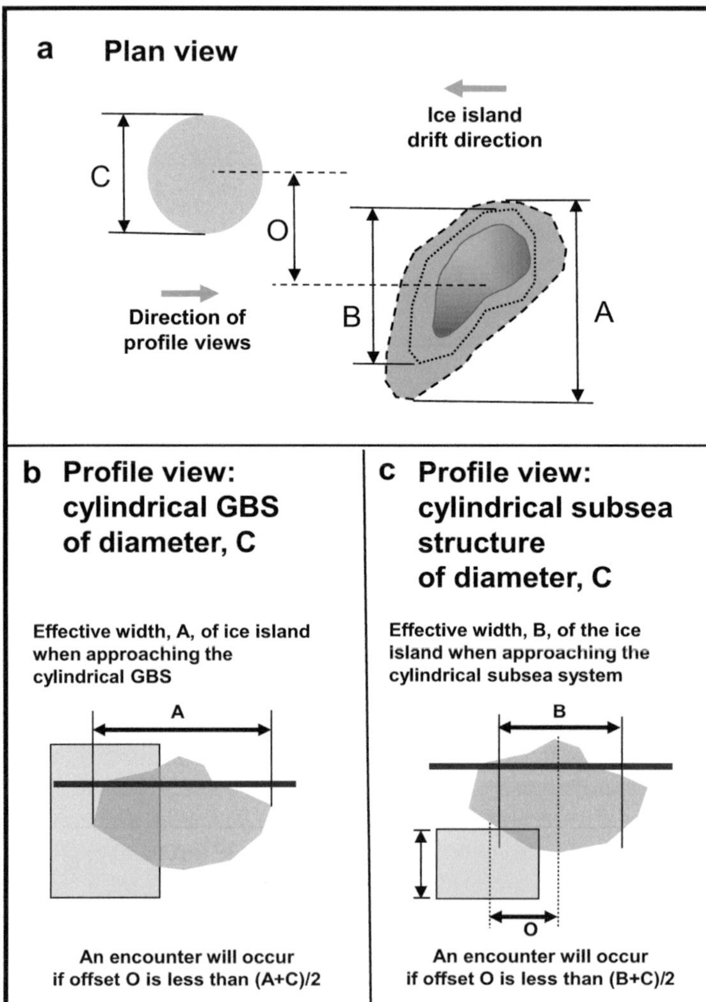

Fig. 15.9 Effective collision widths for cylindrical GBS and cylindrical subsea structure: (**a**) plan view for both cases; (**b**) profile view for cylindrical GBS and (**c**) profile view for cylindrical subsea structure

maximum extent of the ice island perpendicular to its motion. For other structure shapes, the interaction probability can be more complex. For example, for a submerged cylindrical structure, the effective width is defined by the maximum extent of the ice island at depths below the top of the structure. As there are few underwater profiles of ice island fragments (Forrest et al. 2012), judgment may be required. For larger stable ice island fragments; it may be reasonable to assume that the ice island has relatively constant depth and estimate the average effective width over all approach directions. For smaller fragments, it is necessary to consider the underwater profiles. A number of profiles from ice island fragments can be found in Kovacs and Mellor (1971). Work done related to iceberg impacts off the Grand Banks in Canada may serve as guidance, although the icebergs there may be more rounded due to wave action than ice islands in the Arctic. Typically the most common size parameter for describing Grand Banks icebergs is the maximum waterline length, L_W; and based on analyses of underwater shapes, the effective width over all directions (for Grand Banks icebergs impacting a cylindrical GBS) is around 1.04 L_W (Memorial University 1995).

As ice islands break up (for example, in shallower water areas), the risk increases if the remaining pieces that are still large enough to cause failure do not remain grounded. In particular, if a single ice island were to break up into m smaller ice island fragments of approximately equal size, then the areal density ρ of ice island and ice island fragments increases m-fold whereas the average waterline length $\bar{W}_l$ would decrease by $1/\sqrt{m}$ -fold. Based on Eq. 15.1, it is seen that the probability of an encounter in such a situation increases $\sqrt{m}$ -fold.

When determining the average drift velocity in Eq. 15.1 (as well as the distribution of impact velocities), instantaneous velocities should be used. Using velocities that are vector averaged over long periods could result in impact rates and velocities that are underestimated (Fig. 15.10).

Observed distributions of ice island size and velocity will generally not be the same as the corresponding distributions for impacting ice islands. Larger ice islands and fragments are more likely to impact than smaller ones, and features have a higher probability of impact per unit time when they are moving faster. Details on the necessary approach to adjust observed distributions may be found in Dunwoody (1983) and Jordaan (1983).

A key challenge in applying the above methods to determine the probability of encountering an ice island or ice island fragments is the scarcity of data. This is largely a result of a number of factors including the relatively low frequency of ice-island calving events, the variety of trajectories that ice islands can follow, the lack of people and activity in offshore regions to provide observations, and the need to monitor calving, trajectories, breakup and fate of ice islands and fragments by either aerial reconnaissance or the use of satellite imagery. Historically, issues related to detection may have resulted in smaller ice fragments being missed, whereas insufficient coverage could result in ice islands and fragments being missed. These factors are much less of a concern given present technologies. The uncertainty regarding future calving rates associated with climate change adds a new challenge.

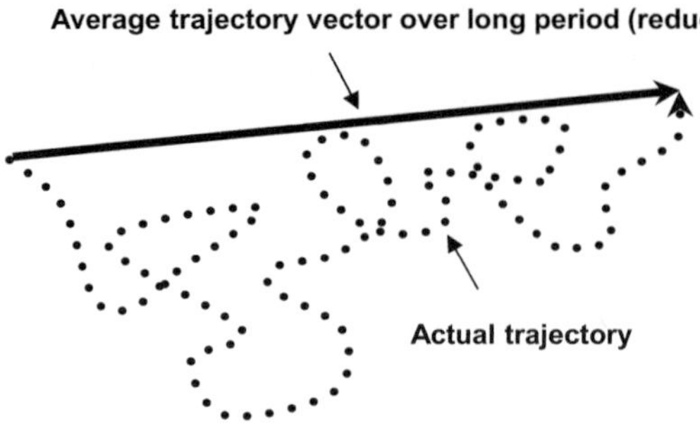

Fig. 15.10 Illustration of reduction in estimated distance travelled, and hence velocity, when using locations at infrequent time intervals

Two studies relating to the probability of iceberg impacts are mentioned here. Dunwoody (1983) based encounter rates on an assumed areal density of two ice islands in the Beaufort Gyre, which has an area of 2×10^6 km^2, giving a density of one ice island per million km^2. Dunwoody indicated that this value is conservative. He estimated that the long term average areal density of ice islands in an exploration area off the Beaufort coast was approximately 0.06 per million km^2, based on the assumption that the probability of an ice island being in the area is roughly proportional to the average concentration of multi-year ice, estimated to be 6%, at the site and 100%, in the Gyre. Ice islands with equivalent diameters ranging from about 4 to 26 km and drift velocities ranging up to 0.8 m s^{-1} were considered. Only ice islands with a draft less than the local water depth (45 m) were considered; the issue of grounded ice islands calving in shallower waters, resulting in additional smaller features, was not discussed. For comparison, within the current lease area in the Canadian Beaufort, water depth increases from 30 to 1000 m and average multi-year ice concentration increases from 5% to 30% from south to north.

Sackinger et al. (1988) present estimates of ice island areal densities over the Arctic based on Monte-Carlo simulations that consider calving events from ice shelves and subsequent trajectory movements. Calving events were assumed to occur once every 4 years at random locations along the northern coast of Ellesmere Island. The length, width and depth distributions of calved fragments were based on observed dimensions of ice islands believed to have been relatively newly formed. The number of fragments calved was determined as a function of the size distributions and observed total volume losses of shelf ice during calving events. Ice island trajectories were then simulated over a 50 km grid covering the Arctic Ocean as far as a line running from northeast Greenland to Severnaya Zemlya. Drift velocities were simulated using mean monthly geostrophic winds with a random component auto-correlated in time.

Grid boundaries included shallow water boundaries adjacent to land masses, plus open water boundaries from the northeast end of Greenland to Severnaya Zemlya, the Bering Strait, the mouth of Nares Strait and Amundsen Gulf. At shallow water boundaries, ice islands were prohibited from moving into water depths less than their drafts, (but could move parallel to or away from the boundaries). At open water boundaries, the ice islands left the grid and did not return. Ice islands thickness was reduced at a rate of 0.82 m year^{-1} based on melt observations of Ice Island T3 over 21 years. Fragmentation into smaller ice island pieces was not considered.

The model calculated an average ice island life of 18 years with some lasting up to 80 years. Causes for the demise of ice islands in the simulation included exit between Greenland and Severnaya Zemlya (75%); exit through Nares Strait (19.5%); exit through Amundsen Gulf (5.4%); and complete melting (0.1%). Of the simulated ice islands, 10% moved directly east after calving and out past Greenland or through Nares Strait. The rest moved west into the Beaufort Gyre of which 47% completed only a single circulation before ejection, with smaller proportions completing more circulations. The time to complete one circuit of the Beaufort Gyre was in the order of 5 years.

Sackinger et al. (1988) present a map with ice island return period contours (Fig. 15.11). The contours are for 5 x 5 km grids. In the Beaufort Gyre, the average return period is somewhere between 10 years and just over 1 year. Closer to shore, the return period increases to over 1000 years. Applying Eq. 15.1 with an assumed average return period of $T = 3$ years for the Beaufort Gyre (i.e. between 1 and 10 years), $\overline{W}_I = 12{,}000$ m, $W_S = 5000$ m (the grid size), $\overline{V}_I = 0.18$ m s^{-1}, and T =

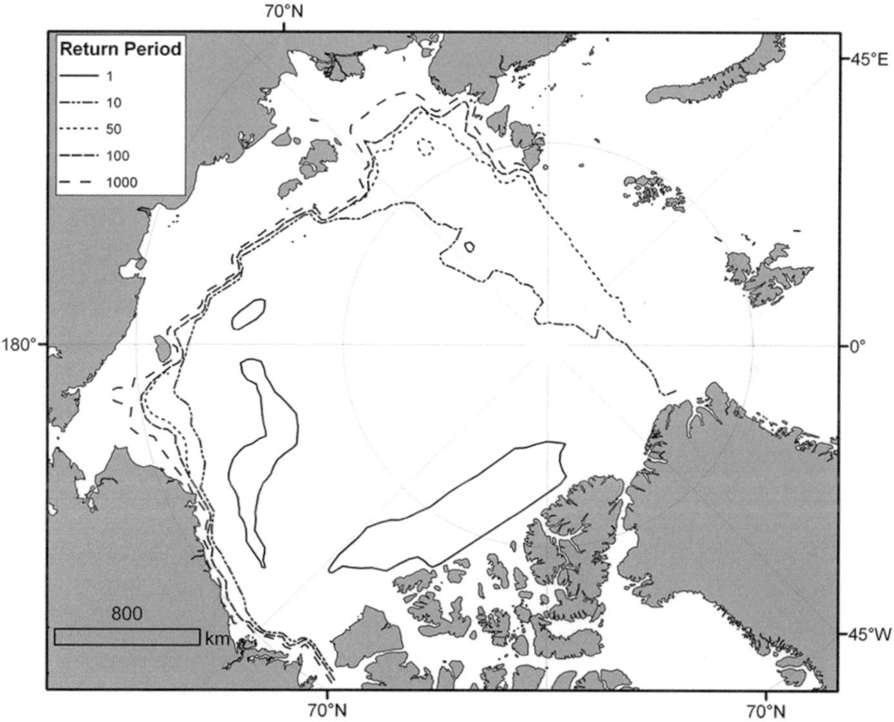

Fig. 15.11 Ice island return period contours (in years) as determined using a simulation model (Reproduction of Figure 146 from Sackinger et al. 1988)

31,536,000 s year^{-1} results in an approximate density ρ of 0.95 ice islands per million km^2. This is in approximate agreement with the estimated ice island density for the Beaufort Gyre given by Dunwoody (1983).

Areas where future ice island trajectory simulation models might be improved include:

- consideration of the movement of ice islands into the Canadian Arctic Archipelago (through Peary Channel, M'Clure Strait and other openings);
- modelling the grounding and breakup of ice islands when they enter shallow water;
- modelling the motions and degradation of smaller ice islands following breakup and calving;
- more accurate modeling of motions along coastlines;
- consideration of additional causes of degradation; and
- consideration of the effect of first and multi-year pack ice on the movement of ice islands.

Detailed observations of the behaviour of recently calved ice islands within the Canadian Arctic Archipelago and along its northwest coastline would be very helpful in validating improved model components.

15.4 Ice Management and Avoidance

Ice management has been used in the Beaufort Sea to break up large sea ice floes and ridges expected to encounter floating platforms, and also has been used on the Grand Banks to tow icebergs away from facilities. Management of ice islands and larger fragments may not be feasible. Management of smaller fragments would require towing operations similar to that on the Grand Banks, but on a larger scale and probably only during open water periods.

If it is determined that an approaching ice island or fragment poses a threat, adequate resources and time to tow the ice island or fragment would be required. The probability of success will depend on the size and shape of the ice island fragment, the sea state and the surrounding ice conditions.

Some guidance is available based on studies for towing icebergs in open water conditions on the Grand Banks. For icebergs with L_W less than 200 m, open water towing success rates are in the order of 75% for sea states with significant wave height less than 4 m, and the success rate decreases significantly for higher sea states. Ralph (pers. comm., 2008) estimates that the number of vessels required to tow icebergs with larger waterline lengths increases to two for L_W between 200 and 400 m, and even higher for larger icebergs. As ice island fragments will be present very infrequently, it is unlikely that multiple equipped vessels and personnel capable of towing operations would be available. Furthermore, the ability to tow in Arctic winter conditions would be extremely limited because of the effect of pack ice on operations. Alternative approaches might be possible, but would need to be tested and success demonstrated. An example would be to clear sea ice from one side of an ice feature, allowing the confinement pressure on the other side to deflect it.

Avoidance is an alternative that can be considered if management cannot adequately reduce risks. Destruction of ice islands and icebergs using explosives has been considered and tested without success (see, for example, Mellor et al. 1977, as well as the review by Crocker et al. 1998). Floating systems can operate seasonally, and can disconnect and move off site when severe ice conditions are expected or large ice islands approach. For oil and gas production, the reliability of ice detection and trajectory forecasting, evacuation procedures and equipment for shutting down production systems and disconnecting and moving off site are critical to ensure safety for personnel and the environment. Good detection and accurate trajectory forecasting of ice islands is required to reduce the probability of an emergency shutdown.

For gouging of linear features such as pipelines, a first step is to determine if the rate of gouges associated with ice islands and fragments is significant enough to pose a significant risk. An alternative to increased burial depth or pipe thickness is to reduce potential consequences by shutting down operations and flushing lines, given that significant warning time will be available that an ice island is approaching.

15.5 Global Loads

Design structural global loads are required to determine foundation strength requirements and the strength of major structural members. The maximum global load during an encounter event depends largely on the ice island mass, moments of inertia and velocity of the impacting ice island, the initial impact offset from the structure center, the local shape of both the ice island and structure at the point of contact, the compliance of the structure, the failure mode and strength of the ice, and the driving forces due to the added mass of the surrounding water, winds, currents and any surrounding sea ice.

Failure of ice island ice is a complex process that includes spalling of large intact ice pieces as well as a number of damage processes that occur during crushing including pressure melting, micro-cracking, extrusion and recrystallization. Large variations in actual contact area and load can occur as a result of major spalling events during interactions (Fig. 15.12). Additional variations in force can occur as a result of minor spalling events and random variations in ice strength (Jordaan 2001). Figure 15.13 illustrates a typical pressure pattern for ice contact over a large contact area. Typically spalling will occur around the peripheries, although larger spalls can occur, especially for very slow velocity impacts. Zones of high pressure have been observed to occur across the contact area in a manner that appears to be random as

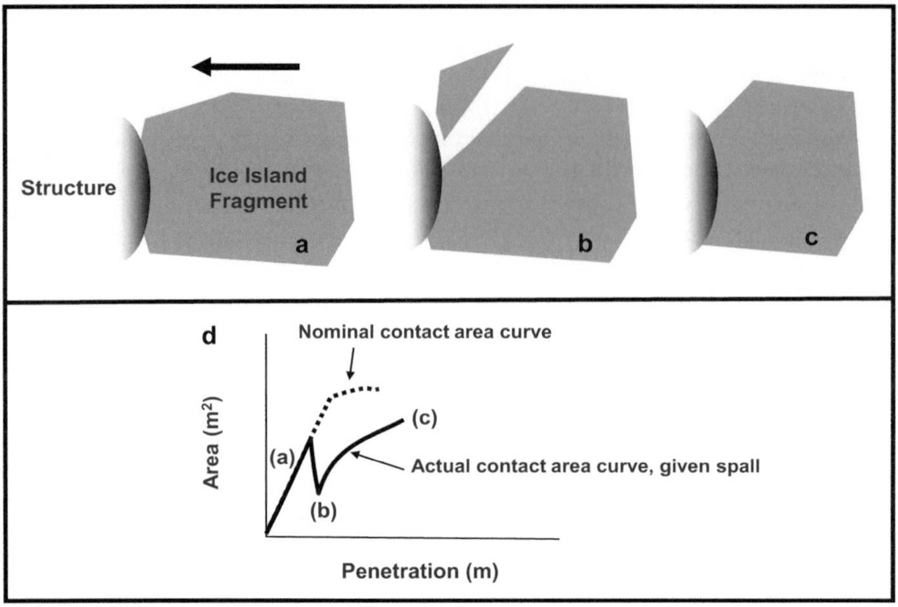

Fig. 15.12 Curve showing the effect of large spall on actual contact area: (**a**), (**b**) and (**c**) plan view showing development of spall and (**d**) resulting development of nominal and actual contact areas with penetration

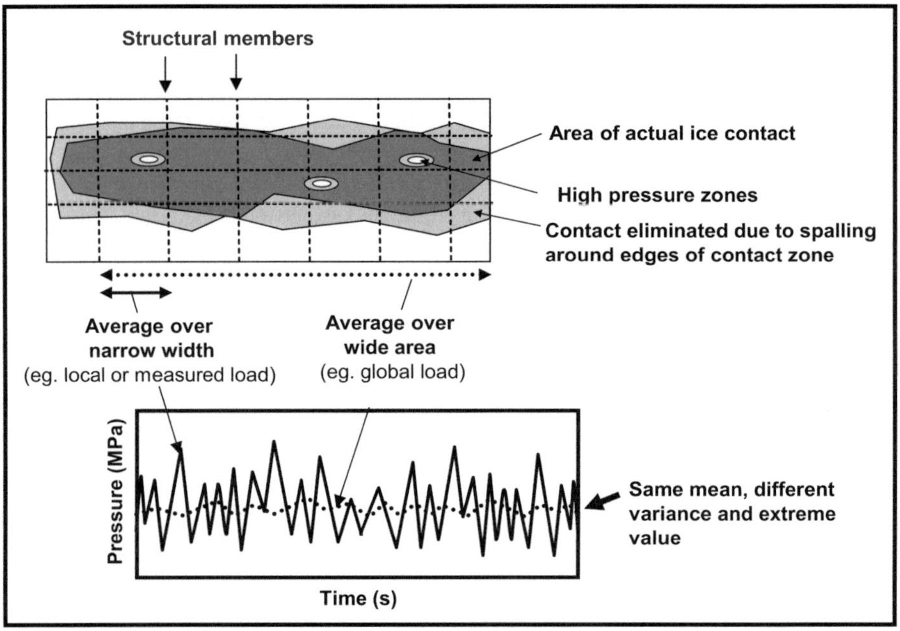

Fig. 15.13 Effect of load measurement area on estimated ice failure pressure

far as location and time. Very high pressure has been observed over very small areas, although the average pressure over large areas is considerably smaller. The random behaviour of crushing failure and scale effect (observed reduction in crushing pressure with contact area) are important considerations.

Measurements of ice pressures are typically based on measurements over a subset of the total structure area as it is very expensive to instrument a complete structure face. It is important to account for the effects of the reduced measurement area when estimating extreme average pressures over the whole structure face. For small panel areas, high pressures are more likely to be observed as a result of either high pressure zones or more coherent ice loading over the panel. These high pressures should not be extrapolated to the whole face without considering averaging processes. As illustrated in Fig. 15.13, the maximum pressure over the area of narrow width, which could represent a measurement panel, is much greater than the maximum pressure over the whole face. If one considers a small area, large pressure variations can occur depending if a high-pressure zone occurs within the area. For larger areas, the variation will be reduced because the pressure associated with any very high pressure zone will be averaged with more typical pressures over a larger surrounding area. If crushing occurs over the structure width, the mean global pressure and force will show much less variation over time than 'local' pressures over small areas. Jordaan et al. (2005) developed a method to account for this 'probabilistic averaging' by considering the ice failure as a first-order autoregressive process as outlined in Vanmarcke (1983); the method is still being enhanced.

As current understanding of the processes involved has not been advanced enough to develop a comprehensive failure model, simpler approaches are required. Approximate analytic models of the impact dynamics and failure of ice developed for icebergs impacting fixed structures may be appropriate for ice island impacts. Matskevitch (1997) and Fuglem et al. (1999) have developed closed form and iterative solutions. The models account for the transfer of the initial iceberg kinetic energy into crushing of ice as well as rotation during indirect impacts. The increase in ice contact area with penetration and the relationship between ice crushing strength and contact area are modeled in terms of power relationships with random coefficients calibrated based on observed iceberg shapes and measured ice loads during ship rams with heavy multi-year ice. In Arctic conditions, the driving force acting behind the ice island from surrounding ice (Fig. 15.14) can be very important. Surrounding ice creates pressure ridges across the back of the ice island as it slows down during impact (the surrounding ice may also form a triangular wedge of rubble behind the ice island in some conditions). If the width of the ice island is much greater than the width of the structure, the ice ridging force (typically given in terms of force per unit contact width) can counteract the crushing force on the structure acting to decelerate the ice island. Models for driving force are provided in ISO 19906 (ISO 2010).

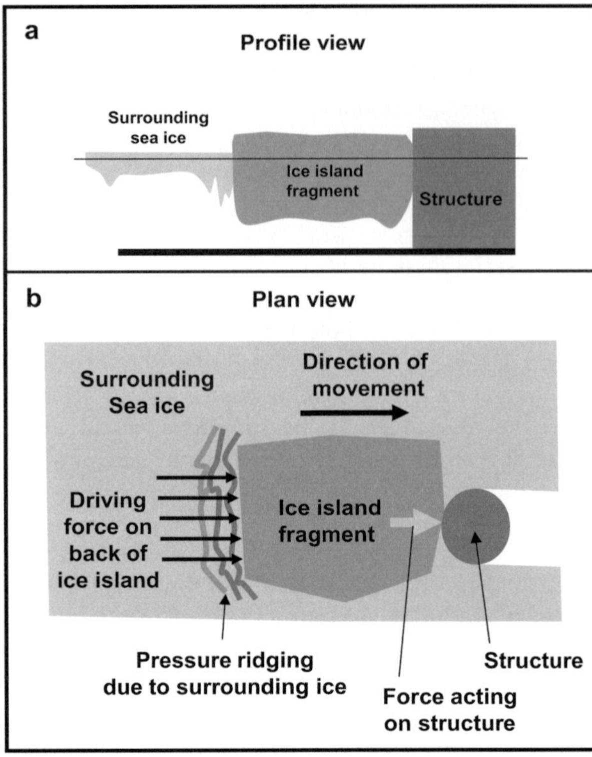

Fig. 15.14 Driving forces from surrounding sea ice: (**a**) profile view and (**b**) plan view

15.6 Local Loads

When the structural designer considers the choice of steel plating (or concrete) thickness and the spacing and strength of internal structural members, the maximum pressure over different fixed areas needs to be considered. As discussed previously with reference to pressure-averaging of global loads, significantly higher pressures can occur over small areas than over the whole contact area. Jordaan et al. (1993) provide a method to model local ice pressures that was developed based on an extreme value analysis of ice pressures over different local contact areas. The method takes into consideration the number of impacts per year in addition to the contact area. In the case of ice island impacts, the increased exposure associated with longer impact duration should also be considered.

15.7 Example Calculation

The probability of an ice island impact with a 100 m diameter structure in 30 m deep water in the Beaufort Sea is considered. Following Dunwoody (1983), it is assumed that the probability of an ice island reaching the location south of the Beaufort Gyre is proportional to multi-year sea ice concentration. A multi-year ice concentration of 5%, is assumed. Applying Eq. 15.1 with ice island areal density $\rho = 1$ ice island per million km^2 for the Beaufort Gyre (with 100% multi-year ice), an average ice island diameter $\overline{W}_I$ of 12,000 m, structure diameter $W_S = 100$ m, average ice drift velocity $\overline{V}_I = 0.18$ m s^{-1}, and T = 31,536,000 s year^{-1} gives an annual probability of impact η equal to 3.4×10^{-3}.

We believe that this result over-estimates the probability of impact for the following reasons. For a water depth of 30 m, approximately 4/5ths of ice islands would ground out before reaching the site (based on the thickness distribution given in Dunwoody 1983). It is likely that the areal density would be further reduced as a result of ice islands with smaller drafts grounding in shallower water to the south of the structure. The mean ice drift velocity of 0.18 m s^{-1} from Dunwoody (1983) may also be high, as winter drift velocities reduce in magnitude as one moves south of the Beaufort Gyre. Given the significant reductions over the last few decades in the extent of ice shelves on Ellesmere Island, one would expect a corresponding reduction in the probability of ice islands reaching the Beaufort Sea as well. It has been noted that the largest ice islands calved during the events of 2008 were 3–5 km wide, which is considerably smaller than the mean value determined based on earlier data. Given the above factors, it may then be possible to show that the probability of impact is less than the ALIE annual exceedance probability of 10^{-4}.

Further north, the probability of impact increases because water depth, multi-year ice concentration and average ice-drift velocity are larger. If the annual probability of impact is too high, consideration might be given to systems that can move off location, if necessary or to systems where human and environmental safety can be ensured by reliable evacuation and shutdown. Ice islands are large enough that detection should not be an issue, allowing sufficient advance warning.

Consideration of possible impact by ice island fragments is also required. Calving and splitting events will be most prevalent where ice islands ground in shallow water. Large fragments may pose similar risks as ice islands. In the case of small fragments, it may be possible to withstand impacts given existing or enhanced ice strengthening, or ice strengthening in combination with ice management. Statistics on the numbers and sizes of fragments created during grounding and calving events would be helpful in evaluating whether existing ice strengthening in place for first and multi-year ice features is sufficient for infrequent impacts from ice island fragments.

Acknowledgments We would like to acknowledge and thank Walt Spring, Derek Mueller and an anonymous reviewer for their detailed review of the chapter and helpful comments.

References

Crocker, G., Wright, B., Thistle, S., & Bruneau, S. (1998). *An assessment of current iceberg management capabilities*. Contract report for National Research Council, Canada, Prepared by C-CORE and B. Wright and Associates Ltd., C-CORE Publication 98-C26, St. John's.

Dunwoody, A. B. (1983, May 2–5). The design ice island for impact against an offshore structure. In *The 15th Annual Offshore Technology Conference*, Houston, pp. 325–330.

Forrest, A. L., Hamilton, A. K., Schmidt, V. E., Laval, B. E., Mueller, D. R., Crawford, A., Brucker, S., & Hamilton, T. (2012). Digital terrain mapping of Petermann Ice Island fragments in the Canadian High Arctic. In *Ice Research for a Sustainable Environment*. International Association for Hydro-Environment Engineering and Research, Dalian, China, pp. 710–721.

Fuglem, M., Muggeridge, K., & Jordaan, I. (1999). Design load calculations for iceberg impacts. *International Journal of Offshore and Polar Engineering, 9*(4), 298–306.

ISO. (2010). *International Standard ISO 19906: Petroleum and natural gas industries – Arctic offshore structures*. International Organization for Standardisation.

Jordaan, I. J. (1983, November). *Risk analysis with application to fixed structures in ice*. Paper presented at the Seminar/Workshop on Sea Ice Management, Memorial University, St. John's.

Jordaan, I. J. (2001). Mechanics of ice-structure interaction. *Engineering Fracture Mechanics, 68*, 1923–1960.

Jordaan, I. J. (2005). *Decisions under uncertainty*. Cambridge: Cambridge University Press.

Jordaan, I. J., Maes, M. A., & Brown, P. W. (1993). Probabilistic analysis of local ice pressures. *Journal of Offshore Mechanics and Arctic Engineering, 115*(1), 83–89. doi:10.1115/1.2920096.

Jordaan, I., Li, C., Mackey, T., Nobahar, A., & Bruce, J. (2005). *Design ice pressure-area relationships; Molikpaq data, PERD/CHC Report 14-121*. Prepared for: National Research Council Canada Canadian Hydraulics Centre and Panel on Energy Research and Development, St. John's.

Kovacs, A., & Mellor, M. (1971). *Sea ice pressure ridges and ice islands, Technical note 122*. Calgary: Creare Inc. for Arctic Petroleum Operators Association.

Matskevitch, D. (1997). Eccentric impact of an ice feature: Non-linear model. *Cold Regions Science and Technology, 26*(1), 55–66. doi:10.1016/S0165-232X(97)00008-6.

Mellor, M., Kovacs, A., & Hnatiuk, J. (1977, September 26–30). Destruction of ice islands with explosives. In *Fourth International Conference on Port and Ocean Engineering Under Arctic Conditions*, Memorial University, St. John's, p. 13.

Memorial University. (1995). *Canadian offshore design for ice environments* (Vol. 1). St. John's: Environment and Routes, Prepared for Department of Industry, Trade and Technology, Canada-Newfoundland Offshore Development Fund, Government of Newfoundland and Labrador.

Mueller, D., Copland, L., & Jeffries, M. O. (2017). Changes in Canadian Arctic ice shelf extent since 1906. In L. Copland & D. Mueller (Eds.), *Arctic ice shelves and ice islands* (p. 109–148). Dordrecht: Springer. doi:10.1007/978-94-024-1101-0_5.

Münchow, A., Padman, L., & Fricker, H. A. (2014). Interannual changes of the floating ice shelf of Petermann Gletscher, North Greenland, from 2000 to 2012. *Journal of Glaciology, 60*, 489–499. doi:10.3189/2014JoG13J135.

Nessim, M.A., Jordan, I.J., Lantos, S. and Cormeau A. 1986. Probability-based design criteria for ice loads on fixed structures in the Beaufort Sea. Det norske Veritas (Canada), Calgary, Joint Industry Report No. 86-CGY-43, 2 volumes. (proprietary)

Sackinger, W. M., Jeffries, M. O., Lu, M. C., & Li, F. C. (1988). *Arctic ice islands*. Fairbanks: University of Alaska.

Sackinger, W. M., Shoemaker, H. D., Serson, H., et al. (1985). Ice islands as hazards to Arctic offshore production structures. In *Proceedings of the 17th Annual Offshore Technology Conference*, Houston, pp. 399–408.

Van Wychen, W., & Copland, L. (2017). Ice island drift mechanisms in the Canadian High Arctic. In L. Copland & D. Mueller (Eds.), *Arctic ice shelves and ice islands* (p. 287–316). Dordrecht: Springer. doi:10.1007/978-94-024-1101-0_11.

Vanmarcke, E. (1983). *Random fields: Analysis and synthesis*. Cambridge: MIT Press.

Erratum to: Holocene History of Arctic Ice Shelves

John H. England, David J.A. Evans, and Thomas R. Lakeman

Erratum to:
Chapter 7 in: L. Copland, D. Mueller (eds.), ***Arctic Ice Shelves and Ice Islands*****, Springer Polar Sciences, DOI 10.1007/978-94-024-1101-0_7**.

The original version of this chapter was inadvertently published with incorrect name of the author (David A. Evans). The correct name should be **David J. A. Evans**.

The incorrect order of reference list was published in the original version of this chapter. It should be organized alphabetically.

The updated online version of this chapter can be found at
DOI 10.1007/978-94-024-1101-0_7

L. Copland, D. Mueller (eds.), *Arctic Ice Shelves and Ice Islands*, Springer Polar Sciences, DOI 10.1007/978-94-024-1101-0_16

Index

L. Copland, D. Mueller (eds.), *Arctic Ice Shelves and Ice Islands*, Springer Polar Sciences, DOI 10.1007/978-94-024-1101-0

S

T

U

V

W

MIX
Papier aus verantwortungsvollen Quellen
Paper from responsible sources
FSC® C105338

If you have any concerns about our products,
you can contact us on
ProductSafety@springernature.com

In case Publisher is established outside the EU,
the EU authorized representative is:
Springer Nature Customer Service Center GmbH
Europaplatz 3, 69115 Heidelberg, Germany

Printed by Libri Plureos GmbH
in Hamburg, Germany